P9-CRN-735

INDEX OF TABLES

UNIVERSITY PHYSICS

SIXTH EDITION

PHYSICS PART II

UNIVERSITY PHYSICS PART II

SIXTH EDITION

Francis W. Sears
Late Professor Emeritus
Dartmouth College

Mark W. Zemansky
Professor Emeritus
City College of the
City University of New York

Hugh D. Young
Professor of Physics
Carnegie-Mellon University

ADDISON-WESLEY PUBLISHING COMPANY

Reading, Massachusetts · Menlo Park, California
London · Amsterdam · Don Mills, Ontario · Sydney

This book is in the
ADDISON-WESLEY SERIES IN PHYSICS

Sponsoring editor: Robert L. Rogers
Production editor: Mary Cafarella
                     Marion E. Howe
Designer: Robert Rose
            Marie E. McAdam
Illustrator: Oxford Illustrators, Ltd.
Cover design: Ann Scrimgeour Rose

Library of Congress Cataloging in Publication Data

Sears, Francis Weston, 1898–
    University physics.

    Includes index.
    1. Physics.   I. Zemansky, Mark Waldo, 1900–
II. Young, Hugh D.   III. Title.
QC21.2.S36   1981   530   81-17551
ISBN 0–201–07195–9   AACR2

PREFACE

University Physics is intended to provide a broad introduction to physics at the beginning college level for students of science and engineering who are taking an introductory calculus course concurrently. Primary emphasis is placed on physical principles and the development of problem-solving ability rather than on historical background or specialized applications. The complete text may be taught in an intensive two- or three-semester course, and the book is also adaptable to a wide variety of shorter courses. Numerous worked-out examples and an extensive collection of problems are included with each chapter. *University Physics* is available as a single volume or as two separate parts. Part I includes mechanics, heat, and sound, and Part II includes electricity and magnetism, optics, and atomic and nuclear physics.

In this new edition, the basic philosophy and outline and the balance between depth of treatment and breadth of subject-matter coverage are unchanged from previous editions. We have tried to preserve those features and characteristics that users of previous editions have found desirable, while incorporating a number of changes, some quite extensive, that should substantially enhance the book's pedagogical usefulness. Here are some of the most important changes:

1. A new single-column two-color format has been adopted, and all the figures have been redrawn for two-color treatment. The second color is used not for cosmetic effect but to add clarity to figures through color-coding of vectors, materials, and other significant features.

   Color is also used to identify important equations and to set off examples. Thus the second color should make a very substantial improvement in the book's usefulness as a learning tool for students.

2. A list of thought-provoking questions, many of them related to everyday experience, has been included with each chapter, about 700 questions in all. These should help students to attain a deeper understanding of principles and to relate these principles to their everyday lives and should also serve as effective springboards for class discussion.

3. The problem collections have been carefully reviewed and expanded, especially with the aims of filling gaps in problem coverage, providing additional straightforward "confidence-builder" problems, and furnishing applications of elementary calculus to physical problems. The book now includes about 1600 problems, an increase of 140 from the previous edition. Worked-out examples in the text have also been reviewed, and about 100 new examples have been included. The authors have resisted the temptation to key problems to specific sections of the text. Learning to select the principles appropriate for a specific problem is, after all, part of learning to solve problems. In addition, many problems require material from more than one section.

4. Unit vectors are introduced in the opening chapter and are used to a limited extent where appropriate. Their use in later chapters of the text is optional, however; instructors who wish to omit unit vectors completely can do so without loss of continuity.

5. There are several minor changes in the outline. The chapter on rigid-body equilibrium has been placed following dynamics of a particle and just before rotational motion. However, users who want to go into this material immediately following equilibrium of a point may still do so. The thermodynamics material has been split into two chapters, one each for the first and second laws. The introductory material on optics has been somewhat condensed and consolidated into a single chapter. The relativity chapter has been moved to the end of the book to accompany the other chapters on twentieth-century physics, which have been considerably expanded. The number of chapters is now 47, but the overall length of the book is about the same as that of the previous edition.

6. Many sections have been extensively rewritten to improve clarity and continuity. New introductory material has been prepared for nearly every chapter; this will provide the student with a sense of perspective and ease his or her entry into the chapter. The first chapter is completely new; it now includes discussion of units, unit conversions, and significant figures, as well as an introduction to vector addition in the context of displacement vectors, preceding the more abstract addition of forces. Products of vectors are also introduced here. All of the material in the areas of magnetic fields and forces has been rewritten with the aim of making it less formal (and less formidable) than in the previous edition.

7. The treatment of modern physics topics has been considerably expanded. There is a completely new chapter on quantum mechanics. The discussion of nuclear fission and fusion has been expanded, as has the treatment of fundamental particles and the associated conservation and symmetry principles. Several topics have been added, including lasers, integrated circuits, superheavy nuclei, quarks, and several others.

8. Some material that is outdated or of peripheral importance has been de-emphasized; a few examples are the Wheatstone bridge, the potentiometer, thermoelectricity, and some aspects of magnetic materials.

The text is adaptable to a wide variety of course outlines. The entire text can be used for an intensive course two or three semesters in length. For a less intensive course, many instructors will want to omit certain chapters or sections to tailor the book to their individual needs. The format of this edition facilitates this kind of flexibility. For example, any or all of the chapters on relativity, hydrostatics, hydrodynamics, acoustics, magnetic properties of matter, electromagnetic waves, optical instruments, and several others can be omitted without loss of continuity. In addition, some sections that are unusually challenging or out of the mainstream have been identified with an asterisk preceding the section title. These too may be omitted without loss of continuity.

Conversely, however, many topics that were regarded a few years ago as of peripheral importance and were purged from introductory courses have now come to the fore again in the life sciences, earth and space sciences, and environmental problems. An instructor who wishes to stress these kinds of applications will find this text a useful source for discussion of the appropriate principles.

In any case, it should be emphasized that instructors should not feel constrained to work straight through the book from cover to cover. Many chapters are, of course, inherently sequential in nature, but within this general limitation instructors are encouraged to select from among the contents those chapters that fit their needs, omitting material that is not relevant for the objectives of a particular course.

Again we wish to thank our many colleagues and students who have contributed suggestions for this new edition. In particular, Professors William M. Cloud (Eastern Illinois University), James R. Gaines (Ohio State University), and A. Lewis Ford (Texas A. and M. University) have read the entire manuscript, and their critical and constructive comments are greatly appreciated. In addition, Professors Malcolm D. Cole and Charles McFarland (University of Missouri at Rolla) have read portions of the manuscript and have provided many valuable suggestions. One of the authors (H.D.Y.) offers special thanks to his students and colleagues at Carnegie-Mellon University for their many helpful comments. He acknowledges a special debt of gratitude to Professors Robert Eisenstein, Robert Kraemer, and Frederick Messing of Carnegie-Mellon for many stimulating discussions about the book and about physics pedagogy generally, and to Professor Kraemer for major contributions to the sections on high-energy physics. The kindness and helpfulness of these people will not soon be forgotten.

As usual, we welcome communications from readers concerning our book, and especially concerning any errors or deficiencies that may remain in this edition.

New York   M.Z.
Pittsburgh   H.D.Y.
November 1981

AVAILABLE SUPPLEMENTS

The following supplementary materials are available for use by students:

Study Guide	James R. Gaines and William F. Palmer (Ohio State University, Columbus)
Solutions Guide	A. Lewis Ford (Texas A&M University, College Station)

The following supplement is available to instructors:

Answers to Even-Numbered Problems	(The answers to the odd-numbered problems are in the back of the book.)

CONTENTS

37
Electromagnetic Waves

38
The Nature and Propagation of Light

39
Images Formed by a Single Surface

40
Lenses and Optical Instruments

41
Interference and Diffraction

42
Polarization

43
Relativistic Mechanics

ABRIDGED CONTENTS

ELECTRICITY AND MAGNETISM, LIGHT, AND ATOMIC PHYSICS
PART

COULOMB'S LAW 24

24-1 Electric charges

It was known to the ancient Greeks as long ago as 600 B.C. that amber, rubbed with wool, acquires the property of attracting light objects. In describing this property today, we say that the amber is *electrified*, or possesses an *electric charge*, or is *electrically charged*. These terms are derived from the Greek word *elektron*, meaning amber. It is possible to impart an electric charge to any solid material by rubbing it with any other material. Thus a person becomes electrified by scuffing his shoes across a nylon carpet, a comb is electrified in passing through dry hair, an electric charge is developed on a sheet of paper moving through a printing press, and so on.

Plastic rods and fur are often used in demonstrations. Suppose a plastic rod is electrified by rubbing it with fur and is then touched to two small, light balls of cork or pith, suspended by thin silk or nylon threads. The balls are found to be *repelled* by the plastic rod, and they also repel each other.

A similar experiment performed with a glass rod that has been rubbed with silk gives rise to the same result; pith balls electrified by contact with such a glass rod are repelled not only by the rod but by each other. On the other hand, when a pith ball that has been in contact with electrified plastic is placed near one that has been in contact with electrified glass, the pith balls *attract* each other. We are therefore led to the conclusion that there are *two kinds* of electric charge—that possessed by plastic after being rubbed with fur, called a *negative* charge, and that possessed by glass after being rubbed with silk, called a *positive* charge. The experiments on pith balls described above lead to the fundamental results that (1) *like charges repel*, (2) *unlike charges attract*.

These repulsive or attractive forces of electrical origin are distinct from *gravitational* attraction, and usually so much larger that the gravitational interaction between two electrically charged bodies may be neglected. Of course, a charged body may be acted on by the gravitational field of a large body such as the earth, and in such cases the electrical and gravitational forces may be of comparable magnitude.

Two bodies may also interact by means of magnetic interaction; the most familiar example is attraction of iron objects to a permanent magnet. Although magnetic forces were once believed to be a fundamentally different type of interaction, it is now known that magnetic interactions are really one aspect of the interactions between charged particles in motion. Indeed, an electromagnet manifests magnetic interactions when an electric current passes through its coils. Magnetic interactions will be discussed in later chapters; for the present we concentrate on charges *at rest,* that is, on *electrostatics.*

Suppose a plastic rod is rubbed with fur and then touched to a suspended pith ball. Both the rod and the pith ball are then negatively charged. If the *fur* is now brought near the pith ball, the ball is *attracted,* indicating that the fur is *positively* charged. It follows that when plastic is rubbed with fur, opposite charges appear on the two materials. This is found to happen whenever any substance is rubbed with any other substance. Thus, glass becomes positive, while the silk with which the glass was rubbed becomes negative. This suggests strongly that electric charges are not generated or created, but that the process of acquiring an electric charge consists of *transferring* something from one body to another, so that one body has an excess and the other a deficiency of that something. It was not until the end of the nineteenth century that this "something" was found to consist of very small, negatively charged particles, known today as *electrons.*

24-2 Atomic structure

The interactions responsible for the structure of atoms and molecules, and hence of all matter, are primarily electrical interactions between electrically charged particles. The fundamental building blocks are three kinds of particles, the negatively charged *electron,* the positively charged *proton,* and the neutral *neutron.* The negative charge of the electron is of the same magnitude as the positive charge of the proton, and no charges of smaller magnitude have ever been observed. The charge of a proton or an electron is the ultimate, natural unit of charge. Measurements of the charge of the electron are discussed in Sec. 26-6.

The particles are arranged in the same general way in all atoms. The protons and neutrons always form a closely packed group called the *nucleus,* which has a net positive charge due to the protons. The diameter of the nucleus, if we think of it as roughly spherical, is of the order of 10^{-14} m. Outside the nucleus, but at relatively large distances from it, are the electrons, whose number is equal to the number of protons within the nucleus. If the atom is undisturbed, and no electrons are removed from or added to the space around the nucleus, the atom as a whole is electrically *neutral.* That is, *the algebraic sum of the positive charges of the nucleus and the negative charges of the electrons is zero,* just as equal positive and negative numbers sum to zero. If one or more electrons are removed, the remaining positively charged structure is called a positive *ion.* A negative ion is an atom that has gained one or more extra electrons. The process of losing or gaining electrons is called *ionization.*

In the atomic model proposed by the Danish physicist Niels Bohr in 1913, the electrons were pictured as whirling about the nucleus in circular or elliptical orbits. More recent research has shown that the electrons are more accurately represented as spread-out distributions of electric charge, governed by the principles of quantum mechanics, which will be discussed in Chapter 45. Nevertheless, the Bohr model is still useful for visualizing the structure of an atom. The diameters of the electron charge distributions, which the Bohr model pictures as orbits, determine the overall size of the atom as a whole, and these are of the order of 10^{-10} m, or about ten thousand times as great as the diameter of the nucleus. A Bohr atom is analogous to a solar system in miniature, with electrical forces taking the place of gravitational forces. The massive, positively charged central nucleus corresponds to the sun, while the electrons, moving around the nucleus under the electrical force of its attraction, correspond to the planets moving around the sun under the influence of its gravitational attraction.

The masses of the proton and neutron are nearly equal, and the mass of the proton is about 1836 times that of the electron. Nearly all the mass of an atom, therefore, is concentrated in its nucleus. Since one mole of monatomic hydrogen consists of 6.022×10^{23} particles (Avogadro's number) and its mass is 1.008 g, the mass of a single hydrogen atom is

$$\frac{1.008 \text{ g}}{6.022 \times 10^{23}} = 1.674 \times 10^{-24} \text{ g} = 1.674 \times 10^{-27} \text{ kg}.$$

The nucleus of a hydrogen atom is a single proton, and around it there is a single electron. Hence, of the total mass of the hydrogen atom, $\frac{1}{1837}$ part is the mass of the electron and the remainder is the mass of a proton. To four significant figures,

$$\text{Mass of electron} = 9.110 \times 10^{-31} \text{ kg,}$$
$$\text{Mass of proton} = 1.673 \times 10^{-27} \text{ kg,}$$
$$\text{Mass of neutron} = 1.675 \times 10^{-27} \text{ kg}$$

After hydrogen, the atom with the next simplest structure is that of helium. Its nucleus consists of two protons and two neutrons, and it has two extranuclear electrons. When these two electrons are removed, the doubly charged helium ion (which is the helium nucleus itself) is often called an *alpha particle,* or α-particle. The next element, lithium, has three protons in its nucleus and has thus a nuclear charge of three units. In the un-ionized state the lithium atom has three extranuclear electrons. Each element has a different number of nuclear protons and therefore a different positive nuclear charge. In the table of elements listed at the end of this book, known as the *periodic table,* each element occupies a box with which is associated a number, called the *atomic number.*

The atomic number is the number of nuclear protons; in the un-ionized state, this is equal to the number of extranuclear electrons.

Every material body contains a tremendous number of charged particles, positively charged protons in the nuclei of its atoms and negatively charged electrons outside the nuclei. When the total number of protons equals the total number of electrons, the body as a whole is electrically neutral.

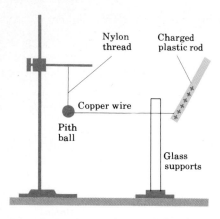

24-1 Copper is a conductor of electricity.

To give a body an excess negative charge, we may either *add* a number of *negative* charges to a neutral body, or *remove* a number of *positive* charges from the body. Similarly, either an *addition* of *positive* charge or a *removal* of *negative* charge results in an excess positive charge. In most instances, it is negative charges (electrons) that are added or removed, and a "positively charged body" is one that has lost some of its normal complement of electrons.

The "charge" of a body refers to its *excess* charge only. The excess charge is always a very small fraction of the total positive or negative charge in the body.

Implicit in the above statements is the *principle of conservation of charge*. This principle states that the algebraic sum of all the electric charges in any closed system is constant. Charge can be transferred from one body to another, but it cannot be created or destroyed. Conservation of charge is believed to be a *universal* conservation law; there is no experimental evidence for any violation of this principle.

24-3 Conductors and insulators

Some materials permit electric charge to move from one region of the material to another, while other materials do not. Suppose we touch one end of a copper wire to an electrified plastic rod and the other end to an initially uncharged pith ball, as in Fig. 24-1. The pith ball is found (by its subsequent interaction with other charged pith balls) to become charged. Thus charge has moved along the wire from the rod to the ball. The copper wire is called a *conductor* of electricity. If the experiment is repeated using a rubber band or nylon thread in place of the wire, *no* transfer of charge is observed; and these materials are called *insulators*. The motion of charge through a material substance will be studied in more detail in Chapter 28, but for the present it is sufficient to state that most substances fall into one or the other of the two classes above. Conductors permit the passage of charge through them, while insulators do not.

Metals in general are good conductors, while most nonmetals are insulators. The positive valency of metals and the fact that they form positive ions in solution indicate that the atoms of a metal will part readily with one or more of their outer electrons. Within a metallic conductor such as a copper wire, a few outer electrons become detached from each atom and can move freely throughout the metal in much the same way that the molecules of a gas can move through the spaces between grains of sand in a sand-filled container. In fact, these free electrons are often referred to as an "electron gas." The positive nuclei and the remainder of the electrons remain fixed in position. Within an insulator, on the other hand, there are no (or at most very few) free electrons.

24-4 Charging by induction

In charging a pith ball by contact with, say, a plastic rod that has been rubbed with fur, some of the extra electrons on the plastic are transferred to the ball, leaving the plastic with a smaller negative charge. There is, however, another way to use the plastic rod to charge other

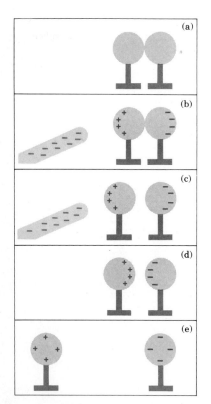

24-2 Two metal spheres are oppositely charged by induction.

bodies, in which the plastic may impart a charge of *opposite* sign and lose none of its own charge. This process, called *charging by induction,* is illustrated in Fig. 24-2.

In Fig. 24-2a, two neutral metal spheres are in contact, both supported on insulating stands. When a negatively charged rod is brought near one of the spheres but without touching it, as in (b), the free electrons in the metal spheres are repelled and drift slightly away from the rod, toward the right. Since the electrons cannot escape from the spheres, an excess negative charge accumulates at the right surface of the right sphere. This leaves a deficiency of negative charge, or an excess positive charge, at the left surface of the left sphere. These excess charges are called *induced* charges.

It should not be inferred that *all* of the free electrons in the spheres are driven to the surface of the right sphere. As soon as any induced charges develop, they also exert forces on the free electrons within the spheres. In this case this force is toward the left (a repulsion by the negative induced charge and an attraction by the positive induced charge). Within an extremely short time the system reaches an equilibrium state in which, at every point in the interior of the spheres, the force on an electron toward the right, exerted by the charged rod, is just balanced by a force toward the left exerted by the induced charges.

The induced charges remain on the surfaces of the spheres as long as the rod is held nearby. When the rod is removed, the electron cloud in the spheres moves to the left and the original neutral condition is restored.

Suppose that the spheres are separated slightly, as shown in (c), while the plastic rod is nearby. If the rod is now removed, as in (d), we are left with two oppositely charged metal spheres whose charges attract each other. When the two spheres are separated by a great distance, as in (e), each of the two charges becomes uniformly distributed over its sphere. It should be noticed that the negatively charged rod has lost none of its charge in the steps from (a) to (e).

The steps from (a) to (e) in Fig. 24-3 should be self-explanatory. In this figure, a single metal sphere (on an insulating stand) is charged by induction. The symbol lettered "ground" in part (c) simply means that the sphere is connected to the earth (a conductor). The earth thus takes the place of the second sphere in Fig. 24-2. In step (c), electrons are repelled to ground either through a conducting wire, or along the moist skin of a person who touches the sphere with his finger. The earth thus acquires a negative charge equal to the induced positive charge remaining on the sphere.

The process taking place in Figs. 24-2 and 24-3 could be explained equally well if the mobile charges in the spheres were *positive* or, in fact, if *both* positive and negative charges were mobile. Although we now know that in a metallic conductor it is actually the *negative* charges that move, it is often convenient to describe a process *as if* the positive charges moved.

24-5 Coulomb's law

The electrical interaction between two charged particles is described in terms of the *forces* they exert on each other. The first quantitative inves-

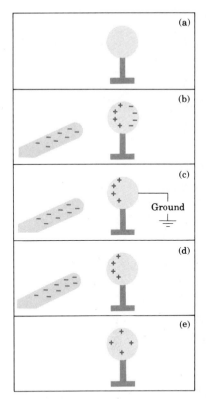

24-3 Charging a single metal sphere by induction.

tigation of the behavior of this force was carried out by Augustin de Coulomb (1736–1806) in 1784, using for force measurements a torsion balance of the type used 13 years later by Cavendish to study (much weaker) gravitational forces (Sec. 4-4). Coulomb studied the force of attraction or repulsion between two "point charges," that is, charged bodies whose dimensions are small compared with the distance between them.

Coulomb found that the force grows weaker with increasing separation between the bodies. When the distance doubles, the force decreases to $\frac{1}{4}$ of its initial value. Thus it varies inversely with the square of the distance. If r is the distance between the particles, then the force is proportional to $1/r^2$.

The force also depends on the quantity of charge on each body, which is usually denoted by q or Q. In Coulomb's time, no *unit* of charge had been defined, nor had any method been developed for comparing a given charge with a unit. Despite this, Coulomb devised an ingenious method of showing how the force exerted on or by a charged body depended on its charge. He reasoned that if a charged spherical conductor were brought in contact with a second identical conductor, originally *uncharged,* the charge on the first would, by symmetry, be shared equally between the conductors. He thus had a method for obtaining one-half, one-quarter, and so on, of any given charge. The results of his experiments were consistent with the conclusion that the force between two point charges q and q' is proportional to each charge and hence proportional to the *product* of the charges. The complete expression for the magnitude of the force between two point charges is therefore

$$F = k\frac{|qq'|}{r^2},$$
(24-1)

where k is a proportionality constant whose magnitude depends on the units in which F, q, q', and r are expressed. Equation (24-1) is the mathematical statement of what is known today as *Coulomb's law:*

> *The force of attraction or repulsion between two point charges is directly proportional to the product of the charges and inversely proportional to the square of the distance between them.*

The *direction* of the force on each particle is always along the line joining the two particles, pulling each particle toward the other in the case of attractive forces on unlike charges, and pushing them apart in the case of repulsive forces on like charges.

The form of Eq. (24-1) is the same as that of the law of gravitation, discussed in Sec. 4-4, but electrical and gravitational interactions are two distinct classes of phenomena. The proportionality of the force to $1/r^2$ has been verified with great precision. There is no reason to suspect, for example, that the electrical force might vary as $1/r^{2.0001}$.

The charges q and q' are *algebraic* quantities, corresponding to the existence of two kinds of charge (positive and negative), but Eq. (24-1) gives the magnitude of the interaction force in all cases. When the charges are of like sign the forces are repulsive; when they are unlike, the forces are attractive. In either case, the forces obey Newton's third law; the force that q exerts on q' is the negative of the force q' exerts on q.

The "absolute value" bars in Eq. (24-1) are needed because F, the magnitude of a vector quantity, is by definition always positive, while the product qq' is negative whenever the two charges have opposite signs.

When two or more charges exert forces simultaneously on a given charge, the total force experienced by that charge is found to be the *vector sum* of the forces that the various charges would exert individually. This important property, called the *principle of superposition,* permits the application of Coulomb's law to arrays of charge of any degree of complexity, although the computational problems can be very great. Example 4 below illustrates the application of the superposition principle.

If there is matter in the space between the charges, the *net* force acting on each is altered because charges are induced in the molecules of the intervening material. This effect will be described later. As a practical matter, the law can be used as stated for point charges in air, since even at atmospheric pressure the effect of the air is to alter the force from its value in vacuum by only about one part in two thousand.

In the chapters of this book dealing with electrical phenomena, we shall use the SI (mks) system of units exclusively. The SI electrical units include all the familiar electrical units such as the volt, the ampere, the ohm, and the watt. The cgs system is also used, more so in scientific work than in commerce and industry, but there is *no* British system of electrical units. This is one of many reasons for abandoning the British system and adopting metric units universally, and there seems little doubt that SI units will eventually receive worldwide adoption.

To the three basic SI units (the meter, kilogram, and second) we now add a fourth, the unit of electric charge. This unit is called one *coulomb* (1 C). The electrical constant k in Eq. (24-1) is, in this system,

$$k = 8.98755 \times 10^9 \, \text{N} \cdot \text{m}^2 \cdot \text{C}^{-2}$$
$$\approx 9.0 \times 10^9 \, \text{N} \cdot \text{m}^2 \cdot \text{C}^{-2}.$$

Later, in connection with the study of electromagnetic radiation, we shall show that k is closely related to the speed of light in vacuum,

$$c = 2.998 \times 10^8 \, \text{m} \cdot \text{s}^{-1}.$$

Specifically,

$$k = 10^{-7} \, c^2.$$

The relationship is not accidental, but results from the definition of the unit of current, which in turn is related to the interaction of electric and magnetic fields, to be studied later.

In the cgs system of electrical units (not used in this book) the constant k is defined to be unity, without units. This defines a unit of electric charge called the *statcoulomb* or the *esu* (electrostatic unit). The conversion factor is

$$1 \, \text{C} = 2.998 \times 10^9 \, \text{esu}.$$

In SI units the constant k in Eq. (24-1) is usually written not as k but as $1/4\pi\epsilon_0$, where ϵ_0 is another constant. This appears to complicate matters, but it actually simplifies some formulas to be encountered later. Thus Coulomb's law is usually written as

$$F = \frac{1}{4\pi\epsilon_0} \frac{|qq'|}{r^2}, \qquad (24\text{-}2)$$

with

$$\frac{1}{4\pi\epsilon_0} = 8.987554 \times 10^9 \text{ N}\cdot\text{m}^2\cdot\text{C}^{-2}$$

and

$$\epsilon_0 = 8.854185 \times 10^{-12} \text{ C}^2\cdot\text{N}^{-1}\cdot\text{m}^{-2}.$$

The "natural" unit of charge is the magnitude of charge of an electron or a proton. This quantity is denoted by e; the most precise measurements to date yield the value

$$e = 1.602192 \times 10^{-19} \text{ C} \approx 1.60 \times 10^{-19} \text{ C}.$$

One coulomb therefore represents the negative of the total charge carried by about 6×10^{18} electrons. For comparison, the population of the earth is estimated to be about 3×10^9 persons, while a cube of copper 1 cm on a side contains about 2.4×10^{24} electrons.

In electrostatics problems, charges as large as one coulomb are unusual; a more typical range of magnitude is 10^{-9} to 10^{-6} C. The microcoulomb (1 μC $= 10^{-6}$ C) is often used as a practical unit of charge.

Example 1 Two charges are located on the positive x-axis of a coordinate system, as shown in Fig. 24-4. Charge $q_1 = 2 \times 10^{-9}$ C is 2 cm from the origin, and charge $q_2 = -3 \times 10^{-9}$ C is 4 cm from the origin. What is the total force exerted by these two charges on a charge $q_3 = 5 \times 10^{-9}$ C located at the origin?

Solution The total force on q_3 is the vector sum of the forces due to q_1 and q_2 individually. Converting distances to meters, we use Eq. (24-2) to find the magnitude F_1 of the force on q_3 due to q_1:

$$F_1 = \frac{(9.0 \times 10^9 \text{ N}\cdot\text{m}^2\cdot\text{C}^{-2})(2 \times 10^{-9} \text{ C})(5 \times 10^{-9} \text{ C})}{(0.02 \text{ m})^2} = 2.25 \times 10^{-4} \text{ N}.$$

This force has a negative x-component because q_3 is repelled (i.e., pushed in the $-x$-direction) by q_1, which has the same sign. Similarly, the force due to q_2 is found to have magnitude

$$F_2 = \frac{(9.0 \times 10^9 \text{ N}\cdot\text{m}^2\cdot\text{C}^{-2})(3 \times 10^{-9} \text{ C})(5 \times 10^{-9} \text{ C})}{(0.04 \text{ m})^2} = 0.84 \times 10^{-4} \text{ N}.$$

This force has a positive x-component because q_3 is attracted (i.e., pulled in the positive x-direction) by the opposite charge q_2. The sum of the x-components is

$$\sum F_x = -2.25 \times 10^{-4} \text{ N} + 0.84 \times 10^{-4} \text{ N} = -1.41 \times 10^{-4} \text{ N}.$$

There are no y- or z-components. Thus the total force on q_3 is directed to the left, with magnitude 1.41×10^{-4} N.

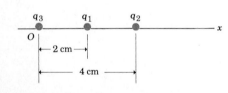

24-4

Example 2 An α-particle is a nucleus of doubly ionized helium. It has a mass m of 6.68×10^{-27} kg and a charge q of $+2e$ or 3.2×10^{-19} C. Compare the force of electrostatic repulsion between two α-particles with the force of gravitational attraction between them.

Solution   The electrostatic force F_e is

$$F_e = \frac{1}{4\pi\epsilon_0}\frac{q^2}{r^2},$$

and the gravitational force F_g is

$$F_g = G\frac{m^2}{r^2}.$$

The ratio of the electrostatic to the gravitational force is

$$\frac{F_e}{F_g} = \frac{1}{4\pi\epsilon_0 G}\frac{q^2}{m^2} = 3.1 \times 10^{35}.$$

The gravitational force is evidently negligible compared with the electrostatic force.

Example 3   The Bohr model of the hydrogen atom (discussed in Chapter 45) consists of a single electron of charge $-e$ revolving in a circular orbit about a single proton of charge $+e$. The electrostatic force of attraction between electron and proton provides the centripetal force that retains the electron in its orbit. Hence if v is the orbital velocity, Newton's second law gives

$$\frac{1}{4\pi\epsilon_0}\frac{e^2}{r^2} = m\left(\frac{v^2}{r}\right).$$

In Bohr's theory, the electron may revolve only in some one of a number of specified orbits. The orbit of smallest radius is that for which the angular momentum L of the electron is $h/2\pi$, where h is a universal constant called *Planck's constant*, equal to 6.625×10^{-34} J $\cdot$ s. Then,

$$L = mvr = \frac{h}{2\pi}. \tag{24-3}$$

When v is eliminated between the preceding equations, we find

$$r = \frac{\epsilon_0 h^2}{\pi k m e^2},$$

and when numerical values are inserted, we find, for the radius of the *first Bohr orbit*,

$$r = 5.29 \times 10^{-11}\,\text{m} = 0.529 \times 10^{-8}\,\text{cm}.$$

This result corresponds reasonably well with other estimates of the "size" of a hydrogen atom obtained from deviations from ideal gas behavior, the density of hydrogen in the liquid and solid states, and other observations.

Example 4   In Fig. 24-5, two equal positive charges $q = 2.0 \times 10^{-6}$ C interact with a third charge $Q = 4.0 \times 10^{-6}$ C. Find the magnitude and direction of the total (resultant) force on Q.

Solution   The key word is *total;* we must compute the force each charge exerts on Q, and then obtain the *vector sum* of the forces. This is

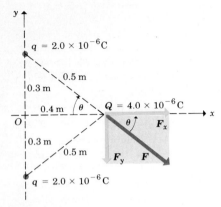

24-5 *F* is the force on *Q* due to the upper charge *q*.

most easily accomplished using components. The figure shows the force on *Q* due to the upper charge *q*. From Coulomb's law,

$$F = (9.0 \times 10^9 \text{ N} \cdot \text{C}^{-2} \cdot \text{m}^{-2}) \frac{(4.0 \times 10^{-6} \text{ C})(2.0 \times 10^{-6} \text{ C})}{(0.5 \text{ m})^2}$$

$$= 0.29 \text{ N}.$$

The components of this force are given by:

$$F_x = F \cos \theta = (0.29 \text{ N}) \left(\frac{0.4 \text{ m}}{0.5 \text{ m}} \right) = 0.23 \text{ N},$$

$$F_y = -F \sin \theta = -(0.29 \text{ N}) \left(\frac{0.3 \text{ m}}{0.5 \text{ m}} \right) = -0.17 \text{ N}.$$

The lower charge *q* exerts a force of the same magnitude, but in a different direction. From symmetry we see that its *x*-component is the same as that due to the upper charge, but its *y*-component is opposite. Hence,

$$\sum F_x = 2(0.23 \text{ N}) = 0.46 \text{ N},$$

$$\sum F_y = 0.$$

The total force on *Q* is horizontal, with magnitude 0.46 N. How would this solution differ if the lower charge were *negative?*

24-6 Electrical interactions

Because matter is made up of charged particles, it is not surprising that electrical interactions play a central and dominant role in all aspects of the structure of matter. The forces that hold atoms together in a molecule or in a solid crystal lattice, the adhesive force of glue, the forces associated with surface tension—all these are basically electrical in nature, arising from the electrical forces between the charged particles making up the interacting atoms. A complete description of the detailed behavior of these forces requires new *mechanical* principles and the introduction of quantum-mechanical concepts, to be discussed in Chapter 45. Nevertheless, Coulomb's law and the additional effects resulting from relative motion of the charges still describe the basic electrical interactions involved.

Electrical interactions alone are *not* sufficient to understand the structure of the *nuclei* of atoms, however. A nucleus is made up of protons, which repel each other, and neutrons, which have no electrical charge. For nuclei to be stable there must be additional forces, attractive in nature, to hold their constituent parts together despite the electrical repulsion. This new kind of interaction, not seen outside the nucleus, is called the *nuclear force;* many phenomena associated with the stability or instability of nuclei pivot around the competition between the repulsive electrical forces and the attractive nuclear forces. These matters will be discussed in greater detail in Chapter 47.

Questions

24-1 Plastic food wraps can be used to cover a container by simply stretching the material across the top and pressing the overhanging material against the sides. What makes it stick? Does it stick to itself with equal tenacity? Why? Does it matter whether the container is metallic or not?

24-2 Bits of paper are attracted to an electrified comb or rod, even though they have no net charge. How is this possible?

24-3 How do we know that the magnitudes of electron and proton charge are *exactly* equal? With what precision is this really known?

24-4 When you walk across a nylon rug and then touch a large metallic object, you may get a spark and a shock. Why does this tend to happen more in the winter than the summer? Why do you not get a spark when you touch a *small* metal object?

24-5 The free electrons in a metal have mass and therefore weight and are gravitationally attracted toward the earth. Why, then, do they not all settle to the bottom of the conductor, as sediment settles to the bottom of a river?

24-6 Simple electrostatics experiments, such as picking up bits of paper with an electrified comb, never work as well on rainy days as on dry days. Why?

24-7 High-speed printing presses sometimes use gas flames to reduce electric charge build-up on the paper passing through the press. Why does this help? (An added benefit is rapid drying of the ink.)

24-8 What similarities do electrical forces have to gravitational forces? What are the most significant differences?

24-9 Given two identical metal objects mounted on insulating stands, describe a procedure for placing charges of equal magnitude and opposite sign on the two objects.

24-10 How do we know that protons have positive charge and electrons negative charge, rather than the reverse?

24-11 Gasoline transport trucks sometimes have chains that hang down and drag on the ground at the rear end. What are these for?

24-12 When a nylon sleeping bag is dragged across a rubberized cloth air mattress in a dark tent, small sparks are sometimes seen. What causes them?

24-13 Atomic nuclei are made of protons and neutrons. This fact by itself shows that there must be another kind of interaction in addition to the electrical forces. Explain.

24-14 When transparent plastic tape is pulled off a roll and one tries to position it precisely on a piece of paper, it often jumps over and sticks where it isn't wanted. Why does it do this?

Problems

24-1 How many excess electrons must be placed on each of two small spheres spaced 3 cm apart if the force of repulsion between the spheres is to be 10^{-19} N?

24-2 Each of two small spheres is positively charged, the combined charge totaling 4×10^{-8} C. What is the charge on each sphere if they are repelled with a force of 27×10^{-5} N when placed 0.1 m apart?

24-3 6.02×10^{23} atoms of monatomic hydrogen have a mass of one gram. How far would the electron of a hydrogen atom have to be removed from the nucleus for the force of attraction to equal the weight of the atom?

24-4 What is the total positive charge, in coulombs, of all the protons in 1 mol of hydrogen atoms?

24-5 If all the positive charges in a mole of hydrogen atoms were lumped into a single charge, and all the negative charges into a single charge, what force would the two lumped charges exert on each other at a distance of

a) 1 m;

b) 10^7 m (comparable to the diameter of the earth)?

24-6 An alpha particle consists of two protons and two neutrons bound together. What is the repulsive force between two alpha particles at a distance of 10^{-15} m, comparable to the sizes of nuclei?

24-7 Two copper spheres, each having mass 1 kg, are separated by 1 m.

a) How many electrons does each sphere contain?

b) How many electrons would have to be removed from one sphere and added to the other to cause an attractive force of 10^4 N (roughly one ton)?

c) What fraction of all the electrons on a sphere does this represent?

24-8 Point charges of 2×10^{-9} C are situated at each of three corners of a square whose side is 0.20 m. What would be the magnitude and direction of the resultant force on a point charge of -1×10^{-9} C if it were placed (a) at the center of the square? (b) at the vacant corner of the square?

24-9 Two charges of $+10^{-9}$ C each are 8 cm apart in air. Find the magnitude and direction of the force exerted by these charges on a third charge of $+5 \times 10^{-11}$ C that is 5 cm distant from each of the first two charges.

24-10 Two positive point charges, each of magnitude q, are located on the y-axis at points $y = +a$ and $y = -a$. A third positive charge of the same magnitude is located at some point on the x-axis.

a) What is the force exerted on the third charge when it is at the origin?

b) What is the magnitude and direction of the force on the third charge when its coordinate is x?

c) Sketch a graph of the force on the third charge as a function of x, for values of x between $+4a$ and $-4a$. Plot forces to the right upward, forces to the left downward.

24-11 A negative point charge of magnitude q is located on the y-axis at the point $y = +a$, and a positive charge of the same magnitude is located at $y = -a$. A third positive charge of the same magnitude is located at some point on the x-axis.

a) What is the magnitude and direction of the force exerted on the third charge when it is at the origin?

b) What is the force on the third charge when its coordinate is x?

c) Sketch a graph of the force on the third charge as a function of x, for values of x between $+4a$ and $-4a$.

24-12 Two small balls, each of mass 10 g, are attached to silk threads 1 m long and hung from a common point. When the balls are given equal quantities of negative charge, each thread makes an angle of 4° with the vertical.

a) Draw a diagram showing all of the forces on each ball.

b) Find the magnitude of the charge on each ball.

24-13 A certain metal sphere of volume 1 cm³ has a mass of 7.5 g and contains 8.2×10^{22} free electrons.

a) How many electrons must be removed from each of two such spheres so that the electrostatic force of repulsion between them just balances the force of gravitational attraction? Assume the distance between the spheres is great enough so that the charges on them can be treated as point charges.

b) Express the number of electrons removed as a fraction of the total number of free electrons.

24-14 In the Bohr model of atomic hydrogen, an electron of mass 9.11×10^{-31} kg revolves about a proton in a circular orbit of radius 5.29×10^{-11} m. The proton has a positive charge equal in magnitude to the negative charge on the electron and its mass is 1.67×10^{-27} kg.

a) What is the radial acceleration of the electron?

b) What is its velocity?

c) What is its angular velocity?

24-15 One gram of monatomic hydrogen contains 6.02×10^{23} atoms, each consisting of an electron with charge -1.60×10^{-19} C and a proton with charge $+1.60 \times 10^{-19}$ C.

a) Suppose all these electrons could be located at the north pole of the earth and all the protons at the south pole. What would be the total force of attraction exerted on each group of charges by the other? The diameter of the earth is 12,800 km.

b) What would be the magnitude and direction of the force exerted by the charges in part (a) on a third positive charge, equal in magnitude to the total charge at one of the poles and located at a point on the equator? Draw a diagram.

24-16 The dimensions of atomic nuclei are of the order of 10^{-14} m. Suppose that two α-particles are separated by this distance.

a) What is the force exerted on each α-particle by the other?

b) What is the acceleration of each? (See Example 2 in Sec. 24–5 for numerical data.)

24-17 The pair of equal and opposite charges in Problem 24–11 is called an *electric dipole*.

a) Show that when the x-coordinate of the third charge in Problem 24–11 is large compared with the distance a, the force on it is inversely proportional to the *cube* of its distance from the midpoint of the dipole.

b) Show that if the third charge is located on the y-axis, at a y-coordinate large compared with the distance a, the force on it is also inversely proportional to the cube of its distance from the midpoint of the dipole.

24-18 A small ball having a positive charge q_1 hangs by an insulating thread. A second ball with a negative charge $q_2 = -q_1$ is kept at a horizontal distance a to the right of the first. (The distance a is large compared with the diameter of the ball.)

a) Show in a diagram all of the forces on the hanging ball in its final equilibrium position.

b) You are given a third ball having a positive charge $q_3 = 2q_1$. Find at least two points at which this ball can be placed so that the first ball will hang vertically.

24-19 Two point charges are located in the xy-plane, as follows: A charge 2.0×10^{-9} C is at the point $(x = 0, y = 4$ cm$)$, and a charge -3.0×10^{-9} C is at the point $(x = 3$ cm, $y = 4$ cm$)$.

a) If a third charge of 4.0×10^{-9} C is placed at the origin, find the x- and y-components of the total force on this third charge.

b) Find the magnitude and direction of the total force on the charge at the origin in (a).

24-20 A charge -3×10^{-9} C is placed at the origin of an xy-coordinate system, and a charge 2×10^{-9} C is placed on the positive y-axis, at $y = 4$ cm. If a third charge 4×10^{-9} C is now placed at the point $(x = 3$ cm, $y = 4$ cm$)$, find the components of the total force exerted on this charge by the other two. Also find the magnitude and direction of this force.

THE ELECTRIC FIELD; GAUSS'S LAW 25

25-1 The electric field

The electrical interaction between charged particles can be reformulated using the concept of *electric field*. This is not only an aid in calculations but also an important concept with fundamental theoretical significance. To introduce the concept, we consider the mutual repulsion of two positively charged bodies A and B, as shown in Fig. 25-1a. In particular, the force on B is labeled F in the figure. This is an "action-at-a-distance" force; it can act across empty space and does not need any matter in the intervening space to transmit the force.

Now let us think of body A as having the effect of modifying some of the properties of the space in its vicinity. We remove body B and label its former position as point P. The charged body A is said to produce or cause an *electric field* at point P (and at all other points in its vicinity). Then when body B is placed at point P and experiences the force F, we take the point of view that the force is exerted on B *by the field*, rather than directly by A. Since B would experience a force at any point in space around A, the electric field exists at all points in the region around A. (One could equally well consider that body B sets up an electric field, and that the force on body A is exerted by the field due to B.)

The experimental test for the existence of an electric field at any point is simply to place a small charged body, which will be called a *test charge,* at the point. If a force (of electrical origin) is exerted on the test charge, then an electric field exists at the point.

> *An electric field is said to exist at a point if a force of electrical origin is exerted on a stationary charged body placed at the point.*

Since force is a vector quantity, the electric field is a *vector quantity* whose properties are determined when both the magnitude and the direction of an electric force are specified. We define the *electric field E* at a point as the quotient obtained when the force F, acting on a positive test charge, is divided by the magnitude q' of the test charge. Thus,

$$E = \frac{F}{q'},$$ (25-1)

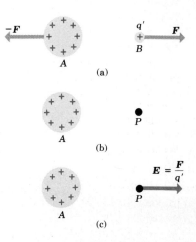

25-1 A charged body creates an electric field in the space around it.

469

and the direction of E is the direction of F. It follows that

$$F = q'E.$$

The force on a *negative* charge, such as an electron, is *opposite* to the direction of the electric field.

The electric field is sometimes called *electric intensity* or *electric field intensity*. In SI (mks) units, where the unit of force is 1 N and the unit of charge is 1 C, the unit of electric field is 1 newton per coulomb (1 N·C^{-1}). Electric field may also be expressed in other units, to be defined later.

The force experienced by the test charge q' varies from point to point, and so the electric field is also different at different points. Thus in general E is not a single vector quantity, but an infinite set of vector quantities, one associated with each point in space. It is an example of a *vector field*. Another example of a vector field is the description of motion of a flowing fluid. In general, different points in the fluid have different velocities, so the velocity is a vector field. If a rectangular coordinate system is used, then in principle each component of E can be expressed as a *function* of the coordinates (x, y, z) of a point in space. Vector fields are an important part of the mathematical language used in many areas of physics, particularly in electricity and magnetism.

One difficulty with our definition of electric field is that in Fig. 25-1 the force exerted by the test charge q' may change the charge distribution A, especially if the body is a conductor on which charge is free to move, so that the electric field around A when q' is present is not the same as when it is absent. However, when q' is very small, the redistribution of charge on body A is also very small; thus the difficulty can be avoided by refining the definition of electric field to be *the limiting value of the force per unit charge on a test charge q' at the point, as the charge q' approaches zero:*

$$E = \lim_{q' \to 0} \frac{F}{q'}.$$

If an electric field exists within a *conductor,* a force is exerted on every charge in the conductor. The motion of the free charges brought about by this force is called a *current.* Conversely, if there is *no* current in a conductor, and hence no motion of its free charges, *the electric field in the conductor must be zero.*

In most instances, the magnitude and direction of an electric field vary from point to point. If the magnitude and direction are constant throughout a certain region, the field is said to be *uniform* in this region.

Example 1   What is the electric field 30 cm from a charge $q = 4 \times 10^{-9}$ C?

Solution   From Coulomb's law, the *force* on a test charge q' 30 cm from q is

$$F = \frac{1}{4\pi\epsilon_0} \frac{|qq'|}{r^2} = \frac{(9 \times 10^9 \text{ N·m}^2\text{·C}^{-2})(4 \times 10^{-9} \text{ C})(q')}{(0.3 \text{ m})^2}$$
$$= (400 \text{ N·C}^{-1})(q').$$

Then from Eq. (25-1), the magnitude of E is

$$E = \frac{F}{q'} = 400 \text{ N·C}^{-1}.$$

The *direction* of E at this point is along the line joining q and q', away from q.

Example 2   When the terminals of a 100-V battery are connected to two large parallel plates 1 cm apart, the field in the region between the plates is very nearly uniform and the electric field magnitude E is 10^4 N·C^{-1}. Suppose the direction of E is vertically upward. Compute the force on an electron in this field and compare with the weight of the electron.

Solution

$$\text{Electron charge } e = 1.60 \times 10^{-19} \text{ C,}$$
$$\text{Electron mass } m = 9.1 \times 10^{-31} \text{ kg.}$$
$$F_{\text{elec}} = eE = (1.60 \times 10^{-19} \text{ C})(10^4 \text{ N·C}^{-1})$$
$$= 1.60 \times 10^{-15} \text{ N;}$$
$$F_{\text{grav}} = mg = (9.1 \times 10^{-31} \text{ kg})(9.8 \text{ m·s}^{-2})$$
$$= 8.9 \times 10^{-30} \text{ N.}$$

The ratio of the electrical to the gravitational force is therefore

$$\frac{1.60 \times 10^{-15} \text{ N}}{8.9 \times 10^{-30} \text{ N}} = 1.8 \times 10^{14}.$$

The gravitational force is negligibly small compared to the electrical force.

Example 3   If released from rest, what speed does the electron of Example 2 acquire while traveling 1 cm? What is then its kinetic energy? How much time is required?

Solution   The force is constant, so the electron moves with a constant acceleration of

$$a = \frac{F}{m} = \frac{eE}{m} = \frac{1.60 \times 10^{-15} \text{ N}}{9.1 \times 10^{-31} \text{ kg}}$$
$$= 1.8 \times 10^{15} \text{ m·s}^{-2}.$$

Its speed after traveling 1 cm, or 10^{-2} m, is

$$v = \sqrt{2ax} = 6.0 \times 10^6 \text{ m·s}^{-1}.$$

Its kinetic energy is

$$\tfrac{1}{2}mv^2 = 1.6 \times 10^{-17} \text{ J.}$$

The time required is

$$t = \frac{v}{a} = 3.3 \times 10^{-9} \text{ s.}$$

Example 4   If the electron of Example 2 is projected into the field with a horizontal velocity, as in Fig. 25–2, find the equation of its trajectory.

Solution   The direction of the field is upward in Fig. 25–2, so the force on the electron is downward. The initial velocity is along the positive x-axis. The x-acceleration is zero, the y-acceleration is $-(eE/m)$. Hence,

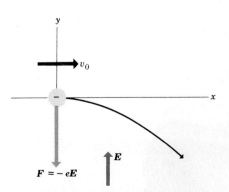

25-2   Trajectory of an electron in an electric field.

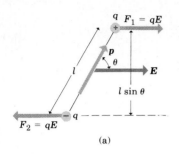

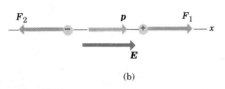

(b)

25-3 (a) The torque on the dipole is $\Gamma = pE \sin \theta$. (b) The dipole is in equilibrium in a uniform field when p and E are parallel. If the field is not uniform, the net force on the dipole is $p(\Delta E / \Delta x)$.

after a time t,

$$x = v_0 t,$$

$$y = -\frac{1}{2}\left(\frac{eE}{m}\right)t^2.$$

Elimination of t gives

$$y = -\left(\frac{eE}{2mv_0{}^2}\right)x^2,$$

which is the equation of a parabola. The motion is the same as that of a body projected horizontally in the earth's gravitational field. The deflection of electrons by an electric field is used to control the direction of an electron stream in many electronic devices, such as the cathode-ray oscilloscope.

Example 5   Figure 25–3 represents two point charges of equal magnitude q but of opposite sign, separated by a distance l. Such a pair of charges is called an *electric dipole*. The dipole is in a *uniform* electric field E, whose direction makes an angle θ with the line joining the two charges, called the *dipole axis*. A force F_1, of magnitude qE, in the direction of the field, is exerted on the positive charge, and a force F_2, of the same magnitude but in the opposite direction, is exerted on the negative charge. The resultant *force* on the dipole is zero, but since the two forces do not have the same line of action, they constitute a *couple* (see Sec. 8–4). The moment of the couple is

$$\Gamma = (qE)(l \sin \theta),$$

since $l \sin \theta$ is the perpendicular distance between the action lines of the forces.

The product ql of the charge q and the distance l is called the *electric dipole moment,* and is represented by p:

$$p = ql.$$

The torque exerted by the couple is therefore

$$\Gamma = pE \sin \theta. \qquad (25\text{–}2)$$

The *vector dipole moment* of the dipole, $p,$ is defined as a vector of magnitude p lying along the dipole axis and pointing from the negative toward the positive charge. With this definition and the concept of vector torque introduced in Sec. 8–5, the torque on an electric dipole in an electric field may be written as

$$\mathbf{\Gamma = p \times E}. \qquad (25\text{–}3)$$

The effect of the torque is to tend to rotate the dipole to a position in which the dipole moment p is parallel to the electric vector E, as in Fig. 25–3b. If the field is *uniform,* the dipole is in equilibrium in this position.

Suppose, however, that the field at each charge has the direction of the vector E but that it is *not* uniform and its magnitude is greater at the position of the $+$ charge than at the position of the $-$ charge. The force F_1 is then *greater* than the force F_2, and there is a net force on the dipole toward the right, urging it toward a region of *stronger* field.

Suppose the field increases uniformly in magnitude along the x-axis, and that it changes by an amount dE in an interval dx. Then, over the length l of the dipole, when it is parallel to the x-axis, the total change of E is $l(dE/dx)$. The difference of the two forces is given by

$$F = F_1 - F_2 = ql\left(\frac{dE}{dx}\right) = p\left(\frac{dE}{dx}\right).$$

The quantity dE/dx, representing the rate of change of E with position, is called the *electric field gradient;* the net force on a dipole oriented parallel to a nonuniform field is equal to the product of the dipole moment and the field gradient.

25-2 Calculation of electric field

The preceding section has described an experimental way to measure the electric field at a point. The method consists of placing a small test charge at the point, *measuring* the force on it, and taking the ratio of the force to the charge. The electric field at a point may also be *computed* from Coulomb's law if the magnitudes and positions of all charges contributing to the field are known. Thus, to find the magnitude of the electric field at a point P, at a distance r from a point charge q, we imagine a test charge q' to be placed at P. The force on the test charge, by Coulomb's law, has magnitude

$$F = \frac{1}{4\pi\epsilon_0}\left(\frac{qq'}{r^2}\right),$$

and hence the electric field at P has magnitude

$$E = \frac{F}{q'} = \frac{1}{4\pi\epsilon_0}\left(\frac{q}{r^2}\right).$$

The direction of the field is away from the charge q if the latter is positive (as in Fig. 25-4a), toward q if it is negative.

The *direction* of the electric field caused by a point charge is represented conveniently by use of *unit vectors,* introduced in Sec. 1-7. Let $\hat{r}$ be a unit vector directed along the line from the charge q to the point P at which the field is to be determined. Then for either a positive or negative charge q,

$$E = \frac{1}{4\pi\epsilon_0}\frac{q\hat{r}}{r^2}, \tag{25-4}$$

where r is the distance from the charge to point P. When q is negative, the direction of E is toward q, opposite to $\hat{r}$.

If a number of point charges q_1, q_2, etc., are at distances r_1, r_2, etc., from a given point P, as in Fig. 25-4b, each exerts a force on a test charge q' placed at the point, and the resultant force on the test charge is the *vector sum* of these forces. The resultant electric field is the *vector sum* of the individual electric fields, and

$$E = E_1 + E_2 + \cdots$$

$$= \frac{1}{4\pi\epsilon_0}\left(\frac{q_1\hat{r}_1}{r_1^2} + \frac{q_2\hat{r}_2}{r_2^2} + \cdots\right). \tag{25-5}$$

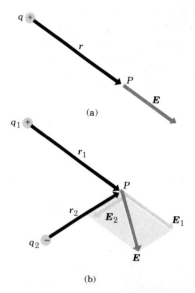

25-4 (a) The electric field E is in the same direction as the vector r when q is positive. (b) The resultant electric field at point P is the vector sum of E_1 and E_2.

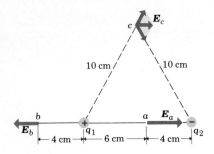

25-5 Electric field at three points, a, b, and c, in the field set up by charges q_1 and q_2.

Because each term in the sum is a vector quantity, the sum is a *vector sum.* The fact that the total field is the sum of the separate fields that would be caused by the individual charges is a direct result of the *principle of superposition,* discussed in Sec. 24–5.

Example 1 Point charges q_1 and q_2 of $+12 \times 10^{-9}$ C and -12×10^{-9} C, respectively, are placed 0.1 m apart, as in Fig. 25–5. Compute the electric fields due to these charges at points a, b, and c.

Solution At point a, the vector due to the positive charge q_1 is directed toward the right, and its magnitude is

$$E_1 = (9 \times 10^9 \text{ N} \cdot \text{m}^2 \cdot \text{C}^{-2}) \frac{(12 \times 10^{-9} \text{ C})}{(0.06 \text{ m})^2}$$

$$= 3.00 \times 10^4 \text{ N} \cdot \text{C}^{-1}.$$

The vector due to the negative charge q_2 is also directed toward the right; its magnitude is

$$E_2 = (9 \times 10^9 \text{ N} \cdot \text{m}^2 \cdot \text{C}^{-2}) \frac{(12 \times 10^{-9} \text{ C})}{(0.04 \text{ m})^2}$$

$$= 6.75 \times 10^4 \text{ N} \cdot \text{C}^{-1}.$$

Hence, at point a,

$$E_a = (3.00 + 6.75) \times 10^4 \text{ N} \cdot \text{C}^{-1}$$
$$= 9.75 \times 10^4 \text{ N} \cdot \text{C}^{-1}, \qquad \text{toward the right.}$$

At point b, the vector due to q_1 is directed toward the left, with magnitude

$$E_1 = (9 \times 10^9 \text{ N} \cdot \text{m}^2 \cdot \text{C}^{-2}) \frac{(12 \times 10^{-9} \text{ C})}{(0.04 \text{ m})^2}$$

$$= 6.75 \times 10^4 \text{ N} \cdot \text{C}^{-1}.$$

The vector due to q_2 is directed toward the right, with magnitude

$$E_2 = (9 \times 10^9 \text{ N} \cdot \text{m}^2 \cdot \text{C}^{-2}) \frac{(12 \times 10^{-9} \text{ C})}{(0.14 \text{ m})^2}$$

$$= 0.55 \times 10^4 \text{ N} \cdot \text{C}^{-1}.$$

Hence, at point b

$$E_b = (6.75 - 0.55) \times 10^4 \text{ N} \cdot \text{C}^{-1}$$
$$= 6.20 \times 10^4 \text{ N} \cdot \text{C}^{-1}, \qquad \text{toward the left.}$$

At point c, the magnitude of each vector is

$$E = (9 \times 10^9 \text{ N} \cdot \text{m}^2 \cdot \text{C}^{-2}) \frac{(12 \times 10^{-9} \text{ C})}{(0.1 \text{ m})^2}$$

$$= 1.08 \times 10^4 \text{ N} \cdot \text{C}^{-1}.$$

The directions of these vectors are shown in the figure; their resultant is easily seen to be

$$E_c = 1.08 \times 10^4 \text{ N} \cdot \text{C}^{-1}, \qquad \text{toward the right.}$$

In practical situations, electric fields are often set up by charges distributed over the surfaces of conductors of finite size, rather than by point charges. The electric field must then be calculated by imagining the charge distribution to be subdivided into many small elements of charge Δq, at varying distances from the point P at which the field is to be calculated, and adding the separate contributions of these elements to the total field. Not all of the charge in each element will be at the same distance from the point P, but if the elements are small compared with the distance to the point, and r represents the distance from any point within the element to the point P, then approximately

$$ \boldsymbol{E} \approx \frac{1}{4\pi\epsilon_0} \sum \frac{\Delta q \hat{\boldsymbol{r}}}{r^2}. $$

The finer the subdivision, the better the approximation, and in the limit as $\Delta q \to 0$,

$$ \boldsymbol{E} = \frac{1}{4\pi\epsilon_0} \lim_{\Delta q \to 0} \sum \frac{\Delta q \hat{\boldsymbol{r}}}{r^2}. $$

The limit of the vector sum, however, is the *vector integral*

$$ \boldsymbol{E} = \frac{1}{4\pi\epsilon_0} \int \frac{\hat{\boldsymbol{r}}\, dq}{r^2}. \tag{25-6} $$

The limits of integration must be assigned so as to include all charges contributing to the field. Like any vector equation, Eq. (25-6) implies three scalar equations, one for each component of the vectors $\boldsymbol{E}$ and $\hat{\boldsymbol{r}}$. To evaluate the vector integral, we evaluate each of the three scalar integrals.

Example 2   A ring-shaped conductor of radius a carries a total charge Q. Find the electric field at a point a distance x from the center, along the line perpendicular to the plane of the ring, through its center.

Solution   The situation is shown in Fig. 25–6. We represent the ring as made up of small segments ds, as shown. The charge of the segment ds is dQ. At point P the element of charge dQ causes an electric field contribution $d\boldsymbol{E}$ having magnitude dE given by

$$ dE = \frac{1}{4\pi\epsilon_0} \frac{dQ}{x^2 + a^2}. $$

The component dE_x of this field along the x-axis is

$$ dE_x = dE \cos\theta = \frac{1}{4\pi\epsilon_0} \frac{dQ}{x^2 + a^2} \frac{x}{\sqrt{x^2 + a^2}} $$

$$ = \frac{1}{4\pi\epsilon_0} \frac{x\, dQ}{(x^2 + a^2)^{3/2}}. $$

To find the total x-component of field, E_x, we integrate this expression:

$$ E_x = \int \frac{1}{4\pi\epsilon_0} \frac{x\, dQ}{(x^2 + a^2)^{3/2}}. $$

On the right side, everything is constant except dQ; x does not vary as we move from point to point on the ring. Thus everything except dQ

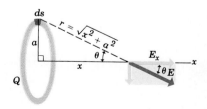

25-6   Electric field due to ring of charge.

may be taken outside the integral. The integral of dQ is simply the total charge Q, and we finally obtain

$$E_x = \frac{1}{4\pi\epsilon_0} \frac{Qx}{(x^2 + a^2)^{3/2}}.$$

(25-7)

In principle, this calculation should also be performed for the components perpendicular to the x-axis, but it is easy to see from symmetry that these add to zero.

Equation (25-7) shows that, at the center of the ring ($x = 0$), the total field is zero, as might be expected; charges on opposite sides pull in opposite directions on a test charge at that point, and their fields cancel. When x is much larger than a, Eq. (25-7) becomes approximately equal to $Q/4\pi\epsilon_0 x^2$, corresponding to the fact that at distances much greater than the dimensions of the ring it appears as a point charge.

Example 3 *Long charged wire.* In Fig. 25-7 a long thin wire of length $2L$ lies along the y-axis and has an electric charge per unit length of λ. We wish to find the electric field caused by this charge at point P, a distance r from the midpoint of the wire, as shown.

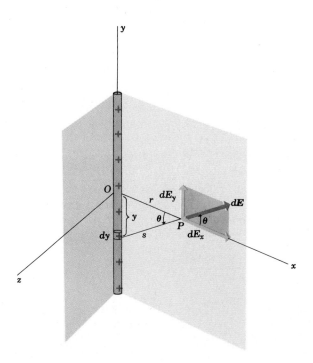

25-7 Electric field of a long thin charged wire.

Solution We imagine dividing the wire into infinitesimal segments dy. The charge dq of a segment dy is then given by $dq = \lambda\, dy$. The distance s from a segment at position y to point P is $s = (y^2 + r^2)^{1/2}$, and the contribution $d\mathbf{E}$ to the field at P due to this segment has magnitude dE given by

$$dE = \frac{1}{4\pi\epsilon_0} \frac{\lambda\, dy}{y^2 + r^2}.$$

To find the total field at P we need to add (integrate) the contributions from *all* the segments. We first express $d\mathbf{E}$ in terms of its x- and y-components. Reference to Fig. 25–7 shows that

$$\cos\theta = \frac{r}{\sqrt{y^2 + r^2}}, \qquad \sin\theta = -\frac{y}{\sqrt{y^2 + r^2}},$$

$$dE_x = dE\cos\theta = \frac{1}{4\pi\epsilon_0}\frac{\lambda r\, dy}{(y^2 + r^2)^{3/2}},$$

$$dE_y = dE\sin\theta = -\frac{1}{4\pi\epsilon_0}\frac{\lambda y\, dy}{(y^2 + r^2)^{3/2}}.$$

The total E_x and E_y caused by the charge on the entire wire are then given by

$$E_x = \int_{-L}^{L} \frac{1}{4\pi\epsilon_0}\frac{\lambda r\, dy}{(y^2 + r^2)^{3/2}},$$

$$E_y = -\int_{-L}^{L} \frac{1}{4\pi\epsilon_0}\frac{\lambda y\, dy}{(y^2 + r^2)^{3/2}}. \tag{25–8}$$

These integrals may be evaluated by trigonometric substitution or by use of a table of integrals. The details are left as an exercise; the final results are

$$E_x = \frac{1}{4\pi\epsilon_0}\frac{2\lambda L}{r\sqrt{L^2 + r^2}},$$

$$E_y = 0. \tag{25–9}$$

The result that $E_y = 0$ could have been predicted from symmetry; for every charge element dq at a given position y there is a corresponding element at $-y$, and the y-components of field caused by these two charges are opposite. Thus the total y-component of field *must* be zero.

If the length $2L$ of the wire is much longer than the distance r (i.e., if the wire is infinitely long), then in Eq. (25–9) r becomes negligible compared to L in the radical, L is canceled by $(L^2 + r^2)^{1/2}$, and we obtain simply

$$E_x = \frac{1}{2\pi\epsilon_0}\frac{\lambda}{r} \qquad \text{(long straight line of charge).} \tag{25–10}$$

The total field is proportional to λ, as expected, and it is inversely proportional to the *first power* of the distance r rather than the *inverse-square* relation for a single point charge.

If the point P had been on the z-axis instead of the x-axis, the only nonzero component of E would have been E_z. The electric field at any point in the xz-plane is directed radially outward from the wire. For an infinitely long wire, the field at *every point* in space is perpendicular to the wire and radially outward from it.

Example 4 *Infinite plane sheet of charge.* In Fig. 25–8, positive charge is distributed uniformly over the entire xy-plane, with a charge *per unit area*, or *surface density of charge*, σ. We wish to calculate the electric field at the point P, at a distance a from the plane.

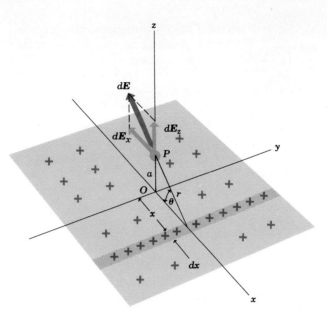

25-8

We divide the charge into narrow strips of width dx, parallel to the y-axis. Each strip can be considered a *line* charge, and we can use the result of the preceding example.

The area of a portion of a strip of length L is $L\,dx$, and the charge dq on the strip is

$$dq = \sigma L\,dx.$$

The charge per unit length, λ, is therefore

$$\lambda = \frac{dq}{L} = \sigma\,dx.$$

From Eq. (25–10), the strip sets up at point P a field $d\mathbf{E}$, lying in the xz-plane, of magnitude

$$dE = \frac{\sigma}{2\pi\epsilon_0}\frac{dx}{r}.$$

The field can be resolved into components $d\mathbf{E}_x$ and $d\mathbf{E}_z$. By symmetry, the components $d\mathbf{E}_x$ will sum to zero when the entire sheet of charge is considered. (Be sure that you understand why.) The resultant field at P is therefore in the z-direction, perpendicular to the sheet of charge. From the diagram,

$$dE_z = dE \sin\theta$$

and hence

$$E = \int dE_z = \frac{\sigma}{2\pi\epsilon_0}\int_{-\infty}^{+\infty}\frac{\sin\theta\,dx}{r}.$$

But

$$\sin\theta = \frac{a}{r}, \qquad r^2 = a^2 + x^2,$$

and therefore

$$E = \frac{\sigma a}{2\pi\epsilon_0} \int_{-\infty}^{+\infty} \frac{dx}{a^2 + x^2} = \frac{\sigma a}{2\pi\epsilon_0} \left[\frac{1}{a} \tan^{-1} \frac{x}{a} \right]_{-\infty}^{+\infty},$$

$$E = \frac{\sigma}{2\epsilon_0}. \tag{25-11}$$

Note that the distance a from the plane to the point P *does not* appear in the final result. This means that the intensity of the field set up by an infinite plane sheet of charge is *independent of the distance from the charge*. In other words, the field is *uniform* and *normal* to the plane of charge.

The same result would have been obtained if point P, in Fig. 25-8, had been taken *below* the xy-plane. That is, a field of the same magnitude but in the opposite sense is set up on the opposite side of the plane.

25-3 Field lines

The concept of field lines was introduced by Michael Faraday (1791-1867) as an aid in visualizing electric (and magnetic) fields. A *field line* (in an electric field) is *an imaginary line drawn in such a way that its direction at any point* (i.e., the direction of its tangent) *is the same as the direction of the field at that point.* (See Fig. 25-9.) Since, in general, the direction of a field varies from point to point, field lines are usually curves. Faraday called these lines "lines of force", but the term "field line" is preferable.

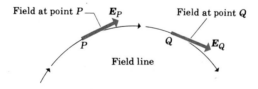

25-9 The direction of the electric field at any point is tangent to the field line through that point.

Figure 25-10 shows some of the field lines in two planes containing (a) a single positive charge; (b) two equal charges, one positive and one negative (an electric dipole); and (c) two equal positive charges. The direction of the resultant field at every point in each diagram is along the tangent to the field line passing through the point. Arrowheads on the field lines indicate the direction in which the tangent is to be drawn.

No field lines originate or terminate in the space surrounding a charge. Every field line in an *electrostatic* field is a continuous line terminated by a positive charge at one end and a negative charge at the other.* While sometimes for convenience we speak of an "isolated" charge and draw its field as in Fig. 25-10a, this simply means that the charges on which the lines terminate are at large distances from the charge under consideration. For example, if the charged body in Fig. 25-10a is a small sphere suspended by a thread from the laboratory ceiling, the negative charges on which its field lines terminate would be

---

* We shall see in a larger chapter that a *changing magnetic field* sets up an electric field whose lines *do not* terminate on electric charges, but close on themselves.

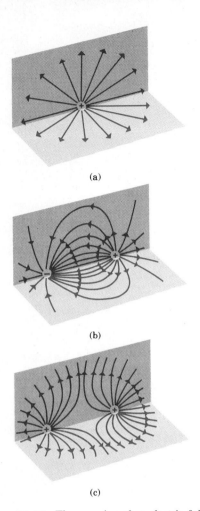

(a)

(b)

(c)

25-10 The mapping of an electric field with the aid of field lines.

found on the walls, floor, or ceiling, as well as on other objects in the laboratory.

At any one point, the resultant field can have but one direction. Hence, only one field line can pass through each point of the field. In other words, field lines never intersect.

If a field line were to be drawn through every point of an electric field, all of space and the entire surface of a diagram would be filled with lines, and no individual line could be distinguished. By suitably limiting the number of field lines one draws to represent a field, the lines can be used to indicate the *magnitude* of a field as well as its *direction*. This is accomplished by spacing the lines in such a way that *the number per unit area crossing a surface at right angles to the direction of the field is at every point proportional to the electric field*. In a region where the field is large, such as that between the positive and negative charges of Fig. 25-10b, the field lines are closely spaced, whereas in a region where the field is small, such as that between the two positive charges of Fig. 25-10c, the lines are widely separated. In a *uniform* field, the field lines are straight, parallel, and uniformly spaced.

25-4 Gauss's Law

Karl Friedrich Gauss (1777–1855) was a German scientist and mathematician who made many contributions to experimental and theoretical physics and to mathematics. The relation known as *Gauss's law* is a statement of an important property of electrostatic fields.

The content of Gauss's law is suggested by consideration of field lines, discussed in Sec. 25-3. The field of an isolated positive point charge q is represented by lines radiating out in all directions. Suppose we imagine this charge as surrounded by a spherical surface of radius R, with the charge at its center. The area of this imaginary surface is $4\pi R^2$, so if the total number of field lines emanating from q is N, then the number of lines *per unit surface area* on the spherical surface is $N/4\pi R^2$. We imagine a second sphere concentric with the first, but with radius $2R$. Its area is $4\pi(2R)^2 = 16\pi R^2$, and the number of lines per unit area on this sphere is $N/16\pi R^2$, one-fourth the density of lines on the first sphere. This corresponds to the fact that, at distance $2R$, the field has only one-fourth the magnitude it has at distance R, and verifies our qualitative statement in Sec. 25-3 that the *density* of lines is proportional to the magnitude of the field.

The fact that the *total* number of lines at distance $2R$ is the same as at R can be expressed another way. The field is inversely proportional to R^2, but the *area* of the sphere is proportional to R^2, so the *product* of the two is independent of R. For a sphere of arbitrary radius r, the magnitude of E on the surface is

$$E = \frac{1}{4\pi\epsilon_0}\frac{q}{r^2},$$

the surface area is

$$A = 4\pi r^2,$$

and the product of the two is

$$EA = \frac{q}{\epsilon_0}. \tag{25-12}$$

This is independent of r and depends *only* on the charge q. As we shall see, this result is of crucial importance in the following development.

What is true of the entire sphere is also true of any portion of its surface. In the construction of Fig. 25-11, an area ΔA is outlined on a sphere of radius R and then projected onto the sphere of radius $2R$ by drawing lines from the center through points on the boundary of ΔA. The area projected on the larger sphere is clearly $4\,\Delta A$; thus again the product $E\,\Delta A$ is independent of the radius of the sphere.

This projection technique shows how this discussion may be extended to nonspherical surfaces. Instead of a second sphere, let us surround the sphere of radius R by a surface of irregular shape, as in Fig. 25-12a. Consider a small areal element ΔA; we note that this area is *larger* than the corresponding element on a spherical surface at the same distance from q. If a normal to the surface makes an angle θ with a radial line from q, two sides of the area projected on the spherical surface are foreshortened by a factor $\cos\theta$, as shown in Fig. 25-12b. Thus, the quantity corresponding to $E\,\Delta A$ for the spherical surface is $E\,\Delta A\,\cos\theta$ for the irregular surface.

Now we may divide the entire irregular surface into small elements ΔA, compute the quantity $E\,\Delta A\,\cos\theta$ for each, and sum the results. Each of these projects onto a corresponding element of area on the sphere, so summing the quantities $E\,\Delta A\,\cos\theta$ over the irregular surface must yield the same result as summing the quantities $E\,\Delta A$ over the sphere. But we have already performed that calculation; the result, given by Eq. (25-12), depends only on the charge q. Thus, for the irregular surface the result is

$$\sum E\,\Delta A\,\cos\theta = \frac{q}{\epsilon_0}, \tag{25-13}$$

no matter what the shape of the surface, provided only that it is a *closed* surface enclosing the charge q. Correspondingly, for a closed surface en-

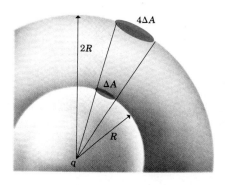

25-11 Projection of an element of area ΔA on a sphere of radius R, onto a sphere of radius $2R$. The projection multiplies each linear dimension by two, so the area element on the larger sphere is $4\,\Delta A$.

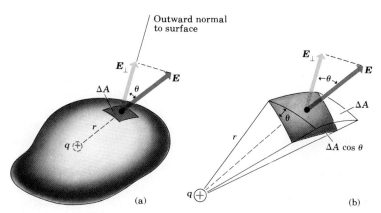

25-12

closing *no* charge,

$$\sum E \, \Delta A \cos \theta = 0.$$

This is a mathematical statement of the fact that when a region contains no charge, any field line that enters on one side must leave again at some other point on the boundary surface. Field lines can begin or end inside a region of space only when there is charge in that region.

Because the field varies from point to point on the irregular surface, Eq. (25-13) is strictly true only in the limit when the area elements become very small. In this limit, the sum becomes an integral called the *surface integral* of $E \cos \theta$, written

$$\oint E \cos \theta \, dA = \frac{q}{\epsilon_0}. \qquad (25\text{-}14)$$

The circle on the integral sign reminds us that the integral is always taken over a *closed* surface enclosing the charge q.

Since $E \cos \theta$ is the component of $\boldsymbol{E}$ perpendicular to the surface at each point, we may use the notation $E_\perp = E \cos \theta$, and write

$$\oint E_\perp \, dA = \frac{q}{\epsilon_0}. \qquad (25\text{-}15)$$

The quantity $E_\perp \, dA = E \cos \theta \, dA$ is also called the *electric flux* through the area dA. Electric flux may be denoted by Ψ, and an element of flux corresponding to a small area dA by $d\Psi$. Then

$$d\Psi = E_\perp \, dA = E \cos \theta \, dA = \boldsymbol{E} \cdot d\boldsymbol{A}. \qquad (25\text{-}16)$$

The total flux Ψ through a finite surface area is

$$\Psi = \int E_\perp \, dA = \int E \cos \theta \, dA = \int \boldsymbol{E} \cdot d\boldsymbol{A}. \qquad (25\text{-}17)$$

(The notation $\boldsymbol{E} \cdot d\boldsymbol{A}$ is explained below.) Thus Eqs. (25-14) and (25-15) state that the total electric flux out of a closed surface is proportional to the charge enclosed:

$$\Psi = \frac{q}{\epsilon_0}. \qquad (25\text{-}18)$$

If the point charge in Fig. 25-12 is negative, the $\boldsymbol{E}$ field is directed radially *inward;* the angle θ is then greater than 90°, its cosine is negative, and the integral in Eq. (25-15) is negative. But since q is also negative, Eq. (25-15) still holds.

If a point charge lies *outside* a closed surface (the reader may construct his own diagram), the electric field is *outward* at some points of the surface and *inward* at others. It is not difficult to show that the positive and negative contributions to the integral over the surface exactly cancel, and the sum is zero. But the charge *inside* the closed surface is also zero, so again Eq. (25-15) is obeyed.

Although we have been concerned only with a single point charge, it is easy to generalize the above results to *any* charge distribution. The total electric field $\boldsymbol{E}$ at a point on the surface is the vector sum of the fields produced by the individual charges, and the quantity $E \, dA \cos \theta$ is therefore the sum of the contributions from these charges. Since Eq. (25–15) holds for each point charge, a corresponding relation holds for the *total* E field and the *total* charge enclosed by the surface. That is,

$$\oint E_{\perp} \, dA = \frac{1}{\epsilon_0} \sum q, \qquad (25\text{–}19)$$

where E is now the total electric field and $\sum q$ represents the *algebraic* sum of all charges enclosed by the surface.

Equation (25–19) is the mathematical statement of *Gauss's law*. It states that when we multiply each element of area of a closed surface by the normal component of E at the element, and sum over the entire surface, the result is a constant times the total charge inside the surface.

The notation can be simplified by use of the *vector area* dA, defined as a vector whose magnitude equals dA and whose direction is that of the *outward* normal at dA. The product $E_{\perp} \, dA = E \cos \theta \, dA$ can then be written as the *scalar product* or *dot product* of the vectors E and dA:

$$E_{\perp} \, dA = \boldsymbol{E} \cdot d\boldsymbol{A}.$$

Denoting the total charge enclosed by $Q = \sum q$, we may write Gauss's law more compactly as

$$\oint \boldsymbol{E} \cdot d\boldsymbol{A} = Q / \epsilon_0. \qquad (25\text{–}20)$$

As mentioned above, the flux of E across a surface, as well as Gauss's law, can be interpreted graphically in terms of field lines. If the number of lines per unit area at right angles to their direction is proportional to E, the surface integral of $E_{\perp}$ over a closed surface is proportional to the total number of lines crossing the surface in an outward direction, and the net charge within the surface is proportional to this number. As an example, consider the field of two equal and opposite point charges shown in Fig. 25–13. Surface A encloses the positive charge only, and 18 lines cross it in an outward direction. Surface B encloses the negative charge only, and it also is crossed by 18 lines, but in an inward direction. Surface C enclosed *both* charges. It is intersected by lines at 16 points, at 8 of which the intersections are outward and at 8 of which they are inward. The *net* number of lines crossing in an outward direction is zero, and the net charge inside the surface is also zero. Surface D is intersected at 6 points, at 3 of which the intersections are outward while at the other 3 they are inward. The net number of lines crossing in an outward direction, and the enclosed charge, are both zero.

In evaluating the surface integral of $E_{\perp}$ over a closed surface, it is often necessary to divide the surface, in imagination, into a number of portions. The integral over the entire surface is the *sum* of the integrals over each portion. There are many cases of practical importance where symmetry considerations simplify the evaluation of the integral enough that only simple algebraic manipulations are necessary. Several examples will be discussed in the next section. The following observations are also useful:

1. If E is at right angles to a surface of area A at all points, and has the

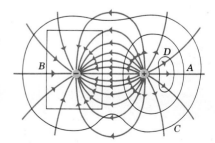

25-13

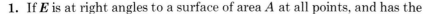

same *magnitude* at all points of the surface, then $E_\perp = E = $ constant, and

$$\int E_\perp \, dA = EA.$$

2. If E is *parallel* to a surface at all points, $E_\perp = 0$ and the integral is zero.

3. If $E = 0$ at all points of a surface, the integral is zero.

4. The surface to which Gauss's law is applied need not be a real physical surface, such as the surface of a solid body. Indeed, in most applications of this law, one considers an imaginary or geometrical surface that may be in empty space, embedded in a solid body, or partly in space and partly within a body.

5. In the integral $\int E \cdot dA$ or $\int E_\perp dA$, E is always the *total* electric field at each point on the surface. In general this field is caused partly be charges within the volume and partly by charges outside. Thus even when there is *no* charge within the volume, the field at points on the surface need not be zero. In that case, however, $\int E \cdot dA$ is always zero.

6. In applications of Gauss's law to field calculations, some judgment is required in choosing a surface. Two useful guiding principles are that the point or points at which the field is to be determined must lie on the surface, and that the surface must have enough symmetry so that it is possible to evaluate the integral. Thus if the problem has spherical or cylindrical symmetry, the gaussian surface will usually be spherical or cylindrical, respectively.

25-5 Applications of Gauss's Law

1. Location of excess charge on a conductor. It has been explained that the electric field E is zero at all points within a conductor when the charges in the conductor are at rest. (If E were *not* zero, the charges would move, and we would not have an electro*static* situation.) We may construct an imaginary surface in the interior of a conductor, such as surface A in Fig. 25-14a. (In applications of Gauss's law, such a surface is often called a *gaussian surface.*) Because $E = 0$ everywhere on this surface, Eq. (25-17) requires that the net charge inside the surface be zero.

If we now imagine the surface to shrink to zero like a collapsing balloon, as suggested in Fig. 25-14a, until it essentially encloses a point, the charge at the point must be zero. Since this process can take place at *any* point in the conductor, *there can be no net charge at any point within the conductor.* It follows that the entire excess charge on the conductor must be located on the *outer surface* of the conductor, as shown.

Now suppose there is a cavity in the conductor, as in Fig. 25-14b, and there are no charges within the cavity. Again a gaussian surface such as A may be used to show that there is no charge within the material of the conductor. Furthermore, consideration of a surface such as B shows that the net charge on the surface of the cavity must be zero. This does not necessarily prove that the entire cavity wall is uncharged, but we shall

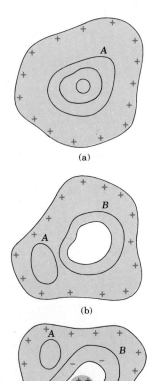

(a)

(b)

(c)

25-14

prove later that this is in fact the case. Thus the entire charge on the conductor lies on its *outer* surface, not on the cavity wall.

Next, suppose that there is a conductor inside the cavity but insulated from it, as in Fig. 25–14c, and that the inner conductor has a charge q. Application of Gauss's law to surface B shows again that the *net* charge inside this surface is zero, so there must be a charge on the cavity wall, equal and opposite in sign to the charge q. If the outer conductor is initially *uncharged* before the charge q is inserted (and is insulated so that the total charge on it cannot change), there must be a charge on its outer surface, equal and opposite to the charge on the cavity wall and therefore equal to, and of the *same* sign as, the charge q. If the outer conductor originally had a charge q', the charge on its outer surface becomes $q + q'$.

It follows that insertion of a charge into a cavity in a hollow conductor results in the appearance of an exactly equal charge on the outer surface of the hollow conductor, whether or not this conductor was originally charged.

2. Coulomb's law. We have considered Coulomb's law as the fundamental equation of electrostatics and have derived Gauss's law from it. An alternative procedure is to consider Gauss's law as a fundamental experimental relation. Coulomb's law can then be derived from Gauss's law, by using this law to obtain the expression for the electric field E due to a point charge.

Consider the electric field of a single positive point charge q, shown in Fig. 25–15. By *symmetry*, the field is everywhere radial (there is no reason why it should deviate to one side of a radial direction rather than to another) and its magnitude is the same at all points at the same distance r from the charge (any point at this distance is like any other). Hence, if we select as a gaussian surface a spherical surface of radius r, $E_\perp = E = constant$ at all points of the surface. Then

$$\oint E_\perp \, dA = E_\perp \oint dA = EA = 4\pi r^2 E.$$

From Gauss's law

$$4\pi r^2 E = \frac{q}{\epsilon_0} \quad \text{and} \quad E = \frac{1}{4\pi\epsilon_0} \frac{q}{r^2}.$$

The force on a point charge q' at a distance r from the charge q is then

$$F = q'E = \frac{1}{4\pi\epsilon_0} \frac{qq'}{r^2},$$

which is Coulomb's law.

3. Field of a charged conducting sphere. Any excess charge on an isolated solid conducting sphere is, by symmetry, distributed *uniformly* over its outer surface. The electric field at any point can be calculated (at least in principle) by summing the contributions from elements of charge on the surface, but it is much simpler to use Gauss's law. At points outside the sphere the field is radial everywhere, for the same reason as for a point charge; because of the spherical symmetry, there is no reason for the field at a given point to deviate one way rather than

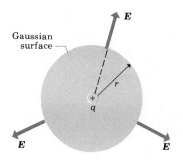

Gaussian surface

25-15

another from the radial direction. Furthermore, the magnitude of the field is again uniform over a spherical surface of radius r, concentric with the conductor.

Thus if we construct a gaussian surface of radius r, where r is greater than the radius R of the sphere, and if q is the total charge on the sphere,

$$4\pi r^2 E = \frac{q}{\epsilon_0},$$

$$E = \frac{1}{4\pi\epsilon_0}\frac{q}{r^2}. \tag{25-21}$$

The field *outside* the sphere is, therefore, the same as though the entire charge were concentrated at a point at its center. Just outside the surface of the sphere, where $r = R$,

$$E = \frac{1}{4\pi\epsilon_0}\frac{q}{R^2}.$$

Inside the sphere, as in the interior of any conductor where no charge is in motion, the field is zero. Thus when r is less than R, $E = 0$.

The same argument may be applied to a conducting, hollow, spherical *shell*, that is, a spherical conductor with a concentric spherical hole in the center, if there is no charge in the hole. This time we take a spherical gaussian surface of radius r less than the radius of the hole. If there *is* a field inside the hole, it must be spherically symmetric (radial) as before, so again $E = q/4\pi\epsilon_0 r^2$. But this time $q = 0$, so E must also be zero.

Because of the relation of Gauss's law to Coulomb's law, this result holds only because the electric field obeys an inverse *square* law. If the field were inversely proportional to r^3, or to $r^{2.147}$, Gauss's law would not hold. Very precise measurements have shown that the internal field of a charged sphere is, in fact, so small that the exponent of r cannot differ from exactly 2 by more than 1 part in 10^7. There seems no reason to doubt that it is exactly 2.

It is left as a problem to find, from Gauss's law, the electric field in the interspace between a charged sphere and a concentric hollow sphere surrounding it.

4. Field of a line charge and of a charged cylindrical conductor. We consider next the electric field set up by a long thin uniformly charged wire. We solved this problem in Sec. 25-2, Example 3, by a straightforward application of Coulomb's law, requiring a somewhat involved integration. Gauss's law makes it possible to find the field by an almost trivial calculation.

If the wire is very long and we are not too near either end, then, by symmetry, the field lines outside the wire are *radial* and lie in planes perpendicular to the wire. Also, the field has the same magnitude at all points at the same radial distance from the wire. This suggests that we use as a gaussian surface a *cylinder* of arbitrary radius r and arbitrary length l, with its ends perpendicular to the wire, as in Fig. 25-16. If λ is the charge *per unit length* on the wire, the charge within the gaussian surface is λl. Since E is at right angles to the wire, the component of E normal to the end faces is zero. Thus the end faces make no contribution

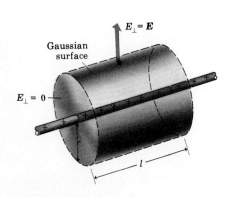

25-16 Cylindrical gaussian surface for calculating the electric field due to a long charged wire.

to the integral in Gauss's law. At all points of the curved surface, $E_\perp = E =$ constant, and since the area of this surface is $2\pi rl$, we have

$$(E)(2\pi rl) = \frac{\lambda l}{\epsilon_0},$$

$$E = \frac{1}{2\pi\epsilon_0}\frac{\lambda}{r}. \tag{25-22}$$

It should be noted that although the *entire* charge on the wire contributes to the field E, only that portion of the total charge lying within the gaussian surface is used when we apply Gauss's law. This feature of the law is puzzling at first; it appears as though we had somehow obtained the right answer by ignoring a part of the charge, and that the field of a *short* wire of length l would be the same as that of a very long wire. The existence of the entire charge on the wire *is*, however, taken into account when we consider the *symmetry* of the problem. Suppose the wire had been a short one, of length l. Then we could *not* conclude by symmetry that the field at one end of the cylinder, say, would equal that at the center, or that the field lines would everywhere be perpendicular to the wire. So the entire charge on the wire actually *is* taken into account, but in an indirect way.

It is left as a problem (1) to show that the field outside a long charged cylinder is the same as though the charge on the cylinder were concentrated in a line along its axis, and (2) to calculate the electic field in the interspace between a charged cylinder and a coaxial hollow cylinder that surrounds it.

5. Field of an infinite plane sheet of charge. To solve this problem, construct the gaussian surface shown by the shaded area in Fig. 25–17, consisting of a cylinder whose ends have an area A and whose walls are perpendicular to the sheet of charge. By symmetry, since the sheet is infinite, the electric field E has the same magnitude E on both sides of the surface, is uniform, and is directed normally away from the sheet of charge. No field lines cross the *side* walls of the cylinder; that is, the component of E normal to these walls is zero. At the ends of the cylinder the normal component of E is equal to E. The integral $\int E_\perp\,dA$, calculated over the entire surface of the cylinder, therefore reduces to $2EA$. If σ is the charge *per unit area* in the plane sheet, the net charge within the gaussian surface is σA. Hence

$$2EA = \frac{\sigma A}{\epsilon_0}, \qquad E = \frac{\sigma}{2\epsilon_0}. \tag{25-23}$$

Note that the magnitude of the field is *independent* of the distance from the sheet and does *not* decrease inversely with the square of the distance. The field lines remain everywhere straight, parallel, and uniformly spaced. This is because the sheet was assumed infinitely large.

Of course, nothing in nature can really be infinitely large; this infinitely large plane sheet is an idealization. Our result is still useful, however; the real content of Eq. (25–23) is that it is a very good approximation to the behavior of the electric field caused by a large but finite sheet of charge, at points that are not near its edge and are close to the plane compared to its dimensions. In such cases the field is very nearly uniform and perpendicular to the plane.

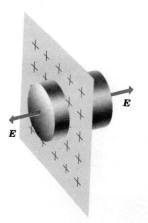

25-17 Gaussian surface in the form of a cylinder for finding the field of an infinite plane sheet of charge.

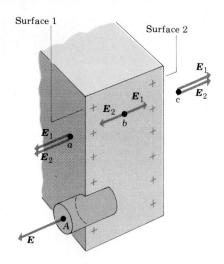

25-18 Electric field inside and outside a charged conducting plate.

6. Field of an infinite plane charged conducting plate. When a flat metal plate is given a net charge, this charge distributes itself over the entire outer surface of the plate and, if the plate is of uniform thickness and is infinitely large (or if we are not too near the edges of a finite plate), the charge per unit area is uniform and is the same on both surfaces. Hence, the field of such a charged plate arises from the super-position of the fields of *two* sheets of charge, one on each surface of the plate. By symmetry, the field is perpendicular to the plate, directed away from it (if the plate has a positive charge) and is uniform. The magnitude of the electric field at any point can be found from Gauss's law, or by using the results already derived for a sheet of charge.

Figure 25–18 shows a portion of a large charged conducting plate. Let σ represent the charge per unit area in the sheet of charge on *each* surface (i.e., the total charge per unit area on both surfaces together is 2σ). At point a, outside the plate at the left, the component of electric field E_1, due to the sheet of charge on the left face of the plate, is directed toward the left, and its magnitude is $\sigma/2\epsilon_0$. The component E_2 due to the sheet of charge on the right face of the plate is also toward the left and its magnitude is also $\sigma/2\epsilon_0$. The magnitude of the *resultant* intensity E is therefore

$$E = E_1 + E_2 = \frac{\sigma}{2\epsilon_0} + \frac{\sigma}{2\epsilon_0} = \frac{\sigma}{\epsilon_0}. \qquad (25\text{-}24)$$

At point b, inside the plate, the two components of electric field are in opposite directions and their resultant is zero, as it must be in any conductor in which the charges are at rest. At point c, the components again add and the magnitude of the resultant is σ/ϵ_0, directed toward the right.

To derive these results directly from Gauss's law, consider the cylinder shown in the figure. Its end faces are of area A; one end lies inside and one outside the plate. The field inside the conductor is zero. The field outside, by symmetry, is perpendicular to the plate, so the normal component of E is zero over the walls of the cylinder and is equal to E over the outside end face. Hence, from Gauss's law,

$$EA = \frac{\sigma A}{\epsilon_0}, \qquad E = \frac{\sigma}{\epsilon_0}. \qquad (25\text{-}25)$$

7. Field between oppositely charged parallel conducting plates. When two plane parallel conducting plates, having the size and spacing shown in Fig. 25–19, are given equal and opposite charges, the field between and around them is approximately as shown in Fig. 25–19a. While most of the charge accumulates at the opposing faces of the plates and the field is essentially uniform in the space between them, there is a small quantity of charge on the outer surfaces of the plates and a certain spreading or "fringing" of the field at the edges of the plates.

As the plates are made larger and the distance between them diminished, the fringing becomes relatively less. Such an arrangement, two oppositely charged plates separated by a distance small compared with their linear dimensions, is encountered in many pieces of electrical equipment, notably in capacitors. Often the fringing is entirely negligible; even if it is not, neglecting it often provides useful approximations in cases where the work of more detailed calculations is not warranted. We shall therefore assume that the field between two oppositely charged plates is uniform, as in Fig. 25–19b, and that the charges are distributed uniformly over the opposing surfaces.

The electric field at any point can be considered as the resultant of that due to two sheets of charge of opposite sign, or it may be found from Gauss's law. Thus at points a and c in Fig. 25–19b, the components $\boldsymbol{E}_1$ and $\boldsymbol{E}_2$ are each of magnitude $\sigma/2\epsilon_0$ but are oppositely directed, so their resultant is zero. At any point b between the plates, the components are in the same direction and their resultant is σ/ϵ_0. It is left as an exercise to show that the same results follow from applying Gauss's law to the surfaces shown by dotted lines.

8. *Field just outside any charged conductor.* Figure 25–20 represents a portion of the surface of a charged conductor of irregular shape. In general, the surface density of charge will vary from point to point of the surface. Let σ represent the surface density at a small area A. We shall show in the next chapter that the electric field *just outside* the surface of any charged conductor is at right angles to the surface.

Let us construct a gaussian surface in the form of a small cylinder, one of whose end faces, of area A, lies within the conductor, while the other lies just outside. The charge within the gaussian surface is σA. The electric field is zero at all points within the conductor. Outside the conductor, the normal component of $\boldsymbol{E}$ is zero at the side walls of the cylinder (since $\boldsymbol{E}$ is normal to the conductor), while over the end face the normal component is equal to E. Hence from Gauss's law,

$$EA = \frac{\sigma A}{\epsilon_0}, \qquad E = \frac{\sigma}{\epsilon_0}. \qquad (25\text{–}26)$$

This agrees with the results already obtained for spherical, cylindrical, and plane surfaces. Just outside the surface of a sphere of radius R carrying a charge q, for example, the electric field is

$$E = \frac{1}{4\pi\epsilon_0}\frac{q}{R^2}.$$

But the surface density of charge on the sphere is $q/4\pi R^2$, so $E = \sigma/\epsilon_0$.

The field outside an infinite charged conducting plate was also shown to equal σ/ϵ_0. In this case, the field is the same at *all* distances from the plate, but in general it decreases with increasing distance from the surface.

The expressions that we have derived for the electric fields set up by a number of simple charge distributions are summarized in Table 25–1.

Finally, we note that, in each of the above examples, Gauss's law could be used to obtain an expression for the field because it enabled us to make effective use of the *symmetry* of the situation. There are many situations where such symmetry is lacking and where Gauss's law is *not* helpful as a tool for practical calculations. To cite a simple example, we consider the electric field produced by a dipole. The field does have axial symmetry about the dipole axis, but there is no simple surface over which $E_\perp$ is constant. Thus there is no practical way to apply Gauss's law to such a configuration, even though the law is still valid in principle in this situation.

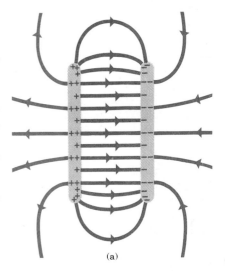

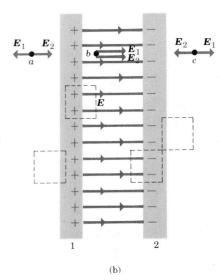

(b)

25–19 Electric field between oppositely charged parallel plates.

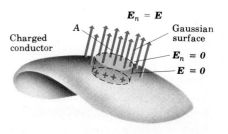

25–20 The field just outside a charged conductor is perpendicular to the surface and is equal to σ/ϵ_0.

Table 25-1   Electric fields around simple charge distributions

Charge distribution responsible for the electric field	Arbitrary point in the electric field	Electric field at this point
Single point charge q	Distance r from q	$E = \dfrac{1}{4\pi\epsilon_0}\dfrac{q\widehat{r}}{r^2}$
Several point charges, $q_1, q_2, \ldots$	Distance r_1 from q_1, r_2 from $q_2 \ldots$	$E = \dfrac{1}{4\pi\epsilon_0}\left(\dfrac{q_1\widehat{r}_1}{r_1{}^2} + \dfrac{q_2\widehat{r}_2}{r_2{}^2} + \cdots\right)$
Charge q uniformly distributed on the surface of a solid conducting sphere of radius R	(a) Outside, $r \geq R$ (b) Inside, $r < R$	(a) $E = \dfrac{1}{4\pi\epsilon_0}\dfrac{q\widehat{r}}{r^2}$ (b) $E = 0$
Long cylinder of radius R, with charge per unit length λ	(a) Outside, $r \geq R$ (b) Inside, $r < R$	(a) $E = \dfrac{1}{2\pi\epsilon_0}\dfrac{\lambda}{r}$ (b) $E = 0$
Two oppositely charged conducting plates with charge per unit area σ	Any point between plates	$E = \dfrac{\sigma}{\epsilon_0}$
Any charged conductor	Just outside the surface	$E = \dfrac{\sigma}{\epsilon_0}$

Questions

25-1 It was shown in the text that the electric field inside a spherical hole in a conductor is zero. Is this also true for a cubical hole? Can the same argument be used?

25-2 Coulomb's law and Newton's law of gravitation have the same *form*. Can Gauss's law be applied to gravitational fields as well as electric fields? If so, what modifications are needed?

25-3 By considering how the lines of force must look near a conducting surface, can you see why the charge density and electric field magnitude at the surface of an irregularly shaped solid conductor must be greatest in regions where the surface curves most sharply, and least in flat regions?

25-4 The electric field and the velocity field in a moving fluid are two examples of vector fields. Think of several other examples. There are also *scalar* fields, which associate a single number with each point in space. Temperature is an example; think of several others.

25-5 Consider the electric field caused by two point charges separated by some distance. Suppose there is a point where the field is zero; what does this tell you about the *signs* of the charges?

25-6 Does an electric charge experience a force due to the field that the charge itself produces?

25-7 A particle having electric charge and mass moves in an electric field. If it starts from rest, does it always move along the field line that passes through its starting point? Explain.

25-8 If the exponent 2 in the r^2 of Coulomb's law were 3 instead, would Gauss's law still be valid?

25-9 A student claimed that an appropriate unit for electric field magnitude is $1\ \text{J}\cdot\text{C}^{-1}\cdot\text{m}^{-1}$. Is this correct?

25-10 A certain region of space bounded by an imaginary closed surface contains no charge. Is the electric field always zero everywhere on the surface? If not, under what circumstances is it zero on the surface?

25-11 A student claimed that the electric field produced by a dipole is represented by field lines that cross each other. Is this correct? Is there any simple rule governing field lines for a superposition of fields due to point charges, if the field lines due to the separate charges are known? Do field lines *ever* cross?

25-12 Nineteenth-century physicists liked to give everything mechanical attributes. Faraday and his contemporaries thought of field lines as elastic strings that repelled each other and arranged themselves in equilibrium under the action of their elastic tension and mutual repulsion. Try this picture on several examples and decide whether it makes any sense. (These mechanical properties are of course now known to be completely fictitious.)

25-13 Are Coulomb's law and Gauss's law *completely* equivalent? Are there any situations in electrostatics where one is valid and the other isn't

25-14 The text states that, in an electrostatic field, every field line must start on a positive charge and terminate on a negative charge. But suppose the field is that of a single positive point charge. Then what?

25-15 Is the total (net) electric charge in the universe positive, negative, or zero?

Problems

25-1 A small object carrying a charge of -5×10^{-9} C experiences a downward force of 20×10^{-9} N when placed at a certain point in an electric field.

a) What is the electric field at the point?

b) What would be the magnitude and direction of the force acting on an electron placed at the point?

25-2 What must be the charge on a particle of mass 2 g for it to remain stationary in the laboratory when placed in a downward-directed electric field of 500 N·C^{-1}?

25-3 A uniform electric field exists in the region between two oppositely charged plane parallel plates. An electron is released from rest at the surface of the negatively charged plate and strikes the surface of the opposite plate, 2 cm distant from the first, in a time interval of 1.5×10^{-8} s.

a) Find the electric field.

b) Find the velocity of the electron when it strikes the second plate.

25-4 An electron is projected with an initial velocity $v_0 = 10^7$ m·s^{-1} into the uniform field between the parallel plates in Fig. 25-21. The direction of the field is vertically downward, and the field is zero except in the space between the plates. The electron enters the field at a point midway between the plates. If the electron just misses the upper plate as it emerges from the field, find the magnitude of the electric field.

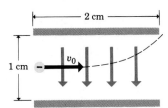

Figure 25-21

25-5 An electron is projected into a uniform electric field of 5000 N·C^{-1}. The direction of the field is vertically upward. The initial velocity of the electron is 10^7 m·s^{-1}, at an angle of 30° above the horizontal.

a) Find the maximum distance the electron rises vertically above its initial elevation.

b) After what horizontal distance does the electron return to its original elevation?

c) Sketch the trajectory of the electron.

25-6 In a rectangular coordinate system a charge of 25×10^{-9} C is placed at the origin of coordinates, and a charge of -25×10^{-9} C is placed at the point $x = 6$ m, $y = 0$. What is the electric field at

a) $x = 3$ m, $y = 0$?

b) $x = 3$ m, $y = 4$ m?

25-7 A charge of 16×10^{-9} C is fixed at the origin of coordinates, a second charge of unknown magnitude is at $x = 3$ m, $y = 0$, and a third charge of 12×10^{-9} C is at $x = 6$ m, $y = 0$. What is the magnitude of the unknown charge if the resultant field at $x = 8$ m, $y = 0$ is 20.25 N·C^{-1} directed to the right?

25-8 In a rectangular coordinate system, two positive point charges of 10^{-8} C each are fixed at the points $x = +0.1$ m, $y = 0$, and $x = -0.1$ m, $y = 0$. Find the magnitude and direction of the electric field at the following points:

a) the origin;

b) $x = 0.2$ m, $y = 0$;

c) $x = 0.1$ m, $y = 0.15$ m;

d) $x = 0$, $y = 0.1$ m.

25-9 Same as Problem 25-8, except that one of the point charges is positive and the other negative.

25-10

a) What is the electric field of a gold nucleus, at a distance of 10^{-12} cm from the nucleus?

b) What is the electric field of a proton, at a distance of 5.28×10^{-9} cm from the proton?

25-11 A small sphere whose mass is 0.1 g carries a charge of 3×10^{-10} C and is attached to one end of a silk fiber 5 cm long. The other end of the fiber is attached to a large vertical conducting plate, which has a surface charge of 25×10^{-6} C·m^{-2} on each side. Find the angle the fiber makes with the vertical.

25-12 How many excess electrons must be added to an isolated spherical conductor 10 cm in diameter to produce a field of intensity 1300 N·C^{-1} just outside the surface?

25-13 What is the magnitude of an electric field in which the force on an electron is equal in magnitude to the weight of the electron?

25-14 Electric charge is distributed uniformly over a disk of radius a, with total charge Q. Find the electric field at a point on the axis of the disk, distance x from its center. [*Hint*: Divide the disk into concentric rings, use the result of Example 2, Sec. 25-2, to find the field due to each ring, and integrate to find the total field.]

25-15 In Example 3 of Sec. 25-2, suppose that the wire is not infinitely long but has total length $2r$, and that point P lies on the perpendicular bisector of the wire. What is the electric field at point P? Compare this with the field of an infinitely long wire.

25-16 Electric charge is uniformly distributed around a semicircle of radius a, with total charge Q. What is the electric field at the center of curvature?

25-17 The electric field in the region between a pair of oppositely charged plane parallel plates, each $100\ \text{cm}^2$ in area, is $10^4\ \text{N·C}^{-1}$. What is the charge on each plate? Neglect edge effects.

25-18 A wire is bent into a ring of radius R and given a charge q.

a) What is the magnitude of the electric field at the center of the ring?

b) Derive the expression for the electric field at a point on a line perpendicular to the plane of the ring and passing through its center, at a distance r from the center of the ring. What is the direction of the E-vector at points on this line?

c) Sketch a graph of the magnitude of E as a function of r, from $r = 0$ to $r = 2R$.

25-19 The electric field E in Fig. 25-22 is everywhere parallel to the x-axis. The field has the same magnitude at all points in any given plane perpendicular to the x-axis (parallel to the yz-plane), but the magnitude is different for various planes. That is, E_x depends on x but not on y and z, and E_y and E_z are zero. At points *in* the yz-plane, $E_x = 400\ \text{N·C}^{-1}$. (The volume shown could be a section of a large insulating slab 1 m thick, with its faces parallel to the yz-plane and with a uniform volume charge distribution imbedded in it.)

a) What is the value of $\int E_\perp\, dA$ over surface I in the diagram?

b) What is the value of the surface integral of E over surface II?

c) There is a positive charge of $26.6 \times 10^{-9}\ \text{C}$ within the volume. What is the magnitude and direction of E at the face opposite I?

25-20 Apply Gauss's law to the dotted gaussian surfaces in Fig. 25-19b to calculate the electric field between and outside the plates.

25-21 A small conducting sphere of radius a, mounted on an insulating handle and having a positive charge q, is inserted through a hole in the walls of a hollow conducting sphere of inner radius b and outer radius c. The hollow sphere is supported on an insulating stand and is initially uncharged, and the small sphere is placed at the center of the hollow sphere. Neglect any effect of the hole.

a) Show that the electric field at a point in the region between the spheres, at a distance r from the center, is equal to

$$E = \frac{1}{4\pi\epsilon_0}\frac{q}{r^2}.$$

b) What is the electric field at a point outside the hollow sphere?

c) Sketch a graph of the magnitude of E as a function of r, from $r = 0$ to $r = 2c$.

d) Represent the charge on the small sphere by four $+$ signs. Sketch the field lines of the system, within a spherical volume of radius $2c$.

e) The small sphere is moved to a point near the inner wall of the hollow sphere. Sketch the field lines.

25-22

a) If the charge per unit length λ on the wire in Fig. 25-16 is finite, and the wire is infinitely long, the *total* charge on the wire is infinite. Explain why this infinite charge does not give rise to an infinite electrical field.

b) Draw a diagram showing an end view of an infinitely long charged wire, and the field lines in a plane perpendicular to the wire, far from either end. Explain, in terms of field lines, why the field decreases with $1/r$, although the field of a *point* charge decreases with $1/r^2$.

25-23 Prove that the electric field outside an infinitely long cylindrical conductor with a uniform surface charge is the same as if all the charge were on the axis.

25-24 A long coaxial cable consists of an inner cylindrical conductor of radius a and an outer coaxial cylinder of inner radius b and outer radius c. The outer cylinder is mounted on insulating supports and has no net charge. The inner cylinder has a uniform positive charge λ per unit length. Calculate the electric field

a) at any point between the cylinders, and

b) at any external point.

c) Sketch a graph of the magnitude of E as a function of the distance r from the axis of the cable, from $r = 0$ to $r = 2c$.

d) Find the charge per unit length on the inner surface of the outer cylinder, and on the outer surface.

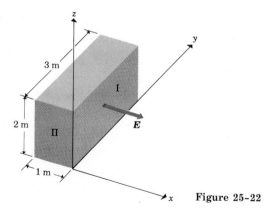

Figure 25-22

25-25 Suppose that positive charge is uniformly distributed throughout a spherical volume of radius R, the charge per unit volume being ρ.

a) Use Gauss's law to prove that the magnitude of the electric field inside the volume, at a distance r from the center, is

$$E = \frac{\rho r}{3\epsilon_0}.$$

b) What is the electric field at a point outside the spherical volume at a distance r from the center? Express your answer in terms of the total charge q within the spherical volume.

c) Compare the answers to (a) and (b) when $r = R$.

d) Sketch a graph of the magnitude of E as a function of r, from $r = 0$ to $r = 3R$.

25-26 Suppose that positive charge is uniformly distributed throughout a very long cylindrical volume of radius R, the charge per unit volume being ρ.

a) Derive the expression for the electric field inside the volume at a distance r from the axis of the cylinder, in terms of the charge density ρ.

b) What is the electric field at a point outside the volume, in terms of the charge per unit length λ in the cylinder?

c) Compare the answers to (a) and (b) when $r = R$.

d) Sketch a graph of the magnitude of E as a function of r, from $r = 0$ to $r = 3R$.

25-27 A conducting spherical shell of inner radius a and outer radius b has a positive point charge Q located at its center. The total charge on the shell is zero, and it is insulated from its surroundings.

a) Derive expressions for the electric field magnitude in terms of the distance r from the center, for the regions $r < a$, $a < r < b$, and $r > b$.

b) What is the surface charge density on the inner surface of the conducting shell?

c) Draw a sketch showing electric field lines and the location of all charges.

d) Draw a graph of E as a function of r.

25-28 The earth has net electric charge that causes a field at points near its surface of the order of $100 \text{ N} \cdot \text{C}^{-1}$. If the earth is regarded as a conducting sphere of radius $6.38 \times 10^6 \text{ m}$, what is the magnitude of its charge?

26 POTENTIAL

26-1 Electrical potential energy

In Chapter 6 the concepts of *work* and *energy* were introduced, and we saw that they provide a very important method of analysis for many mechanical problems. Energy plays an equally fundamental role in electricity and magnetism; in this chapter we shall apply work and energy considerations to the electric field. When a charged particle moves in an electric field, the field does *work* on the particle. We shall see that the work can always be expressed in terms of a potential energy, which in turn is associated with a new concept called *electrical potential* or simply *potential*.

In Sec. 6-6 the concept of a *conservative force field* was introduced; the reader would do well to review that discussion. Whenever the work done on a body undergoing a displacement can be expressed in terms of a potential energy function, the corresponding force is said to be *conservative*. If the potential energy function U has the value U_a at point a and the value U_b at point b, then the work $W_{a \to b}$ done by the force during any displacement from a to b, along any path, is given by

$$W_{a \to b} = U_a - U_b. \tag{26-1}$$

That is, the work done on the body equals its *loss* in potential energy. In order for this statement to have meaning, this work must be the same for all possible paths from a to b. If all the forces acting on the body are conservative, then the total mechanical energy of the body, kinetic and potential, is *conserved;* as we have seen, this principle simplifies the analysis of many mechanical systems.

We shall now show that the force on a charged particle in an electric field, due to other charges at rest, is a conservative force field. To begin, we consider only the interaction of two point charges. Specifically, consider a particle with a positive charge q' moving in the field produced by a stationary point charge q, as shown in Fig. 26-1. We shall first calculate the work done on the charge q' during a displacement along a radial line from point a to point b.

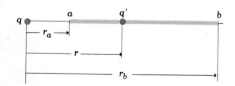

26-1 Charge q' moves along a straight line extending radially from charge q. As it moves from a to b, the distance varies from r_a to r_b.

The magnitude of the force on q' is given by Coulomb's law:

$$F = \frac{1}{4\pi\epsilon_0}\frac{qq'}{r^2}. \qquad (26\text{-}2)$$

The work done by this force on q' as it moves from r_a to r_b is given by

$$W_{a\to b} = \int_{r_a}^{r_b} F\,dr = \int_{r_a}^{r_b} \frac{1}{4\pi\epsilon_0}\frac{qq'}{r^2}\,dr$$

$$= \frac{qq'}{4\pi\epsilon_0}\left(\frac{1}{r_a} - \frac{1}{r_b}\right). \qquad (26\text{-}3)$$

Thus the work for this particular path depends only on the end-points. We have not yet proved that the work is the same for *all possible* paths from a to b. To do this, we consider a more general displacement, in which a and b *do not* lie on the same radial line, as in Fig. 26-2. To find the work in the displacement from point a to point b we must now use the more general definition of work given by Eq. (6-9):

$$W = \int_a^b F\cos\theta\,dl = \int_a^b \boldsymbol{F}\cdot d\boldsymbol{l}. \qquad (26\text{-}4)$$

But from the figure, $\cos\theta\,dl = dr$. That is, the work done during a small displacement dl depends only on the change dr in the distance r between the charges, which is the *radial component* of the displacement. In any displacement in which r does not change, no work is done because $\boldsymbol{F}$ and $d\boldsymbol{l}$ are perpendicular, $\cos\theta = 0$, and $\boldsymbol{F}\cdot d\boldsymbol{l} = 0$.

Thus Eq. (26-3) is equal to the work even for this more general displacement. This result shows that the work done on q' by the $\boldsymbol{E}$ field depends only on the initial and final distances r_a and r_b and not on the

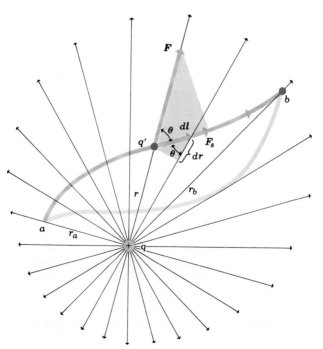

26-2 The work done by the electric-field force on charge q' depends only on the distances r_a and r_b.

path connecting these points. The work is the same for *any* path between these points, or between any pair of points at distances r_a and r_b from the charge q. Also, if the test charge is returned from b to a along any path, the work done by the $\boldsymbol{E}$ field is the *negative* of that done in the displacement from a to b. Thus, the total work done during a *loop* displacement returning to the starting point is always zero.

Comparing Eqs. (26–1) and (26–3), we see that the term $qq'/4\pi\epsilon_0 r_a$ is the potential energy U_a when q' is at point a, at distance r_a from q, and $qq'/4\pi\epsilon_0 r_b$ the potential energy U_b when it is at point b, at distance r_b from q. Thus the potential energy U of the test charge q' at *any* distance r from charge q is given by

$$U = \frac{1}{4\pi\epsilon_0}\frac{qq'}{r}. \tag{26-5}$$

If the field in which charge q' moves is not that of a single point charge but is due to a more general charge distribution, we may always divide the charge distribution into small elements and treat each element as a point charge. Then, since the total field is the vector sum of the fields due to the individual elements, and since the total work on q' is the sum of the contributions from the individual charge elements, we may conclude that *every* electric field due to a static charge distribution is a conservative force field. Furthermore, the potential energy of a test charge q' at point a in Fig. 26–3, due to a collection of charges q_1, q_2, q_3, etc., at distances r_1, r_2, r_3, etc., from the test charge q', is given by

$$U = \frac{q'}{4\pi\epsilon_0}\left(\frac{q_1}{r_1} + \frac{q_2}{r_2} + \frac{q_3}{r_3} + \cdots\right),$$

$$U = \frac{q'}{4\pi\epsilon_0}\sum\frac{q_i}{r_i}. \tag{26-6}$$

At a second point b, the potential energy is given by the same expression except that r_1, r_2, ... now represent the distances from the respective charges to point b. The work of the electric force in moving the test charge from a to b along any path is equal to the difference $U_a - U_b$ between its potential energies at a and at b.

In the discussion of potential energy in Chapter 6, it was observed that it is always possible to add an arbitrary constant C to a potential-energy function U, since in Eq. (26–1) adding the same constant C to both U_a and U_b does not change the physically significant quantity, namely the difference $U_a - U_b$. Thus we are always free to choose the constant C so that U is zero at some convenient reference position. In Chapter 6, the potential energy of a body in a uniform gravitational field could be taken to be zero at some convenient reference level, often the surface of the earth. When the body is above this reference level, its potential energy is positive; when below, negative.

In Eqs. (26–5) and (26–6), the reference position for electrical potential energy, the position at which $U = 0$, has been chosen implicitly to be that at which all the distances r_1, r_2, ... are *infinite*. That is, the potential energy of the test charge is zero when it is very far removed from all the charges setting up the field. This is the most convenient reference level for most electrostatic problems. When dealing with electrical *circuits,* other reference levels are more convenient, which simply means

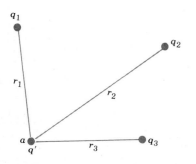

26-3 Potential energy of a charge q' at point a depends on charges q_1, q_2, and q_3, and on their distances r_1, r_2, r_3 from point a.

that a constant term is added to the potential energy. This is not significant, since it is only *differences* in potential energy that are of practical significance or even measurable.

From the above definitions, *the potential energy of a test charge at any point in an electric field is equal to the work of the electric force when the test charge is brought from the point in question to a zero reference level, often taken at infinity.*

26-2 Potential

Instead of dealing directly with the potential energy U of a charged particle, it is useful to introduce the more general concept of *potential energy per unit charge*. This quantity is called the *potential*, and *the potential at any point of an electrostatic field is defined as the potential energy per unit charge* at the point. Potential is represented by the letter V:

$$V = \frac{U}{q'}, \quad \text{or} \quad U = q'V. \tag{26-7}$$

Potential energy and charge are both *scalars*, so potential is a scalar quantity. Its basic SI unit is 1 *joule per coulomb* (1 J · C^{-1}). For brevity, a potential of 1 J · C^{-1} is called 1 *volt* (1 V). The unit is named in honor of the Italian scientist Alessandro Volta (1745–1827). Commonly used multiples of the volt are the kilovolt (1 kV = 10^3 V), the megavolt (1 MV = 10^6 volts), and the gigavolt (1 GV = 10^9 V). Submultiples include the millivolt (1 mV = 10^{-3} V) and the microvolt (1 μV = 10^{-6} V).

To put Eq. (26–1) on a "work per unit charge" basis, we divide both sides by q', obtaining

$$\frac{W_{a \to b}}{q'} = \frac{U_a}{q'} - \frac{U_b}{q'} = V_a - V_b, \tag{26-8}$$

where $V_a = U_a/q'$ is the potential energy per unit charge at point a, and similarly for V_b. V_a and V_b are called the *potential at point a* and *potential at point b*, respectively. From Eq. (26–6), the potential V at a point due to an arbitrary collection of point charges is given by

$$V = \frac{U}{q'} = \frac{1}{4\pi\epsilon_0} \sum \frac{q_i}{r_i} \tag{26-9}$$

The difference $V_a - V_b$ is called the *potential of a with respect to b*, and is often abbreviated V_{ab}. It should be noted that potential, like electric field, is independent of the test charge q' used to define it.

The potential due to a collection of point charges is usually obtained most easily by use of Eq. (26–9), but in some problems where the E field is known it is easier to work with it directly. The force F on the test charge q' can be written as $F = q'E$. When this relation is used in Eq. (26–4) and the result combined with Eq. (26–8), one obtains the useful result

$$V_{ab} = V_a - V_b = \int_a^b \boldsymbol{E} \cdot d\boldsymbol{l} = \int_a^b E \cos \theta \, dl. \tag{26-10}$$

Either integral on the right side is called the *line integral* of E. It represents the conceptual process of dividing a path into small elements *dl,* multiplying each magnitude *dl* by the component of E parallel to *dl* at that point, and summing the results for the entire path. The concept of line integral is very useful in formulating principles of physics, particularly in electricity and magnetism.

Equation (26–10) states that when a positive test charge moves from a region of high potential to one of lower potential (that is, $V_b < V_a$), the electric field does positive work on it. Thus a positive charge tends to "fall" from a high-potential region to a lower-potential one. The opposite is true for a negative charge.

A *voltmeter* is an instrument that measures the potential *difference* between the points to which its terminals are connected. The principle of the common type of moving-coil voltmeter will be described later. There are also much more sensitive potential-measuring devices that make use of electronic amplification, and devices that can measure a potential difference of $1 \mu V$ are used routinely.

Example 1   A particle having a charge $q = 3 \times 10^{-9}\,C$ moves from point a to point b along a straight line, a total distance $d = 0.5\,m$. The electric field is uniform along this line, in the direction from a to b, with magnitude $E = 200\,N \cdot C^{-1}$. Determine the force on q, the work done on it by the field, and the potential difference $V_a - V_b$.

Solution   The force is in the same direction as the electric field, and its magnitude is given by

$$F = qE = (3 \times 10^{-9}\,C)(200\,N \cdot C^{-1}) = 600 \times 10^{-9}\,N.$$

The work done by this force is

$$W = Fd = (600 \times 10^{-9}\,N)(0.5\,m) = 300 \times 10^{-9}\,J.$$

The potential difference is the work per unit charge, which is

$$V_a - V_b = \frac{W}{q} = \frac{300 \times 10^{-9}\,J}{3 \times 10^{-9}\,C} = 100\,J \cdot C^{-1} = 100\,V.$$

Alternatively, since E is force per unit charge, the work per unit charge is obtained by multiplying E by the distance d:

$$V_a - V_b = Ed = (200\,N \cdot C^{-1})(0.5\,m) = 100\,J \cdot C^{-1} = 100\,V.$$

Example 2   Point charges of $+12 \times 10^{-9}\,C$ and $-12 \times 10^{-9}\,C$ are placed 10 cm apart, as in Fig. 26–4. Compute the potentials at points a, b, and c.

Solution   We must evaluate the *algebraic* sum $(1/4\pi\epsilon_0)\,\Sigma(q_i/r_i)$ at each point. At point a, the potential due to the positive charge is

$$(9 \times 10^9\,N \cdot m^2 \cdot C^{-2})\frac{12 \times 10^{-9}\,C}{0.06\,m} = 1800\,N \cdot m \cdot C^{-1}$$

$$= 1800\,J \cdot C^{-1} = 1800\,V,$$

and the potential due to the negative charge is

$$(9 \times 10^9\,N \cdot m^2 \cdot C^{-2})\frac{-12 \times 10^{-9}\,C}{0.04\,m} = -2700\,J \cdot C^{-1} = -2700\,V.$$

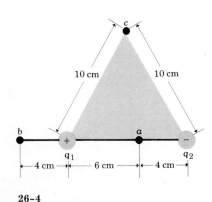

26–4

Hence

$$V_a = 1800 \text{ V} - 2700 \text{ V} = -900 \text{ V}$$
$$= -900 \text{ J} \cdot \text{C}^{-1}.$$

At point b, the potential due to the positive charge is $+2700$ V and that due to the negative charge is -770 V. Hence,

$$V_b = 2700 \text{ V} - 770 \text{ V} = 1930 \text{ V}$$
$$= 1930 \text{ J} \cdot \text{C}^{-1}.$$

At point c the potential is

$$V_c = 1080 \text{ V} - 1080 \text{ V} = 0.$$

Example 3 Compute the potential energy of a point charge of $+4 \times 10^{-9}$ C if placed at points a, b, and c in Fig. 26-4.

Solution   First,

$$U = qV.$$

Hence at point a,

$$U = qV_a = (4 \times 10^{-9} \text{ C})(-900 \text{ J} \cdot \text{C}^{-1})$$
$$= -36 \times 10^{-7} \text{ J}.$$

At point b,

$$U = qV_b = (4 \times 10^{-9} \text{ C})(1930 \text{ J} \cdot \text{C}^{-1})$$
$$= 77 \times 10^{-7} \text{ J}.$$

At point c,

$$U = qV_c = 0.$$

(All relative to a point at infinity.)

Example 4   In Fig. 26-5, a particle having mass $m = 5$ g and charge $q' = 2 \times 10^{-9}$ C starts from rest at point a and moves in a straight line to point b. What is its speed v at point b?

Solution   Conservation of energy gives

$$K_a + U_a = K_b + U_b.$$

For this situation, $K_a = 0$ and $K_b = \frac{1}{2}mv^2$. The potential energies are given in terms of the potentials by Eq. (26-6): $U_a = q'V_a$ and $U_b = q'V_b$. Thus the energy equation becomes

$$0 + q'V_a = \frac{1}{2}mv^2 + q'V_b.$$

When solved for v, this yields

$$v = \sqrt{\frac{2q'(V_a - V_b)}{m}}.$$

We obtain the potentials just as we did in the preceding examples:

$$V_a = (9.0 \times 10^9 \text{ N} \cdot \text{m}^2 \cdot \text{C}^{-2})\left(\frac{3 \times 10^{-9} \text{ C}}{0.01 \text{ m}} + \frac{-3 \times 10^{-9} \text{ C}}{0.02 \text{ m}}\right) = 1350 \text{ V},$$

$$V_b = (9.0 \times 10^9 \text{ N} \cdot \text{m}^2 \cdot \text{C}^{-2})\left(\frac{3 \times 10^{-9} \text{ C}}{0.02 \text{ m}} + \frac{-3 \times 10^{-9} \text{ C}}{0.01 \text{ m}}\right) = -1350 \text{ V}.$$

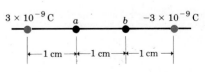

26-5

Finally,

$$v = \sqrt{\frac{2(2 \times 10^{-9}\,\text{C})(2700\,\text{V})}{5 \times 10^{-3}\,\text{kg}}} = 4.65 \times 10^{-2}\,\text{m·s}^{-1} = 4.65\,\text{cm·s}^{-1}.$$

Consistency of units may be checked by noting that $1\,\text{V} = 1\,\text{J·C}^{-1}$, and thus the numerator under the radical has units of J or $\text{kg·m}^2\text{·s}^{-2}$.

26-3 Calculation of potential differences

In principle the potential difference between any two points can be calculated by using Eq. (26–9) to find the potential at each point, but in many cases, especially where the electric field is easily obtained, it is easier to use Eq. (26–10), calculating from the known electric field the work done on a test charge during a displacement from one point to the other. In some problems a combination of these two approaches is useful. The following problems illustrate these remarks.

Example 1 *Charged spherical conductor.* We consider first a solid conducting sphere of radius R, with total charge q. From Gauss's law, as discussed in Chapter 25, we conclude that, at all points *outside* the sphere, the field is the same as that of a point charge q at the center of the sphere. *Inside* the sphere the field is zero everywhere; otherwise charge would move within the sphere.

As mentioned above, it is convenient to take the reference level of potential (the point where $V = 0$) at a very large distance from all charges. Since the field at any point r greater than R is the same as for a point charge, the work on a test charge moving from a finite value of r to infinity is also the same as for a point charge, so the potential V at a radial distance r, relative to a point at infinity, is

$$V = \frac{1}{4\pi\epsilon_0}\frac{q}{r}. \tag{26–11}$$

The potential is positive if q is positive, negative if q is negative.

Equation (26–11) applies to the field of a charged spherical conductor only when r is greater than or equal to the radius R of the sphere. The potential *at* the surface is

$$V = \frac{1}{4\pi\epsilon_0}\frac{q}{R}. \tag{26–12}$$

Inside the sphere the field is zero everywhere, and no work is done on a test charge displaced from any point to any other in this region. Thus the potential is the same at all points inside the sphere, and is equal to its value $q/4\pi\epsilon_0 R$ at the surface. The field and potential are shown as functions of r in Fig. 26–6.

The electric field $\boldsymbol{E}$ at the surface has a magnitude

$$E = \frac{1}{4\pi\epsilon_0}\frac{q}{R^2}. \tag{26–13}$$

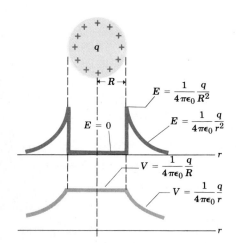

26-6 Electric field magnitude E and potential V at points inside and outside a charged spherical conductor.

The maximum potential to which a conductor in air can be raised is limited by the fact that air molecules become ionized, and hence the air becomes a conductor, at an electric field of about $3 \times 10^6\,\text{N·C}^{-1}$. Comparing Eqs. (26–12) and (26–13), we note that at the surface the field and potential are related by $V = ER$. Thus if E_m represents the upper limit of electric field, known as the *dielectric strength,* the maximum potential to which a spherical conductor can be raised is

$$V_m = RE_m.$$

For a sphere 1 cm in radius, in air,

$$V_m = (10^{-2}\,\text{m})(3 \times 10^6\,\text{N}\cdot\text{C}^{-1}) = 30{,}000\,\text{V},$$

and no amount of "charging" could raise the potential of a sphere of this size, in air, higher than about 30,000 V. It is this fact that necessitates the use of large spherical terminals on high-voltage machines. If we make $R = 2\,\text{m}$, then

$$V_m = (2\,\text{m})(3 \times 10^6\,\text{N}\cdot\text{C}^{-1}) = 6 \times 10^6\,\text{V} = 6\,\text{MV}.$$

At the other extreme is the effect produced by sharp points, a "point" being a surface of very *small* radius of curvature. Since the maximum potential is proportional to the radius, even relatively small potentials applied to sharp points in air will produce sufficiently high fields just outside the point to result in ionization of the surrounding air.

Example 2 *Parallel plates.*　The electric field between oppositely charged parallel plates, as derived in Sec. 25–5, is

$$E = \frac{\sigma}{\epsilon_0}, \tag{26–14}$$

where σ is the magnitude of the surface charge density (charge per unit area) on either plate. To obtain the potential difference between points a and x in Fig. 26–7, we note that the force on a test charge q' is $F = Eq'$, and the work done during a displacement from a to x is

$$W_{ax} = Fx = Eq'x.$$

Thus, the potential difference, the work per unit charge, also called the potential at a with respect to x, is

$$V_a - V_x = \frac{W_{ax}}{q'} = Ex,$$

or

$$V_x = V_a - Ex = V_a - \frac{\sigma}{\epsilon_0}x. \tag{26–15}$$

The potential therefore decreases *linearly* with x. At point b, where $x = l$ and $V_x = V_b$,

$$V_b = V_a - El,$$

and hence

$$E = \frac{V_a - V_b}{l} = \frac{V_{ab}}{l}. \tag{26–16}$$

That is, *the electric field equals the potential difference between the plates divided by the distance between them.* This is a more useful ex-

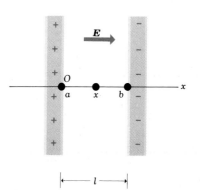

26-7

pression for E than Eq. (26–14) is, because the potential difference V_{ab} can readily be measured with a voltmeter, while there are no instruments that read surface density of charge directly. It must be understood, however, that Eqs. (26–15) and (26–16) are valid only when $\boldsymbol{E}$ is *uniform*.

Equation (26–16) also shows that the unit of electric field can be expressed as 1 *volt per meter* (1 V·m^{-1}), as well as 1 N·C^{-1}. In practice, the volt per meter is most commonly used as the unit of E.

Example 3   A simple type of vacuum tube known as a *diode* consists essentially of two electrodes within a highly evacuated enclosure. One electrode, the cathode, is maintained at a high temperature and emits electrons from its surface. A potential difference of a few hundred volts is maintained between the cathode and the other electrode, known as the *anode,* with the anode at the higher potential. Suppose that, in a certain diode, the anode potential V_a is 250 V above that (V_c) of the cathode, and an electron is emitted from the cathode with no initial velocity. What is its velocity when it reaches the anode?

Solution   Let V_c and V_a represent the cathode and anode potentials, respectively, and let the charge on the electron be $-e$. Since this charge is negative, the electron goes from the cathode, where the potential is lower, to the anode, where the potential is higher. Hence the work done on the elctron by the field is its charge ($-e$), where e is the (positive) magnitude of the electron charge, multiplied by the (negative) potential difference $V_c - V_a = V_{ca}$. This work in turn equals the change of kinetic energy of the electron.

Letting v_c and v_a be the speeds at cathode and anode, respectively, we have

$$(-e)(V_c - V_a) = e(V_a - V_c) = eV_{ac} = \tfrac{1}{2}mv_a{}^2 - \tfrac{1}{2}mv_c{}^2.$$

Since $v_c = 0$,

$$
\begin{aligned}
v_a &= \sqrt{2eV_{ac}/m} \\
&= \sqrt{\frac{2(1.6 \times 10^{-19}\ \text{C})(250\ \text{J·C}^{-1})}{9.1 \times 10^{-31}\ \text{kg}}} \\
&= 9.4 \times 10^6 (\text{J·kg}^{-1})^{1/2} = 9.4 \times 10^6\ \text{m·s}^{-1}.
\end{aligned}
$$

Note that the shape or separation of the electrodes need not be known. The final velocity depends only on the difference of potential between cathode and anode. Of course, the time of transit from cathode to anode depends on the geometry of the tube.

Example 4   *Line charge and charged conducting cylinder.*   The field of a line charge, and the field outside a charged conducting cylinder, are both given by

$$E = \frac{1}{2\pi\epsilon_0}\frac{\lambda}{r},$$

where λ is the charge per unit length.

The potential difference between any two points a and b at radial distances r_a and r_b is

$$V_a - V_b = \frac{\lambda}{2\pi\epsilon_0}\int_{r_a}^{r_b}\frac{dr}{r} = \frac{\lambda}{2\pi\epsilon_0}\ln\frac{r_b}{r_a}. \tag{26–17}$$

If we take point b at infinity and set $V_b = 0$, we find for the potential V_a,

$$V_a = \frac{\lambda}{2\pi\epsilon_0} \ln \frac{\infty}{r_a} = \infty.$$

Hence a reference point at infinity is not suitable for this field! We can, however, set $V = 0$ at some arbitrary radius r_0. Then at any radius r,

$$V = \frac{\lambda}{2\pi\epsilon_0} \ln \frac{r_0}{r}. \tag{26-18}$$

Equations (26–17) and (26–18) give the potential in the field of a cylinder only for values of r equal to or greater than the radius R of the cylinder. If r_0 is taken as the cylinder radius R, so that the potential of the cylinder is considered zero, the potential at any external point, relative to that of the cylinder, is

$$V = \frac{\lambda}{2\pi\epsilon_0} \ln \frac{R}{r}. \tag{26-19}$$

26-4 Equipotential surfaces

The potential distribution in an electric field may be represented graphically by *equipotential surfaces*. An equipotential surface is a surface such that the potential has the same value at all points on the surface. While an equipotential surface may be constructed through every point of an electric field, it is customary to show only a few of the equipotentials in a diagram.

Since the potential energy of a charged body is the same at all points of a given equipotential surface, it follows that no (electrical) work is needed to move a charged body over such a surface. Hence, the equipotential surface through any point must be at right angles to the direction of the field at that point. If this were not so, the field would have a component lying in the surface and work would have to be done against electrical forces to move a charge in the direction of this component. The field lines and the equipotential surfaces thus form a mutually perpendicular network. In general, the field lines of a field are curves and the equipotentials are curved surfaces. For the special case of a uniform field, where the field lines are straight and parallel, the equipotentials are parallel planes perpendicular to the field lines.

Figure 26–8 shows several arrangements of charges. The field lines are represented by colored lines, and cross sections of the equipotential surfaces as black lines. The actual field is of course three-dimensional. At each crossing of an equipotential and a field line, the two are perpendicular.

When charges at rest reside on the surface of a conductor, the electric field just outside the conductor must be everywhere perpendicular to the surface. To prove this, we imagine transporting a test charge q' around the loop $abcda$ in Fig. 26–9a. The segments bc and da (and the associated work) may be made arbitrarily small, and no work is done in the segment cd because, as already shown, the field is zero everywhere inside the conductor. If the field just outside the conductor has a component $E_\parallel$ parallel to the surface, this component does work equal to $q'E_\parallel l$.

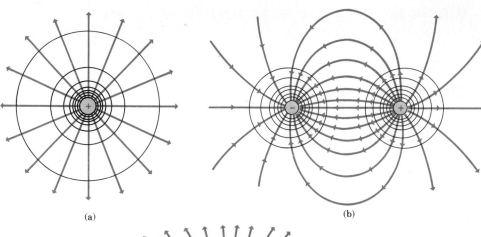

(a)

(b)

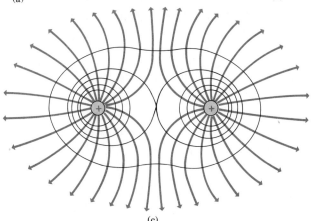

(c)

26-8 Equipotential surfaces (black lines) and field lines (colored lines) in the neighborhood of point charges.

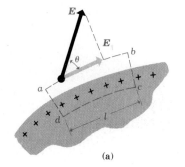

(a)

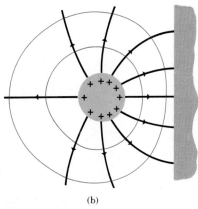

(b)

26-9 (a) Construction for finding the direction of E outside the surface of a conductor. (b) Field lines always meet charged conducting surfaces at right angles.

But then the net work done in the displacement ab is different from that in the displacement $adcb$, and the force field is not conservative. To avoid this contradiction, we must conclude that there *cannot* be a component of E parallel to the surface, and that E is therefore perpendicular to the surface. It also follows that, when all charges are at rest, *the surface of a conductor is always an equipotential surface.* Figure 26–9b illustrates these conclusions. Field lines (color) and equipotentials (black) are shown.

26-5 Potential gradient

The electric field E and the potential V in a region of space can always be regarded as functions of the coordinates of the point under consideration. These functions are of course closely related, and Eq. (26–10) is one form of this relation. A different form which is often useful relates the components of E to *derivatives* of V.

In Eq. (26–10), $V_{ab} = V_a - V_b$ is the change of potential when the point moves from b to a. This may be written as

$$V_{ab} = \int_b^a dV = -\int_a^b dV,$$

where dV is the infinitesimal change of potential accompanying a small displacement dl. As usual, $E \cos \theta$ is the component of E parallel to dl.

For brevity we denote this by $E_\parallel$; then, using the above expression, for V_{ab}, we may rewrite Eq. (26–10) as

$$-\int_a^b dV = \int_a^b E_\parallel \, dl.$$

Since these two integrals are equal for any pair of limits a and b, the *integrands* must be equal. Thus for an *infinitesimal* displacement dl,

$$-dV = E_\parallel \, dl,$$

or   (26–20)

$$E_\parallel = -\frac{dV}{dl}.$$

The derivative dV/dl is the rate of change of V for a displacement in the direction of dl. In particular, if dl is parallel to the x-axis, then the component of E parallel to dl is just the x-component of E, that is, E_x. Thus $E_x = -dV/dx$. Because V is in general also a function of y and z, the appropriate notation for a derivative in which only x varies is $\partial V/\partial x$. The y- and z-components of E are related to the corresponding derivatives of V in the same way, so we have

$$E_x = -\frac{\partial V}{\partial x},$$

$$E_y = -\frac{\partial V}{\partial y},\qquad (26\text{–}21)$$

$$E_z = -\frac{\partial V}{\partial z}.$$

In terms of unit vectors, E may be written as

$$E = -\left(i\frac{\partial V}{\partial x} + j\frac{\partial V}{\partial y} + k\frac{\partial V}{\partial z} \right)$$

(26–22)

$$= -\left(i\frac{\partial}{\partial x} + j\frac{\partial}{\partial y} + k\frac{\partial}{\partial z} \right) V.$$

In vector notation the operation

$$i\frac{\partial}{\partial x} + j\frac{\partial}{\partial y} + k\frac{\partial}{\partial z}$$

is called the *gradient* and is denoted by the symbol ∇. Thus in vector notation, Eqs. (26–21) are summarized compactly as

$$E = -\nabla V. \qquad (26\text{–}23)$$

This is read "E is the negative of the gradient of V."

Clearly, the units of potential gradient (volts per meter) must be the same as those of electric field (newtons per coulomb). This may be verified by recalling that potential is potential energy per unit charge and that $1\ \text{V} = 1\ \text{J} \cdot \text{C}^{-1}$. Then

$$1\ \text{V} \cdot \text{m}^{-1} = 1\ \text{J} \cdot \text{C}^{-1} \cdot \text{m}^{-1} = 1\ \text{N} \cdot \text{m} \cdot \text{C}^{-1} \cdot \text{m}^{-1}$$
$$= 1\ \text{N} \cdot \text{C}^{-1}.$$

Example 1   We have shown that the potential at a radial distance r from a point charge q is

$$V = \frac{1}{4\pi\epsilon_0} \frac{q}{r}.$$

By symmetry, the electric field is in the radial direction, so

$$E = E_r = -\frac{dV}{dr} = -\frac{d}{dr}\left(\frac{1}{4\pi\epsilon_0}\frac{q}{r}\right) = \frac{1}{4\pi\epsilon_0}\frac{q}{r^2},$$

in agreement with Coulomb's law.

Example 2   The potential outside a charged conducting cylinder of radius R and charge per unit length λ (Example 4 in Sec. 26-3) is

$$V = \frac{\lambda}{2\pi\epsilon_0} \ln \frac{R}{r} = \frac{\lambda}{2\pi\epsilon_0} (\ln R - \ln r).$$

The electric field is radial, and its magnitude is given by

$$E = -\frac{dV}{dr} = \frac{\lambda}{2\pi\epsilon_0 r},$$

in agreement with our previous result.

26-6 The Millikan oil-drop experiment

We have now developed the theory of electrostatics to a point where one of the classical physical experiments of all time can be described. In a brilliant series of investigations carried out at the University of Chicago in the period 1909–1913, Robert Andrews Millikan not only demonstrated conclusively the discrete nature of electric charge, but actually measured the charge of an individual electron.

Millikan's apparatus is shown schematically in Fig. 26-10a. Two accurately parallel horizontal metal plates A and B, are insulated from each other and separated by a few millimeters. Oil is sprayed in fine droplets from an atomizer above the upper plate and a few of the droplets are allowed to fall through a small hole in this plate. A beam of light is directed horizontally between the plates, and a telescope is set up with its axis at right angles to the light beam. The oil drops, illuminated by the light beam and viewed through the telescope, appear like tiny bright stars, falling slowly with a terminal velocity determined by their weight and by the viscous air-resistance force opposing their motion.

It is found that some of the oil droplets are electrically charged, presumably because of frictional effects. Charges can also be given the drops if the air in the apparatus is ionized by x-rays or a bit of radioactive material. Some of the electrons or ions then collide with the drops and stick to them. The drops are usually negatively charged, but occasionally one with a positive charge is found.

The simplest method, in principle, for measuring the charge on a drop is as follows. Suppose a drop has a negative charge and the plates are maintained at a potential difference such that a downward electric field of intensity $E\ (= V_{AB}/l)$ is set up between them. The forces on the drop are then its weight mg and the upward force qE. By adjusting the

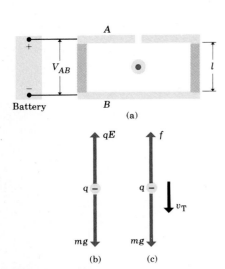

26-10   (a) Schematic diagram of Millikan apparatus. (b) Forces on a drop at rest. (c) Forces on a drop falling with its terminal velocity v_T.

field E, qE can be made just equal to mg, so that the drop remains at rest, as indicated in Fig. 26-10b. Under these circumstances,

$$q = \frac{mg}{E}.$$

The mass of the drop equals the product of its density ρ and its volume, $4\pi r^3/3$, and $E = V_{AB}/l$, so

$$q = \frac{4\pi}{3} \frac{\rho r^3 g l}{V_{AB}}. \qquad (26\text{-}24)$$

All the quantities on the right are readily measurable with the exception of the drop radius r, which is of the order of 10^{-5} cm and is much too small to be measured directly. It can be calculated, however, by cutting off the electric field and measuring the terminal velocity v_T of the drop as it falls through a known distance d defined by reference lines in the ocular of the telescope.

The terminal velocity is that at which the weight mg is just balanced by the viscous force f. The viscous force on a sphere of radius r, moving with a velocity v through a fluid of viscosity η, is given by Stokes' law, discussed in Sec. 13–7:

$$f = 6\pi\eta r v.$$

If Stokes' law applies, and the drop is falling with its terminal velocity v_T,

$$mg = f,$$

$$\frac{4}{3}\pi r^3 \rho g = 6\pi\eta r v_T,$$

and

$$r = 3\sqrt{\eta v_T/2\rho g}.$$

When this expression for r is inserted in Eq. (26-24), we have

$$q = 18\pi \frac{l}{V_{AB}} \sqrt{\frac{\eta^3 v_T^3}{2\rho g}}, \qquad (26\text{-}25)$$

which expresses the charge q in terms of measurable quantities.

In actual practice this procedure is modified somewhat. To correct for the buoyant force of the air through which the drop falls, the density ρ of the oil should be replaced by $(\rho - \rho_g)$, where ρ_g is the density of air. A correction to Stokes' law is also required, because air is not a continuous fluid but a collection of molecules separated by distances that are of the same order of magnitude as the dimensions of the drops.

Millikan and his co-workers measured the charges of thousands of drops, and found that, within the limits of their experimental error, every drop had a charge equal to some small integer multiple of a basic charge e. That is, drops were observed with charges of e, $2e$, $3e$, etc., but never with such values as $0.76e$ or $2.49e$. The evidence is conclusive that electric charge is not something that can be divided indefinitely, but that it exists in nature only in units of magnitude e. When a drop is observed with charge e, we conclude it has acquired one extra electron; if its charge is $2e$, it has two extra electrons, and so on.

As stated in Sec. 24–5, the best experimental value of the charge e is

$$e = 1.602192 \times 10^{-19}\,\text{C} \approx 1.60 \times 10^{-19}\,\text{C}.$$

The quark theory of fundamental particle structure, discussed in Sec. 47–11, incorporates particles called *quarks* having fractional charges $\pm e/3$ and $\pm 2e/3$. There are conflicting views among physicists as to whether properties (such as charge) of an individual quark should be directly observable. In 1979 the American physicist William Fairbank reported an experiment in which fractionally charged particles were observed. His results have not been duplicated in other laboratories, and their significance is still uncertain.

26-7 The electronvolt

The change in potential energy of a particle having a charge q, when it moves from a point where the potential is V_a to a point where the potential is V_b, is

$$\Delta U = q(V_a - V_b) = qV_{ab}.$$

In particular, if the charge q is the electronic charge $e = 1.60 \times 10^{-19}$ C, and the potential difference $V_{ab} = 1$ V, the change in energy is

$$\Delta U = (1.60 \times 10^{-19}\,\text{C})(1\,V) = 1.60 \times 10^{-19}\,\text{J}.$$

This quantity of energy is called 1 *electronvolt* (1 eV):

$$1\,\text{eV} = 1.60 \times 10^{-19}\,\text{J}.$$

Other commonly used units are:

$$1\,\text{keV} = 10^3\,\text{eV},$$
$$1\,\text{MeV} = 10^6\,\text{eV},$$
$$1\,\text{GeV} = 10^9\,\text{eV}.$$

The electronvolt is a convenient energy unit when one is dealing with the motions of electrons and ions in electric fields, because the change in potential energy between two points on the path of a particle having a charge e, when expressed in electronvolts, is *numerically* equal to the potential difference between the points, in volts. If the charge is some multiple of e, say Ne, the change in potential energy in electronvolts is, numerically, N times the potential difference in volts. For example, if a particle having a charge $2e$ moves between two points for which the potential difference is 1000 V, the change in its potential energy is

$$\begin{aligned}
\Delta U = qV_{ab} &= (2)(1.6 \times 10^{-19}\,\text{C})(10^3\,V) \\
&= 3.2 \times 10^{-16}\,\text{J} \\
&= 2000\,\text{eV}.
\end{aligned}$$

Although the electronvolt was defined above in terms of *potential* energy, energy of *any* form, such as the kinetic energy of a moving particle, can be expressed in terms of the electronvolt. Thus one may speak of a "one-million-volt electron," meaning an electron having a kinetic energy of one million electronvolts (1 MeV).

One of the principles of the special theory of relativity, to be developed in Chapter 43, is that the mass m of a particle is equivalent to a quantity of energy mc^2, where c is the speed of light. The rest mass of an electron is 9.108×10^{-31} kg, and the energy equivalent to this is

$$E_0 = mc^2 = (9.108 \times 10^{-31} \text{ kg})(3 \times 10^8 \text{ m} \cdot \text{s}^{-1})^2$$
$$= 82 \times 10^{-15} \text{ J}.$$

Since $1 \text{ eV} = 1.60 \times 10^{-19}$ J, this is equivalent to

$$E_0 = 511{,}000 \text{ eV} = 0.511 \text{ MeV}.$$

When the kinetic energy of a particle becomes of comparable magnitude to its rest energy, Newton's laws of motion are not strictly valid and must be replaced by the more general relations of relativistic mechanics, to be developed in detail in Chapter 43. For example, an electron accelerated through a potential difference of 500 kV acquires a kinetic energy approximately *equal* to its rest energy, and a correct analysis of the motion of the particle requires the use of relativistic mechanics.

*26–8 The cathode-ray oscilloscope

Figure 26–11 is a schematic diagram of the elements of a cathode-ray oscilloscope tube. The interior of the tube is highly evacuated. The *cathode* at the bottom is raised to a high temperature by the *heater,* and electrons evaporate from its surface. (Before the nature of this process of electron emission was fully understood, these electrons were given the name "cathode rays.") The *accelerating anode,* which has a small hole at its center, is maintained at a high positive potential V_1 relative to the cathode, so that there is an electric field, directed downward between the anode and cathode. This field is confined to the cathode-anode region, and electrons passing through the hole in the anode travel with *constant* vertical velocity from the anode to the *fluorescent screen.*

The function of the *control grid* is to regulate the number of electrons that reach the anode (and hence the brightness of the spot on the screen). The *focusing anode* ensures that electrons leaving the cathode in slightly different directions all arrive at the same spot on the screen. These two electrodes need not be considered in the following analysis. The complete assembly of cathode, control grid, focusing anode, and accelerating electrode is referred to as an *electron gun.*

The accelerated electrons pass between two pairs of *deflecting plates.* An electric field between the first pair of plates deflects them to the front or back, and a field between the second pair deflects them right or left. In the absence of such fields the electrons travel in a straight line from the hole in the accelerating anode to the *fluorescent screen* and produce a bright spot on the screen where they strike it.

We first calculate the speed v imparted to the electrons by the electron gun, just as in Example 4 of Sec. 26–2. We assume the electrons leave the cathode with zero initial speed; in fact, the electrons do have some small initial speed when they evaporate from the cathode, but it is very small compared to the final speed and can be neglected. As in the example, we find

$$v = \sqrt{\frac{2eV_1}{m}}. \qquad (26\text{–}26)$$

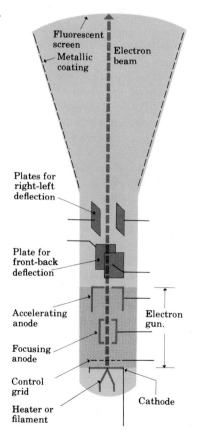

26–11 Basic elements of a cathode-ray tube.

As a numerical example, if $V_1 = 2000$ V,

$$v = \sqrt{\frac{2(1.6 \times 10^{-19}\,\text{C})(2 \times 10^3\,\text{V})}{9.11 \times 10^{-31}\,\text{kg}}}$$

$$= 2.65 \times 10^7\,\text{m}\cdot\text{s}^{-1}.$$

We note that the kinetic energy of an electron at the anode depends only on the *potential difference* between anode and cathode, and not at all on the details of the fields within the electron gun or on the shape of the electron trajectory within the gun.

If there is no electric field between the front–back deflection plates, the electrons enter the region between the other plates with a speed equal to v and represented by v_x in Fig. 26–12. (This view is turned 90° from that of Fig. 26–11.) If there is a potential difference V_2 between the plates, and the upper plate is positive, a downward electric field of magnitude $E = V_2/l$ is set up between the plates. A constant upward force eE then acts on the electrons, and their upward acceleration is

$$a_y = \frac{eE}{m} = \frac{eV_2}{ml}. \tag{26–27}$$

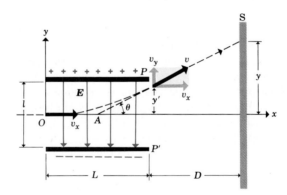

26-12 Electrostatic deflection of cathode rays.

The *horizontal* component of velocity v_x remains constant, so the time required for the electrons to travel the length L of the plates is

$$t = \frac{L}{v_x}. \tag{26–28}$$

In this time, they acquire an upward velocity component given by

$$v_y = a_y t, \tag{26–29}$$

and are displaced upward by an amount

$$y' = \frac{1}{2}a_y t^2.$$

On emerging from the deflecting field, their velocity v makes an angle θ with the x-axis, where

$$\tan\theta = \frac{v_y}{v_x};$$

and from this point on they travel in a straight line to the screen. It is not difficult to show that this straight line, if projected backward, intersects the x-axis at a point A that is midway between the ends of the plates. Then if y is the vertical coordinate of the point of impact with screen S,

$$\tan \theta = \frac{y}{D + (L/2)}.$$

When this is combined with Eqs. (26–27), (26–28), and (26–29), we get finally

$$y = \left[\frac{L}{2l} \left(D + \frac{L}{2} \right) \right] \frac{V_2}{V_1}. \qquad (26\text{–}30)$$

The term in brackets is a purely geometrical factor. If the accelerating voltage V_1 is held constant, *the deflection y is proportional to the deflecting voltage V_2.*

If a field is also set up between the *horizontal* deflecting plates, the beam is also deflected in the horizontal direction, perpendicular to the plane of Fig. 26–12. The coordinates of the luminous spot on the screen are then proportional, respectively, to the horizontal and vertical deflecting voltages. This device, with its associated circuitry, forms the basis of the *oscilloscope,* a laboratory instrument widely used for the observation and measurement of rapidly varying voltages.

The picture tube in a television set is similar in its operation to the oscilloscope tube, except that the beam is deflected by magnetic fields (to be discussed in later chapters) rather than by electric fields. Graphic display devices on some computer terminals also are similar, using a deflected electron beam and a fluorescent screen. The accelerating voltage in TV picture tubes (V_1 in the above discussion) is typically 20 to 25 kV.

Questions

26-1 Are there cases in electrostatics where a conducting surface is *not* an equipotential surface? If so, give an example.

26-2 If the electric potential at a single point is known, can the electric field at that point be determined?

26-3 If two points are at the same potential, is the electric field necessarily zero everywhere between them?

26-4 If the electric field is zero throughout a certain region of space, is the potential also zero in this region? If not, what *can* be said about the potential?

26-5 A student said: "Since electrical potential is always proportional to potential energy, why bother with the concept of potential at all?" How would you respond?

26-6 Is potential gradient a scalar or vector quantity?

26-7 A conducting sphere is to be charged by bringing in positive charge, a little at a time, until the total charge is Q. The total work required for this process is alleged to be proportional to Q^2. Is this correct? Why or why not?

26-8 The potential (relative to a point at infinity) midway between two charges of equal magnitude and opposite sign is zero. Can you think of a way to bring a test charge from infinity to this midpoint in such a way that no work is done in any part of the displacement?

26-9 A high-voltage dc power line falls on a car, so the entire metal body of the car is at a potential of 10,000 V with respect to the ground. What happens to the occupants

a) when they are sitting in the car;

b) when they step out of the car?

26-10 In electronics it is customary to define the potential of ground (thinking of the earth as a large conductor) as zero. Is this consistent with the fact that the earth has a net electric charge that is not zero? (Cf. Problem 25–28.)

26-11 A positive point charge is placed near a very large conducting plane. A professor of physics asserted that the field caused by this configuration is the same as would be obtained by removing the plane and placing a negative point charge of equal magnitude in the mirror-image position behind the initial position of the plane. Is this correct?

26-12 It is easy to produce a potential of several thousand volts on your body by scuffing your shoes across a nylon carpet, yet contact with a power line of comparable voltage would probably be fatal. What's the difference?

Problems

26-1 Two large parallel metal sheets carrying equal and opposite electric charges are separated by a distance of 0.05 m. The electric field between them is approximately uniform and has magnitude 600 N·C^{-1}. What is the potential difference between the plates? Which plate is at higher potential?

26-2 Two large parallel metal plates carry opposite charges. They are separated by 0.1 m, and the potential difference between them is 500 V.

a) What is the magnitude of the electric field, if it is uniform, in the region between the plates?

b) Compute the work done by this field on a charge of 2.0×10^{-9} C as it moves from the higher-potential plate to the lower.

c) Compare the result of (b) to the change of potential energy of the same charge, computed from the electrical potential.

26-3 A total electric charge of 4×10^{-9} C is distributed uniformly over the surface of a sphere of radius 0.20 m. If the potential is zero at a point at infinity, what is the value of the potential

a) at a point on the surface of the sphere;

b) at a point inside the sphere, 0.1 m from the center?

26-4 A particle of charge $+3 \times 10^{-9}$ C is in a uniform electric field directed to the left. It is released from rest and moves a distance of 5 cm, after which its kinetic energy is found to be $+4.5 \times 10^{-5}$ J.

a) What work was done by the electrical force?

b) What is the magnitude of the electric field?

c) What is the potential of the starting point with respect to the endpoint?

26-5 In Problem 26-4, suppose that another force in addition to the electrical force acts on the particle, so that when it is released from rest it moves to the right. After it has moved 5 cm, the additional force has done 9×10^{-5} J of work and the particle has 4.5×10^{-5} J of kinetic energy.

a) What work was done by the electrical force?

b) What is the magnitude of the electric field?

c) What is the potential of the starting point with respect to the endpoint?

26-6 A charge of 2.5×10^{-8} C is placed in an upwardly directed uniform electric field having magnitude 5×10^4 N·C^{-1}. What is the work of the electrical force when the charge is moved

a) 45 cm to the right?

b) 80 cm downward?

c) 260 cm at an angle of 45° upward from the horizontal?

26-7

a) Show that $1 \, \text{N·C}^{-1} = 1 \, \text{V·m}^{-1}$.

b) A potential difference of 2000 V is established between parallel plates in air. If the air becomes electrically conducting when the electric field exceeds 3×10^6 N·C^{-1}, what is the minimum separation of the plates?

26-8 A small sphere of mass 0.2 g hangs by a thread between two parallel vertical plates 5 cm apart. The charge on the sphere is 6×10^{-9} C. What potential difference between the plates will cause the thread to assume an angle of 30° with the vertical?

26-9 Two point charges $q_1 = +40 \times 10^{-9}$ C and $q_2 = -30 \times 10^{-9}$ C are 10 cm apart. Point A is midway between them, point B is 8 cm from q_1 and 6 cm from q_2. Find

a) the potential at point A;

b) the potential at point B;

c) the work required to carry a charge of 25×10^{-9} C from point B to point A.

26-10 Three equal point charges of 3×10^{-7} C are placed at the corners of an equilateral triangle whose side is one meter. What is the potential energy of the system? Take as zero potential energy the energy of the three charges when they are infinitely far apart.

26-11 The potential at a certain distance from a point charge is 600 V, and the electric field is 200 N·C^{-1}.

a) What is the distance to the point charge?

b) What is the magnitude of the charge?

26-12 Two point charges whose magnitudes are $+20 \times 10^{-9}$ C and -12×10^{-9} C are separated by a distance of 5 cm. An electron is released from rest between the two charges, 1 cm from the negative charge, and moves along the line connecting the two charges. What is its velocity when it is 1 cm from the positive charge?

26-13 Two positive point charges, each of magnitude q, are fixed on the y-axis at the points $y = +a$ and $y = -a$.

a) Draw a diagram showing the positions of the charges.

b) What is the potential V_0 at the origin?

c) Show that the potential at any point on the x-axis is

$$V = \frac{1}{4\pi\epsilon_0} \frac{2q}{\sqrt{a^2 + x^2}}.$$

d) Sketch a graph of the potential on the x-axis as a function of x over the range from $x = +4a$ to $x = -4a$.

e) At what value of x is the potential one-half that at the origin?

26–14 Consider the same distribution of charges as in Problem 26–13.

a) Sketch a graph of the potential on the y-axis as a function of y, over the range from $y = +4a$ to $y = -4a$.

b) Discuss the physical meaning of the graph at the points $+a$ and $-a$.

c) At what point or points on the y-axis is the potential equal to that at the origin?

d) At what points on the y-axis is the potential equal to half its value at the origin?

26–15 Consider the same charge distribution as in Problem 26–13.

a) Suppose a positively charged particle of charge q' and mass m is placed precisely at the origin and released from rest. What happens?

b) What will happen if the charge in part (a) is displaced slightly in the direction of the y-axis?

c) What will happen if it is displaced slightly in the direction of the x-axis?

26–16 Again consider the charge distribution in Problem 26–13. Suppose a positively charged particle of charge q' and mass m is displaced slightly from the origin in the direction of the x-axis.

a) What is its speed at infinity?

b) Sketch a graph of the velocity of the particle as a function of x.

c) If the particle is projected toward the left along the x-axis from a point at a large distance to the right of the origin, with a velocity half that acquired in part (a), at what distance from the origin will it come to rest?

d) If a negatively charged particle were released from rest on the x-axis, at a very large distance to the left of the origin, what would be its velocity as it passed the origin?

26–17 A positive charge $+q$ is located at the point $x = -a$, $y = -a$, and an equal negative charge $-q$ is located at the point $x = +a$, $y = -a$.

a) Draw a diagram showing the positions of the charges.

b) What is the potential at the origin?

c) What is the expression for the potential at a point on the x-axis, as a function of x?

d) Sketch a graph of the potential as a function of x, in the range from $x = +4a$ to $x = -4a$. Plot positive potentials upward, negative potentials downward.

26–18 A potential difference of 1600 V is established between two parallel plates 4 cm apart. An electron is released from the negative plate at the same instant that a proton is released from the positive plate.

a) How far from the positive plate will they pass each other?

b) How do their velocities compare when they strike the opposite plates?

c) How do their energies compare when they strike the opposite plates?

26–19 Consider the same charge distribution as in Problem 26–13.

a) Construct a graph of the potential energy of a positive point charge on the x-axis, as a function of x.

b) Construct a graph of the potential energy of a negative point charge on the axis, as a function of x.

c) What is the x-component of the potential gradient at the origin?

26–20 In the Bohr model of the hydrogen atom, a single electron revolves around a single proton in a circle of radius R.

a) By equating the electrical force to the electron mass times its acceleration, derive an expression for the electron's speed.

b) Obtain an expression for the electron's kinetic energy, and show that its magnitude is just half that of the electrical potential energy.

c) Obtain an expression for the total energy, and evaluate it using $R = 0.528 \times 10^{-10}$ m.

26–21 A vacuum diode consists of a cylindrical cathode of radius 0.05 cm, mounted coaxially within a cylindrical anode 0.45 cm in radius. The potential of the anode is 300 V higher than that of the cathode. An electron leaves the surface of the cathode with zero initial speed. Find its speed when it strikes the anode.

26–22 A vacuum triode may be idealized as follows. A plane surface (the cathode) emits electrons with negligible initial velocities. Parallel to the cathode and 3 mm away from it is an open grid of fine wire at a potential of 18 V above the cathode. A second plane surface (the anode) is 12 mm beyond the grid and is at a potential of 15 V above the cathode. Assume that the plane of the grid is an equipotential surface, and that the potential gradients between cathode and grid, and between grid and anode, are uniform. Assume also that the structure of the grid is sufficiently open for electrons to pass through it freely.

a) Draw a diagram of potential vs. distance, along a line from cathode to anode.

b) With what speed will electrons strike the anode?

26-23 A ring-shaped conductor of radius a carries a total charge Q; it is placed with its plane perpendicular to the x-axis and its center at the origin. (Cf. Sec. 25-2, Example 2.)

a) Derive an expression for the potential at any point on the x-axis.

b) Using Eq. (26-21) and the result of (a), derive an expression for the x-component of electric field at any point on the x-axis. Compare your result to that of the example cited above.

c) Construct graphs of the potential V and the x-component of field E_x as functions of x, and show the geometric relationship between the two graphs.

26-24 A metal sphere of radius r_a is supported on an insulating stand at the center of a hollow metal sphere of inner radius r_b. There is a charge $+q$ on the inner sphere and a charge $-q$ on the outer. (See Problem 25-21.)

a) Show that the potential of the inner sphere with respect to the outer is

$$V_{ab} = \frac{q}{4\pi\epsilon_0}\left(\frac{1}{r_a} - \frac{1}{r_b}\right).$$

b) Show that the electric field at any point between the spheres has magnitude

$$E = \frac{V_{ab}}{(1/r_a - 1/r_b)} \cdot \frac{1}{r^2}.$$

c) Find the electric field at a point outside the larger sphere, at a distance r from the center, where $r > r_b$.

d) Suppose the charge on the outer sphere is not $-q$ but a negative charge of different magnitude, say $-Q$. Show that the answers for (a) and (b) are the same as before but (c) is different.

26-25 A long metal cylinder of radius r_a is supported on an insulating stand on the axis of a long hollow metal cylinder of inner radius r_b. The positive charge per unit length on the inner cylinder is λ and there is an equal negative charge per unit length on the outer cylinder. (See Problem 25-24.)

a) Show that the potential of the inner cylinder with respect to the outer is

$$V_{ab} = \frac{\lambda}{2\pi\epsilon_0}\ln\frac{r_b}{r_a}.$$

b) Show that the electric field at any point between the cylinders has magnitude

$$E = \frac{V_{ab}}{\ln(r_b/r_a)} \cdot \frac{1}{r}$$

c) What is the potential difference if the outer cylinder has no net charge?

26-26 Refer to Problem 25-25.

a) Find the expression for the potential V as a function of r, both inside and outside the sphere, relative to a point at infinity.

b) Sketch graphs of V and E as functions of r from $r = 0$ to $r = 3R$, and compare with Fig. 26-6.

26-27 Refer to Problem 25-26.

a) Find the expressions for the potential V as a function of r, both inside and outside the cylinder. Let $V = 0$ at the surface of the cylinder.

b) Sketch graphs of V and E as functions of r, from $r = 0$ to $r = 3R$.

26-28 Refer to Problem 24-19.

a) Calculate the potential at the origin, and at the point (3 cm, 0) due to the first two point charges.

b) Calculate the work the electric field would do on the third charge if it moved from the origin to the point (3 cm, 0).

26-29 Refer to Problem 24-20.

a) Calculate the potential at the point (3 cm, 0) and at the point (3 cm, 4 cm) due to the first two charges.

b) If the third charge moves from the point (3 m, 0) to the point (3 cm, 4 cm), calculate the work done on it by the field of the first two charges. Comment on the *sign* of this work; is your result reasonable?

26-30 In an apparatus for measuring the electronic charge e by Millikan's method, an electric intensity of 6.34×10^4 V·m^{-1} is required to maintain a certain charged oil drop at rest. If the plates are 1.5 cm apart, what potential difference between them is required?

26-31 An oil droplet of mass 3×10^{-11} g and of radius 2×10^{-4} cm carries 10 excess electrons. What is its terminal velocity

a) when falling in a region in which there is no electric field?

b) When falling in an electric field of magnitude 3×10^5 N·C^{-1} directed downward?

The viscosity of air is 180×10^{-7} N·s·m^{-2}. (Neglect the buoyant force of the air.)

26-32 A charged oil drop, in a Millikan oil-drop apparatus, is observed to fall through a distance of 1 mm in a time of 27.4 s, in the absence of any external field. The same drop can be held stationary in a field of 2.37×10^4 N·C^{-1}. How many excess electrons has the drop acquired? The viscosity of air is 180×10^{-7} N·s·m^{-2}. The density of oil is 824 kg·m^{-3}, and the density of air is 1.29 kg·m^{-3}.

26-33 Find the energy equivalent of the rest mass of the proton; express your result in MeV.

26-34 Find the potential energy of the interaction of two protons at a distance of 10^{-15} m, typical of the dimensions of atomic nuclei. Express your result in MeV.

26-35 An alpha particle with kinetic energy 10 MeV makes a head-on collision with a gold nucleus at rest.

What is the distance of closest approach of the two particles?

26-36

a) Prove that when a particle of constant mass and charge is accelerated from rest in an electric field, its final velocity is proportional to the square root of the potential difference through which it is accelerated.

b) Find the magnitude of the proportionality constant if the particle is an electron, the velocity is in meters per second, and the potential difference is in volts.

c) What is the final velocity of an electron accelerated through a potential difference of 1136 V if it has an initial velocity of 10^7 m·s⁻¹?

26-37

a) What is the maximum potential difference through which an electron can be accelerated if its kinetic energy is not to exceed 1% of the rest energy?

b) What is the speed of such an electron, expressed as a fraction of the speed of light, c?

c) Make the same calculations for a *proton*.

26-38 The electric field in the region between the deflecting plates of a certain cathode-ray oscilloscope is 30,000 N·C⁻¹.

a) What is the force on an electron in this region?

b) What is the acceleration of an electron when acted on by this force?

26-39 In Fig. 26-13, an electron is projected along the axis midway between the plates of a cathode-ray tube with an initial velocity of 2×10^7 m·s⁻¹. The uniform electric field between the plates has an intensity of 20,000 N·C⁻¹ and is upward.

a) How far below the axis has the electron moved when it reaches the end of the plates?

b) At what angle with the axis is it moving as it leaves the plates?

c) How far below the axis will it strike the fluorescent screen S?

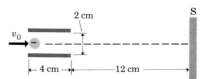

Figure 26-13

26-40 An electron with kinetic energy 100 MeV collides head-on with a gold nucleus at rest. Assuming that the gold nucleus can be treated as a uniform distribution of charge through a sphere of radius 7×10^{-15} m and that the electron can penetrate into the nucleus, what is its kinetic energy when it reaches the center of the nucleus?

CAPACITANCE.
PROPERTIES
OF DIELECTRICS

27

27–1 Capacitors

Any two conductors separated by an insulator are said to form a *capacitor*. In most cases of practical interest the conductors usually have charges of equal magnitude and opposite sign, so that the *net* charge on the capacitor as a whole is zero. The electric field in the region between the conductors is then proportional to the magnitude of this charge, and it follows that the *potential difference* V_{ab} between the conductors is also proportional to the charge magnitude Q.

The *capacitance C* of a capacitor is defined as the ratio of the magnitude of the charge Q on *either* conductor to the magnitude of the potential difference V_{ab} between the conductors:

$$C = \frac{Q}{V_{ab}}. \tag{27–1}$$

It follows from its definition that the unit of capacitance is one *coulomb per volt* (1 C·V^{-1}). A capacitance of one coulomb per volt is called one *farad* (1 F) in honor of Michael Faraday. A capacitor is represented by the symbol

When one speaks of a capacitor as having charge Q, what is really meant is that the conductor at higher potential has a charge Q, and the conductor at lower potential a charge $-Q$ (assuming Q is a positive quantity). This interpretation should be kept in mind in the following discussion and examples.

Capacitors find many applications in electrical circuits. Capacitors are used for tuning radio circuits and for "smoothing" the rectified current delivered by a power supply. A capacitor is used to eliminate sparking when a circuit containing inductance is suddenly opened. The ignition system of every automobile engine contains a capacitor to eliminate sparking of the "points" when they open and close. The efficiency of alternating-current power transmission can often be increased by the use of large capacitors.

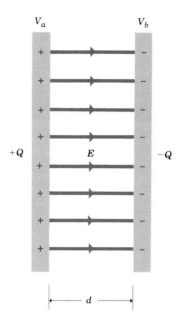

27-1 Parallel-plate capacitor.

27-2 The parallel-plate capacitor

The most common type of capacitor consists in principle of two conducting plates parallel to each other and separated by a distance that is small compared with the linear dimensions of the plates (see Fig. 27–1). Practically the entire field of such a capacitor is localized in the region between the plates, as shown. There is a slight "fringing" of the field at its outer boundary (bending outward of field lines near the edges), but if the plates are sufficiently close, this may be neglected. The field between the plates is then uniform, and the charges on the plates are uniformly distributed over their opposing surfaces. This arrangement is known as a *parallel-plate capacitor.*

Let us assume first that the plates are in vacuum. It has been shown that the electric field magnitude between a pair of closely spaced parallel plates in vacuum is

$$E = \frac{\sigma}{\epsilon_0} = \frac{Q}{\epsilon_0 A},$$

where σ is the magnitude of surface charge density on either plate, A is the area of each plate, and Q is the magnitude of total charge on each plate. Since the electric field (potential gradient) between the plates is uniform, the potential difference between the plates is

$$V_{ab} = Ed = \frac{1}{\epsilon_0}\frac{Qd}{A},$$

where d is the separation of the plates. Hence the capacitance of a parallel-plate capacitor in vacuum is

$$C = \frac{Q}{V_{ab}} = \epsilon_0 \frac{A}{d}. \qquad (27\text{-}2)$$

Since ϵ_0, A, and d are constants for a given capacitor, the capacitance is a constant independent of the charge on the capacitor, and is directly proportional to the area of the plates and inversely proportional to their separation. If SI units are used, A is expressed in square meters and d in meters. The capacitance C is then in farads.

As an example, let us compute the area of the plates of a 1-F parallel-plate capacitor if the separation of the plates is 1 mm and the plates are in vacuum:

$$C = \epsilon_0 \frac{A}{d},$$

$$A = \frac{Cd}{\epsilon_0} = \frac{(1\ \text{F})(10^{-3}\ \text{m})}{8.85 \times 10^{-12}\ \text{C}^2\cdot\text{N}^{-1}\cdot\text{m}^{-2}}$$
$$= 1.13 \times 10^8\ \text{m}^2.$$

This corresponds to a square 10,600 m, or 34,800 ft, or about $6\frac{1}{2}$ miles on a side!

27-2 Variable air capacitor. (Courtesy of General Radio Company.)

Since the farad is such a large unit of capacitance, units of more convenient size are the *microfarad* (1 μF = 10^{-6} F), and the *picofarad* (1 pF = 10^{-12} F). For example, a common radio set contains in its power supply several capacitors whose capacitances are of the order of ten or more microfarads, while the capacitances of the tuning capacitors are of the order of ten to one hundred picofarads.

Variable capacitors whose capacitance may be varied (between limits) are widely used in the tuning circuits of radio receivers. These usually have a number of fixed parallel metal plates connected together and constituting one "plate" of the capacitor, while a second set of movable plates (also connected together) forms the other "plate" (Fig. 27-2). The movable plates are mounted on a shaft and may be caused to interleave the fixed plates to a greater or lesser extent. The effective area of the capacitor is that of the interleaved portion of the plates.

A variable capacitor is represented by the symbol

Capacitors consisting of concentric spheres and of coaxial cylinders are sometimes used in standards laboratories, since the corrections for the "fringing" fields can be made more readily and the capacitance can be accurately calculated from the dimensions of the apparatus. Problems involving spherical and cylindrical capacitors will be found at the end of the chapter.

Example   The plates of a parallel-plate capacitor are 5 mm apart and 2 m² in area. The plates are in vacuum. A potential difference of 10,000 V is applied across the capacitor. Compute (a) the capacitance, (b) the charge on each plate, and (c) the electric field in the space between them.

Solution

a)
$$C = \epsilon_0 \frac{A}{d} = \frac{(8.85 \times 10^{-12}\ \text{C}^2 \cdot \text{N}^{-1} \cdot \text{m}^{-2})(2\ \text{m}^2)}{5 \times 10^{-3}\ \text{m}}$$
$$= 3.54 \times 10^{-9}\ \text{C}^2 \cdot \text{N}^{-1} \cdot \text{m}^{-1}.$$

But
$$1\ \text{C}^2 \cdot \text{N}^{-1} \cdot \text{m}^{-1} = 1\ \text{C}^2 \cdot \text{J}^{-1} = 1\ \text{C}\left(\frac{\text{J}}{\text{C}}\right)^{-1}$$
$$= 1\ \text{C} \cdot \text{V}^{-1} = 1\ \text{F},$$

so
$$C = 3.54 \times 10^{-9}\ \text{F} = 0.00354\ \mu\text{F}.$$

b) The charge on the capacitor is
$$Q = CV_{ab} = (3.54 \times 10^{-9}\ \text{C} \cdot \text{V}^{-1})(10^4\ \text{V})$$
$$= 3.54 \times 10^{-5}\ \text{C}.$$

That is, the plate at higher potential has charge $+3.54 \times 10^{-5}$ C, and the other plate has charge -3.54×10^{-5} C.

c) The electric field is
$$E = \frac{\sigma}{\epsilon_0} = \frac{Q}{\epsilon_0 A} = \frac{3.54 \times 10^{-5}\ \text{C}}{(8.85 \times 10^{-12}\ \text{C}^2 \cdot \text{N}^{-1} \cdot \text{m}^{-2})(2\ \text{m}^2)}$$
$$= 20 \times 10^5\ \text{N} \cdot \text{C}^{-1};$$

or, since the electric field equals the potential gradient,

$$E = \frac{V_{ab}}{d} = \frac{10^4 \text{ V}}{5 \times 10^{-3} \text{ m}}$$
$$= 20 \times 10^5 \text{ V} \cdot \text{m}^{-1}.$$

Of course, the newton per coulomb and the volt per meter are equivalent units.

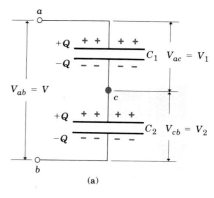

27-3 Capacitors in series and parallel

In Fig. 27–3a, two capacitors are connected in *series* between points a and b, maintained at a constant potential difference V_{ab}. The capacitors are both initially uncharged. In this connection, both capacitors always have the same charge Q. One might ask whether the lower plate of C_1 and the upper plate of C_2 might have charges different from those on the remaining two plates, but in this case the net charge on each capacitor would not be zero, and the resulting electric field in the conductor connecting the two capacitors would cause a current to flow until the total charge on each capacitor is zero. Hence, *in a series connection the magnitude of charge on all plates is the same.*

Referring again to Fig. 27–3a, we have

$$V_{ac} \equiv V_1 = \frac{Q}{C_1}, \qquad V_{cb} \equiv V_2 = \frac{Q}{C_2},$$

$$V_{ab} \equiv V = V_1 + V_2 = Q\left(\frac{1}{C_1} + \frac{1}{C_2}\right),$$

and

$$\frac{V}{Q} = \frac{1}{C_1} + \frac{1}{C_2}. \tag{27-3}$$

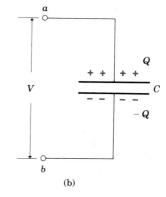

27-3 (a) Two capacitors in series, and (b) their equivalent.

The *equivalent* capacitance C of the series combination is defined as the capacitance of a *single* capacitor for which the charge Q is the same as for the combination, when the potential difference V is the same. For such a capacitor, shown in Fig. 27–3b,

$$Q = CV, \qquad \frac{V}{Q} = \frac{1}{C}. \tag{27-4}$$

Hence, from Eqs. (27–3) and (27–4),

$$\frac{1}{C} = \frac{1}{C_1} + \frac{1}{C_2}.$$

Similarly, for any number of capacitors in *series*,

$$\frac{1}{C} = \frac{1}{C_1} + \frac{1}{C_2} + \frac{1}{C_3} + \cdots \tag{27-5}$$

The reciprocal of the equivalent capacitance equals the sum of the reciprocals of the individual capacitances.

In Fig. 27–4a, two capacitors are connected in *parallel* between points a and b. In this case, the potential difference $V_{ab} \equiv V$ is the same for both, and the charges Q_1 and Q_2, not necessarily equal, are

$$Q_1 = C_1 V, \qquad Q_2 = C_2 V.$$

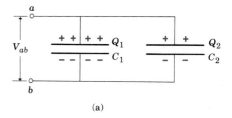

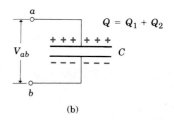

27-4 (a) Two capacitors in parallel, and (b) their equivalent.

The *total* charge Q supplied by the source was

$$Q = Q_1 + Q_2 = V(C_1 + C_2),$$

and

$$\frac{Q}{V} = C_1 + C_2. \qquad (27\text{-}6)$$

The *equivalent* capacitance C of the parallel combination is defined as that of a single capacitor, shown in Fig. 27-4b, for which the total charge is the same as in part (a). For this capacitor,

$$\frac{Q}{V} = C,$$

and hence

$$C = C_1 + C_2.$$

In the same way, for any number of capacitors in *parallel*

$$C = C_1 + C_2 + C_3 + \cdots \qquad (27\text{-}7)$$

The equivalent capacitance equals the *sum* of the individual capacitances.

Example In Figs. 27-3 and 27-4, let $C_1 = 6\,\mu\text{F}$, $C_2 = 3\,\mu\text{F}$, $V_{ab} = 18$ V.

The equivalent capacitance of the series combination in Fig. 27-3a is given by

$$\frac{1}{C} = \frac{1}{6\,\mu\text{F}} + \frac{1}{3\,\mu\text{F}}, \qquad C = 2\,\mu\text{F}.$$

The charge Q is

$$Q = CV = (2\,\mu\text{F})(18\text{ V}) = 36\,\mu\text{C}.$$

The potential differences across the capacitors are

$$V_{ac} \equiv V_1 = \frac{Q}{C_1} = \frac{36\,\mu\text{C}}{6\,\mu\text{F}} = 6\text{ V}, \qquad V_{cb} \equiv V_2 = \frac{Q}{C_2} = \frac{36\,\mu\text{C}}{3\,\mu\text{F}} = 12\text{ V}.$$

The *larger* potential difference appears across the *smaller* capacitor.

The equivalent capacitance of the parallel combination in Fig. 27-4a is

$$C = C_1 + C_2 = 9\,\mu\text{F}.$$

The charges Q_1 and Q_2 are

$$Q_1 = C_1 V = (6\,\mu\text{F})(18\text{ V}) = 108\,\mu\text{C},$$
$$Q_2 = C_2 V = (3\,\mu\text{F})(18\text{ V}) = 54\,\mu\text{C}.$$

27-4 Energy of a charged capacitor

The process of charging a capacitor consists of transferring charge from the plate at lower potential to the plate at higher potential. The charging process therefore requires the expenditure of *energy*. We may imagine charging the capacitor by starting with both plates completely uncharged, and then repeatedly removing small positive charges dq from one plate and transferring them to the other plate. The final charge Q

and the final potential difference V are related by

$$Q = CV.$$

At a stage of the process at which the magnitude of the net charge on either plate is q, the potential difference v between the plates is $v = q/C$. The work dW required to transfer the next charge dq is

$$dW = v \, dq = \frac{q \, dq}{C}.$$

The total work W to increase the charge from zero to a final value Q is

$$W = \int dW = \frac{1}{C} \int_0^Q q \, dq = \frac{Q^2}{2C}.$$

The final potential difference V between the plates is $V = Q/C$ and we can also write

$$W = \frac{Q^2}{2C} = \frac{1}{2} CV^2 = \frac{1}{2} QV. \tag{27–8}$$

When Q is expressed in coulombs and V in volts (joules per coulomb), W is expressed in joules.

The last form of Eq. (27–8) also shows that the total work is equal to the *average* potential $V/2$ during the charging process, multiplied by the total charge Q transferred.

A charged capacitor is the electrical analog of a stretched spring, whose elastic potential energy equals $\frac{1}{2}kx^2$. The charge Q is analogous to the elongation x, and the *reciprocal* of the capacitance, $1/C$, is analogous to the force constant k. The energy supplied to a capacitor in the charging process is stored by the capacitor and is released when the capacitor discharges.

It is often useful to consider the stored energy to be localized in the *electric field* between the capacitor plates. The capacitance of a parallel-plate capacitor in vacuum is

$$C = \epsilon_0 \frac{A}{d}.$$

The electric field fills the space between the plates, of volume Ad, and is given by

$$E = \frac{V}{d}.$$

The energy per unit volume, or the *energy density,* denoted by u, is

$$u = \text{Energy density} = \frac{\frac{1}{2}CV^2}{Ad}.$$

Making use of the preceding equations, we can express this as

$$u = \frac{1}{2}\epsilon_0 E^2. \tag{27–9}$$

Example In Fig. 27–5 capacitor C_1 is initially charged by connecting it to a source of potential difference V_0 (not shown in the figure). Let $C_1 = 8\,\mu\text{F}$ and $V_0 = 120$ V. The charge Q_0 is

$$Q_0 = C_1 V_0 = 960\,\mu\text{C},$$

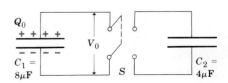

27–5 When the switch S is closed, the charged capacitor C_1 is connected to an uncharged capacitor C_2.

and the energy of the capacitor is

$$\frac{1}{2} Q_0 V_0 = 0.0576 \text{ J}.$$

After the switch S is closed, the positive charge Q_0 is distributed over the upper plates of both capacitors, and the negative charge $-Q_0$ is distributed over the lower plates of both. Let Q_1 and Q_2 represent the magnitudes of the final charges on the respective capacitors. Then

$$Q_1 + Q_2 = Q_0.$$

When the motion of charges has ceased, both upper plates are at the same potential and both lower plates are at the same potential (different from that of the upper plates). The final potential difference between the plates, V, is therefore the same for both capacitors, and

$$Q_1 = C_1 V, \qquad Q_2 = C_2 V.$$

When this is combined with the preceding equation, we find that

$$V = \frac{Q_0}{C_1 + C_2} = \frac{960 \ \mu\text{C}}{12 \ \mu\text{F}} = 80 \text{ V},$$
$$Q_1 = 640 \ \mu\text{C}, \qquad Q_2 = 320 \ \mu\text{C}.$$

The final energy of the system is

$$\frac{1}{2} Q_0 V = 0.0384 \text{ J}.$$

This is less than the original energy of 0.0576 J; the difference is converted to energy of some other form. If the resistance of the connecting wires is large, most of the energy is converted to heat. If the resistance is small, most of the energy is radiated in the form of electromagnetic waves.

This process is exactly analogous to an inelastic collision of a moving car with a stationary car. In the electrical case, the charge $Q = CV$ is conserved. In the mechanical case, the momentum $p = mv$ is conserved. The electrical energy $\frac{1}{2}CV^2$ is *not* conserved, and the mechanical energy $\frac{1}{2}mv^2$ is *not* conserved.

* 27-5 Effect of a dielectric

Most capacitors have a solid, nonconducting material or *dielectric* between their plates. A common type of capacitor incorporates strips of metal foil, forming the plates, separated by strips of wax-impregnated paper or plastic sheet such as Mylar, which serves as the dielectric. A sandwich of these materials is rolled up, forming a compact unit that can provide a capacitance of several microfarads in a relatively small volume.

Electrolytic capacitors have as their dielectric an extremely thin layer of nonconducting oxide between a metal plate and a conducting solution. Because of the thinness of the dielectric, electrolytic capacitors of relatively small dimensions may have a capacitance of the order of 100 to 1000 μF.

The function of a solid dielectric between the plates of a capacitor is threefold. First, it solves the mechanical problem of maintaining two large metal sheets at an extremely small separation but without actual contact.

Second, any dielectric material, when subjected to a sufficiently large electric field, experiences *dielectric breakdown,* a partial ionization, which permits conduction through a material that is supposed to insulate. Many insulating materials can tolerate stronger electric fields without breakdown than can air.

Third, the capacitance of a capacitor of given dimensions is *larger* when there is dielectric material between the plates than when the plates are separated only by air or vacuum. This effect can be demonstrated with the aid of a sensitive electrometer, a device that can measure the potential difference between two conductors without permitting any charge to flow from one to the other. In Fig. 27-6a, a capacitor is charged, with magnitude of charge Q on each plate and potential difference V_0. When a sheet of dielectric, such as glass, paraffin, or polystyrene, is inserted between the plates, the potential difference is found to *decrease* to a smaller value V. When the dielectric is removed, the potential difference returns to its original value, showing that the original charges on the plates have not been affected by insertion of the dielectric.

The original capacitance of the capacitor, C_0, was

$$C_0 = \frac{Q}{V_0}.$$

Since Q does not change and V is observed to be less than V_0, it follows that C is *greater* than C_0. The ratio of C to C_0 is called the *dielectric constant* of the material, K.

$$K = \frac{C}{C_0}. \tag{27-10}$$

Since C is always greater than C_0, the dielectric constants of all dielectrics are greater than unity. Some representative values of K are given in Table 27-1. For vacuum, of course, $K = 1$ by definition, and K for air is so nearly equal to 1 that for most purposes an air capacitor is equivalent to one in vacuum; the original measurement of V_0 in Fig. 27-6a could have been made with the plates in air instead of in vacuum.

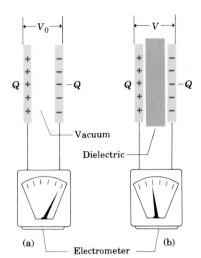

27-6 Effect of a dielectric between the plates of a parallel-plate capacitor. (a) With a given charge, the potential difference is V_0. (b) With the same charge, the potential difference V is smaller than V_0.

Table 27-1   Dielectric constant K at 20°C

Material	K	Material	K
Vacuum	1	Strontium titanate	310
Glass	5–10	Titanium dioxide (rutile)	173($\perp$),
Mica	3–6		86($\parallel$)
Mylar	3.1	Water	80.4
Neoprene	6.70	Glycerin	42.5
Plexiglas	3.40	Liquid ammonia (-78°C)	25
Polyethylene	2.25	Benzene	2.284
Polyvinyl chloride	3.18	Air (1 atm)	1.00059
Teflon	2.1	Air (100 atm)	1.0548
Germanium	16		

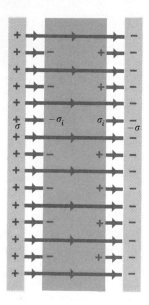

27-7 Induced charges on the faces of a dielectric in an external field.

With vacuum (or air) between the plates, the electric field E_0 in the region between the plates of a parallel-plate capacitor is

$$E_0 = \frac{V_0}{d} = \frac{\sigma}{\epsilon_0}.$$

The observed reduction in potential difference, when a dielectric is inserted between the plates, implies a reduction in the electric field, which in turn implies a reduction in the charge per unit area. Since no charge has leaked off the plates, such a reduction could be caused only by induced charges of opposite sign appearing on the two surfaces of the *dielectric.* That is, the dielectric surface adjacent to the positive plate must have an *induced negative charge* and that adjacent to the negative plate an *induced positive charge of equal magnitude,* as shown in Fig. 27-7.

If σ_i is the magnitude of induced charge per unit area on the surfaces of the dielectric, then the *net* surface charge on each side that contributes to the electric field within the dielectric has magnitude $(\sigma - \sigma_i)$, and the electric field in the dielectric is

$$E = \frac{V}{d} = \frac{\sigma - \sigma_i}{\epsilon_0}. \tag{27-11}$$

But

$$K = \frac{C}{C_0} = \frac{Q/V}{Q/V_0} = \frac{V_0}{V} = \frac{E_0}{E} = \frac{\sigma}{\sigma - \sigma_i}; \tag{27-12}$$

and therefore

$$\sigma - \sigma_i = \frac{\sigma}{K}. \tag{27-13}$$

Substituting Eq. (27-13) into Eq. (27-11), we get

$$E = \frac{\sigma}{K\epsilon_0}. \tag{27-14}$$

The product $K\epsilon_0$ is called the *permittivity* of the dielectric and is represented by ϵ.

$$\epsilon = K\epsilon_0. \tag{27-15}$$

The electric field within the dielectric may therefore be written

$$E = \frac{\sigma}{\epsilon}. \tag{27-16}$$

Also,

$$C = KC_0 = K\epsilon_0 \frac{A}{d},$$

and the capacitance of a parallel-plate capacitor with a dielectric between its plates is therefore

$$C = \epsilon \frac{A}{d}. \tag{27-17}$$

In empty space, where $K = 1$, $\epsilon = \epsilon_0$, and therefore ϵ_0 may be described as the "permittivity of empty space" or the "permittivity of vacuum." Since K is a pure number, the units of ϵ and ϵ_0 are evidently the same, $C^2 \cdot N^{-1} \cdot m^{-2}$.

Example The parallel plates in Fig. 27–7 have an area of 2000 cm² or 2×10^{-1} m², and are 1 cm or 10^{-2} m apart. The original potential difference between them, V_0, is 3000 V, and it decreases to 1000 V when a sheet of dielectric is inserted between the plates. Compute (a) the original capacitance C_0, (b) the charge Q on each plate, (c) the capacitance C after insertion of the dielectric, (d) the dielectric constant K of the dielectric, (e) the permittivity ϵ of the dielectric, (f) the induced charge Q_i on each face of the dielectric, (g) the original electric field E_0 between the plates, and (h) the electric field E after insertion of the dielectric.

Solution

a) $C_0 = \epsilon_0 \dfrac{A}{d} = (8.85 \times 10^{-12}\,C^2 \cdot N^{-1} \cdot m^{-2})\,\dfrac{2 \times 10^{-1}\,m^2}{10^{-2}\,m}$

$\qquad = 17.7 \times 10^{-11}\,F = 177\,pF.$

b) $Q = C_0 V_0 = (17.7 \times 10^{-11}\,F)(3 \times 10^3\,V) = 53.1 \times 10^{-8}\,C.$

c) $C = \dfrac{Q}{V} = \dfrac{53.1 \times 10^{-8}\,C}{10^3\,V} = 53.1 \times 10^{-11}\,F = 531\,pF.$

d) $K = \dfrac{C}{C_0} = \dfrac{53.1 \times 10^{-11}\,F}{17.7 \times 10^{-11}\,F} = 3.$

The dielectric constant could also be found from Eq. (27–12),

$$K = \frac{V_0}{V} = \frac{3000\,V}{1000\,V} = 3.$$

e) $\epsilon = K\epsilon_0 = (3)(8.85 \times 10^{-12}\,C^2 \cdot N^{-1} \cdot m^{-2})$

$\qquad = 26.6 \times 10^{-12}\,C^2 \cdot N^{-1} \cdot m^{-2}.$

f) $\qquad Q_i = A\sigma_i, \qquad Q = A\sigma,$

$\qquad \sigma - \sigma_i = \dfrac{\sigma}{K}, \qquad \sigma_i = \sigma\left(1 - \dfrac{1}{K}\right),$

$\qquad\qquad Q_i = Q\left(1 - \dfrac{1}{K}\right) = (53.1 \times 10^{-8}\,C)\left(1 - \dfrac{1}{3}\right)$

$\qquad\qquad\qquad = 35.4 \times 10^{-8}\,C.$

g) $E_0 = \dfrac{V_0}{d} = \dfrac{3000\,V}{10^{-2}\,m} = 3 \times 10^5\,V \cdot m^{-1}.$

h) $E = \dfrac{V}{d} = \dfrac{1000\,V}{10^{-2}\,m} = 1 \times 10^5\,V \cdot m^{-1};$

or

$$E = \frac{\sigma}{\epsilon} = \frac{Q}{A\epsilon} = \frac{53.1 \times 10^{-8}\,C}{(2 \times 10^{-1}\,m^2)(26.6 \times 10^{-12}\,C^2 \cdot N^{-1} \cdot m^{-2})}$$

$\qquad\qquad = 1 \times 10^5\,V \cdot m^{-1};$

or

$$E = \frac{\sigma - \sigma_i}{\epsilon_0} = \frac{Q - Q_i}{A\epsilon_0}$$

$$= \frac{(53.1 - 35.4) \times 10^{-8}\,\text{C}}{(2 \times 10^{-1}\,\text{m}^2)(8.85 \times 10^{-12}\,\text{C}^2 \cdot \text{N}^{-1} \cdot \text{m}^{-2})}$$

$$= 1 \times 10^5\,\text{V} \cdot \text{m}^{-1};$$

or, from Eq. (27–12),

$$E = \frac{E_0}{K} = \frac{3 \times 10^5\,V \cdot \text{m}^{-1}}{3} = 1 \times 10^5\,\text{V} \cdot \text{m}^{-1}.$$

Any dielectric material, when subjected to a sufficiently strong electric field, becomes a conductor, a phenomenon known as *dielectric breakdown*. The onset of conduction, associated with cumulative ionization of molecules of the material, is often quite sudden, and may be characterized by spark or arc discharges. When a capacitor is subjected to excessive voltage, an arc may be formed through a layer of dielectric, burning or melting a hole in it, permitting the two metal foils to come in contact, creating a short circuit, and rendering the device permanently useless as a capacitor.

The maximum electric field a material can withstand without the occurrence of breakdown is called the *dielectric strength*. Because dielectric breakdown is affected significantly by impurities in the material, small irregularities in the metal electrodes, and other factors difficult to control, only approximate figures can be given for dielectric strengths. The dielectric strength of dry air is about $0.8 \times 10^6\,\text{V} \cdot \text{m}^{-1}$. Typical values for plastic and ceramic materials commonly used to insulate capacitors and current-carrying wires are of the order of $10^7\,\text{V} \cdot \text{m}^{-1}$. For example, a layer of such a material, $10^{-4}\,\text{m}$ in thickness, could withstand a maximum voltage of 1000 V, for $1000\,\text{V}/10^{-4}\,\text{m} = 10^7\,\text{V} \cdot \text{m}^{-1}$.

*27–6 Molecular theory of induced charges

It remains to be discussed how a dielectric, which supposedly has no electric charges that are free to move within the material, can nevertheless acquire a surface charge distribution as described in the preceding section. When a *conductor* is placed in an electric field, the free charges within it are displaced by the forces exerted on them by the field. In the final steady state, the conductor has an induced charge on its surface, distributed in such a way that the field of this induced charge neutralizes the original field at all internal points, and the net electric field within the conductor is reduced to zero. A *dielectric*, however, contains no free charges. How can induced charge appear on the surfaces of a dielectric when it is inserted in the electric field between the plates of a charged capacitor?

The explanation involves the redistribution of electric charge in the material at the molecular level. First, the molecules of a dielectric may be classified as either *polar* or *nonpolar*. A nonpolar molecule is one in which the "centers of gravity" of the positive nuclei and the electrons normally coincide, while a polar molecule is one in which they do not.

Symmetrical molecules such as H_2, N_2, and O_2 are nonpolar. In the molecules N_2O and H_2O, on the other hand, both nitrogen atoms or both hyrogen atoms lie on the same side of the oxygen atom. These molecules are *polar,* and each is a tiny electric dipole.

Under the influence of an electric field, the charges of a *nonpolar* molecule become displaced relative to each other, as indicated schematically in Fig. 27-8. The molecules are said to become *polarized* by the field and are called *induced dipoles.* When a nonpolar molecule becomes polarized, restoring forces come into play on the displaced charges, pulling them together much as if they were connected by a spring. Under the influence of a given external field, the charges separate until the restoring force is equal and opposite to the force exerted on the charges by the field. Naturally, the restoring forces vary in magnitude from one kind of molecule to another, with corresponding differences in the displacement produced by a given field.

When a dielectric consists of polar molecules or *permanent dipoles,* these dipoles are oriented at random when no electric field is present, as in Fig. 27-9a. When an electric field is present, as in Fig. 27-9b, the forces on a dipole give rise to a torque whose effect is to tend to orient the dipole in the same direction as the field. The stronger the field, the greater is the aligning effect.

Whether the molecules of a dielectric are polar or nonpolar, the net effect of an external field is substantially the same, as shown in Fig. 27-10. Within the two extremely thin surface layers indicated by light color lines there is an excess charge, negative in one layer and positive in the other. It is these layers of charge that give rise to the induced charge on the surface of a dielectric. The charges are not free, but each is *bound* to a molecule lying in or near the surface. Within the remainder of the dielectric the net charge per unit volume remains zero.

The four parts of Fig. 27-11 illustrate the behavior of a sheet of dielectric when inserted in the field between a pair of oppositely charged plane parallel plates. Part (a) shows the original field. Part (b) is the situation after the dielectric has been inserted but before any rearrangement of charges has occurred. Part (c) shows by dotted lines the field set up in the dielectric by its induced surface charges. This field is opposite to the original field but, since the charges in the dielectric are not free to move indefinitely, their displacement does not proceed to such an extent that the induced field is equal in magnitude to the original field. The field in the dielectric is therefore *weakened* but not reduced to zero, as it would be in the interior of a conductor.

The resultant field is shown in Fig. 27-11d. Some of the field lines leaving the positive plate penetrate the dielectric; others terminate on the induced charges on the faces of the dielectric.

The charges induced on the surface of a dielectric in an external field afford an explanation of the attraction of an *uncharged* pith ball or bit of paper by a charged rod of rubber or glass. Figure 27-12 shows an uncharged dielectric sphere B in the radial field of a positive charge A. The induced positive charges on B experience a force toward the right, while the force on the negative charges is toward the left. Since the negative charges are closer to A and therefore in a stronger field than are the positive, the force toward the left exceeds that toward the right, and B, although its net charge is zero, experiences a resultant force toward

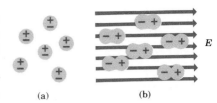

27-8 Behavior of nonpolar molecules (a) in the absence and (b) in the presence of an electric field.

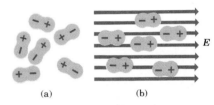

27-9 Behavior of polar molecules (a) in the absence and (b) in the presence of an electric field.

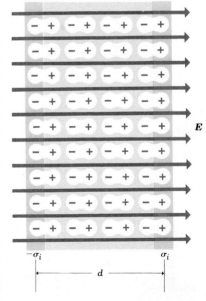

27-10 Polarization of a dielectric in an electric field gives rise to thin layers of bound charges on the surfaces.

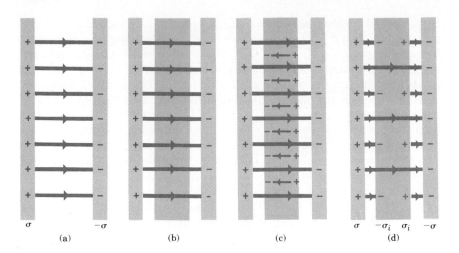

27-11 (a) Electric field between two charged plates. (b) Introduction of a dielectric. (c) Induced surface charges and their field. (d) Resultant field when a dielectric is between charged plates.

σ $-\sigma$ σ $-\sigma_i$ σ_i $-\sigma$

(a)   (b)   (c)   (d)

A. The sign of A's charge does not affect the conclusion, as may readily be seen. Furthermore, the effect is not limited to dielectrics; a conducting sphere would be similarly attracted.

More general arguments based on energy considerations show that a dielectric body in a nonuniform field *always* experiences a force urging it from a region where the field is weak toward a region where it is stronger, provided the dielectric constant of the body is greater than that of the medium in which it is immersed. If the dielectric constant is less, the reverse is true.

*27-7 Polarization and displacement

The extent to which the molecules of a dielectric become polarized by an electric field, or oriented in the direction of the field, is described by a vector quantity called the *polarization P.* If *p* is the component of the vector dipole moment (defined in Sec. 25–1, Example 5) of each molecule in the direction of the applied field, and there are n molecules per unit volume, the polarization is defined as

$$P = np. \tag{27-18}$$

Polarization is therefore *dipole moment per unit volume.* The polarization vector has the same direction as the molecular dipole moments, from left to right in Fig. 27–10. For the special case in Fig. 27–10, the magnitude of *P* is the same at all points of the dielectric. In other cases it can vary from point to point and the quantities n and *p* then refer to a small volume including the point. The SI unit of *P* is one *coulomb meter per cubic meter,* or one *coulomb per square meter* (1 C·m^{-2}).

The polarized dielectric in Fig. 27–10 can be considered a single large dipole, consisting of the induced bound charges Q_i at the opposite faces, separated by the thickness d of the dielectric. The dipole moment of the dielectric is then $Q_i d$, and since the volume of the dielectric is the product of its cross-sectional area A and thickness d, the dipole moment per unit volume, or the polarization P, is

$$P = \frac{Q_i d}{Ad} = \frac{Q_i}{A} = \sigma_i, \tag{27-19}$$

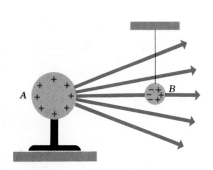

27-12 An uncharged dielectric sphere B in the radial field of a positive charge A.

where σ_i is the surface density of bound charge. In this special case, the polarization is numerically equal to the surface density of bound charge. More generally, the surface density of bound charge equals the normal component of P at the surface.

Figure 27–13 shows in cross section a sheet of polarized dielectric between two oppositely charged conducting plates. The thickness is exaggerated; we assume the sheet is very thin compared with its other dimensions, so that the electric field is uniform throughout the dielectric. The broken line is the outline of a Gaussian surface in the form of a cylinder with its ends parallel to the sheet and having cross-sectional area A. We consider the surface integral $\int P\cdot dA$ of the polarization vector P over this closed surface. P is zero at the left face, which is inside the left conducting plate. On the sides of the cylinder, P is parallel to the surface and makes no contribution to the integral. Only the right face, perpendicular to P, contributes, and gives simply PA. But, from Eq. (27–19), this is equal to Q_i, the total induced bound charge enclosed by the surface.

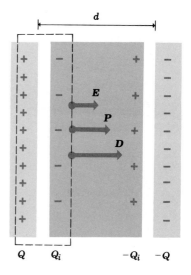

27–13

In the general case, we have

$$\oint P\cdot dA = -Q_i.\qquad(27\text{–}20)$$

The minus sign must be included because, from the diagram, the flux of P is *outward* (and hence positive) when the enclosed bound charge is negative. Equation (27–20) is Gauss's law for the polarization vector P: *the surface integral of P over any closed surface (the flux of P) equals the negative of the total bound charge within the surface.*

The resultant electric field E at any point, when bound charges are present, is due to *both* the free and bound charges. The general form of Gauss's law for E is therefore

$$\oint E\cdot dA = \frac{1}{\epsilon_0}(Q + Q_i).\qquad(27\text{–}21)$$

We may use Eq. (27–20) to eliminate the induced charge Q_i from Eq. (27–21); the result is

$$\oint E\cdot dA = \frac{1}{\epsilon_0}\left(Q - \oint P\cdot dA\right)\qquad(27\text{–}22)$$

or

$$\oint (\epsilon_0 E + P)\cdot dA = Q.\qquad(27\text{–}23)$$

Let us define a new quantity D called the *displacement* as the vector sum

$$D = \epsilon_0 E + P.\qquad(27\text{–}24)$$

Equation (27–23) then takes the simple form

$$\oint D\cdot dA = Q,\qquad(27\text{–}25)$$

which is Gauss's law for the displacement vector: *the surface integral of D over any closed surface (the flux of D) is equal to the free charge **only** within the surface.*

In summary, for any closed surface, the flux of E equals the *total* enclosed charge (divided by ϵ_0), the flux of P equals the (negative of the) *bound* charge, and the flux of D equals the *free* charge.

Applying Eq. (27–25) to the Gaussian surface of Fig. 27–13, we see that the magnitude of D is given simply by $D = \sigma$, where σ is the surface density of free charge on the conducting plate. Similarly, applying Gauss's law for E to this same surface gives $E = (\sigma - \sigma_i)/\epsilon_0$, where σ_i is the magnitude of the induced surface charge. Thus we find

$$\frac{D}{E} = \frac{\epsilon_0 \sigma}{\sigma - \sigma_i}.$$

Comparing this with Eq. (27–12), we obtain the simple relation

$$D = K\epsilon_0 E = \epsilon E. \tag{27–26}$$

More generally, in any dielectric where the polarization is proportional to the electric field (so σ_i is proportional to σ), the vector quantities D and E are related by

$$D = \epsilon E.$$

This relation will be useful in our study of electromagnetic waves in later chapters.

Questions

27-1 Two identical capacitors are connected in series. Is the resulting capacitance greater or less than that of each individual capacitor? What if they are connected in parallel?

27-2 Could one define a capacitance for a single conductor? What would be a reasonable definition?

27-3 Suppose the two plates of a capacitor have different areas. When the capacitor is charged by connecting it to a battery, do the charges on the two plates have equal magnitude, or may they be different?

27-4 Can you think of a situation in which the two plates of a capacitor *do not* have equal magnitudes of charge?

27-5 A capacitor is charged by connecting it to a battery, and is then disconnected from the charging agency. The plates are then pulled apart a little. How does the electric field change? The potential difference? The total energy?

27-6 According to the text, the energy in a charged capacitor can be considered to be located in the field between the plates. But suppose there is vacuum between the plates; can there be energy in vacuum?

27-7 The charged plates of a capacitor attract each other. To pull the plates farther apart therefore requires work by some external force. What becomes of the energy added by this work?

27-8 A solid slab of metal is placed between the plates of a capacitor without touching either plate. Does the capacitance increase, decrease, or remain the same?

27-9 The two plates of a capacitor are given charges $\pm Q$, and then they are immersed in a tank of oil. Does the electric field between them increase, decrease, or remain the same? How might this field be measured?

27-10 Is dielectric strength the same thing as dielectric constant?

27-11 Liquid dielectrics having polar molecules (such as water) always have dielectric constants that decrease with increasing temperature. Why?

27-12 A capacitor made of aluminum foil strips separated by Mylar film was subjected to excessive voltage, and the resulting dielectric breakdown melted holes in the Mylar. After this the capacitance was found to be about the same as before, but the breakdown voltage was much less. Why?

27-13 Two capacitors have equal capacitance, but one has a higher maximum voltage rating than the other. Which one is likely to be bulkier? Why?

27-14 A capacitor is made by rolling a sandwich of aluminum foil and Mylar, as described in Sec. 27–5. A student claimed that the capacitance when the sandwich is rolled up is twice the value when it is flat. Discuss his allegation.

Problems

27-1 An air capacitor is made from two flat parallel plates 0.5 mm apart. The magnitude of charge on each plate is 0.01 μC when the potential difference is 200 V.

a) What is the capacitance?

b) What is the area of each plate?

c) What maximum voltage can be applied without dielectric breakdown?

d) When the charge is 0.01 μC, what total energy is stored?

27-2 A parallel-plate air capacitor has a capacitance of 0.001 μF.

a) What potential difference is required for a charge of 0.5 μC on each plate?

b) In (a), what is the total stored energy?

c) If the plates are 1.0 mm apart, what is the area of each plate?

d) What potential difference is required for dielectric breakdown?

27-3 An air capacitor consisting of two closely spaced parallel plates has a capacitance of 1000 pF. The charge on each plate is 1 μC.

a) What is the potential diference between the plates?

b) If the charge is kept constant, what will be the potential difference between the plates if the separation is doubled?

c) How much work is required to double the separation?

27-4 A capacitor has a capacitance of 8.5 μF. How much charge must be removed to lower the potential difference of its plates by 50 V?

27-5 The capacitance of a variable capacitor can be changed from 50 pF to 950 pF by turning the dial from 0° to 180°. With the dial set at 180°, the capacitor is connected to a 400-V battery. After charging, the capacitor is disconnected from the battery and the dial is turned to 0°.

a) What is the charge on the capacitor?

b) What is the potential difference across the capacitor when the dial reads 0°.

c) What is the energy of the capacitor in this position?

d) How much work is required to turn the dial, if friction is neglected?

27-6 A 20-μF capacitor is charged to a potential difference of 1000 V. The terminals of the charged capacitor are then connected to those of an uncharged 5-μF capacitor. Compute

a) the original charge of the system,

b) the final potential difference across each capacitor,

c) the final energy of the system, and

d) the decrease in energy when the capacitors are connected.

27-7 In Fig. 27–14, each capacitance C_3 is 3 μF and each capacitance C_2 is 2 μF.

a) Compute the equivalent capacitance of the network between points a and b.

b) Compute the charge on each of the capacitors nearest a and b when V_{ab} = 900 V.

c) With 900 V across a and b, compute V_{cd}.

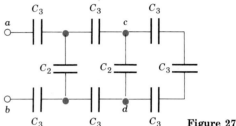

Figure 27–14

27-8 A number of 0.5-μF capacitors are available. The voltage across each is not to exceed 400 V. A capacitor of capacitance 0.5 μF is required to be connected across a potential difference of 600 V.

a) Show in a diagram how an equivalent capacitor having the desired properties can be obtained.

b) No dielectric is a perfect insulator, of infinite resistance. Suppose that the dielectric in one of the capacitors in your diagram is a moderately good conductor. What will happen?

27-9 A 1-μF capacitor and a 2-μF capacitor are connected in series across a 1200-V supply line.

a) Find the charge on each capacitor and the voltage across each.

b) The charged capacitors are disconnected from the line and from each other, and reconnected with terminals of like sign together. Find the final charge on each and the voltage across each.

27-10 A 1-μF capacitor and a 2-μF capacitor are connected in parallel across a 1200-V supply line.

a) Find the charge on each capacitor and the voltage across each.

b) The charged capacitors are then disconnected from the line and from each other, and reconnected with terminals of unlike sign together. Find the final charge on each and the voltage across each.

27-11 In Fig. 27–4a, let $C_1 = 6\,\mu$F, $C_2 = 3\,\mu$F, and $V_{ab} = 18$V. Suppose that the charged capacitors are disconnected from the source and from each other, and reconnected with plates of *opposite* sign connected together. By how much does the energy of the system decrease?

27-12 Three capacitors having capacitances of 8, 8, and 4 μF are connected in series across a 12-V line.

a) What is the charge on the 4-μF capacitor?

b) What is the total energy of all three capacitors?

c) The capacitors are disconnected from the line and reconnected in parallel with the positively charged plates connected together. What is the voltage across the parallel combination?

d) What is the energy of the combination?

27-13 A 500-μF capacitor is charged to 120 V. How many calories are produced on discharging the capacitor if all of the energy goes into heating the wire?

27-14 The capacitors in Fig. 27–15 are initially uncharged, and are connected as in the diagram with switch S open.

a) What is the potential difference V_{ab}?

b) What is the potential of point b after switch S is closed?

c) How much charge flowed through the switch when it was closed?

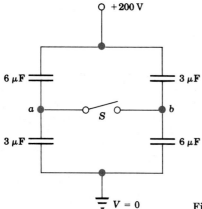

Figure 27–15

27-15 The plates of a parallel-plate capacitor in vacuum have charges $+Q$ and $-Q$ and the distance between the plates is x. The plates are disconnected from the charging voltage and pulled apart a short distance dx.

a) What is the change dC in the capacitance of the capacitor?

b) What is the change dW in its energy?

c) Equate the work $F\,dx$ to the increase in energy dW and find the force of attraction F between the plates.

d) Explain why F is not equal to QE, where E is the electric field between the plates.

27-16 A parallel-plate capacitor is to be constructed using, as a dielectric, rubber having a dielectric constant of 3 and a dielectric strength of 2×10^5 V·cm^{-1}. The capacitor is to have a capacitance of 0.15 μF and must be able to withstand a maximum potential difference of 6000 V.

What is the minimum area the plates of the capacitor may have?

27-17 The paper dielectric in a paper and foil capacitor is 0.005 cm thick. Its dielectric constant is 2.5 and its dielectric strength is 50×10^6 V·m^{-1}.

a) What area of paper, and of tinfoil, is required for a 0.1-μF capacitor?

b) If the electric intensity in the paper is not to exceed one-half the dielectric strength, what is the maximum potential difference that can be applied across the capacitor?

27-18 A Mylar–aluminum foil capacitor is to be designed for a capacitance of 0.1 μF and a voltage rating of 500 V. It is to be made from strips of aluminum foil and Mylar about 4 cm wide, and the foil has a thickness of 0.02 mm.

a) What thickness Mylar film is required?

b) What total area of Mylar and aluminum is required?

c) What is the diameter of the rolled-up capacitor?

27-19 A parallel-plate air capacitor is made using two plates 0.2 m square, spaced 1 cm apart. It is connected to a 50-V battery.

a) What is the capacitance?

b) What is the charge on each plate?

c) What is the electric field between the plates?

d) What is the energy stored in the capacitor?

e) If the battery is disconnected and then the plates are pulled apart to a separation of 2 cm, what are the answers to parts (a), (b), (c), and (d)?

27-20 In Problem 27–19, suppose the battery remains connected while the plates are pulled apart. What are the answers to parts (a), (b), (c), and (d)?

27-21 A parallel-plate capacitor with plate area A and separation x is charged to a charge of magnitude q on each plate.

a) What is the total energy stored in the capacitor?

b) The plates are now pulled apart an additional distance dx; now what is the total energy?

c) If F is the force with which the plates attract each other, then the difference in the two energies above must equal the work $dW = F\,dx$ done in pulling the plates apart. Hence show that $F = q^2/2\epsilon_0 A$.

27-22 A parallel-plate capacitor has the space between the plates filled with a slab of dielectric with constant K_1 and one with constant K_2, each of thickness $d/2$, where d is the plate separation. Show that the capacitance is

$$C = \frac{2\epsilon_0 A}{d}\left(\frac{K_1 K_2}{K_1 + K_2}\right).$$

27-23 Three square metal plates A, B, and C, each 10 cm on a side and 3 mm thick, are arranged as in Fig. 27–16. The plates are separated by sheets of paper 0.5 mm thick and of

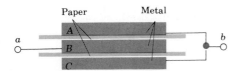

Paper   Metal

a

b

Figure 27-16

dielectric constant 5. The outer plates are connected together and connected to point *b*. The inner plate is connected to point *a*.

a) Copy the diagram, and show by + and − signs the charge distribution on the plates when point *a* is maintained at a positive potential relative to point *b*?

b) What is the capacitance between points *a* and *b*?

27-24 Two parallel plates have equal and opposite charges. When the space between the plates is evacuated, the electric field is 2×10^5 V·m^{-1}. When the space is filled with dielectric, the electric field is 1.2×10^5 V·m^{-1}.

a) What is the charge density on the surface of the dielectric?

b) What is its dielectric constant?

27-25 Two oppositely charged conducting plates, having numerically equal quantities of charge per unit area, are separated by a dielectric 5 mm thick, of dielectric constant 3. The resultant electric field in the dielectric is 10^6 V·m^{-1}. Compute

a) the charge per unit area on the conducting plate and

b) the charge per unit area on the surfaces of the dielectric.

27-26 A capacitor consists of two parallel plates of area 25 cm^2 separated by a distance of 0.2 cm. The material between the plates has a dielectric constant of 5. The plates of the capacitor are connected to a 300-V battery.

a) What is the capacitance of the capacitor?

b) What is the charge on either plate?

c) What is the energy in the charged capacitor?

d) What is the energy density in the dielectric?

27-27 Two parallel plates of 100 cm^2 area are given equal and opposite charges of 10^{-7} C. The space between the plates is filled with a dielectric material, and the electric field within the dielectric is 3.3×10^5 V·m^{-1}.

a) What is the dielectric constant of the dielectric?

b) What is the total induced charge on either face of the dielectric?

27-28 An air capacitor is made using two flat plates of area A separated by a distance d. Then a metal slab having thickness a (less than d) and the same shape and size as the plates is inserted between them, parallel to the plates and not touching either plate.

a) What is the capacitance of this arrangement?

b) Express the capacitance as a multiple of the capacitance when the metal slab is not present.

27-29 A spherical capacitor consists of an inner metal sphere of radius r_a supported on an insulating stand at the center of a hollow metal sphere of inner radius r_b; there is a charge $+Q$ on the inner sphere and a charge $-Q$ on the outer. (See Problem 26-24.)

a) What is the potential difference V_{ab} between the spheres?

b) Prove that the capacitance is

$$C = 4\pi\epsilon_0 \cdot \frac{r_b r_a}{r_b - r_a}.$$

27-30 A coaxial cable consists of an inner solid cylindrical conductor of radius r_a supported by insulating disks on the axis of a conducting tube of inner radius r_b. The two cylinders are oppositely charged with a charge λ per unit length. (See Problem 26-25.)

a) What is the potential difference between the two cylinders?

b) Prove that the capacitance of a length L of the cable is

$$C = \frac{2\pi\epsilon_0 L}{\ln(r_b/r_a)}.$$

Neglect any effect of the supporting disks.

CURRENT, RESISTANCE, AND ELECTROMOTIVE FORCE

28

28-1 Current

When there is a net flow of charge perpendicular to any area (passing through it from one side to the other), we say there is a *current* across the area. If an isolated conductor is placed in an electrostatic field, the charges in the conductor rearrange themselves so as to make the interior of the conductor a field-free region throughout which the potential is constant. This motion of the charges constitutes a *transient* current, of short duration only, and the current ceases when the field in the conductor becomes zero. To maintain a *continuous* current, we must in some way maintain a force on the mobile charges in a conductor. The force may result from an electrostatic field or from other causes that will be described later. For the present, we assume that within a conductor an electric field *E* is maintained such that a charged particle in the conductor is acted on by a force $F = qE$. We shall refer to this force as the *driving force* on the particle.

The motion of a free charged particle in a conductor is very different from that of a particle in empty space. After a momentary acceleration, the particle makes an inelastic collision with one of the fixed particles in the conductor and makes a fresh start. Thus *on the average* it moves in the direction of the driving force with an average velocity called its *drift velocity*. The inelastic collisions with the fixed particles result in a transfer of energy to them. This increases their energy of vibration and causes a rise in temperature if the conductor is thermally insulated, or results in a flow of heat from the conductor to its surroundings if it is not.

The current across an area is defined quantitatively as *the net charge flowing across the area per unit time*. Thus if a net charge ΔQ flows across a certain area in a time interval Δt, the average current I_{av} across this area is

$$I_{av} = \frac{\Delta Q}{\Delta t}.$$

The rate of flow of charge may not be constant, in which case we generalize the definition of current in a natural way, using the derivative. The instantaneous current I is defined as

$$I = \frac{dQ}{dt}. \tag{28-1}$$

Current is a *scalar* quantity.

The SI unit of current, *one coulomb per second,* is called *one ampere* (1 A), in honor of the French scientist André Marie Ampère (1775–1836). Small currents are more conveniently expressed in *milliamperes* (1 mA $= 10^{-3}$ A) or in *microamperes* (1 $\mu A = 10^{-6}$ A).

The current across an area can be expressed in terms of the drift velocity of the moving charges as follows. Consider a portion of a conductor of cross-sectional area A within which there is a resultant electric field E from left to right. We suppose first that the conductor contains free *positively* charged particles; these move in the same direction as the field. A few positive particles are shown in Fig. 28-1. Suppose there are n such particles per unit volume, all moving with a drift velocity v. In a time Δt each advances a distance $v \, \Delta t$. Hence, all of the particles within the shaded cylinder of length $v \, \Delta t$, and only those particles, will flow across the end of the cylinder in time Δt. The volume of the cylinder is $Av \, \Delta t$, the number of particles within it is $nAv \, \Delta t$, and if each has a charge q, the charge ΔQ flowing across the end of the cylinder in time Δt is

$$\Delta Q = nqvA \, \Delta t,$$

The current carried by the positively charged particles is therefore

$$I = \frac{\Delta Q}{\Delta t} = nqvA. \tag{28-2}$$

If the moving charges are negative rather than positive, the electric-field force is opposite to E, and so the drift velocity is right to left, opposite to the direction shown in Fig. 28-1. Positive particles crossing from left to right *increase* the *positive* charge at the right of the section, while negative particles crossing from right to left *decrease* the *negative* charge at the right of the section. But a *decrease* of *negative* charge is equivalent to an *increase* of *positive* charge, so the motion of *both* kinds of charge has the same effect, namely, to increase the positive charge at the right of the section. In either case, particles flowing out through an end of the cylindrical section are continuously replaced by particles flowing *in* through the oposite end.

In general, a conductor may contain a number of different kinds of charged particles having charges q_i, densities n_i, and drift velocities v_i; the total current is then

$$I = A \sum n_i q_i v_i. \tag{28-3}$$

In metals the moving charges are always (negative) electrons, while an ionized gas has moving electrons and positively charged ions. In a semiconductor material such as germanium or silicon, conduction is partly by electrons and partly by motion of vacancies, also known as *holes;* these are sites of missing electrons and act like positive charges.

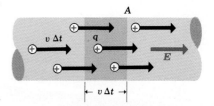

28-1 All of the particles, and only those particles, within the shaded cylinder will cross its base in time Δt.

The *current per unit cross-section area* is called the *current density J*:

$$J = \frac{I}{A} = \sum nqv. \tag{28–4}$$

The *vector current density J* is defined by the equation

$$\boldsymbol{J} = \sum n_i q_i \boldsymbol{v}_i. \tag{28–5}$$

Current, by definition, is a scalar quantity and thus it is not correct to speak of the "direction of a current." This expression is often used, however, for brevity, meaning thereby the direction of the *vector* current density *J*.

The direction of the drift velocity *v* of a positive charge is the same as that of the electric field *E,* and the direction of the velocity of a negative charge is opposite to *E*. But since the charge *q* is negative, each of the vectors *nqv* is in the same direction as *E,* and hence the *vector current density J always has the same direction as the field E.* Thus even in a metallic conductor, where the moving charges are negative electrons only and move in the *opposite* direction to *E*, the *vector* current density *J* is in the *same* direction as *E*.

The above discussion has shown that the effect of a current is the same, whether it consists of positive charges moving in the direction of *E,* of negative charges moving in the opposite direction, or a combination of the two. In describing circuit behavior it is customary to describe currents as though they consisted entirely of positive charge flow, even in cases where the actual current is known to be due to electrons. This convention will be followed consistently in the following sections. In Chapter 31, in studying the effect of a *magnetic* field on a moving charge, we shall consider a phenomenon, the Hall effect, in which the sign of the moving charges *is* important.

When there is a steady current in a closed loop (a "complete circuit"), the total charge in every portion of a conductor remains constant. Hence, if we consider a portion between two fixed cross sections, the rate of flow of charge *out* of the portion at one end equals the rate of flow of charge *into* the portion at the other end. In other words, the current is the same at any two cross sections and hence is *the same at all cross sections.* Current is *not* something that squirts out of the positive terminal of a battery and gets all used up by the time it reaches the negative terminal! These statements are direct consequences of the principle of conservation of charge, introduced in Sec. 24–2.

Example A copper conductor of square cross section 1 mm on a side carries a constant current of 20 A. The density of free electrons is 8×10^{28} electrons per cubic meter. Find the current density and the drift velocity.

Solution The current density in the wire is

$$J = \frac{I}{A} = 20 \times 10^6 \, \text{A} \cdot \text{m}^{-2}.$$

From Eq. (28–4),

$$v = \frac{J}{nq} = \frac{(20 \times 10^6 \, \text{A} \cdot \text{m}^{-2})}{(8 \times 10^{28} \, \text{m}^{-3})(1.6 \times 10^{-19} \, \text{C})}$$
$$= 1.6 \times 10^{-3} \, \text{m} \cdot \text{s}^{-1},$$

or about 1.6 mm·s^{-1}. At this speed, an electron would require 625 s or about 10 min to travel the length of a wire 1 m long. Thus the drift velocity is *very* small compared with the velocity of propagation of a current pulse along a wire, about 3×10^8 m·s^{-1}.

28-2 Resistivity

The current density J in a conductor depends on the electric field E, and on the nature of the conductor. In general the dependence of J on E can be quite complex but, for some materials, especially the metals, it can be represented quite well by a direct proportionality. For such materials the ratio of E to J is constant; we define the *resistivity* ρ of a particular material as the ratio of electric field to current density:

$$\rho = \frac{E}{J}. \tag{28-6}$$

That is, the resistivity is the *electric field per unit current density*. The greater the resistivity, the greater the field needed to establish a given current density, or the smaller the current density for a given field. Representative values are given in Table 28-1. The unit Ω·m (ohm·meter) will be explained in the following section. A "perfect" conductor would have zero resistivity and a "perfect" insulator an infinite resistivity. Metals and alloys have the lowest resistivities and are the best conductors. The resistivities of insulators exceed those of the metals by a factor of the order of 10^{22}.

Table 28-1   Resistivities at room temperature

Substance		ρ, $\Omega \cdot$ m	Substance		ρ, $\Omega \cdot$ m
Conductors			Semiconductors		
Metals	Silver	1.47×10^{-8}	Pure	Carbon	3.5×10^{-5}
	Copper	1.72×10^{-8}		Germanium	0.60
	Gold	2.44×10^{-8}		Silicon	2300
	Aluminum	2.63×10^{-8}	Insulators		
	Tungsten	5.51×10^{-8}		Amber	5×10^{14}
	Steel	$20 \quad \times 10^{-8}$		Glass	10^{10}–10^{14}
	Lead	$22 \quad \times 10^{-8}$		Lucite	$> 10^{13}$
	Mercury	$95 \quad \times 10^{-8}$		Mica	10^{11}–10^{15}
Alloys	Manganin	$44 \quad \times 10^{-8}$		Quartz (fused)	75×10^{16}
	Constantan	$49 \quad \times 10^{-8}$		Sulfur	10^{15}
	Nichrome	$100 \quad \times 10^{-8}$		Teflon	$> 10^{13}$
				Wood	10^{8}–10^{11}

Comparison with Table 16-1 shows that *thermal* insulators have thermal resistivities (the reciprocals of their thermal conductivities) that differ from those of good thermal conductors by factors of only about 10^3. By the use of electrical insulators, electric currents can be confined to well-defined paths in good electrical conductors, while it is impossible to confine heat currents to a comparable extent. It is also interesting to note that the metals, as a class, are also the best *thermal* conductors. The free electrons in a metal that carry charge in electrical conduction also play an important role in the conduction of heat; hence, a correlation can be expected between electrical and thermal conductivity. It is a familiar fact that good electrical conductors, such as the metals, are also good conductors of heat, while poor electrical conductors, such as ceramic and plastic materials, are also poor thermal conductors.

The *semiconductors* form a class intermediate between the metals and the insulators. They are of importance not primarily because of their resistivities, but because of the way in which these are affected by temperature and by small amounts of impurities.

The discovery that ρ is a constant for a metallic conductor at constant temperatures was made by G. S. Ohm (1789–1854) and is called *Ohm's law*. A material obeying Ohm's law is called an *ohmic* conductor or a *linear* conductor. If Ohm's law is *not* obeyed, the conductor is called *nonlinear*. Thus Ohm's law, like the ideal gas equation, Hooke's law, and many other relations describing the properties of materials, is an *idealized model* that describes the behavior of certain materials reasonably well but is by no means a general property of all matter.

The resistivity of all *metallic* conductors increases with increasing temperature, as shown in Fig. 28–2a. Over a temperature range that is not too great, the resistivity of a metal can be represented approximately by the equation

$$\rho_T = \rho_0[1 + \alpha(T - T_0)], \tag{28-7}$$

where ρ_0 is the resistivity at a reference temperature T_0 (often taken as 0°C or 20°C) and ρ_T the resistivity at temperature T°C. The factor α is called the *temperature coefficient of resistivity*. Some representative values are given in Table 28–2. The resistivity of carbon (a nonmetal) *decreases* with increasing temperature and its temperature coefficient of resistivity is negative. The resistivity of the alloy manganin is practically independent of temperature.

A number of materials have been found to exhibit the property of *superconductivity*. As the temperature is decreased, the resistivity at first decreases regularly, like that of any metal. At the so-called *critical temperature*, usually in the range 0.1 K to 20 K, a phase transition occurs, and the resistivity suddenly drops to zero, as shown in Fig. 28–2b. A current once established in a superconducting ring will continue of itself, apparently indefinitely, without the presence of any driving field.

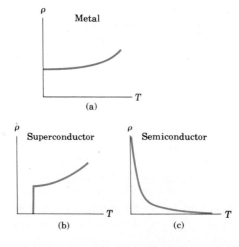

28-2 Variation of resistivity with temperature for three conductors: (a) an ordinary metal, (b) a superconducting metal, alloy, or compound, and (c) a semiconductor.

Table 28-2   Temperature coefficients of resistivity
(approximate values near room temperature)

Material	α, °C^{-1}	Material	α, °C^{-1}
Aluminum	0.0039	Lead	0.0043
Brass	0.0020	Manganin (Cu 84,	0.000000
Carbon	−0.0005	Mn 12, Ni 4)	
Constantan (Cu 60, Ni 40)	+0.000002	Mercury	0.00088
Copper (Commercial	0.00393	Nichrome	0.0004
annealed)		Silver	0.0038
Iron	0.0050	Tungsten	0.0045

The resistivity of a *semiconductor* decreases rapidly with increasing temperature, as shown in Fig. 28–2c. A tiny bead of semiconducting material, called a *thermistor,* serves as a sensitive thermometer.

28-3 Resistance

The current density J, at a point within a conductor where the electric field is E, is given by Eq. (28–6):

$$E = \rho J.$$

It is often difficult to measure E and J directly, and it is useful to put this relation in a form involving readily measured quantities such as total current and potential difference. To do this we consider a conductor with uniform cross-section area A and length l, as shown in Fig. 28–3. Assuming a constant current density over a cross section, and a uniform electric field along the length of the conductor, the total current I is given by

$$I = JA,$$

and the potential difference V between the ends is

$$V = El. \tag{28-8}$$

Solving these equations for J and E, respectively, and substituting the results in Eq. (28–6), we obtain

$$\frac{V}{l} = \frac{\rho I}{A}. \tag{28-9}$$

Thus the total current is proportional to the potential difference.

The quantity $\rho l/A$ for a particular specimen of material is called its *resistance R*:

$$R = \frac{\rho l}{A}. \tag{28-10}$$

Equation (28–9) then becomes

$$V = IR. \tag{28-11}$$

This relation is often referred to as *Ohm's law;* in this form it refers to a specific piece of material, not to a general property of the material as with Eq. (28–6).

Equation (28–10) shows that the resistance of a wire or other conductor of uniform cross section is directly proportional to its length and

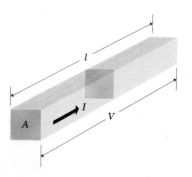

28-3   A conductor of uniform cross section. The current density is uniform over any cross section, and the electric field is constant along the length.

inversely proportional to its cross-section area. It is of course also proportional to the resistivity of the material of which the conductor is made.

The SI unit of resistance is one *volt per ampere* ($1 \text{ V}\cdot\text{A}^{-1}$). A resistance of $1 \text{ V}\cdot\text{A}^{-1}$ is called 1 *ohm* (1Ω). The unit of resistivity is therefore one *ohm·meter* ($1 \Omega\cdot\text{m}$).

Large resistances are conveniently expressed in *kilohms* ($1 \text{ k}\Omega = 10^3 \Omega$) or *megohms* ($1 \text{ M}\Omega = 10^6 \Omega$), and small resistances in microhms ($1 \mu\Omega = 10^{-6} \Omega$). Resistivities are also expressed in a variety of hybrid units, most common of which is the *ohm·centimeter* ($1 \Omega\cdot\text{cm} = 10^{-2} \Omega\cdot\text{m}$).

Because the resistance of any specimen of material is proportional to its resistivity, which varies with temperature, resistance also varies with temperature. For temperature ranges that are not too great, this variation may be represented approximately as a linear relation analogous to Eq. (28–7):

$$R_T = R_0[1 + \alpha(T - T_0)]. \qquad (28\text{--}12)$$

Here R_T is the resistance at temperature T, and R_0 is the resistance at the temperature T_0, often taken to be 20°C or 0°C. Within the limits of validity of Eq. (28–12), the *change* in resistance resulting from a temperature change $T - T_0$ is given by $\alpha(T - T_0)$, where α is the temperature coefficient of resistivity, given for several common materials in Table 28–2.

Example 1   For the example at the end of Sec. 28–1, find the electric field, and the potential difference between two points 100 m apart.

Solution   From Eq. (28–6), the electric field is given by

$$E = \rho J = (1.72 \times 10^{-8} \Omega\cdot\text{m})(20 \times 10^6 \text{ A}\cdot\text{m}^{-2})$$
$$= 0.344 \text{ V}\cdot\text{m}^{-1}.$$

The potential difference is given by

$$V = El = (0.344 \text{ V}\cdot\text{m}^{-1})(100 \text{ m}) = 34.4 \text{ V}.$$

Thus the resistance of a piece of this wire 100 m in length is

$$R = \frac{V}{I} = \frac{34.4 \text{ V}}{20 \text{ A}} = 1.72 \, \Omega.$$

This result can also be obtained directly from Eq. (28–10):

$$R = \frac{\rho l}{A} = \frac{(1.72 \times 10^{-8} \Omega\cdot\text{m})(100 \text{ m})}{(1 \times 10^{-3} \text{ m})^2} \times 1.72 \, \Omega.$$

Example 2   In the previous example, suppose the resistance is $1.72 \, \Omega$ at a temperature of 20°C. Find the resistance at 0°C and at 100°C.

Solution   We use Eq. (28–12). In this instance $T_0 = 20°\text{C}$ and $R_0 = 1.72 \, \Omega$. From Table 28–2, the temperature coefficient of resistivity of copper is $\alpha = 0.00393(°\text{C})^{-1}$. Thus at $T = 0°\text{C}$,

$$R = 1.72 \, \Omega[1 + (0.00393°\text{C}^{-1})(0°\text{C} - 20°\text{C})]$$
$$= 1.58 \, \Omega,$$

and at $T = 100°C$,

$$R = 1.72 \, \Omega[1 + (0.00393°C^{-1})(100°C - 20°C)]$$
$$= 2.26 \, \Omega.$$

Example 3   The space between two metallic coaxial cylinders of radii r_a and r_b is filled with a material of resistivity ρ. What is the resistance between the cylinders?

Solution   Equation (28–10) cannot be used directly because the cross section through which the charge travels varies from $2\pi r_a l$ at the inner cylinder to $2\pi r_b l$ at the outer cylinder. Instead, we consider a cylindrical shell of inner radius r and thickness dr. The area A is then $2\pi r l$, and the length of the current path through the shell is dr. Thus, the resistance dR of the shell is

$$dR = \frac{\rho \, dr}{2\pi r l}$$

and the total resistance between the cylinders is

$$R = \frac{\rho}{2\pi l} \int_{r_a}^{r_b} \frac{dr}{r} = \frac{\rho}{2\pi l} \ln \frac{r_b}{r_a}.$$

28-4 Electromotive Force

In order for a steady current to exist in a conducting path, that path must form a closed loop or *complete circuit*. Otherwise charge would accumulate at the ends of the conductor, the resulting electric field would change with time, and the current could not be constant.

However, such a path cannot consist solely of resistance. Current in a resistor requires an electric field and an associated potential. The field always does *positive* work on the charge, which moves always in the direction of *decreasing* potential. But after a complete trip around the loop, a charge returns to its starting point, and the potential then must be the same as when it left that point. This cannot be so if its travel around the loop involves only *decreases* in potential.

Thus there must be some portion of the loop where a charge travels from lower to higher potential, despite the electrostatic force trying to push it from higher to lower potential. The influence that makes charge move from lower to higher potential is called *electromotive force*. Every complete circuit in which there is a steady current must have some device which provides electromotive force.

Batteries, generators, photovoltaic cells, and thermocouples are examples of such devices, which are called *seats of electromotive force*. Any such device can transfer energy into a circuit in which it is connected; thus such a device is sometimes called a *source*, although the term *energy converter* is preferable. Electromotive force is usually abbreviated emf, pronounced "ee-em-eff."

Figure 28–4 is a schematic representation of a seat of emf, such as a battery or generator. Such a device has the property that it can maintain a potential difference between conductors a and b, called the *terminals* of the device. In Fig. 28–4, there is no conducting path *outside* the device connecting a and b, and the device is said to be on *open circuit*.

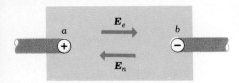

28-4 Schematic diagram showing the general directions of the electrostatic field E_e and the non-electrostatic field E_n within a source on open circuit. In this case, $E_n = -E_e$ and $V_{ab} = \mathcal{E}$.

Terminal a, marked $+$, is maintained by the source at a *higher* potential than terminal b, marked $-$. Associated with this potential difference is an electrostatic field E_e, at all points between and around the terminals, both inside and outside the source. The electrostatic field E_e inside the device is directed from a toward b, as shown. The source is itself a conductor, however, and if the *only* force on the free charges within it were that exerted by the electrostatic field, the positive charges would move from a toward b (or the negative charges from b toward a). The excess charges on the terminals would decrease, and the potential difference between them would decrease and eventually become zero.

But this is not the way batteries and generators actually act; in fact, they maintain a potential difference even when there is a steady current. From this we conclude that there must be some additional force on the charges within the source, tending to push them from a lower-potential point to a higher, opposite to the tendency of the electrostatic force. The origin of the non-electrostatic force depends on the nature of the source. In a generator it results from the action of a *magnetic* field on moving charges. In a battery it is associated with varying electrolyte concentrations arising from chemical reactions. In an electrostatic machine such as a Van de Graaff or Wimshurst generator, it is a mechanical force applied by a moving belt or wheel.

Whatever the origin of the non-electrostatic force, which we may call F_n, its effect is the same as though there were an additional electric field E_n, of non-electrostatic origin, related to the force by $F_n = qE_n$. That is, the non-electrostatic force is the same *as if* there were a non-electrostatic field E_n in addition to the purely electrostatic field E_e.

When the source is on open circuit, as in Fig. 28-4, the charges are in equilibrium, and the *resultant* field E, the vector sum of E_e and E_n, must be zero at every point:

$$E_e + E_n = 0.$$

Now the electrostatic potential difference V_{ab} is defined as the work per unit charge performed by the *electrostatic* field E_e on a charge moving from a to b. Similarly, we may consider the work done by the *non-electrostatic* field E_n. It is customary to speak of the (positive) work of this field during a displacement from b to a, rather than the reverse. Specifically, the work performed by E_n, per unit charge, when a charge moves from b to a, is called the *electromotive force* $\mathcal{E}$ *of the source.*

When $E_e = -E_n$, we have $V_{ab} = \mathcal{E}$. Hence *for a source on open circuit,* the potential difference V_{ab}, or the *open-circuit terminal voltage,* is equal to the electromotive force:

$$V_{ab} = \mathcal{E} \qquad \text{(source on open circuit).} \qquad (28\text{-}13)$$

The term *electromotive force,* although widely used, is somewhat unfortunate, in that the concept to which it refers is not a *force* but a *work per unit charge.* The term *electromotance* is sometimes used, but the concept is usually referred to simply as *emf,* pronounced "ee-em-eff."

The SI unit of E_n is the same as that of E_e, namely, one volt per meter, so the unit of emf is the same as that of potential or potential difference, namely, 1 V. However, an electromotive force is not the *same thing* as a potential difference, since the latter is the work of an *electrostatic* field and the former is the work of a *non-electrostatic* field.

As we shall see later, the *electrostatic* field within a source, and hence the potential difference between its terminals, depend on the current in the source. The *non*-electrostatic field, and hence the emf of the source, is in many cases a *constant* independent of the current, and hence the emf represents a definite property of a source. Unless stated otherwise, we shall assume, in what follows, that the emf of a source is constant.

Now suppose that the terminals of a source are connected by a wire, as shown schematically in Fig. 28–5, forming a *complete circuit*. The driving force on the free charges *in the wire* is due solely to the *electrostatic* field E_e set up by the charged terminals a and b of the source. This field sets up a current *in the wire* from a toward b. The charges on the terminals decrease slightly and the electrostatic fields, both within the wire and within the source, decrease also. As a result, the *electrostatic* field within the source becomes smaller than the (constant) *non-electrostatic* field. Hence positive charges within the source are driven toward the positive terminal, and there is a current within the source from b toward a. The circuit settles down to a steady state in which the current is the same at all cross sections.

If current could travel through the source without impediment (that is, if the source had no *internal* resistance), charge entering the external circuit through terminal a would be replaced immediately by charge flow through the source. In this case the internal electrostatic field in the source would not change under complete-circuit conditions, and the terminal potential difference V_{ab} would still be equal to $\mathcal{E}$. Since V_{ab} is also related to the current and resistance in the external circuit by Eq. (28–11), we would then have

$$\mathcal{E} = IR, \tag{28–14}$$

where R is the resistance of the external circuit. This relation determines the current in the circuit, once $\mathcal{E}$ and R are specified.

We say "if" in the above paragraph because every real source has some *internal* resistance, which we may denote as r. Under closed-circuit conditions the total electric field $E_e + E_n$ inside the source cannot then be precisely zero because some net field is required in order to push the charge through the internal resistance. Thus E_e must have somewhat smaller magnitude than E_n, and correspondingly V_{ab} is less than $\mathcal{E}$; the difference is equal to the work per unit charge done by the total field, which is simply Ir. Thus the terminal potential difference under closed-circuit conditions is given by

$$V_{ab} = \mathcal{E} - Ir, \tag{28–15}$$

where r is the *internal resistance* of the source. The equation determining the current in the complete circuit is then

$$\mathcal{E} - Ir = IR,$$

or

$$I = \frac{\mathcal{E}}{R + r}. \tag{28–16}$$

That is, the current equals the source emf divided by the *total* circuit resistance, external plus internal.

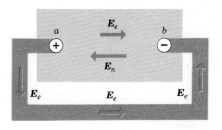

28-5 Schematic diagram of a source with a complete circuit. The vectors E_n and E_e represent the directions of the corresponding fields. The current is everywhere the same and is in the direction of the *resultant* field, that is, from a to b in the external circuit and from b to a within the source. $V_{ab} = IR = \mathcal{E} - Ir$.

If the terminals of a source are connected by a conductor of zero (or negligible) resistance, the source is said to be *short circuited*. (This would be an *extremely* dangerous procedure to carry out with the storage battery of your car, or with the terminals of the power line!) Then $R = 0$, and from the circuit equation the *short-circuited current* I_s is

$$I_s = \frac{\mathcal{E}}{r}. \tag{28-17}$$

The terminal voltage is then zero:

$$V_{ab} = \mathcal{E} - \left(\frac{\mathcal{E}}{r}\right)r = 0. \tag{28-18}$$

The *electrostatic* field within the source is zero, and the driving force on the charges within it is due to the *non*-electrostatic field only.

A source is completely described by its emf $\mathcal{E}$ and its internal resistance r. These properties may be found (at least in principle) from measurements of the open-circuit terminal voltage, which equals $\mathcal{E}$, and the short-circuit current, which enables r to be calculated from Eq. (28-17).

We must consider one more special case. If a source is connected to an external circuit containing other sources, it is possible that the electrostatic field within the given source will be *greater* than the non-electrostatic field, as in Fig. 28-6. When this is the case, the current *within* the source is from terminal a toward terminal b. This is the case when the storage battery of an automobile is being "charged" by the alternator. Equation (28-15), then becomes:

$$V_{ab} = \mathcal{E} + Ir. \tag{28-19}$$

The terminal voltage is then *greater* than the emf $\mathcal{E}$.

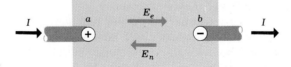

28-6

The diagrams in the preceding sections have been more or less representational, so as to show the electric fields within sources and conductors. Every conductor (except a superconductor) has resistance, and is therefore a *resistor,* also. In fact, the terms "conductor" and "resistor" may be used interchangeably. Resistance units constructed to introduce into a circuit lumped resistances that are large compared with those of leads and contacts are called *resistors*. A resistor is represented by the symbol

Conductors having negligible resistance are shown by straight lines.

A variable resistor is called a *rheostat* or, particularly in electronics, a *potentiometer*. A common type consists of a resistor with a sliding contact that can be moved along its length and is represented by the symbol

Connections are made to either end of the resistor and to the sliding contact. The symbol

is also used for a variable resistor.

A source is represented by the symbol

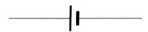

The longer vertical line always corresponds to the + terminal. We shall modify this in the following examples to

or

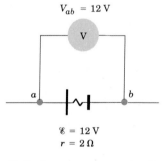

so as to show explicitly that a source has an internal resistance.

Example 1   Consider a source whose emf $\mathscr{E}$ is constant and equal to 12 V, and whose internal resistance r is 2 Ω. (The internal resistance of a commercial 12-V lead storage battery is only a few thousandths of an ohm.) Figure 28–7 represents the source with a voltmeter V connected between its terminals a and b. A voltmeter reads the potential difference between its terminals. If it is of the conventional type, the voltmeter provides a conducting path between the terminals and so there is a current in the source (and through the voltmeter). We shall assume, however, that the resistance of the voltmeter is so large (essentially infinite) that it draws no appreciable current. The source is then on *open circuit,* corresponding to the source in Fig. 28–4, and the voltmeter reading V_{ab} equals the emf $\mathscr{E}$ of the source, or 12 V.

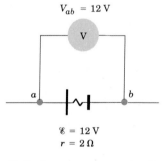

V_{ab} = 12 V

$\mathscr{E}$ = 12 V
$r = 2\,\Omega$

28-7   A source on open circuit.

Example 2   In Fig. 28–8, an ammeter A and a resistor of resistance $R = 4\,\Omega$ have been connected to the terminals of the source to form a complete circuit. The total resistance of the circuit is the sum of the resistance R, the internal resistance r, and the resistance of the ammeter. The ammeter resistance, however, can be made very small, and we shall assume it so small (essentially zero) that it can be neglected. The ammeter (whatever its resistance) reads the current I through it. The circuit corresponds to that in Fig. 28–5.

The wires connecting the resistor to the source and the ammeter, shown by the straight lines, have zero resistance and hence there is no potential difference between their ends. Thus, points a and a' are at the

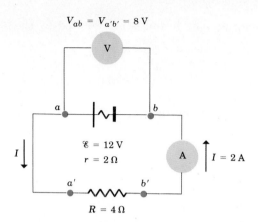

$$V_{ab} = V_{a'b'} = 8 \text{ V}$$

28-8   A source in a complete circuit.

same potential and are electrically equivalent, as are points b and b'. The potential differences V_{ab} and $V_{a'b'}$ are therefore equal. In the future, we shall use the same symbol to represent all points in a circuit that are connected by resistanceless conductors and are at the same potential.

The current I in the resistor (and hence at all points of the circuit) could be found from the relation $I = V_{ab}/R$, if the potential difference V_{ab} were known. However, V_{ab} is the terminal voltage of the source, equal to $\mathcal{E} - Ir$, and since this depends on I it is unknown at the start. We can, however, calculate the current from the circuit equation:

$$I = \frac{\mathcal{E}}{R + r} = \frac{12 \text{ V}}{4\Omega + 2\,\Omega} = 2 \text{ A}.$$

The potential difference V_{ab} can now be found by considering a and b either as the terminals of the resistor or as those of the source. If we consider them as the terminals of the resistor,

$$V_{a'b'} = IR = (2 \text{ A})(4\,\Omega) = 8 \text{ V}.$$

If we consider them as the terminals of the source,

$$V_{ab} = \mathcal{E} - Ir = 12 \text{ V} - (2 \text{ A})(2\,\Omega) = 8 \text{ V}.$$

The voltmeter therefore reads 8 V and the ammeter reads 2 A.

Example 3   In Fig. 28–9, the source is short circuited. The current is

$$I = \frac{\mathcal{E}}{r} = \frac{12 \text{ V}}{2\,\Omega} = 6 \text{ A}.$$

The terminal voltage is

$$V_{ab} = \mathcal{E} - Ir = 12 \text{ V} - (6 \text{ A})(2\,\Omega) = 0.$$

The ammeter reads 6 A and the voltmeter reads zero.

The rule represented by Eq. (28–16), for finding the current in a circuit, can be recast in a form which is particularly useful in more complex circuits having several seats of emf or several branches. The electrostatic field is a *conservative* force field. Suppose we go around a loop, measuring potential differences across successive circuit elements. When we return to the starting point we must find that the *algebraic sum* of

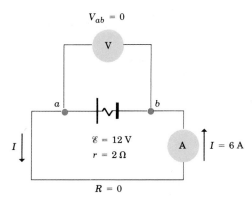

28-9   A source on short circuit.

these differences is zero; otherwise we could not say that the potential at this point has a definite value.

Thus, in Fig. 28-8, if we start at point b and travel counterclockwise (the same direction as I) we find a *rise* in potential (a *positive* change) due to the battery emf, a *drop* (a *negative* change) due to the battery's internal resistance, and an additional drop due to the 4-Ω resistor. The algebraic sum of these must be zero. Thus

$$12\text{ V} - I(2\,\Omega) - I(4\,\Omega) = 0, \qquad I = 2\text{ A},$$

in agreement with the result obtained in Example 2.

To allow for the possibility of several seats of emf in the loop, we generalize this procedure as follows: *The algebraic sum of the emf's around a series circuit, minus the sum of the IR products for the resistive elements in the circuit, must equal zero.* Formally,

$$\sum \mathcal{E} - \sum IR = 0. \qquad (28\text{-}20)$$

This is called *Kirchhoff's loop rule.* In applying it, we need some sign conventions. We first assume a direction for the current, and mark it on the diagram. Then, starting at any point in the circuit, we go around the circuit in the direction of the assumed current, adding emf's and IR products as we come to them. When a source is traversed in the direction from ($-$) to ($+$) (the direction of E_n in that source), the emf is considered *positive,* and when from ($+$) to ($-$), negative. The IR terms are all negative because the direction of current is always that of *decreasing* potential. Of course, we could also traverse the loop in the direction *opposite* to that of the assumed current. In that case all the emf's have opposite sign, and all the IR terms are positive because in going in the opposite direction to the current we are going "up-hill" from lower to higher potential.

For the circuit in Fig. 28-8, if we start at point b and go clockwise, the resulting equation is

$$I(4\,\Omega) + I(2\,\Omega) - 12\text{ V} = 0,$$

which is the same as the previous equation except for an overall factor of (-1) which does not change the value of I. But if we assume that I is clockwise, then starting at b and going counterclockwise around the circuit yields the equation

$$12\text{ V} + I(2\,\Omega) + I(4\,\Omega) = 0, \qquad I = -2\text{ A}.$$

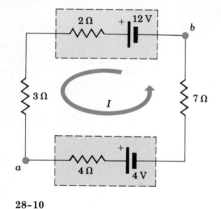

28-10

In this case the negative sign on the result shows that our initial assumption about the current direction was wrong; the actual direction is counterclockwise.

This same bookkeeping system may be used to determine the potential difference between any two points a and b in a circuit. To find $V_{ab} = V_a - V_b$ (the potential at a with respect to b), we start at b and add the potential changes encountered in going from b to a. Again an emf is considered positive when we go from $(-)$ to $(+)$, negative otherwise. An IR term is positive when we go "up-hill," against the current direction, negative when in the same direction as the current.

Example 4   The circuit shown in Fig. 28-10 contains two batteries, each having an emf and an internal resistance, and two resistors. Find the current in the circuit and the potential difference V_{ab}.

Solution   We assume a direction for the current, as shown; then, starting at a and going counterclockwise, we add potential increases and decreases to compute $\Sigma \mathcal{E} - \Sigma IR$. The resulting equation is

$$-I(4\,\Omega) - 4\,\text{V} - I(7\,\Omega) + 12\,\text{V} - I(2\,\Omega) - I(3\,\Omega) = 0.$$

Collecting terms containing I and solving for I, we find

$$8\,\text{V} = I(16\,\Omega) \qquad \text{and} \qquad I = 0.5\,\text{A}.$$

The result for I is positive, showing that our assumed current direction is correct. The reader may verify that if one assumes the opposite direction the result is $I = -0.5\,\text{A}$, indicating that the actual current is opposite to this assumption.

To find V_{ab}, the potential at a with respect to b, we start at b and go toward a, adding potential changes. There are two possible paths from b to a; taking the lower one first, we find

$$V_{ab} = (0.5\,\text{A})(7\,\Omega) + 4\,\text{V} + (0.5\,\text{A})(4\,\Omega) = 9.5\,\text{V}.$$

Point a is at 9.5 V higher potential than b. All the terms in this sum are positive because each represents an *increase* in potential as we go from b toward a. If instead we use the upper path, the resulting equation is

$$V_{ab} = 12\,\text{V} - (0.5\,\text{A})(2\,\Omega) - (0.5\,\text{A})(3\,\Omega) = 9.5\,\text{V}.$$

Here the IR terms are negative because our path goes in the direction of the current, with potential decreases through the resistors. The result is the same as for the other path, as it must be in order for the total potential change around the complete loop to be zero. In each case potential rises are taken as positive, and drops as negative.

In the foregoing analysis, emf's have been treated on the same basis as potential differences in circuit elements. This procedure might well be questioned; an emf is, strictly speaking, *not* a potential difference, inasmuch as it represents work done (per unit charge) by a force of *non*-electrostatic origin. But as we have seen the (electrostatic) potential difference between terminals of a source can always be expressed in terms of its emf and the drop across its internal resistance. Thus the emf terms in Eq. (28-20) really *do* represent true potential differences.

28-5 Current-voltage relations

The current through a device such as a resistor depends on the potential difference between its terminals. For a device obeying Ohm's law, the current is directly proportional to voltage, as shown in Eq. (28-11). But as mentioned in Sec. 28-2, there are many devices for which this simple model is not an adequate description. Current may depend on voltage in a more complicated way, and the current resulting from a given potential difference may depend on the polarity of the potential difference. This is the case with *diodes,* devices constructed deliberately to conduct much better in one direction than the other.

It is convenient to represent the current–voltage relation as a graph, and Fig. 28-11 shows several examples. Part (a) shows the behavior of a resistor that obeys Ohm's law, for which the graph is a straight line. Part (b) shows the relation for a vacuum diode. For positive potentials of anode with respect to cathode, I is approximately proportional to $V^{3/2}$, while for negative potentials the current is several orders of magnitude smaller and for most purposes may be assumed to be zero. Germanium diode behavior (c) is somewhat different but still strongly asymmetric, acting as a one-way valve in a circuit. Diodes are used to convert alternating current to direct and to perform a wide variety of logic functions in computer circuitry. The microscopic basis of diode behavior will be explored in some detail in later chapters.

An additional consideration is that for nearly all materials the current–voltage relation is temperature-dependent. Thus at low temperatures the curve in Fig. 28-11c rises more steeply for positive V than at higher temperatures, and at successively higher temperatures the asymmetry in the curve becomes less and less pronounced.

The current–voltage relation for a source may also be represented graphically. For a source represented by Eq. (28-15), that is,

$$V = \mathcal{E} - Ir,$$

the graph appears as in Fig. 28-12a. The intercept on the V-axis, corresponding to the open-circuit condition ($I = 0$), is at $V = \mathcal{E}$, and the intercept on the I-axis, corresponding to a short-circuit situation ($V = 0$), is at $I = \mathcal{E}/r$.

This line may be used to find the current in a circuit containing a nonlinear device, as in Fig. 28-12b. Its current-voltage relation is shown in Fig. 28-12c, and Eq. (28-15) is also plotted on this graph. Each curve represents a current-voltage relation that must be satisfied, so the intersection represents the only possible values of V and I. This amounts to a graphical solution of two simultaneous equations for V and I, one of which is nonlinear.

When a device has a nonlinear voltage–current relation, the quantity V/I is not constant. This ratio may still be called resistance, but now it varies with current; it is constant only for a device obeying Ohm's law. Often a more useful quantity is dV/dI, which expresses the relation between a small change in current and the resulting voltage change. This is called the *dynamic* or *incremental* resistance, and in general it too depends on current. For the voltage-current relation shown in Fig. 28-13, the dynamic resistance at current I_0 is the slope of the tangent line to the curve at I_0, while the ordinary resistance V/I is the slope of the line from the origin to this point.

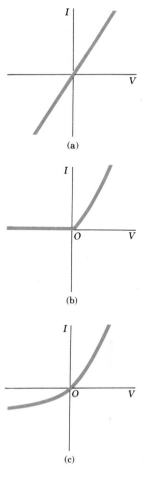

28-11 Current-voltage relations for (a) a resistor obeying Ohm's law; (b) a vacuum diode; (c) a semiconductor diode.

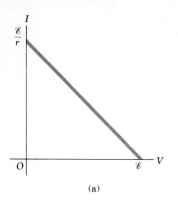

(a)

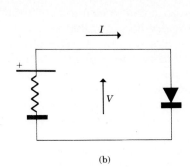

(b)

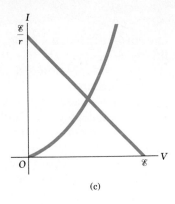

(c)

28-12 (a) Current–voltage relation for a source with emf $\mathcal{E}$ and internal resistance r; (b) a circuit containing a source and a nonlinear element; (c) simultaneous solution of I–V equations for this circuit.

Finally, we remark that Eq. (28–15) is not always an adequate representation of the behavior of a source. What we have described as an internal resistance may actually be a more complex voltage–current relation, and the emf may vary somewhat. Nevertheless, the concept of internal resistance frequently provides an adequate description of batteries, generators, and other energy converters. The difference between a fresh flashlight battery and an old one is not in the emf, which decreases only slightly with use, but principally in the internal resistance, which may increase from a few ohms when fresh to as much as $1000 \, \Omega$ or more after long use. Similarly, the current a car battery can deliver to the starter motor on a cold morning is less than when the battery is warm, partly because the emf is somewhat less but principally because the internal resistance is temperature-dependent, decreasing with increasing temperature.

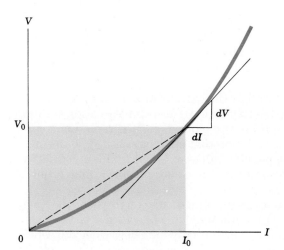

28-13 Dynamic resistance is the slope of the voltage–current curve.

28-6 Work and power in electrical circuits

We are now ready to consider in detail the energy relations in electric circuits. The rectangle in Fig. 28-14 represents a portion of a circuit

28-14 The power input P to the portion of the circuit between a and b is $P = V_{ab}I$.

having current I and potential difference $V_a - V_b = V_{ab}$ between the two conductors leading to and from this point of the circuit. The detailed nature of this circuit element need not be specified at present. As charge passes through, the electric field does work on it. In a time interval Δt, an amount of charge $\Delta Q = I \Delta t$ passes through, and the work ΔW done by the electric field is given by the product of the potential difference (work per unit charge) and the quantity of charge:

$$\Delta W = V_{ab} \Delta Q = V_{ab} I \Delta t.$$

By means of this work the electric field transfers energy into this portion of the circuit.

 Rate of transfer of energy is *power,* denoted by P; dividing the above relation by Δt, we obtain the rate at which energy enters this part of the circuit:

$$P = \frac{\Delta W}{\Delta t} = V_{ab}I. \tag{28–21}$$

It may happen that the potential at b is higher than that at a; in this case V_{ab} is negative. The charge then *gains* potential energy (at the expense of some other form of energy), and there is a corresponding transfer of electrical energy *out of* this portion of the circuit.

 Equation (28–21) is the general expression for the magnitude of the electrical power input to (or the power output from) any portion of an electrical circuit. The unit of V_{ab} is one volt, or one joule per coulomb, and the unit of I is one ampere or one coulomb per second. The SI unit of power is therefore

$$(1 \text{ J}\cdot\text{C}^{-1})(1 \text{ C}\cdot\text{s}^{-1}) = 1 \text{ J}\cdot\text{s}^{-1} = 1 \text{ W} = 1 \text{ watt}.$$

We now consider some special cases.

1. Pure resistance. If the portion of the circuit in Fig. 28–14 is a pure resistance, the potential difference is given by $V_{ab} = IR$ and

$$P = V_{ab}I = I^2R = \frac{V^2}{R}. \tag{28–22}$$

 The potential at a is necessarily higher than at b and there is a power *input* to the resistor. The circulating charges give up energy to the atoms of the resistor when they collide with them, and the temperature of the resistor increases unless there is a flow of heat out of it. We say that energy is *dissipated* in the resistor at a rate I^2R.

 Because of this heat, every resistor has a maximum power rating, the maximum power that can be dissipated without overheating the device. When this rating is exceeded the resistance may change unpredictably; in more extreme cases the resistor may melt or even explode. In practical applications, the power rating of a resistor is just as important a characteristic as its resistance value.

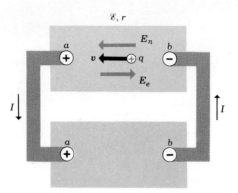

28-15 The rate of conversion of non-electrical to electrical energy in the source equals $\mathcal{E}I$. The rate of energy dissipation in the source is I^2r. The difference $\mathcal{E}I - I^2r$ is the power output of the source.

2. *Power output of a source.* The upper rectangle in Fig. 28–15 represents a source having emf $\mathcal{E}$ and internal resistance r, connected by ideal (resistanceless) conductors to an external circuit represented by the lower rectangle; the precise nature of the external circuit does not matter. We assume only that there is a current I in the circuit in the direction shown, from a to b in the external circuit and from b to a within the source. The letters a and b can be considered to represent either the terminals of the source or those of the external circuit, and $V_a > V_b$. The external circuit then corresponds to the rectangle in Fig. 28–14, and the power input to it is

$$P = V_{ab}I.$$

If a and b are considered as the terminals of the source, then as we have shown,

$$V_{ab} = \mathcal{E} - Ir$$

and hence,

$$P = V_{ab}I = \mathcal{E}I - I^2r. \tag{28–23}$$

The terms $\mathcal{E}I$ and I^2r have direct significance. The emf $\mathcal{E}$ has been defined as the work per unit charge performed on the charges by the non-electrostatic field $\boldsymbol{E}_n$ as the charges move from b to a in the source. Hence, if a charge ΔQ flows in time Δt, this field does work $\Delta W = \mathcal{E}\,\Delta Q$, and its *rate* of doing work, or power, is

$$P = \frac{\Delta W}{\Delta t} = \mathcal{E}\frac{\Delta Q}{\Delta t} = \mathcal{E}I. \tag{28–24}$$

Hence, the product $\mathcal{E}I$ is the rate at which work is done on the circulating charges by the agency that maintains the non-electrostatic field.

The term I^2r is the rate at which energy is *dissipated* in the internal resistance of the source, and the difference $\mathcal{E}I - I^2r$ is the rate at which energy is delivered by the source to the remainder of the circuit. In other words, the power P in Eq. (28–23) represents the *power output* of the source, or the *power input* to the remainder of the circuit.

3. *Power input to a source.* Suppose that the lower rectangle in Fig. 28–15 is itself a source of emf larger than that of the upper source and with its emf opposite to that of the upper source. The current I in the circuit is then *opposite* to that shown in Fig. 28–15. The lower output of

the lower source is

$$P = V_{ab}I.$$

Considering a and b as the terminals of the upper source, we have

$$V_{ab} = \mathcal{E} + Ir,$$

and

$$P = V_{ab}I = \mathcal{E}I + I^2r. \tag{28-25}$$

The charges in the upper source now move from left to right, in a direction opposite the field $\boldsymbol{E}_n$, and the work done by the non-electrostatic force $\boldsymbol{F}_n$ is *negative*. That is, work is done *on* the agent maintaining the non-electrostatic field. The product $\mathcal{E}I$ equals the rate at which work is done on this agent, and the term I^2r again equals the rate of dissipation of energy in the internal resistance of the source. The sum $\mathcal{E}I + I^2r$ is therefore the total *power input* to the upper source. The source converts electrical energy to non-electrical energy.

To illustrate these relations, let us apply them to the circuits in Figs. 28–8 through 28–10.

Example 1 The rate of energy conversion in the source in Fig. 28–8 is

$$\mathcal{E}I = (12\ \text{V})(2\ \text{A}) = 24\ \text{W}.$$

The rate of dissipation of energy in the source is

$$I^2r = (2\ \text{A})^2(2\ \Omega) = 8\ \text{W}.$$

The power *output* of the source is the difference between these, or 16 W.

The power output is also given by

$$V_{ab}I = (8\ \text{V})(2\ \text{A}) = 16\ \text{W}.$$

The power input to the resistor is

$$V_{a'b'}I = (2\ \text{A})(8\ \text{V}) = 16\ \text{W}.$$

This equals the rate of dissipation of energy in the resistor:

$$I^2R = (2\ \text{A})^2(4\ \Omega) = 16\ \text{W}.$$

Example 2 The rate of energy conversion in the source in Fig. 28–9 is

$$\mathcal{E}I = (12\ \text{V})(6\ \text{A}) = 72\ \text{W}.$$

The rate of dissipation of energy in the source is

$$I^2r = (6\ \text{A})^2(2\ \Omega) = 72\ \text{W}.$$

The power *output* of the source (also given by $V_{ab}I$) equals zero. *All* of the energy converted is dissipated within the source.

Example 3 The rate of energy conversion in the upper source in Fig. 28–10 is

$$\mathcal{E}I = (12\ \text{V})(0.5\ \text{A}) = 6\ \text{W}.$$

The rate of dissipation of energy in the source is

$$I^2r = (0.5\ \text{A})^2(2\ \Omega) = 0.5\ \text{W}.$$

The power output of the source is $6\,W - 0.5\,W = 5.5\,W$.

The rates of dissipation of energy in the 3-Ω and 7-Ω resistors are, respectively,

$$(0.5\,A)^2(3\,\Omega) = 0.75\,W,$$

$$(0.5\,A)^2(7\,\Omega) = 1.75\,W.$$

The rate of energy conversion in the lower source is

$$\mathcal{E}I = (4\,V)(0.5\,A) = 2\,W,$$

and the rate of energy dissipation in this source is

$$I^2r = (0.5\,A)^2(4\,\Omega) = 1\,W.$$

The power *input* to this source is $2\,W + 1\,W = 3\,W$. Thus, of the 5.5-W output of the upper source, 2.5 W is dissipated in the two resistors, and the remaining 3 W is partly converted, partly dissipated in the lower source. Conversion of energy in the sources is often reversible, as in a storage battery where electrical energy is converted to chemical for later retrieval as electrical energy. Energy dissipated in resistors is converted irreversibly to heat.

28-7 Physiological effects of currents

Electrical potential differences and currents play a vital role in the nervous systems of animals. Conduction of nerve impulses is basically an electrical process, although the mechanism of conduction is much more complex than in simple material such as metals. A nerve fiber or *axon,* along which an electrical impulse can travel, includes a cylindrical membrane with one conducting fluid (electrolyte) inside and another outside. By mechanisms similar to those in batteries, a potential difference of the order of 0.1 V is maintained between these fluids.

When a pulse is initiated, the membrane temporarily becomes more permeable to the ions in the fluids, leading to a local drop in potential. As the pulse passes, with a typical speed of the order of $30\,m\cdot s^{-1}$, the membrane recovers and the potential returns to its initial value. There are several aspects of this process that are not yet well understood.

The basically electrical nature of nerve-impulse conduction is responsible for the great sensitivity of the body to externally supplied electrical currents. Currents through the body as small as 0.1 A, much too small to produce significant heating, are fatal because they interfere with nerve processes essential for vital functions such as heartbeat. The *resistance* of the human body is highly variable, principally because of the fairly low conductivity of the skin, but the resistance between two electrodes grasped by dry hands is of the order 5 kΩ to 10 kΩ. For $R = 10\,k\Omega$, a current of 0.1 A requires a potential difference $V = IR = (0.1\,A)(10\,k\Omega) = 1000\,V$.

Even much smaller currents can be very dangerous. A current of 0.01 A causes strong, convulsive muscle action and considerable pain, and with 0.02 A the person typically is unable to release a conductor inflicting the shock. Currents of this magnitude and even as small as 0.001 A can cause ventricular fibrillation, a disorganized twitching of heart muscles that occurs instead of regular beating, and that unfortunately pumps very little blood. Surprisingly, very large currents (over

0.1 A) are somewhat *less* likely to cause fatal fibrillation because the heart muscle is "clamped" in one position and is more likely to resume normal beating when the current is removed. Severe burns are, of course, more likely with large currents.

The moral of this rather morbid story, if there is one, is that under certain conditions voltages as small as 10 V can be dangerous, and should not be regarded with anything but respect and caution.

On the positive side, rapidly alternating currents can have beneficial effects. Alternating currents with frequencies the order of 10^6 Hz do not interfere appreciably with nerve processes, and can be used for therapeutic heating for arthritic conditions, sinusitis, and a variety of other disorders. If one electrode is made very small, the resulting concentrated heating can be used for local destruction of tissue, such as tumors, or even for cutting tissue in certain surgical procedures.

Study of particular nerve impulses is also an important *diagnostic* tool in medicine. The most familiar examples are electrocardiography (EKG) and electroencephalography (EEG). Electrocardiograms, obtained by attaching electrodes to the chest and back and recording the regularly varying potential differences, are used to study heart function. Similarly, electrodes attached to the scalp permit study of potentials in the brain, and the resulting patterns can be helpful in diagnosing epilepsy, brain tumors, and other disorders.

* 28-8 The electric field of the earth

If the molecules of the earth's atmosphere were electrically neutral, the atmosphere would be a nonconductor or insulator. Actually, because of the bombardment of the earth by cosmic rays (e.g., high-speed protons from outer space), a few ions, both positive and negative, are always present. With increasing elevation, the ionization produced by cosmic rays increases; and since the density of the atmosphere decreases, ions can move more freely and the conductivity increases. Above an elevation of about 50 km, the atmosphere is a relatively good conductor; and the earth itself is also a relatively good conductor. An extremely simplified model of the earth and its lower atmosphere consists of a conducting sphere surrounded by a conducting spherical shell, the two being separated by a poorly conducting layer about 50 km thick.

Near the earth's surface, there is found to be a radial electrostatic field or potential gradient of about 100 V·m^{-1}. The field becomes weaker with increasing elevation, and the total potential difference between the earth's surface and the outer conducting layer is about 400,000 V. The direction of the field is downward, so the earth has a negative surface charge and the outer conducting layer has a positive charge. The surface charge density and the electric field are related by Eq. (25–22),

$$\sigma = \epsilon_0 E;$$

and hence, if $E = 100$ V·m^{-1}, the surface density is, roughly, 10^{-9} C·m^{-2}.

Approximating the earth by a sphere of radius 5000 km, we find its surface area is about 3×10^{14} m^2, so the total surface charge is about 3×10^5 C.

Because the atmosphere is not a perfect insulator, there is a current in it, directed downward in the same direction as the electrostatic field. (Positive ions drift downward; negative ions drift upward.) The total current over the entire earth's surface is fairly constant and is about 1800 A or 1800 C·s⁻¹. The time required to neutralize the entire surface charge of 3×10^5 C, if it were not replenished in some way, would therefore be only about two or three minutes.

The charge, however, remains constant, so there must be some mechanism that pumps positive charges upward and negative charges downward, opposite to the directions of the forces exerted on them by the electrostatic field. The nature of this mechanism is not fully understood, but it is believed to be the processes that go on in the development of thunderstorms. In some way, positive and negative ions are developed in a thunderstorm. The positive ions are carried upward and the negative ions downward by the air currents in the storm. These currents, in turn, are driven by pressure differences resulting from nonuniform temperatures in the atmosphere, so that the ultimate source of the energy input is the radiant energy reaching the earth from the sun.

The earth's atmosphere can be compared to an enormous Van de Graaff generator immersed in a conducting medium. Instead of a moving belt, the motion of the atmosphere transports charge carriers in a direction opposite to the electrostatic force acting on them. The charges then flow back by conduction through the lower atmosphere. The terminal voltage of the generator is 400,000 V and the current is 1800 A, so the power is about 700 MW. By way of comparison, a modern power plant may develop on the order of 1000 MW, and others to be built are in the GW range.

*28-9 Theory of metallic conduction

Additional insight into the phenomenon of conduction can be gained by examining the microscopic mechanisms of conductivity. Here we consider only a crude and primitive model that treats the electrons as classical particles and ignores their inherently quantum-mechanical behavior in solids. Thus this model is not correct conceptually; yet it is useful in helping to develop an intuitive idea of the microscopic basis of conduction.

In the simplest microscopic model of metallic conduction, each atom in the crystal lattice is assumed to give up one or more of its outer electrons. These electrons are then free to move through the crystal lattice, colliding at intervals with the stationary positive ions. Their motion is like that of the molecules of gas in a container and they are often referred to as an "electron gas." In the absence of an electric field, the electrons move in straight lines between collisions, but if there is an electric field, the paths are slightly curved, as in Fig. 28–16, which represents schematically a few free paths of an electron in an electric field directed from right to left. At each collision, the electron is assumed to lose any energy it may have acquired from the field and to make a fresh start. The energy given up in these collisions increases the thermal energy of vibration of the positive ions.

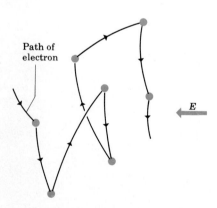

Path of electron

E

28–16 Random motion of an electron, with superimposed drift under action of electric field.

A force $F = eE$ is exerted on each electron by the field, and produces an acceleration a in the direction of the force given by

$$a = \frac{F}{m} = \frac{eE}{m},$$

where m is the electron mass. Let u represent the average *random* speed of an electron, and λ the mean free path. The average time t between collisions, called the *mean free time,* is

$$t = \frac{\lambda}{u}.$$

In this time, the electron acquires a final velocity component v_f in the direction of the force, given by

$$v_f = at = \frac{eE}{m}\frac{\lambda}{u}.$$

Its *average* velocity v in the direction of the force, which is superposed on its random velocity and which we interpret as the *drift* velocity, is one-half the final velocity, so

$$v = \frac{1}{2}v_f = \frac{1}{2}\frac{e\lambda}{mu}E.$$

The drift velocity is therefore proportional to the electric field E.

The current density is

$$J = nev = \frac{ne^2\lambda}{2mu}E,$$

and the resistivity is

$$\rho = \frac{E}{J} = \frac{2mu}{ne^2\lambda}. \tag{28-26}$$

This theoretical expression for resistivity is in *qualitative* agreement with experiment. At a given temperature, the quantities m, u, n, e, and λ are constant. The resistivity is then constant, and Ohm's law is obeyed. When the temperature is increased, the random speed u increases and the theory predicts that the resistivity of a metal increases with increasing temperature.

In a semiconductor, the number of charge carriers per unit volume, n, increases rapidly with increasing temperature. The increase in n far outweighs any increase in u, and the resistivity *decreases*. At low temperatures, n is very small and the resistivity becomes so large that the material can be considered an insulator.

The modern theory of superconductivity predicts that, in effect, at temperatures below the critical temperature the electrons move freely throughout the lattice. The mean free path λ then becomes very large and the resistivity very small.

Questions

28-1 A rule of thumb used to determine the internal resistance of a source is that it is the open-circuit voltage divided by the short-circuit current. Is this correct?

28-2 The energy that can be extracted from a storage battery is always less than the energy that goes into it while it is being charged. Why?

28-3 In circuit analysis one often assumes that a wire connecting two circuit elements has no potential difference between its ends; yet there must be an electric field within the wire to make the charges move, and so there must be a potential difference. How do you resolve this discrepancy?

28-4 Long-distance electric-power transmission lines always operate at very high voltage, sometimes as much as 750 kV. What are the advantages of such high voltages? The disadvantages?

28-5 Ordinary household electric lines usually operate at 110 V. Why is this a desirable voltage, rather than a value considerably larger or smaller? What about cars, which usually have 12-V electrical systems?

28-6 What is the difference between an emf and a potential difference?

28-7 Electric power for household and commercial use always uses *alternating current,* which reverses direction 120 times each second. A student claimed that the power conveyed by such a current would have to average out to zero, since it is going one way half the time and the other way the other half. What is your response?

28-8 As discussed in the text, the drift velocity of electrons in a good conductor is very slow. Why then does the light come on so quickly when the switch is turned on?

28-9 A fuse is a device designed to break a circuit, usually by melting, when the current exceeds a certain value. What characteristics should the material of the fuse have?

28-10 What considerations determine the maximum current-carrying capacity of household wiring?

28-11 The text states that good thermal conductors are also good electrical conductors. If so, why don't the cords used to connect toasters, irons, and similar heat-producing appliances get hot by conduction of heat from the heating element?

28-12 Eight flashlight batteries in series have an emf of about 12 V, about the same as that of a car battery. Could they be used to start a car with a dead battery?

28-13 High-voltage power supplies are sometimes designed intentionally to have rather large internal resistance, as a safety precaution. Why is such a power supply with a large internal resistance safer than one with the same voltage but lower internal resistance?

28-14 How would you expect the resistivity of a good insulator such as glass or polystyrene to vary with temperature? Why?

Problems

28-1 A silver wire 1 mm in diameter transfers a charge of 90 C in 1 hr and 15 min. Silver contains 5.8×10^{28} free electrons per m^3.

a) What is the current in the wire?

b) What is the drift velocity of the electrons in the wire?

28-2 When a sufficiently high potential difference is applied between two electrodes in a gas, the gas ionizes; electrons move toward the positive electrode and positive ions toward the negative electrode.

a) What is the current in a hydrogen discharge if, in each second, 4×10^{18} electrons and 1.5×10^{18} protons move in opposite directions past a cross section of the tube?

b) What is the direction of the current?

28-3 A vacuum diode can be approximated by a plane cathode and a plane anode, parallel to each other and 5 mm apart. The area of both cathode and anode is 2 cm^2. In the region between cathode and anode the current is carried solely by electrons. If the electron current is 50 mA, and the electrons strike the anode surface with a speed of 1.2×10^7 m·s^{-1}, find the number of electrons per cubic millimeter in the space just outside the surface of the anode.

28-4 In the Bohr model of the hydrogen atom the electron makes about 6×10^{15} rev·s^{-1} around the nucleus. What is the average current at a point on the orbit of the electron?

28-5 A copper wire has a square cross section, 2.0 mm on a side. It is 4 m long and carries a current of 10 A. The density of free electrons is 8×10^{28} m^{-3}.

a) What is the current density in the wire?

b) What is the electric field?

c) How much time is required for an electron to travel the length of the wire?

28-6 A wire 100 m long and 2 mm in diameter has a resistivity of 4.8×10^{-8} Ω·m.

a) What is the resistance of the wire?

b) A second wire of the same material has the same mass as the 100-m length, but twice its diameter. What is its resistance?

28-7 A certain electrical conductor has a square cross section, 2.0 mm on a side, and is 12 m long. The resistance between its ends is 0.072 Ω.

a) What is the resistivity of the material?

b) If the electric field magnitude in the conductor is $0.12 \, \text{V} \cdot \text{m}^{-1}$, what is the total current?

c) If the material has 8.0×10^{28} free electrons per cubic meter, find the average drift velocity under conditions of part (b).

28-8 The following measurements of current and potential difference were made on a resistor constructed of Nichrome wire:

I, A	V_{ab}, V
0.5	2.18
1.0	4.36
2.0	8.72
4.0	17.44

a) Make a graph of V_{ab} as a function of I.

b) Does the Nichrome obey Ohm's law?

c) What is the resistance of the resistor, in ohms?

28-9 The following measurements were made on a Thyrite resistor:

I, A	V_{ab}, V
0.5	4.76
1.0	5.81
2.0	7.05
4.0	8.56

a) Make a graph of V_{ab} as a function of I. Does Thyrite have constant resistance?

b) Construct a graph of the resistance R as a function of I.

c) Find the dynamic resistance at 2.0 A, and compare with the resistance values found in part (b).

28-10 The current in a wire varies with time according to the relation

$$i = 4 \, \text{A} + (2 \, \text{A} \cdot \text{s}^{-2})t^2.$$

a) How many coulombs pass a cross section of the wire in the time interval between $t = 5 \, \text{s}$ and $t = 10 \, \text{s}$?

b) What constant current would transport the same charge in the same time interval?

28-11 The current in a wire varies with time according to the relation

$$i = (20 \, \text{A}) \sin (377 \, \text{s}^{-1}t).$$

a) How many coulombs pass a cross section of the wire in the time interval between $t = 0$ and $t = 1/120 \, \text{s}$?

b) In the interval between $t = 0$ and $t = 1/60 \, \text{s}$?

c) What constant current would transport the same charge in each of the intervals above?

28-12 An aluminum bar 2.5 m long has a rectangular cross section 1 cm by 5 cm.

a) What is its resistance?

b) What would be the length of an iron wire 15 mm in diameter having the same resistance?

28-13 A solid cube of silver has a mass of 84.0 g. What is its resistance between opposite faces?

28-14 The two parallel plates of a capacitor have equal and opposite charges Q. The dielectric has a dielectric constant K and a resistivity ρ. Show that the "leakage" current carried by the dielectric is given by the relationship $i = Q/K\epsilon_0\rho$.

28-15

a) What is the resistance of a Nichrome wire at $0 \, ^\circ\text{C}$, if its resistance is $100.00 \, \Omega$ at $12 \, ^\circ\text{C}$?

b) What is the resistance of a carbon rod at $30 \, ^\circ\text{C}$, if its resistance is $0.0150 \, \Omega$ at $0 \, ^\circ\text{C}$?

28-16 A copper wire is initially at $0 \, ^\circ\text{C}$. How much must its temperature rise for the resistance to increase by 10 percent?

28-17 Refer to the third example in Sec. 28-3. Let the resistivity of the material between the cylinders be $10 \, \Omega \cdot \text{m}$ and let $r_a = 10 \, \text{cm}$, $r_b = 20 \, \text{cm}$, and $l = 5 \, \text{cm}$.

a) Find the resistance between the cylinders.

b) Find the current between the cylinders if $V_{ab} = 8 \, \text{V}$.

28-18 The region between two concentric conducting spheres of radii r_a and r_b is filled with a conducting material of resistivity ρ.

a) Show that the resistance between the spheres is given by

$$R = \frac{\rho}{4\pi}\left(\frac{1}{r_a} - \frac{1}{r_b}\right).$$

b) Derive an expression for the current density as a function of radius, if the potential difference between the spheres is V_{ab}.

28-19 A carbon resistor is to be used as a thermometer. On a winter day when the temperature is $0 \, ^\circ\text{C}$, its resistance is $217.3 \, \Omega$. What is the temperature on a hot summer day when the resistance is $214.2 \, \Omega$?

28-20 The resistance of a coil of copper wire is $200 \, \Omega$ at $20 \, ^\circ\text{C}$. What is its resistance at $50 \, ^\circ\text{C}$?

28-21 A toaster using a Nichrome heating element operates on 120 volts. When it is switched on at $0 \, ^\circ\text{C}$, it carries an initial current of 1.5 A. A few seconds later the current reaches the steady value of 1.33 A. What is the final temperature of the element? The average value of the temperature coefficient of Nichrome over the temperature range is $0.00045 \, (\text{C}^\circ)^{-1}$.

28-22 A certain resistor has a resistance of $150.4 \, \Omega$ at $20 \, ^\circ\text{C}$ and a resistance of $162.4 \, \Omega$ at $40 \, ^\circ\text{C}$. What is its temperature coefficient of resistivity?

28-23 A resistance thermometer using a platinum wire is used to measure the temperature of a liquid. The resistance is 2.42 Ω at 0°C, and when immersed in the liquid it is 2.98 Ω. The temperature coefficient of resistivity of platinum is 0.0038 (C°)$^{-1}$. What is the temperature of the liquid?

28-24 A piece of wire has a resistance R. It is cut into three pieces of equal length, and the pieces are twisted together in parallel. What is the resistance of the resulting wire?

28-25 What diameter must an aluminum wire have if it is to have the same resistance as an equal length of copper wire of diameter 2.0 mm?

28-26 When switch S is open, the voltmeter V, connected across the terminals of the dry cell in Fig. 28-17, reads 1.52 V. When the switch is closed, the voltmeter reading drops to 1.37 V and the ammeter A reads 1.5 A. Find the emf and internal resistance of the cell. Neglect meter corrections.

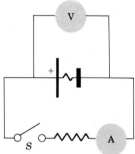

Figure 28-17

28-27 The potential difference across the terminals of a battery is 8.5 V when there is a current of 3 A in the battery from the negative to the positive terminal. When the current is 2 A in the reverse direction, the potential difference becomes 11 V.

a) What is the internal resistance of the battery?

b) What is the emf of the battery?

28-28

a) What is the potential difference V_{ad} in the circuit of Fig. 28-18?

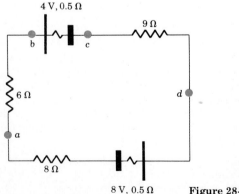

Figure 28-18

b) What is the terminal voltage of the 4-V battery?

c) A battery of emf 17 V and internal resistance 1 Ω is inserted in the circuit at d, with its positive terminal connected to the positive terminal of the 8-V battery. What is now the difference of potential V_{bc} between the terminals of the 4-V battery?

28-29 A closed circuit consists of a 12-V battery, a 3.7-Ω resistor, and a switch. The internal resistance of the battery is 0.3 Ω. The switch is opened. What would a high-resistance voltmeter read when placed

a) across the terminals of the battery,

b) across the resistor,

c) across the switch?

Repeat (a), (b), and (c) for the case when the switch is closed.

28-30 The internal resistance of a dry cell increases gradually with age, even though the cell is not used. The emf, however, remains fairly constant at about 1.5 V. Dry cells may be tested for age at the time of purchase by connecting an ammeter directly across the terminals of the cell and reading the current. The resistance of the ammeter is so small that the cell is practically short circuited.

a) The short-circuit current of a fresh No. 6 dry cell (1.5-V emf) is about 30 A. Approximately what is the internal resistance?

b) What is the internal resistance if the short-circuit current is only 10 A?

c) The short-circuit current of a 6-V storage battery may be as great as 1000 A. What is its internal resistance?

28-31 The open-circuit terminal voltage of a source is 10 V and its short-circuit current is 4.0 A.

a) What will be the current when the source is connected to a linear resistor of resistance 2 Ω?

b) What will be the current in the Thyrite resistor of Problem 28-9 when connected across the terminals of this source?

c) What is the terminal voltage at this current?

28-32 A "660-W" electric heater is designed to operate from 120-V lines.

a) What is its resistance?

b) What current does it draw?

c) What is the rate of dissipation of energy, in calories per second?

d) If the line voltage drops to 110 V, what power does the heater take, in watts? (Assume the resistance constant. Actually, it will change because of the change in temperature.)

28-33 A resistor develops heat at the rate of 360 W when the potential difference across its ends is 180 V. What is its resistance?

28-34 A motor operating on 120 V draws a current of 2 A. If heat is developed in the motor at the rate of 9 cal·s⁻¹, what is its efficiency?

28-35

a) Express the rate of dissipation of energy in a resistor in terms of (1) potential difference and current, (2) resistance and current, (3) potential difference and resistance.

b) Energy is dissipated in a resistor at the rate of 40 W when the potential difference between its terminals is 60 V. What is its resistance?

28-36 A storage battery whose emf is 12 V and whose internal resistance is 0.1 Ω is to be charged from a 112-V dc supply. [*Caution:* Don't try to charge a battery directly from a power line, either dc or the more usual ac in household wiring. Real 12-V car batteries have internal resistance of only a few milliohms, and a serious explosion is possible.]

a) Should the + or the − terminal of the battery be connected to the + side of the line?

b) What will be the charging current if the battery is connected directly across the line?

c) Compute the resistance of the series resistor required to limit the current to 10 A.

With this resistor in the circuit, compute

d) the potential difference between the terminals of the battery,

e) the power taken from the line,

f) the power dissipated in the series resistor, and

g) the *useful* power input to the battery.

h) If electrical energy costs 3 cents per kWh, what is the cost of operating the circuit for 2 hr when the current is 10 A?

28-37 In the circuit in Fig. 28–19, find

a) the rate of conversion of internal energy to electrical energy within the battery,

b) the rate of dissipation of energy in the battery,

c) the rate of dissipation of energy in the external resistor.

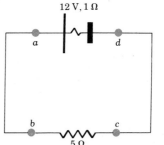

12 V, 1 Ω

a   *d*

b   *c*

5 Ω   **Figure 28–19**

28-38 A source whose emf is $\mathcal{E}$ and whose internal resistance is r is connected to an external circuit.

a) Show that the power output of the source is maximum when the current in the circuit is one-half the short-circuit current of the source.

b) If the external circuit consists of a resistance R, show that the power output is maximum when $R = r$, and that the maximum power is $\mathcal{E}^2/4r$.

DIRECT-CURRENT CIRCUITS AND INSTRUMENTS

29

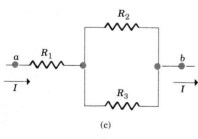

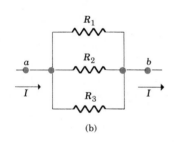

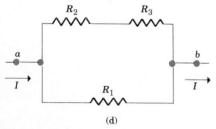

29–1 Four different ways of connecting three resistors.

29-1 Resistors in series and in parallel

Most electrical circuits consist not merely of a single source and a single external resistor, but comprise a number of sources, resistors, or other elements such as capacitors, motors, etc., interconnected in a more or less complicated manner. The general term applied to such a circuit is a *network*. We shall next consider a few of the simpler types of network.

Figure 29–1 illustrates four different ways in which three resistors having resistances R_1, R_2, and R_3 might be connected between points a and b. In part (a), the resistors provide only a single path between the points, and are said to be connected in *series* between these points. Any number of circuit elements such as resistors, cells, motors, etc., are similarly said to be in series with one another between two points if connected as in (a) so as to provide only a single path between the points. The *current* is the same in each element.

The resistors in Fig. 29–1b are said to be in *parallel* between points a and b. Each resistor provides an alternative path between the points, and any number of circuit elements similarly connected are in parallel with one another. The *potential difference* is the same across each element.

In Fig. 29–1c, resistors R_2 and R_3 are in parallel with each other, and this combination is in series with the resistor R_1. In Fig. 29–1d, R_2 and R_3 are in series, and this combination is in parallel with R_1.

It is always possible to find a single resistor that could replace a combination of resistors in any given circuit and leave unaltered the potential difference between the terminals of the combination and the current in the rest of the circuit. The resistance of this single resistor is called the *equivalent* resistance of the combination. If any one of the networks in Fig. 29–1 were replaced by its equivalent resistance R, we could write

$$V_{ab} = IR \quad \text{or} \quad R = \frac{V_{ab}}{I},$$

where V_{ab} is the potential difference between the terminals a and b of the network and I is the current at point a or b. Hence the method of computing an equivalent resistance is to assume a potential difference V_{ab} across the actual network, compute the corresponding current I, and take the ratio V_{ab}/I. The simple series and parallel connections of resistors are sufficiently common so that it is worthwhile to develop formulas for these two special cases.

If the resistors are in *series,* as in Fig. 29–1a, the current in each must be the same and equal to the line current I. Hence

$$V_{ax} = IR_1, \qquad V_{xy} = IR_2, \qquad V_{yb} = IR_3,$$

and

$$V_{ab} = V_{ax} + V_{xy} + V_{yb} = I(R_1 + R_2 + R_3),$$
$$\frac{V_{ab}}{I} = R_1 + R_2 + R_3.$$

But V_{ab}/I is, by definition, the equivalent resistance R. Therefore

$$R = R_1 + R_2 + R_3. \tag{29–1}$$

The equivalent resistance of any number of resistors in series equals the *sum* of their individual resistances.

If the resistors are in *parallel,* as in Fig. 29–1b, the potential difference between the terminals of each must be the same and equal to V_{ab}. If the currents in each are denoted by I_1, I_2, and I_3, respectively,

$$I_1 = \frac{V_{ab}}{R_1}, \qquad I_2 = \frac{V_{ab}}{R_2}, \qquad I_3 = \frac{V_{ab}}{R_3}.$$

Charge is delivered to point a by the line current I, and removed from a by the currents I_1, I_2, and I_3. Since charge is not accumulating at a, it follows that

$$I = I_1 + I_2 + I_3 = V_{ab}\left(\frac{1}{R_1} + \frac{1}{R_2} + \frac{1}{R_3}\right),$$

or

$$\frac{I}{V_{ab}} = \frac{1}{R_1} + \frac{1}{R_2} + \frac{1}{R_3}.$$

But

$$\frac{I}{V_{ab}} = \frac{1}{R},$$

so

$$\frac{1}{R} = \frac{1}{R_1} + \frac{1}{R_2} + \frac{1}{R_3}. \tag{29–2}$$

For any number of resistors in parallel, the *reciprocal* of the equivalent resistance equals the *sum of the reciprocals* of their individual resistances.

For the special case of *two* resistors in parallel,

$$\frac{1}{R} = \frac{1}{R_1} + \frac{1}{R_2} = \frac{R_2 + R_1}{R_1 R_2}$$

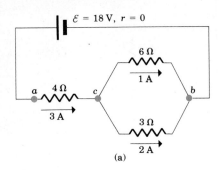

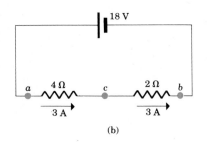

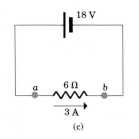

29-2 Reduction of a network of resistors to a single equivalent resistor.

and

$$R = \frac{R_1 R_2}{R_1 + R_2}.$$

Also, since $V_{ab} = I_1 R_1 = I_2 R_2$,

$$\frac{I_1}{I_2} = \frac{R_2}{R_1}, \tag{29-3}$$

and the currents carried by two resistors in parallel are *inversely proportional* to their resistances.

The equivalent resistances of the networks in Figs. 29-1c and 29-1d could be found by the same general method, but it is simpler to consider them as combinations of series and parallel arrangements. Thus, in (c) the combination of R_2 and R_3 in parallel is first replaced by its equivalent resistance, which then forms a simple series combination with R_1. In (d), the combination of R_2 and R_3 in series forms a simple parallel combination with R_1. Not all networks, however, can be reduced to simple series–parallel combinations, and special methods must be used for handling such networks.

Example   Compute the equivalent resistance of the network in Fig. 29-2, and find the current in each resistor.

Solution   Successive stages in the reduction to a single equivalent resistance are shown in parts (b) and (c). The 6-Ω and the 3-Ω resistors in part (a) are equivalent to the single 2-Ω resistor in part (b), and the series combination of this with the 4-Ω resistor results in the single equivalent 6-Ω resistor in part (c).

In the simple series circuit of part (c), the current is 3 A, and hence the current in the 4-Ω and 2-Ω resistors in part (b) is 3 A also. The potential difference V_{cb} is therefore 6 V, and since it must be 6 V in part (a) as well, the currents in the 6-Ω and 3-Ω resistors in part (a) are 1 A and 2 A, respectively.

29-2 Kirchhoff's rules

Not all networks can be reduced to simple series–parallel combinations. An example is a resistance network with a cross connection, as in Fig. 29-3a. A circuit like that in Fig. 29-3b, which contains sources in parallel paths, is another example. No new *principles* are required to compute the currents in these networks, but there are a number of techniques that enable such problems to be handled systematically. We shall describe only one of these, first developed by Gustav Robert Kirchhoff (1824–1887).

We first define two terms. A *branch point* in a network is a point where three or more conductors are joined. A *loop* is any closed conducting path. In Fig. 29-3a, for example, points a, d, e, and b are branch points but c and f are not. In Fig. 29-3b there are only two branch points, a and b.

Some possible loops in Fig. 29-3a are the closed paths *aceda*, *defbd*, *hadbgh*, and *hadefbgh*.

Kirchhoff's rules consist of the following two statements:

Point rule   *The algebraic sum of the currents **toward** any branch point is zero:*

$$\sum I = 0. \qquad (29\text{-}4)$$

Loop rule   *The algebraic sum of the potential differences in any loop, including those associated with emfs and those of resistive elements, must equal zero:*

$$\sum \mathcal{E} - \sum IR = 0. \qquad (29\text{-}5)$$

The point rule is an application of the principle of conservation of electric charge. Since no charge can accumulate at a branch point, the total current entering the point must equal the total leaving, or (considering those entering as positive and those leaving as negative) the algebraic sum of currents into a point must be zero.

The loop rule, which we have already encountered in Sec. 28–4, is an expression of conservation of *energy*; as a charge traverses a loop and returns to its starting point, the sum of the *rises* in potential associated with emf's in the loop must equal the sum of *drops* in potential associated with resistors (or, in more general problems, other circuit elements).

These basic rules are sufficient for the solution of a wide variety of network problems. Usually in such problems some of the emf's, currents, and resistances are known and others are unknown. The number of equations obtained from Kirchhoff's rules must always be equal to the number of unknowns, to permit simultaneous solution of the equations. The principal difficulty is not in understanding the basic ideas but in keeping track of algebraic signs! The following procedures should be followed carefully.

First, all quantities, known and unknown, should be labeled carefully, including an assumed sense of direction for each unknown current and emf. Often one does not know in advance the *actual* direction of an unknown current or emf, but this does not matter. The solution is carried out using the assumed directions, and if the actual direction of a particular quantity is opposite to the assumed direction, the value of the quantity will emerge from the analysis with a negative sign. Hence Kirchhoff's rules, correctly used, give the directions as well as the magnitudes of unknown currents and emf's. This point will be illustrated in the examples to follow.

Usually in labeling currents it is advantageous to use the point rule immediately to express the currents in terms of as few quantities as possible. For example, Fig. 29–4a shows a circuit correctly labeled, and Fig. 29–4b shows the same circuit, relabeled by applying the point rule to point *a* to eliminate I_3.

The following guidelines will help with the problem of signs:

1. Choose any closed loop in the network, and designate a direction (clockwise or counterclockwise) to traverse the loop in applying the loop rule.

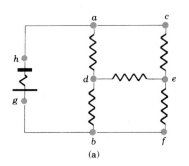

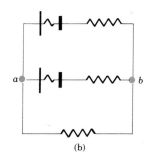

29-3   Two networks that cannot be reduced to simple series–parallel combinations of resistors.

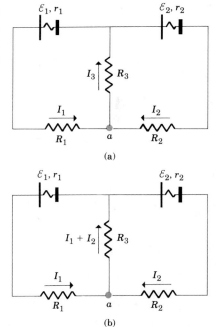

29-4   Application of the point rule to point *a* reduces the number of unknown currents from three to two.

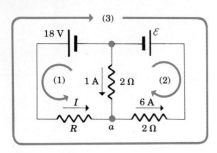

29-5

2. Go around the loop in the designated direction, adding emf's and potential differences. An emf is counted as positive when it is traversed from $(-)$ to $(+)$ (in the direction of the non-electrostatic field E_n in the source) and negative when from $(+)$ to $(-)$. An IR term is counted as negative if the resistor is traversed in the *same* direction as the assumed current, positive if in the opposite direction.

3. Equate the sum of step (2) to zero.

4. If necessary, choose another loop to obtain a different relation between the unknowns, and continue until there are as many equations as unknowns, or until every circuit element has been included in at least one of the chosen loops.

Example 1   In the circuit shown in Fig. 29–5, find the unknown current I, resistance R, and emf $\mathcal{E}$.

Solution   Application of the point rule to point a yields the relation

$$I + 1\,\mathrm{A} - 6\,\mathrm{A} = 0; \qquad I = 5\,\mathrm{A}.$$

To determine R we apply the loop rule to the loop labeled (1), obtaining

$$18\,\mathrm{V} - (5\,\mathrm{A})R + (1\,\mathrm{A})(2\,\Omega) = 0,$$
$$R = 4\,\Omega.$$

The term for resistance R is negative because our loop traverses that element in the same direction as the current and hence finds a potential *drop*, while the term for the 2-Ω resistor is positive because in traversing it in the direction opposite to the current we find a potential *rise*. If we had chosen to traverse loop (1) in the opposite direction, every term would have had the opposite sign, and the result for R would have been the same.

To determine $\mathcal{E}$, we apply the loop rule to loop (2);

$$\mathcal{E} + (6\,\mathrm{A})(2\,\Omega) + (1\,\mathrm{A})(2\,\Omega) = 0,$$
$$\mathcal{E} = -14\,\mathrm{V}.$$

This shows that the actual polarity of this emf is opposite to that assumed, and that the positive terminal of this source is really on the left side. Alternatively, one could use loop (3), obtaining the equation

$$\mathcal{E} + (6\,\mathrm{A})(2\,\Omega) + (5\,\mathrm{A})(4\,\Omega) - 18\,\mathrm{V} = 0,$$

from which again $\mathcal{E} = -14\,\mathrm{V}$.

Example 2   In Fig. 29–6, find the current in each resistor, and find the equivalent resistance.

Solution   As pointed out at the beginning of this section, it is not possible to represent this network in terms of series and parallel combinations. There are five different currents to determine, but by applying the point rule to junctions a and b we represent them in terms of three unknown currents, as indicated in the figure. The current in the battery is of course $(I_1 + I_2)$.

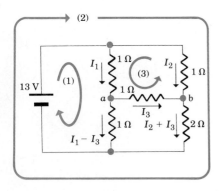

29-6

We now apply the loop rule to the three loops shown, obtaining the following three equations:

$$13 \text{ V} - I_1(1\,\Omega) - (I_1 - I_3)(1\,\Omega) = 0; \tag{1}$$

$$- I_2(1\,\Omega) - (I_2 + I_3)(2\,\Omega) + 13 \text{ V} = 0; \tag{2}$$

$$- I_1(1\,\Omega) - I_3(1\,\Omega) + I_2(1\,\Omega) = 0. \tag{3}$$

This is a set of three simultaneous equations for the three unknown currents. They may be solved by various methods; one straightforward procedure is to solve the third for I_2, obtaining $I_2 = I_1 + I_3$, and then substitute this expression into the first two equations to eliminate I_2. When this is done one obtains the two equations

$$13 \text{ V} = I_1(2\,\Omega) - I_3(1\,\Omega), \tag{1'}$$

$$13 \text{ V} = I_1(3\,\Omega) + I_3(5\,\Omega). \tag{2'}$$

Now I_3 may be eliminated by multiplying the first of these by 5 and adding the two to obtain

$$78 \text{ V} = I_1(13\,\Omega),$$

from which $I_1 = 6$ A. This result may be substituted back into (1') to obtain $I_3 = -1$ A, and finally from (3) we find $I_2 = 5$ A. We note that the direction of I_3 is opposite that of the initial assumption.

The total current through the network is $I_1 + I_2 = 11$ A, and the potential drop across it is equal to the battery emf, namely, 13 V. Thus the equivalent resistance of the network is

$$R = \frac{13 \text{ V}}{11 \text{ A}} = 1.18\,\Omega.$$

29-3 Ammeters and voltmeters

The most common instruments for measuring potential or current use a device called a *d'Arsonval galvanometer*. A pivoted coil of fine wire is placed in the magnetic field of a permanent magnet, as shown in Fig. 29-7. When there is a current in the coil, the magnetic field exerts on the coil a *torque* that is proportional to the current. (This magnetic interaction is discussed in detail in Chapter 31.) This torque is opposed by a spring, similar to the hairspring on the balance wheel of a watch, which exerts a restoring torque proportional to the angular displacement.

Thus the angular deflection of the indicator needle attached to the pivoted coil is directly proportional to the coil current, and the device can be calibrated to measure current. The maximum deflection for which the meter is designed, typically 90° to 120°, is called *full-scale deflection*. The current required to produce full-scale deflection (typically of the order of 10 μA to 10 mA) and the resistance of the coil (typically of the order of 10 to 1000 Ω) are the essential electrical characteristics of the meter.

The meter deflection is proportional to the *current* in the coil, but if the coil obeys Ohm's law, the current is proportional to the *potential difference* between the terminals of the coil. Thus the deflection is also proportional to this potential difference. For example, consider a meter whose coil has a resistance of 20 Ω, and which deflects full scale with a

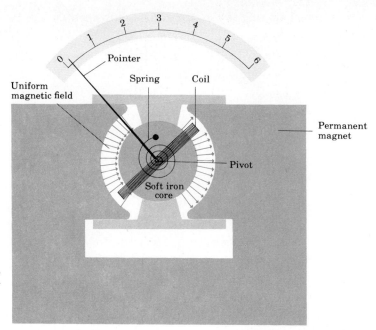

29-7 A d'Arsonval meter movement, showing pivoted coil with attached pointer, permanent magnet supplying uniform magnetic field, and spring to provide restoring torque, which opposes magnetic-field torque.

current of 1 mA in its coil. The corresponding potential difference is

$$V_{ab} = IR = (10^{-3}\,\text{A})(20\,\Omega) = 0.020\,\text{V} = 20\,\text{mV}.$$

Let us consider further the use of the d'Arsonval meter as a current-measuring instrument, often called an *ammeter*. To measure the current in a circuit, an ammeter must be inserted in *series* in the circuit so that the current to be measured actually passes *through* the meter. If the galvanometer above is inserted in this way, it will measure any current from zero to 1 mA. However, the resistance of the *coil* adds to the total resistance of the circuit, with the result that the current *after* the galvanometer is inserted, although it is correctly indicated by the instrument, may be less than it was *before* insertion of the instrument. It is evidently desirable that the resistance of the instrument should be much *smaller* than that of the remainder of the circuit, so that when the instrument is inserted it does not change the very thing we wish to measure. An *ideal* ammeter would have *zero* resistance.

Furthermore, the *range* of the galvanometer, if it is used without modification, is limited to a maximum current of 1 mA. The range can be extended, and at the same time the equivalent resistance can be reduced, by connecting a low resistance R_{sh} in parallel with the moving coil, as in Fig. 29-8a. The parallel resistor is called a *shunt*; its effect is to permit some of the circuit current I to bypass the meter and pass through the shunt instead. Usually the meter and the shunt are mounted inside a case, with binding posts or plug connectors for external connections at a and b. Occasionally, when several interchangeable shunts are used, they are mounted outside the case.

Suppose we wish to modify the meter described above for use as an ammeter with a range of 0 to 10 A. That is, the coil is to deflect full-scale when the current I in the *circuit* in which the ammeter is inserted equals 10 A. The *coil* current I_c must then be 1 mA, so the current I_{sh} in the

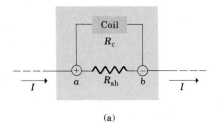

(a)

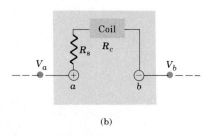

(b)

29-8 (a) Internal connections of an ammeter. (b) Internal connections of a voltmeter.

shunt is 9.999 A. The potential difference V_{ab} is

$$V_{ab} = I_c R_c = I_{sh} R_{sh}.$$

Hence

$$R_{sh} = R_c \left(\frac{I_c}{I_{sh}} \right) = 20\,\Omega \left(\frac{0.001}{9.999} \right) = 0.00200\,\Omega.$$

The equivalent resistance R of the instrument is

$$\frac{1}{R} = \frac{1}{R_c} + \frac{1}{R_{sh}},$$

and

$$R = 0.00200\,\Omega.$$

Thus we have a low-resistance instrument with the desired range of 0 to 10 A. Of course, if the current I is *less* than 10 A, the coil current (and the deflection) is correspondingly less, also.

Now let us consider the construction of a *voltmeter*. A voltmeter measures the potential difference beween two points, and its terminals must be connected to these points. A moving-coil meter cannot be used to measure the potential differences between, say, two charged spheres. When its terminals are connected to the spheres, the coil provides a conducting path from one sphere to the other. There will be a *momentary* current in the coil, but the charges on the sphere will change until the entire system is at the *same* potential. Only if the resistance of the instrument is so great that a very long time is required to reach equilibrium can a voltmeter be used for this purpose. Thus an *ideal* voltmeter has an *infinite* resistance, but a moving-coil galvanometer can be deflected only by a *current* in its coil; its resistance must be finite.

A moving-coil galvanometer *can* be used to measure the potential difference between the terminals of a *source,* or between *two points of a circuit containing a source,* because the source *maintains* a difference of potential between the points. However, the *range* of the galvanometer in our example, if used without modification, is limited to a maximum value of 20 mV. The range can be extended, and at the same time the equivalent resistance can be increased, by connecting a resistance R_s in *series* with the moving coil, as in Fig. 29–8b.

Suppose we wish to modify the galvanometer for use as a voltmeter with a range of 0 to 10 V. The coil is to deflect full-scale when the potential difference between the terminals of the instrument is 10 V. In other words, the current in the instrument is to be 1 mA when the potential difference between its terminals is 10 V.

The terminal potential difference is

$$V_{ab} = I(R_c + R_s),$$

and the necessary series resistance is

$$R_s = \frac{V_{ab}}{I} - R_c = \frac{10\,\text{V}}{0.001\,\text{A}} - 20\,\Omega = 9980\,\Omega.$$

The equivalent resistance R is

$$R = R_c + R_s = 10{,}000\,\Omega.$$

Thus we have a high-resistance instrument with the desired range of 0 to 10 V.

A voltmeter, like an ammeter, can disturb the circuit to which it is connected. For example, when a source is on *open* circuit, the potential difference between its terminals equals its emf. It would seem, therefore, that to measure the emf we need only measure this potential difference. But when the terminals of a voltmeter are connected to those of a source, the meter and source form a *complete* circuit in which there is a current. The potential difference *after* the meter is connected, although it is correctly indicated by the instrument, is not equal to $\mathcal{E}$, but to $\mathcal{E} - Ir$, and is *less* than it was *before* the instrument was connected. Again, the measuring instrument alters the quantity it is intended to measure. It is evidently desirable that the resistance of a voltmeter, even though not infinite, should be as *large* as possible.

A voltmeter and an ammeter can be used together to measure *resistance* and *power*. The resistance of a resistor equals the potential difference V_{ab} between its terminals, divided by the current I:

$$R = \frac{V_{ab}}{I},$$

and the power input to any portion of a circuit equals the product of the potential difference across this portion and the current:

$$P = V_{ab}I.$$

The most straightforward method of measuring R or P is therefore to measure V_{ab} and I simultaneously.

In Fig. 29–9a, ammeter A reads correctly the current I in the resistor R. Voltmeter V, however, reads the *sum* of the potential difference V_{ab} across the resistor and the potential difference V_{bc} across the ammeter.

If we transfer the voltmeter terminal from c to b, as in Fig. 29–9b, the voltmeter reads correctly the potential difference V_{ab} but the ammeter now reads the *sum* of the current I in the resistor and the current I_V in the voltmeter. Thus, whichever connection is used, we must correct the reading of one instrument or the other to obtain the true values of V_{ab} or I (unless, of course, the corrections are small enough to be neglected).

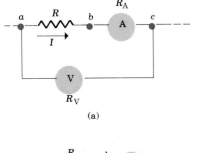

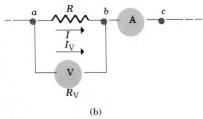

29-9   Ammeter–voltmeter method for measuring resistance or power.

Example 1   The circuit of Fig. 29–9a is to be used to measure an unknown resistor R. The meter resistances are $R_V = 10{,}000\ \Omega$ and $R_A = 2.0\ \Omega$. If the voltmeter reads 12.0 V and the ammeter reads 0.10 A, what is the true resistance?

Solution   If the meters were ideal (i.e., $R_V = \infty$ and $R_A = 0$), the resistance would be simply $R = V/I = (12.0\ \text{V})/(0.10\ \text{A}) = 120\ \Omega$. But the voltmeter reading includes the potential V_{bc} across the ammeter as well as that (V_{ab}) across the resistor. We have $V_{bc} = IR_A = (0.10\ \text{A})(2.0\ \Omega) = 0.2\ \text{V}$, so the actual potential drop V_{ab} across the resistor is 12.0 V − 0.2 V = 11.8 V, and the resistance is

$$R = \frac{V_{ab}}{I} = \frac{11.8\ \text{V}}{0.10\ \text{A}} = 118\ \Omega.$$

Example 2   Suppose the same meters are connected instead as shown in Fig. 29-9b, and the above readings are obtained. What is the true resistance?

Solution   In this case the voltmeter measures the potential across the resistor correctly; the difficulty is that the ammeter measures the voltmeter current I_V as well as the current I in the resistor. We have $I_V = V/R_V = (12.0\text{ V})/(10{,}000\ \Omega) = 1.2\text{ mA}$; so the actual current I in the resistor is $I = 0.10\text{ A} - 0.0012\text{ A} = 0.0988\text{ A}$. Thus the resistance is

$$R = \frac{V_{ab}}{I} = \frac{12.0\text{ V}}{0.0988\text{ A}} = 121.5\ \Omega.$$

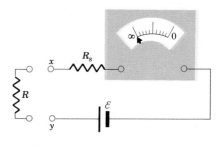

29-10

29-4 The ohmmeter

Although not a precision instrument, the ohmmeter is a useful device for rapid measurement of resistance. It consists of a meter, a resistor, and a source (often a flashlight cell) connected in series, as in Fig. 29-10. The resistance R to be measured is connected between terminals x and y.

The series resistance R_s is chosen so that, when terminals x and y are short-circuited (that is, when $R = 0$), the galvanometer deflects full scale. When the circuit between x and y is open (that is, when $R = \infty$), the galvanometer shows no deflection. For a value of R between zero and infinity, the galvanometer deflects to some intermediate point depending on the value of R, and hence the galvanometer scale can be calibrated to read the resistance R.

*29-5 The *R-C* series circuit

In the circuits considered thus far, we have assumed that the emf's and resistances are constant, so that all potentials and currents are constant, independent of time. Figure 29-11 shows a simple example of a circuit in which the current and voltages are *not* constant. When the double-pole, double-throw (dpdt) switch is in the upper position, points a and b are connected to the battery, which charges the capacitor through resistor R.

If the capacitor is initially uncharged, then the initial potential difference across the capacitor is zero, and the entire battery voltage appears across the resistor, causing an initial current $I = V/R$. As the capacitor charges, its voltage increases, and the potential difference across the resistor decreases, corresponding to a decrease in current. After a long time the capacitor has become fully charged, the entire battery voltage V appears across the capacitor, there is no potential difference across the resistor, and the current becomes zero.

Let q represent the charge on the capacitor and i the charging current at some instant after the switch is thrown in the "up" position. (As is customary in circuit analysis, we use small letters for quantities that vary with time, and capital letters for constant quantities.) The instantaneous potential differences v_{ac} and v_{cb} are

$$v_{ac} = iR, \qquad v_{cb} = \frac{q}{C}. \tag{29-6}$$

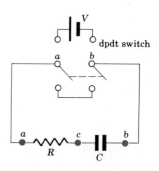

29-11

Therefore,

$$V_{ab} = V = v_{ac} + v_{cb} = iR + \frac{q}{C}, \tag{29–7}$$

where $V = $ constant. The current i is then

$$i = \frac{V}{R} - \frac{q}{RC}. \tag{29–8}$$

At the instant connections are made, $q = 0$ and the *initial current* $I_0 = V/R$, which would be the steady current if the capacitor were not present.

As the charge q increases, the term q/RC becomes larger, and the current decreases and eventually becomes zero. When $i = 0$,

$$\frac{V}{R} = \frac{q}{RC}, \qquad q = CV = Q_f,$$

where Q_f is the final charge on the capacitor.

To obtain a quantitative description of the variation with time of the charge and current in this circuit, we replace i in Eq. (29–8) with dq/dt, obtaining

$$\frac{dq}{dt} = \frac{V}{R} - \frac{q}{RC}, \tag{29–9}$$

which may be rearranged as

$$\frac{dq}{VC - q} = \frac{dt}{RC}.$$

Integrating both sides, we obtain

$$-\ln(VC - q) = t/RC + \text{constant}.$$

To evaluate the constant, we note that at $t = 0$, $q = 0$, so

$$-\ln(VC - 0) = 0 + \text{constant}.$$

Rearranging again, we obtain

$$\ln(VC - q) - \ln VC = \ln \frac{VC - q}{VC} = -\frac{t}{RC},$$

$$1 - \frac{q}{VC} = e^{-t/RC},$$

$$q = VC(1 - e^{-t/RC}) = Q_f(1 - e^{-t/RC}). \tag{29–10}$$

The time derivative of this expression is the current:

$$i = \frac{dq}{dt} = \frac{V}{R}e^{-t/RC} = I_0 e^{-t/RC} \tag{29–11}$$

The charge and current are therefore both *exponential* functions of time. Figure 29–12a is a graph of Eq. (29–11), and Fig. 29–12b is a graph of Eq. (29–10). At a time $t = RC$, the current has decreased to $1/e$ (about 0.368) of its initial value and the charge has increased to *within* $1/e$ of its final value. The product RC is called the *time constant*, or the *relaxation*

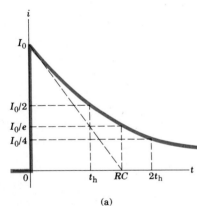

(a)

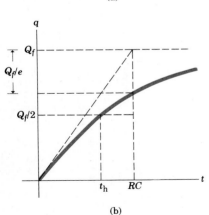

(b)

29–12

time, of the circuit, denoted by τ:

$$\tau = RC. \tag{29-12}$$

The *half-life* of the circuit, t_h, is the time for the current to decrease to half its initial value, or for the capacitor to acquire half its final charge. Setting $i = I_0/2$ in Eq. (29–11), we find

$$t_h = RC \ln 2 = 0.693\,RC.$$

The half-life depends only on the time constant RC, and not on the initial current. Thus if the current decreases from I_0 to $I_0/2$ in a time t_h, as shown in Fig. 29–12a, it decreases to half this value, or to $I_0/4$, in another half-life, and so on.

Example   A resistor of resistance $R = 10\ \mathrm{M\Omega}$ is connected in series with a capacitor of capacitance $1\ \mu\mathrm{F}$. The time constant is

$$\tau = RC = (10 \times 10^6\ \Omega)(10^{-6}\ \mathrm{F}) = 10\ \mathrm{s},$$

and the half-life is

$$t_h = (10\ \mathrm{s})(\ln 2) = 6.9\ \mathrm{s}.$$

On the other hand, if $R = 10\ \Omega$, the time constant is only $10 \times 10^{-6}\ \mathrm{s}$, or $10\ \mu\mathrm{s}$.

Suppose next that the capacitor has acquired a charge Q_0 and that the switch is thrown to the "down" position. The capacitor then *discharges* through the resistor and its charge eventually decreases to zero. (We designate the charge by Q_0 because this is the *initial* charge in the discharge process. It is not necessarily equal to the charge Q_f defined above.)

Again let i and q represent the time-varying current and charge at some instant after the switch is thrown. Since V_{ab} is now zero, we have, from Eq. (29–7),

$$0 = v_{ac} + v_{cb}.$$

The direction of the current in the resistor is now from c to a, so $v_{ca} = -v_{ac} = iR$ and

$$i = \frac{q}{RC}. \tag{29-13}$$

When $t = 0$, $q = Q_0$ and the initial current I_0 is

$$I_0 = \frac{Q_0}{RC} = \frac{V_0}{R},$$

where V_0 is the initial potential difference across the capacitor. As the capacitor discharges, both q and i decrease.

The same procedures as above can be followed to obtain $i(t)$ and $q(t)$. If we replace i in Eq. (29–13) by $-dq/dt$ (the charge q is now *decreasing*), we get

$$\frac{dq}{dt} = -\frac{q}{RC}. \tag{29-14}$$

Integration of this equation gives $q(t)$, and by differentiation we find $i(t)$.

Alternatively, differentiation of Eq. (29–13) gives

$$\frac{di}{dt} = -\frac{i}{RC}, \tag{29–15}$$

from which we can get $i(t)$ and, by a second integration, get $q(t)$. It is left as a problem to show that

$$i = I_0 e^{-t/RC}, \tag{29–16}$$

$$q = Q_0 e^{-t/RC}. \tag{29–17}$$

Both the current and the charge decrease exponentially with time.

*29-6 Displacement current

For a *conducting* circuit, in the steady state, the total current *into* any given portion must be equal to the current *out of* that portion. This statement forms the basis of Kirchhoff's point rule, discussed in Sec. 29-2. However, this rule is *not* obeyed for a capacitor that is being charged. In Fig. 29–13, there is a conduction current *into* the left plate but no conduction current *out of* this plate; similarly, there is conduction current out of the right plate, but none into it.

James Clerk Maxwell (1831–1879) showed that it is possible to generalize the definition of current so that one can still say that the current out of each plate is equal to the current into it. As the capacitor charges, the conduction current increases the charge on each plate, and this in turn increases the electric field between the plates. The *rate* of increase of field is proportional to the conduction current; Maxwell's scheme was to associate an equivalent current density with this rate of increase of field.

Assuming, for simplicity, that the capacitor consists of two parallel plates with uniform charge density σ, the field E between the plates is given by

$$E = \frac{\sigma}{\epsilon_0} = \frac{Q}{\epsilon_0 A}. \tag{29–18}$$

In a time interval dt, the charge Q increases by $dQ = I\,dt$; the corresponding change in E is

$$dE = \frac{dQ}{\epsilon_0 A} = \frac{I\,dt}{\epsilon_0 A},$$

and the *rate* of change of E is

$$\frac{dE}{dt} = \frac{I}{\epsilon_0 A}. \tag{29–19}$$

We now define an *equivalent current density* J_D between the plates as

$$J_D = \epsilon_0 \frac{dE}{dt}. \tag{29–21}$$

Then the total equivalent current between the plates, which we may call I_D is

$$I_D = J_D A = \epsilon_0 \left(\frac{dE}{dt}\right) A = I.$$

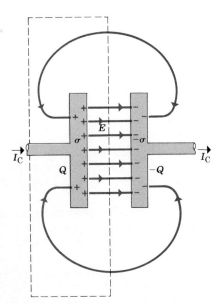

29-13 The conduction current into the left plate of the capacitor equals the displacement current between the plates.

Thus, if we include the "effective current" as well as the conduction current, the current *into* the region bounded by the broken line in Fig. 29-13 equals the current *out of* this region. The subscript D is chosen because Maxwell originally called this effective current the *displacement current;* this term is still used, although the reasons for its use are dubious.

Displacement current may also be expressed in terms of *electric flux,* defined in Sec. 25-4, Eqs. (25-16) and (25-17). For a uniform field E perpendicular to an area A, $\Psi = EA$, and Eq. (29-21) may be rewritten as

$$I_D = \epsilon_0 \frac{d\Psi}{dt}. \tag{29-22}$$

If the space between capacitor plates contains a dielectric, Eq. (29-18) must be replaced by the more general relation derived in Sec. 27-5:

$$E = \frac{\sigma}{\epsilon} = \frac{Q}{\epsilon A} = \frac{Q}{k\epsilon_0 A}. \tag{29-23}$$

The corresponding modification of the definition of displacement current density is

$$J_D = \epsilon \left(\frac{dE}{dt} \right) = K\epsilon_0 \left(\frac{dE}{dt} \right). \tag{29-24}$$

Maxwell's generalized view of current and current density may appear to be merely an artifice introduced to preserve Kirchhoff's current rule even in cases where charge is accumulating in a certain region of space, such as a capacitor plate. However, it is much more than this. In a later chapter we shall study the role of current as a source of *magnetic* field, and we shall see that a *displacement* current sets up a magnetic field in exactly the same way as an ordinary *conduction* current. Thus displacement current, far from being an artifice, is a fundamental fact of nature, and Maxwell's discovery of it was the bold step of an extraordinary genius. As we shall see, the concept of displacement current provided the necessary basis for the understanding of electromagnetic waves in the last third of the nineteenth century.

Questions

29-1 Can the potential difference between terminals of a battery ever be opposite in direction to the emf?

29-2 Why do the lights on a car become dimmer when the starter is operated?

29-3 What determines the maximum current that can be carried safely by household wiring? (Typical limits are 15 A for 14-gauge wire, 20 A for 12-gauge, and so on.)

29-4 Lights in a house often dim momentarily when a motor, such as a washing machine or a power saw, is turned on. Why does this happen?

29-5 Compare the formulas for resistors in series and parallel with those for capacitors in series and parallel. What similarities and differences do you see? Sometimes in cir-

cuit analysis one uses the quantity *conductance,* denoted as g and defined as the reciprocal of resistance: $g = 1/R$. What is the corresponding comparison for conductance and capacitance?

29-6 Is it possible to connect resistors together in a way that cannot be reduced to some combination of series and parallel combinations? If so, give examples; if not, state why not.

29-7 In a two-cell flashlight, the batteries are usually connected in series. Why not connect them in parallel?

29-8 Some Christmas-tree lights have the property that, when one bulb burns out, all the lights go out, while with others only the burned-out bulb goes out. Discuss this difference in terms of series and parallel circuits.

29-9 What possible advantage could there be in connecting several identical batteries in parallel?

29-10 Two 110-V lightbulbs, one 25-W and one 200-W, were connected in series across a 220-V line. It seemed like a good idea at the time, but one bulb burned out almost instantaneously. Which one burned out, and why?

29-11 When the direction of current in a battery reverses, does the direction of its emf also reverse?

29-12 Under what conditions would the terminal voltage of a battery be zero?

29-13 What sort of meter should be used to test the condition of a dry cell (such as a flashlight battery) having constant emf but internal resistance that increases with age and use?

29-14 For very large resistances, it is easy to construct RC circuits having time constants of several seconds or minutes. How might this fact be used to measure very large resistances, too large to measure by more conventional means?

Problems

29-1 Prove that, when two resistors are connected in parallel, the equivalent resistance of the combination is always smaller than that of either resistor.

29-2

a) A resistance R_2 is connected in parallel with a resistance R_1. What resistance R_3 must be connected in series with the combination of R_1 and R_2 so that the equivalent resistance is equal to the resistance R_1? Draw a diagram.

b) A resistance R_2 is connected in series with a resistance R_1. What resistance R_3 must be connected in parallel with the combination of R_1 and R_2 so that the equivalent resistance is equal to R_1? Draw a diagram.

29-3

a) Calculate the equivalent resistance of the circuit of Fig. 29-14 between x and y.

b) What is the potential difference between x and a if the current in the 8-Ω resistor is 0.5 A?

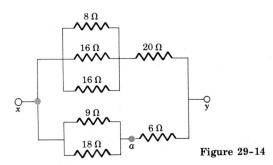

29-4 Each of three resistors in Fig. 29-15 has a resistance of 2 Ω and can dissipate a maximum of 18 W without becoming excessively heated. What is the maximum power the circuit can dissipate?

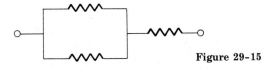

Figure 29-15

29-5 Two lamps, marked "60 W, 120 V" and "40 W, 120 V," are connected in series across a 120-V line. What power is consumed in each lamp? Assume that the resistance of the filaments does not vary with current.

29-6 Three equal resistors are connected in series. When a certain potential difference is applied across the combination, the total power dissipated is 10 W. What power would be dissipated if the three resistors were connected in parallel across the same potential difference?

29-7

a) The power rating of a 10,000-Ω resistor is 2 W. (The power rating is the maximum power the resistor can safely dissipate without too great a rise in temperature.) What is the maximum allowable potential difference across the terminals of the resistor?

b) A 20,000-Ω resistor is to be connected across a potential difference of 300 V. What power rating is required?

c) It is desired to connect an equivalent resistance of 1000 Ω across a potential difference of 200 V. A number of 10-W, 1000-Ω resistors are available. How should they be connected?

29-8 A 60-Ω resistor and a 90-Ω resistor are connected in parallel and the combination is connected across a 120-V dc line.

a) What is the resistance of the parallel combination?

b) What is the total current through the parallel combination?

c) What is the current through each resistor?

29-9 A 1000-Ω 2-W resistor is needed, but only several 1000-Ω 1-W resistors are available.

a) How can the required resistance and power rating be obtained by a combination of the available units?

b) What power is then dissipated in each resistor?

29-10 A 25-W, 120-V lightbulb and a 100-W, 120-V lightbulb are connected in series across a 240-V line. Assume that the resistance of each bulb does not vary with current.

a) Find the current through the bulbs.

b) Find the power dissipated in each bulb.

c) One bulb burns out very quickly. Which one? Why?

29-11 An electric heating element was designed for a power rating of 550 W when connected to a 110-V line. If the line voltage is 120 V, what power is consumed?

29-12 Prove that the resistance of the infinite network shown in Fig. 29-16 is equal to $(1 + \sqrt{3})r$.

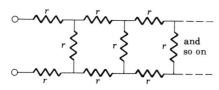

Figure 29-16

29-13

Note: Figure 29-17 employs a convention often used in circuit diagrams. The battery (or other power supply) is not shown explicitly. It is understood that the point at the top, labeled "36 V," is connected to the positive terminal of a 36-V battery having negligible internal resistance, and the "ground" symbol at the bottom to its negative terminal. The circuit is completed through the battery, even though it is not shown on the diagram.

a) In Fig. 29-17a, what is the potential difference V_{ab} when switch S is open?

b) What is the current through switch S, when it is closed?

c) In Fig. 29-17b, what is the potential difference V_{ab} when switch S is open?

d) What is the current through switch S when it is closed?

What is the equivalent resistance in Fig. 29-17b,

e) when switch S is open?   f) when switch S is closed?

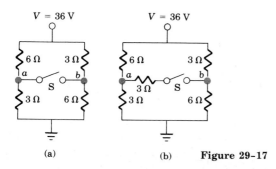

(a)   (b)   **Figure 29-17**

29-14 (See note with Problem 29-13.)

a) What is the potential difference between points a and b in Fig. 29-18 when switch S is open?

b) Which point, a or b, is at the higher potential?

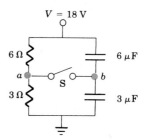

Figure 29-18

c) What is the final potential of point b when switch S is closed?

d) How much charge flows through switch S when it is closed?

29-15 (See note with Problem 29-13.)

a) What is the potential difference between points a and b in Fig. 29-19 when switch S is open?

b) Which point, a or b, is at the higher potential?

c) What is the final potential of point b when switch S is closed?

d) How much does the charge on each capacitor change when S is closed?

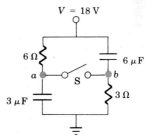

Figure 29-19

29-16 Calculate the three currents indicated in the circuit diagram of Fig. 29-20.

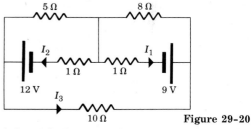

Figure 29-20

29-17 Find the emf's $\mathcal{E}_1$ and $\mathcal{E}_2$ in the circuit of Fig. 29-21, and the potential difference between points a and b.

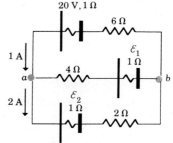

Figure 29-21

29-18

a) Find the potential difference between points a and b in Fig. 29-22.

b) If a and b are connected, find the current in the 12-V cell.

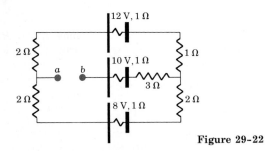

Figure 29-22

29-19 Find the current in each branch of the circuit shown in Fig. 29-23.

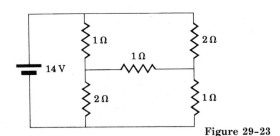

Figure 29-23

29-20 In the circuit shown in Fig. 29-24, find

a) the current in resistor R;

b) the resistance R;

c) the unknown emf $\mathcal{E}$.

d) If the circuit is broken at point x, what is the current in the 28-V battery?

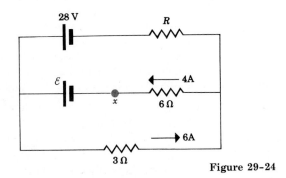

Figure 29-24

29-21 In the circuit shown in Fig. 29-25, find

a) the current in each branch;

b) the potential difference V_{ab}.

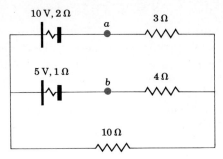

Figure 29-25

29-22 Suppose a resistor R lies along each edge of a cube (12 resistors in all) with connections at the corners. Find the equivalent resistance between two diagonally opposite corners of the cube.

29-23 A 600-Ω resistor and a 400-Ω resistor are connected in series across a 90-V line. A voltmeter across the 600-Ω resistor reads 45 V.

a) Find the voltmeter resistance.

b) Find the reading of the same voltmeter if connected across the 400-Ω resistor.

29-24 Point a in Fig. 29-26 is maintained at a constant potential of 300 V above ground. (See note with Problem 29-13.)

a) What is the reading of a voltmeter of the proper range, and of resistance $3 \times 10^4 \Omega$, when connected between point b and ground?

b) What would be the reading of a voltmeter of resistance $3 \times 10^6 \Omega$?

c) Of a voltmeter of infinite resistance?

Figure 29-26

29-25 Two 150-V voltmeters, one of resistance 15,000 Ω and the other of resistance 150,000 Ω, are connected in series across a 120-V dc line. Find the reading of each voltmeter.

29-26 A 150-V voltmeter has a resistance of 20,000 Ω. When connected in series with a large resistance R across a 110-V line, the meter reads 5 V. Find the resistance R. (This problem illustrates one method of measuring large resistances.)

29-27 A 100-V battery has an internal resistance of 5 Ω.

a) What is the reading of a voltmeter having a resistance of 500 Ω when placed across the terminals of the battery?

b) What maximum value may the ratio r/R_V have if the error in the reading of the emf of a battery is not to exceed 5 percent?

29-28 The resistance of a galvanometer coil is 50 Ω and the current required for full-scale deflection is 500 μA.

a) Show in a diagram how to convert the galvanometer to an ammeter reading 5 A full-scale, and compute the shunt resistance.

b) Show how to convert the galvanometer to a voltmeter reading 150 V full-scale, and compute the series resistance.

29-29 The resistance of the coil of a pivoted-coil galvanometer is 10 Ω, and a current of 0.02 A causes it to deflect full-scale. It is desired to convert this galvanometer to an ammeter reading 10 A full-scale. The only shunt available has a resistance of 0.03 Ω. What resistance R must be connected in series with the coil? (See Fig. 29–27.)

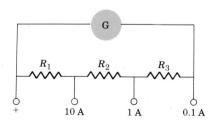

Figure 29-27

29-30 The resistance of the moving coil of the galvanometer G in Fig. 29–28 is 25 Ω; the meter deflects full-scale with a current of 0.010 A. Find the magnitudes of the resistances R_1, R_2, and R_3, required to convert the galvanometer to a multirange ammeter deflecting full-scale with currents of 10 A, 1 A, and 0.1 A.

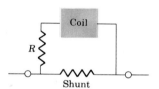

Figure 29-28

29-31 Figure 29–29 shows the internal wiring of a "three-scale" voltmeter whose binding posts are marked +, 3 V,

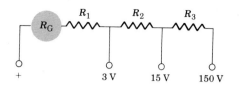

Figure 29-29

15 V, 150 V. The resistance of the moving coil, R_G, is 15 Ω, and a current of 1 mA in the coil causes it to deflect full-scale. Find the resistances R_1, R_2, R_3, and the overall resistance of the meter on each of its ranges.

29-32 A certain dc voltmeter is said to have a resistance of "one thousand ohms per volt of full-scale deflection." What current, in milliamperes, is required for full-scale deflection?

29-33 Let V and I represent the readings of the voltmeter and ammeter, respectively, shown in Fig. 29–9, and R_V and R_A their equivalent resistances.

a) When the circuit is connected as in Fig. 29–9a, show that

$$R = \frac{V}{I} - R_A.$$

b) When the connections are as in Fig. 29–9b, show that

$$R = \frac{V}{I - (V/R_V)}.$$

c) Show that the power delivered to the resistor in part (a) is $IV - IR_A$, and in part (b) is $IV - (V^2/R_V)$.

29-34 A certain galvanometer has a resistance of 200 Ω and deflects full-scale with a current of 1 mA in its coil. It is desired to replace this with a second galvanometer whose resistance is 50 Ω and which deflects full-scale with a current of 50 μA in its coil. Devise a circuit incorporating the second galvanometer such that

a) the equivalent resistance of the circuit equals the resistance of the first galvanometer, and

b) the second galvanometer will deflect full-scale when the current into and out of the circuit equals the full-scale current of the first galvanometer.

29-35 Suppose the galvanometer of the ohmmeter in Fig. 29–10 has a resistance of 50 Ω and deflects full-scale with a current of 1 mA in its coil. The emf $\mathcal{E} = 1.5$ V.

a) What should be the value of the series resistance R_S?

b) What values of R correspond to galvanometer deflections of $\frac{1}{4}$, $\frac{1}{2}$, and $\frac{3}{4}$ full-scale?

c) Does the ohmmeter have a linear scale?

29-36 In the ohmmeter in Fig. 29–30, M is a 1-mA meter having a resistance of 100 Ω. The battery B has an emf of 3 V and negligible internal resistance. R is so chosen that, when the terminals a and b are shorted ($R_x = 0$), the meter reads full scale. When a and b are open ($R_x = \infty$), the meter reads zero.

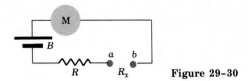

Figure 29-30

a) What should be the value of the resistor R?

b) What current would indicate a resistance R_x of 600 Ω?

c) What resistances correspond to meter deflections of $\frac{1}{4}$, $\frac{1}{2}$, and $\frac{3}{4}$ full-scale?

29-37 In Fig. 29-31, a resistor of resistance 75 Ω is connected between points a and b. The resistance of the galvanometer G is 90 Ω. What should be the resistance between b and the sliding contact c, if the galvanometer current I_G is to be $\frac{1}{3}$ of the current I?

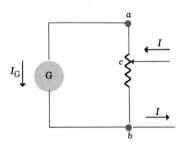

Figure 29-31

29-38 The circuit shown in Fig. 29-32 is called a *Wheatstone bridge;* it is used to determine the value of an unknown resistor X by comparison with three precisely known resistors M, N, and P. With switches K_1 and K_2 closed, these resistors are varied until the current in the galvanometer G is zero; the bridge is then said to be *balanced.* Show that under this condition the unknown resistance is given by $X = MP/N$. (This method permits very high precision in comparing resistors.)

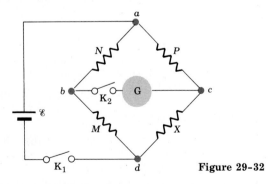

Figure 29-32

29-39 In Fig. 29-32, the galvanometer G shows zero deflection when $M = 1000 \, \Omega$, $N = 10.00 \, \Omega$, and $P = 27.49 \, \Omega$. What is the unknown resistance X?

29-40 The circuit shown in Fig. 29-33 is called a *potentiometer.* It permits measurements of potential difference

without drawing current from the circuit being measured, and hence acts as an infinite-resistance voltmeter. The resistor between a and b is a uniform wire of length l, with a sliding contact c at a distance x from b. An unknown potential difference V_x is measured by sliding the contact until the galvanometer G reads zero.

a) Show that under this condition the unknown potential difference is given by $V_x = (x/l)\mathcal{E}$.

b) Why is the internal resistance of the galvanometer not important?

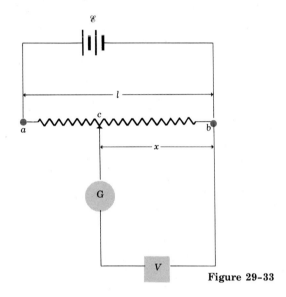

Figure 29-33

29-41 In Fig. 29-33, suppose $\mathcal{E} = 12.00$ V and $l = 1.000$ m. The galvanometer G reads zero when $x = 0.793$ m.

a) What is the potential difference V?

b) Suppose V is the emf of a battery; can its internal resistance be determined by this method?

29-42 A 0.05-μF capacitor is charged to a potential of 200 V and is then permitted to discharge through a 10-MΩ resistor. How much time is required for the charge to decrease to

a) $1/e$,   b) $1/e^2$

of its initial value?

29-43 A capacitor is charged to a potential of 10 V and is then connected to a voltmeter having an internal resistance of 1.0 MΩ. After a time of 5 s, the voltmeter reads 5 V. What is the capacitance?

29-44 A 10-μF capacitor is connected through a 1-MΩ resistor to a constant potential difference of 100 V.

a) Compute the charge on the capacitor at the following times after the connections are made: 0, 5 s, 10 s, 20 s, 100 s.

b) Compute the charging current at the same instants.

c) How long a time would be required for the capacitor to acquire its final charge if the charging current remained constant at its initial value?

d) Find the time required for the charge to increase from zero to 5×10^{-4} C.

e) Construct graphs of the results of parts (a) and (b) for a time interval of 20 s.

29-45 Two capacitors are charged in series by a 12-V battery (Fig. 29-34).

a) What is the time constant of the charging circuit?

b) After being closed for the length of time determined in (a), the switch S is opened. What is the voltage across the 6-μF capacitor?

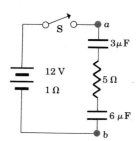

Figure 29-34

29-46 A capacitor of capacitance C is charged by connecting it through a resistance R to the terminals of a battery of emf $\mathcal{E}$ and of negligible internal resistance.

a) How much energy is supplied by the battery in the charging process?

b) What fraction of this energy appears as heat in the resistor?

29-47 The current in a discharging capacitor is given by Eq. (29-16).

a) Using Eq. (29-16), derive an expression for the instantaneous power $P = i^2R$ dissipated in the resistor.

b) Integrate the expression for P to find the total energy dissipated in the resistor, and show that this is equal to the total energy initially stored in the capacitor.

29-48 The current in a charging capacitor is given by Eq. (29-11).

a) The instantaneous power supplied by the battery is Vi. Integrate this to find the total energy supplied by the battery.

b) The instantaneous power dissipated in the resistor is i^2R. Integrate this to find the total energy dissipated in the resistor.

c) Find the final energy stored in the capacitor, and show that this equals the total energy supplied by the battery, less the energy dissipated in the resistor, as obtained in parts (a) and (b).

d) What fraction of the energy supplied by the battery is stored in the capacitor? How does this fraction depend on R?

29-49

a) The differential equation for the instantaneous charge q of a capacitor a moment after its terminals have been disconnected from a source and connected to a resistance R is given by Eq. (29-14), namely,

$$\frac{dq}{dt} = -\frac{q}{RC}.$$

Show that

$$q = Q_0 e^{-t/RC}.$$

b) The current in the circuit of part (a) is given by Eq. (29-15), namely,

$$\frac{di}{dt} = -\frac{i}{RC}.$$

Show that

$$i = I_0 e^{-t/RC}.$$

29-50 Suppose that the parallel plates in Fig. 29-13 have an area of 2 m² and are separated by a sheet of dielectric 1 mm thick, of dielectric constant 3. (Neglect edge effects.) At a certain instant, the potential difference between the plates is 100 V and the current I_C equals 2 mA.

a) What is the charge Q on each plate?

b) What is the rate of change of charge on the plates?

c) What is the displacement current in the dielectric?

29-51 In a certain copper conductor ($\rho = 2 \times 10^{-8}\,\Omega\cdot\text{m}$) carrying a current, the electric field varies sinusoidally with time according to $E = E_0 \sin \omega t$, where $E_0 = 0.1$ V m^{-1} and $\omega = (2\pi)(60$ Hz).

a) Find the magnitude of the maximum conduction current density in the wire.

b) Assuming $\epsilon = \epsilon_0$, find the maximum displacement current density in the wire, and compare with the result of part (a).

29-52

a) Repeat the calculations of Problem 29-51 for a rod of pure silicon having $\rho = 2000\,\Omega\cdot\text{m}$.

b) At what frequency would the maximum conduction and displacement densities become equal, if $\epsilon = \epsilon_0$ (which is not actually the case)?

c) At the frequency determined in part (b), what is the relative *phase* of the conduction and displacement currents?

29-53 A capacitor is made from two cylindrical rods of radius R, placed end to end with a gap of width d between them, where d is much smaller than R. There is a current i in each rod, and the charge q on each end surface changes at a rate given by $dq/dt = i$.

a) Find the conduction current density in each rod.

b) Find the electric field between the surfaces, in terms of q.

c) Find the displacement current density in the gap, and show that it is equal to the conduction current density in the rods.

29-54 A parallel-plate capacitor is made from two plates, each having area A, spaced a distance d apart. The space between plates is filled with a material having dielectric constant K. This material is not a perfect insulator but has resistivity ρ. The capacitor is initially charged with charge of magnitude Q_0 on each plate, which gradually discharges by conduction through the dielectric. Show that at any instant the displacement current density in the dielectric is equal in magnitude to the conduction current density but opposite in direction, so that the *total* current density is zero at every instant.

THE MAGNETIC FIELD 30

30-1 Magnetism

The first magnetic phenomena observed were those associated with naturally occurring magnets, fragments of iron ore found near the ancient city of Magnesia (whence the term "magnet"). These natural magnets attract unmagnetized iron; the effect is most pronounced at certain regions of the magnet known as its *poles*. It was known to the Chinese as early as 121 A.D. that an iron rod, after being brought near a natural magnet, would acquire and retain this property of the natural magnet, and that such a rod, when freely suspended about a vertical axis, would set itself approximately in the north–south direction. The use of magnets as aids to navigation can be traced back at least to the eleventh century.

The study of magnetic phenomena was confined for many years to magnets made in this way. Not until 1819 was any connection between electrical and magnetic phenomena shown. In that year, the Danish scientist Hans Christian Oersted (1777–1851) observed that a pivoted magnet (a compass needle) was deflected when in the neighborhood of a wire carrying a current. Twelve years later, Michael Faraday (1791–1867), an English physicist, found that a momentary current existed in a circuit while the current in a nearby circuit was being *started or stopped*. Shortly afterward it was discovered that the *motion* of a magnet toward or away from the circuit would produce the same effect. Joseph Henry (1797–1878), an American scientist who later became the first director of the Smithsonian Institution, discovered some of the same phenomena independently. The work of Oersted thus demonstrated that *magnetic* effects could be produced by moving *electric* charges, and that of Faraday and Henry showed that currents could be produced by moving magnets.

It is now known that all magnetic phenomena result from forces between electric charges in motion. That is, charges in motion relative to an observer set up a *magnetic* field as well as an *electric* field, and this magnetic field exerts a force on a second charge in motion relative to the observer. The electrons in atoms are in motion about the atomic nuclei, and each electron has additional motion ("spin") that can be visualized roughly as rotation about an axis passing through it. Thus all atoms can be expected to exhibit magnetic effects, and in fact they do. The possibility that the magnetic properties of matter are the result of tiny atomic currents was first suggested by Ampère in 1820. Not until recent years has the verification of these ideas been possible.

The material through which the charges are moving may have a significant effect on the observed magnetic forces between them. In the present chapter we shall assume that the charges or conductors are in otherwise empty space. For all practical purposes, the results will apply equally well to other charges and conductors in air.

30-2 The magnetic field

In describing the interaction of two charges *at rest,* we found it useful to introduce the concept of *electric field,* and to describe the interaction in two stages:

1. One charge sets up or creates an electric field E in the space surrounding it;
2. The electric field E exerts a force $F = qE$ on a charge q placed in the field.

We shall follow the same pattern in describing the interactions of charges in motion:

1. A *moving* charge or a current sets up or creates a *magnetic field* in the space surrounding it;
2. The magnetic field exerts a *force* on a *moving* charge or a current in the field.

Like electric field, magnetic field is a *vector field,* that is, a vector quantity associated with each point in space. We shall use the symbol B for magnetic field.

In this chapter and the next we consider the *second* aspect of the interaction: Given the presence of a magnetic field, what force does it exert on a moving charge or a current? Then in Chapter 32 we return to the problem of how magnetic fields are *created* by moving charges and currents.

Some aspects of the magnetic force on a moving charge are analogous to corresponding properties of the electric-field force. Both have magnitude proportional to the charge. If a 1-μC charge and a 2-μC charge move through a given magnetic field with the same velocity, the force on the 2-μC charge is twice as great as that on the 1-μC charge. Also, both are proportional to the magnitude or "strength" of the magnetic field; if a given charge moves with the same velocity in two magnetic fields, one having twice the magnitude (and the same direction) as the other, it experiences twice as great a force in the larger field as in the smaller.

The dependence of the magnetic force on the particle's *velocity* is quite different from the electric-field case. The *electric* force on a charge *does not* depend on velocity; it is the same whether the charge is moving or not. The *magnetic* force, conversely, is found to have a magnitude that increases with speed. Furthermore, the *direction* of the force depends on the directions of the magnetic field B and the velocity v in an interesting way. The force F *does not* have the same direction as B, but instead of always *perpendicular* to both B and v. The magnitude F of the force is found to be proportional to the component of v perpendicular to the field; when that component is zero, that is, when v and B are parallel or antiparallel, there is *no* force!

These characteristics of the magnetic force can be summarized, with reference to Fig. 30–1, as follows: The direction of F is perpendicular to the plane containing v and B; its magnitude is given by

$$F = qv_\perp B = qvB \sin \phi, \qquad (30\text{-}1)$$

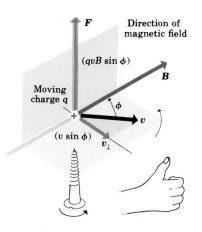

where q is the charge and ϕ is the angle between the vectors v and B, as shown in the figure.

This prescription does not specify the direction of F completely; there are always two opposite directions both perpendicular to the plane of v and B. To complete the description we use the same righthand-thread rule used to define the vector product in Sec. 1–8 and for angular mechanical quantities in Sec. 9–14. We imagine turning v into B, through the smaller of the two possible angles; the direction of F is the direction in which a righthand-thread screw would advance if turned the same way. Alternatively, one can wrap the fingers of the right hand around the line perpendicular to the plane of v and B so that they curl around with this sense of rotation; the thumb then points in the direction of F.

30-1 The magnetic force F acting on a charge q moving with velocity v is perpendicular to both the magnetic field B and to v.

Thus the force on a charge q moving with velocity v in a magnetic field B is given, both in magnitude and in direction, by

$$F = qv \times B. \qquad (30\text{-}2)$$

The vector product is extremely useful in formulating relationships involving magnetic fields; this is our first example of its application in this area.

Spatial relations for magnetic forces can be troublesome; the following review of the definition of the vector product may help. First, always draw the vectors v and B from a common point, that is, with their tails together. Second, visualize and if possible draw the *plane* in which these two vectors lie. The force vector always lies along a line perpendicular to this plane. To determine its direction along this line, imagine rotating v into B, as in Fig. 30–1. The direction of F is the direction of advance of a righthand-thread screw rotated in this direction, or the direction of the extended thumb of the right hand if the fingers are curled around the perpendicular line with this direction of rotation.

Finally, we note that Eq. (30–1) can be interpreted in a different but equivalent way. Recalling that ϕ is the angle between the direction of vectors v and B, we may interpret $(B \sin \phi)$ as the component of B perpendicular to v, that is, $B_\perp$. With this notation, the force expression becomes

$$F = qvB_\perp. \qquad (30\text{-}3)$$

Although equivalent to Eq. (30–1), this is sometimes more convenient, especially in problems involving currents rather than individual particles. Forces on currents in conductors are discussed in Chapter 31.

The *units* of B can be deduced from Eq. (30–1); they must be the same as the units of F/qv. Therefore the SI unit of B is one newton second per coulomb, or $1 \text{ N} \cdot \text{s} \cdot \text{C}^{-1} \cdot \text{m}^{-1}$, or, since one ampere is one coulomb per second, $1 \text{ N} \cdot \text{A}^{-1} \cdot \text{m}^{-1}$. This unit is called *one tesla* (1 T). The cgs unit of B, the *gauss* ($1 \text{ G} = 10^{-4} \text{ T}$) is also in common use. To summarize,

$$1 \text{ T} = 1 \text{ N} \cdot \text{A}^{-1} \cdot \text{m}^{-1} = 10^4 \text{ G}.$$

In this discussion we have assumed that q is a *positive* charge. If q is negative, the direction of F is opposite to that shown in Fig. 30–1 and given by the righthand rule. Thus if two charges of equal magnitude and opposite sign move in the same B field with the same velocity, the forces on the two charges have equal magnitude and opposite direction.

An unknown *electric* field can be "explored" by measuring the magnitude and direction of the force on a test charge. To "explore" an unknown *magnetic* field, we must measure the magnitude and direction of the force on a *moving* test charge. The cathode-ray tube, discussed in Sec. 26–8, is a convenient device for studying the behavior of moving charges in a magnetic field. At one end of this tube an electron gun shoots out a narrow electron beam at a speed that can be controlled and calculated. At the other end is a fluorescent screen that emits light at the small spot where the electron beam strikes it. In the absence of any deflecting force, the beam strikes the center of the screen. If the spot does not move when the tube is moved or its orientation changed, we may conclude that there is no detectable magnetic field. But if, as we move the tube, the spot of light changes position because the electron beam is deflected, we conclude that a magnetic field is present.

At a given point in a magnetic field, the electron beam will, in general, be deflected. By rotating the cathode-ray tube, however, one finds that there are always two directions for which *no* deflection takes place. *One of the directions of motion of a charge on which a magnetic field exerts **no** force is the direction of the **B**-vector.* Thus, in Fig. 30–2, the electron beam is undeflected when its direction is parallel to the z-axis and hence the B-vector must point either up or down.

Let us now place the cathode-ray tube so that the electron beam moves in a plane perpendicular to this direction. Experiment shows that a deflection always takes place in such a manner as to indicate a force acting in this plane, but at right angles to the velocity of the electron beam. Thus, in Fig. 30–2, when the electron beam lies in the xy-plane and its direction is initially along the positive x-axis, the beam is deflected in the direction of the positive y-axis. That is, *when the velocity of the moving charge is perpendicular to the magnetic field, the force is perpendicular to both the magnetic field and the velocity.* Thus this observation corroborates our previous discussion. Additional experiments in which B and v are at various angles can be performed, and the results are always consistent with Eq. (30–2) and the accompanying discussion.

Although this experiment uses electrons, which have negative charge, similar experiments can be carried out with beams of positively charged particles, typically ionized atoms, and again the results are consistent with our general statements about magnetic forces.

Finally, when a charged particle moves through a region of space where *both* electric and magnetic fields are present, both fields exert forces on the particle, and the total force is the vector sum of the electric-field and magnetic-field forces:

$$F = q(E + v \times B). \tag{30-4}$$

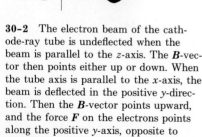

30-2 The electron beam of the cathode-ray tube is undeflected when the beam is parallel to the z-axis. The B-vector then points either up or down. When the tube axis is parallel to the x-axis, the beam is deflected in the positive y-direction. Then the B-vector points upward, and the force F on the electrons points along the positive y-axis, opposite to $v \times B$ because q is negative.

Example   A proton beam moves through a region of space where there is a uniform magnetic field of magnitude 2.0 T, with direction along the positive z-axis, as in Fig. 30-3. The protons have velocity of magnitude 3×10^5 m·s^{-1}, in the xz-plane, at an angle of 30° to the positive z-axis. Find the force on a proton. ($q = 1.6 \times 10^{-19}$ C.)

Solution   The righthand rule shows that the direction of the force is along the negative y-axis. The magnitude of the force, from Eq. (30-1), is

$$F = (1.6 \times 10^{-19}\text{ C})(3 \times 10^5 \text{ m·s}^{-1})(2.0\text{ T})(\sin 30°)$$
$$= 4.8 \times 10^{-14}\text{ N}.$$

Alternatively, in vector language, with Eq. (30-2),

$$\boldsymbol{v} = (3 \times 10^5 \text{ m·s}^{-1})(\sin 30°)\boldsymbol{i} + (3 \times 10^5 \text{ m·s}^{-1})(\cos 30°)\boldsymbol{k},$$
$$\boldsymbol{B} = (2.0\text{ T})\boldsymbol{k},$$
$$\boldsymbol{F} = q\boldsymbol{v} \times \boldsymbol{B} = (1.6 \times 10^{-19}\text{ C})(3 \times 10^5 \text{ m·s}^{-1})(2.0\text{ T})$$
$$(\sin 30° \ \boldsymbol{i} + \cos 30° \ \boldsymbol{k}) \times \boldsymbol{k}$$
$$= (-4.8 \times 10^{-14}\text{ N})\boldsymbol{j},$$

since $\boldsymbol{i} \times \boldsymbol{k} = -\boldsymbol{j}$ and $\boldsymbol{k} \times \boldsymbol{k} = \boldsymbol{0}$.

If the beam consists of *electrons* rather than protons, the charge is negative ($q = -1.6 \times 10^{-19}$ C), and the direction of the force is reversed. The solution proceeds just as before; the force is now directed along the *positive* y-axis: $\boldsymbol{F} = +(4.8 \times 10^{-14}\text{ N})\boldsymbol{j}$.

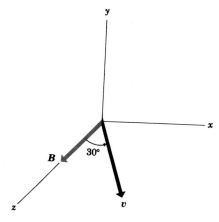

30-3

30-3 Magnetic field lines; magnetic flux

A magnetic field can be represented by lines, such that the direction of a line through a given point is the same as that of the magnetic field vector $\boldsymbol{B}$ at that point. We shall call these *magnetic field lines;* they are sometimes called magnetic lines of force, but this term is unfortunate because, unlike electric field lines, they *do not* point in the direction of the force on a charge.

In a uniform magnetic field, where the $\boldsymbol{B}$ vector has the same magnitude and direction at every point in a region, the field lines are straight and parallel. If the poles of an electromagnet are large, flat, and close together, there is a region between them where the magnetic field is approximately uniform.

The *magnetic flux* across a surface is defined in analogy to the electric field flux used with Gauss's law (Sec. 25-4). Any surface can be divided into elements of area dA, as shown in Fig. 30-4. For each element we obtain the components of $\boldsymbol{B}$ normal and tangent to the surface at the position of that element, as shown. In general these components will vary from point to point on the surface. From the figure, $B_\perp = B \cos\theta$. The magnetic flux $d\Phi$ through this area is defined as

$$d\Phi = B_\perp\, dA = B \cos\theta\, dA = \boldsymbol{B}\cdot d\boldsymbol{A}. \qquad (30\text{-}5)$$

The total magnetic flux through the surface is the sum of the contributions from the individual area elements, given by

$$\Phi = \int B_\perp\, dA = \int \boldsymbol{B}\cdot d\boldsymbol{A}. \qquad (30\text{-}6)$$

In the special case where $\boldsymbol{B}$ is uniform over a plane surface with total

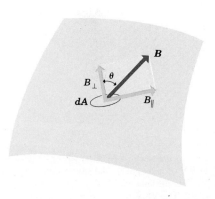

30-4 The magnetic flux through an area element dA is defined to be $\Phi = B_\perp\, dA$.

area A,

$$\Phi = B_\perp A = BA \cos \theta. \qquad (30\text{-}7)$$

If B happens to be perpendicular to the surface, $\cos \theta = 1$, and this expression reduces to $\Phi = BA$. The chief usefulness of the concept of magnetic flux is in the study of electromagnetic induction, to be considered in Chapter 33.

The above definitions of magnetic flux have an ambiguity of sign associated with the direction of the vector area element dA. In Gauss's law we always took dA as pointing *out of* a closed surface; some applications of magnetic flux involve *open* surfaces with an edge, and then we shall have to take care to define which is the positive side of the surface.

The SI unit of magnetic field B is one tesla = one newton per ampere meter; hence the unit of magnetic flux is *one newton meter per ampere* $(1\,\text{N}\cdot\text{m}\cdot\text{A}^{-1})$. In honor of Wilhelm Weber (1804–1890), $1\,\text{N}\cdot\text{m}\cdot\text{A}^{-1}$ is called one *weber* (1 Wb).

If the element of area dA in Eq. (30-5) is at right angles to the field lines, $B_\perp = B$, and hence

$$B = \frac{d\Phi}{dA}. \qquad (30\text{-}8)$$

That is, the magnetic field equals the *flux per unit area* across an area at right angles to the magnetic field. Since the unit of flux is 1 weber, the unit of field, 1 tesla, is equal to one *weber per square meter* $(1\,\text{Wb}\cdot\text{m}^{-2})$. The magnetic field B is sometimes called *flux density*.

The total flux across a surface can then be pictured as proportional to the number of field lines crossing the surface, and the field (the flux density) as the number of lines *per unit area*.

In the cgs system, the unit of magnetic flux is one *maxwell* and the corresponding unit of flux density, one maxwell per square centimeter, is called one *gauss* (1 G). (Instruments for measuring flux density are often referred to as *gaussmeters*.)

The largest values of steady magnetic field that have been achieved in the laboratory are of the order of 30 T = 300,000 G; some pulsed-current electromagnets can produce fields of the order of 120 T = 1.2×10^6 G for short time intervals of the order of a millisecond. By comparison, the magnetic field of the earth is of the order of 10^{-4} T or 1 G.

30-4 Motion of charged particles in magnetic fields

Let a positively charged particle at point O in a uniform magnetic field B be given a velocity v in a direction at right angles to the field (Fig. 30-5). An upward force F, equal in magnitude to qvB, is exerted on the particle at this point. As explained in Sec. 5-4, when the force acting on a particle (and thus also its acceleration) has a direction perpendicular to the particle's velocity, the velocity changes only in direction, not in magnitude. Thus the particle moves with constant speed. At points such as P and Q the directions of force and velocity will have changed as shown; the magnitude of the force is constant, since the magnitudes of q, v, and B are constant. The particle therefore moves under the influence of a force whose *magnitude* is constant but whose *direction* is always at right angles to the velocity of the particle. The orbit of the particle is therefore a

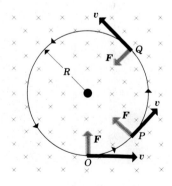

30-5 The orbit of a charged particle in a uniform magnetic field is a circle when the initial velocity is perpendicular to the field. The crosses represent a uniform magnetic field directed *away from* the reader.

circle described with constant tangential speed v. Since the centripetal acceleration is v^2/R, as shown in Sec. 5-6, we have, from Newton's second law,

$$F = qvB = m\left(\frac{v^2}{R}\right),$$

where m is the mass of the particle. The radius of the circular orbit is

$$R = \frac{mv}{Bq}. \tag{30-9}$$

If the direction of the initial velocity is *not* perpendicular to the field, the velocity component parallel to the field remains constant and the particle moves in a helix. In this case, v in Eq. (30-9) is the component of velocity perpendicular to the field.

Note that the radius is proportional to the *momentum* of the particle, mv. Note also that the *work* of the magnetic force acting on a charged particle is always *zero,* because the force is always at *right angles* to the motion. The only effect of a magnetic force is to change the *direction* of motion, never to increase or decrease the *magnitude* of the velocity. Thus, *motion of a charged particle under the action of a magnetic field alone is always motion with constant speed.*

Figure 30-6 is a photograph of the circular tracks made in a cloud chamber by charged particles moving in a magnetic field perpendicular to the plane of the paper. The photograph shows three pairs of tracks originating at common points but curving in opposite directions. A study of the density of droplets along the paths shows that both particles ionize like electrons, but since the tracks curve in opposite directions, the charges must be of opposite sign. These tracks are in fact made by *electron–positron pairs,* created at the points from which the tracks originate by the annihilation of a high-energy gamma ray in the process known as *pair production.* Photographs such as this provided, in the early 1930s, the first experimental evidence of the existence of positrons.

The other tracks at the top of the photograph are portions of the circular paths of photoelectrons ejected from the lead sheet in the chamber. One complete circular track appears in the lower part of the photograph.

The spiral in Fig. 30-7 is the track made by a 4-MeV electron as it slows down in a liquid hydrogen bubble chamber in which there is a magnetic field. Many more ions are produced per unit length in a liquid than in a gas, so that, although an electron may pass completely through a cloud chamber without appreciable energy loss, as in Fig. 30-6, the electron in Fig. 30-7 loses all its energy and comes to rest at the end of the spiral. The shape of the path shows in a striking way how the radius of curvature decreases as the velocity decreases.

30-5 Thomson's measurement of *e/m*

The charge-to-mass ratio of an electron, e/m, was first measured by Sir J. J. Thomson in 1897 at the Cavendish Laboratory in Cambridge, England. The discovery that this ratio is constant provided the best experimental evidence available at that time of the *existence* of electrons, particles with definite mass and charge. Thomson's term for these particles was "cathode corpuscles."

30-6 An early photograph showing cloud-chamber tracks of three electron–positron pairs in a magnetic field. The gamma-ray photons entering at the top materialize into pairs within a lead sheet. The coiled tracks are due to low-energy photoelectrons ejected from the lead. (Courtesy of Lawrence Berkeley Laboratory, University of California.)

30-7 A 4-MeV electron slowing down in a liquid-hydrogen bubble chamber traversed by a magnetic field. The shape of the path shows how the radius of curvature decreases with velocity. (Courtesy of Lawrence Berkeley Laboratory, University of California.)

Thomson's apparatus (Fig. 30–8) was very similar in principle to the cathode-ray tube discussed in Sec. 26–8. It consisted of a highly evacuated glass tube into which several metal electrodes were sealed. Electrode C is the hot cathode from which the electrons emerged. Electrode A is the anode, which was maintained at a high positive potential. Most of the electrons accelerated toward A by this potential hit electrode A, but there was a small hole in A through which some of them passed. They were further restricted by an electrode A′ in which there was another hole. Thus a narrow *beam* of electrons, all having the same speed, passed into the region between the two plates P and P′. After passing between the plates, the electrons struck the end of the tube, where they caused fluorescent material at S to glow. The speed of the electrons is related to the accelerating potential V; we simply equate the kinetic energy $\frac{1}{2}mv^2$ to the loss of potential energy eV (where e is the magnitude of the electron charge):

$$\tfrac{1}{2}mv^2 = eV \qquad \text{or} \qquad v = \sqrt{\frac{2eV}{m}}. \qquad (30\text{–}10)$$

If now a potential difference is established between the two deflecting plates P and P′, as shown, the resulting downward electric field deflects the negatively charged electrons upward. Alternatively, we may impose a *magnetic* field directed into the plane of the figure, as shown by the x's; this field results in a downward deflection of the beam; the reader should verify this direction. Finally, if *both* $\boldsymbol{E}$ and $\boldsymbol{B}$ fields are applied simultaneously, it is possible to adjust their relative magnitudes so that the two forces cancel and the beam is undeflected. The condition that must be satisfied, obtained by equating the two force magnitudes, is

$$eE = evB, \qquad \text{or} \qquad v = \frac{E}{B}. \qquad (30\text{–}11)$$

Finally, we may combine this with Eq. (30–10) to eliminate v and obtain an expression for the charge-to-mass ratio e/m in terms of the other quantities:

$$\frac{E}{B} = \sqrt{\frac{2eV}{m}},$$
$$\frac{e}{m} = \frac{E^2}{2B^2 V}. \qquad (30\text{–}12)$$

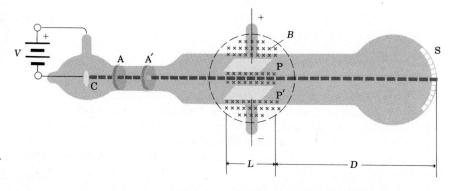

30-8 Thomson's apparatus for measuring the ratio e/m for cathode rays.

All the quantities on the right side can be measured, so e/m can be determined. We note that it is *not* possible to measure e or m separately by this method, but only their ratio.

Thomson measured e/m for his "cathode corpuscles" and found a unique value for this quantity that was independent of the cathode material and the residual gas in the tube. This independence indicated that cathode corpuscles are a common constituent of all matter. The modern accepted value of e/m is $(1.758803 \pm 0.000003) \times 10^{11}$ C·kg^{-1}. Thus Thomson is credited with discovery of the first subatomic particle, the electron. He also found that the speed of the electrons in the beam was about one-tenth the speed of light, much larger than any previously measured material particle speed.

Fifteen years after Thomson's experiments, Millikan succeeded in measuring the charge of the electron with his famous oil-drop experiment, described in Sec. 26-6. The magnitude of the electron charge e is 1.602×10^{-19} C. Thus the *mass* of the electron can be obtained:

$$m = \frac{1.602 \times 10^{-19}\,\text{C}}{1.759 \times 10^{11}\,\text{C·kg}^{-1}} = 9.109 \times 10^{-31}\,\text{kg}.$$

*30-6 Isotopes and mass spectroscopy

Thomson devised a method similar to the above e/m measurement, for measuring the charge–mass ratio for positive ions. An added difficulty was that in Thomson's day it was difficult to produce a beam of ions all having the same speed. Because the e/m electron experiment depends on the particles having a common speed (in order for the electric and magnetic field forces to balance), this method is not directly applicable for a beam of particles having various velocities. Thomson's method was to make the electric and magnetic fields *parallel,* so that the deflections due to these fields are in perpendicular directions. The net deflection can then never be zero, but it turns out that the relation between the x- and y-deflections for a beam permits determination of the charge–mass ratio of the particles.

Thomson assumed that each positive ion had a charge equal in magnitude to that of the electron because each was an atom that had lost one electron. He could then identify particular values of q/m with particular ions. Positive particles move more slowly than electrons and have lower values of q/m because they are much more massive. The *largest* q/m for positive particles is that for the *lightest* element, hydrogen. From the value of q/m it was found that the mass of the *hydrogen ion* or *proton* is 1836.13 ± 0.01 times the mass of an electron. Electrons contribute only a small amount of the mass of material objects.

The most striking result of these experiments was that certain chemically pure gases have *more than one* value of q/m. Most notable was the case of neon, which has atomic mass 20.2 g·mol^{-1}. Thomson obtained *two* values of q/m, corresponding to 20 and 22 g·mol^{-1}, and after trying and discarding various explanations he concluded that there must be two kinds of neon atoms with different masses.

Soon afterward Aston, a student of Thomson, succeeded in actually separating these two atomic species. Aston permitted the gas to diffuse repeatedly through a porous plug between two containers. At a given temperature T, each atom has an average kinetic energy $3kT/2$, as discussed in Sec. 20-4. Thus the more massive atoms have, on the average,

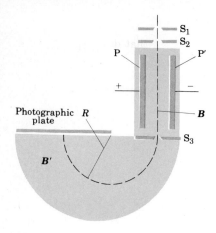

30-9 Bainbridge's mass spectrograph, utilizing a velocity selector.

somewhat smaller speeds; because of this speed difference, the gas emerging from the plug had a slightly greater concentration of the less massive atoms than the gas entering the plug. Thus Aston demonstrated directly the existence of two species of neon atoms with different masses.

Since these experiments in the early years of the twentieth century, many other elements have been shown to have several kinds of atoms, identical in their chemical behavior but differing in mass. Such forms of an element are called *isotopes*. We shall see in a later chapter that the mass differences are due to differing numbers of neutrons in the *nuclei* of the atoms.

A detailed search for the isotopes of all the elements required precise experimental technique. Aston built the first of many instruments called *mass spectrometers* in 1919. His instrument had a precision of one part in 10,000, and he found that many elements have isotopes. We shall describe here a variation built by Bainbridge. The Bainbridge mass spectrometer (Fig. 30–9) has a source of ions (not shown) situated above S_1. The ions under study pass through slits S_1 and S_2 and move down into the electric field between the two plates P and P'. In the region of the electric field there is also a magnetic field B, perpendicular to the paper. Thus the ions enter a region of crossed electric and magnetic fields like those used by Thomson to measure the velocity of electrons in his determination of e/m. Only those ions whose speed is equal to E/B, as in Eq. (30–11), pass undeviated through this region; ions with other speeds are stopped by the slit S_3. Thus all ions that emerge from S_3 have the same velocity. The region of crossed fields is called a *velocity selector*.

Below S_3 the ions enter a region where there is another magnetic field B', perpendicular to the page, but no electric field. Here the ions move in circular paths of radius R. From Eq. (30–9), we find that

$$m = \frac{qB'R}{v}. \tag{30–13}$$

Assuming equal charges on each ion, the mass of each ion is proportional to the radius of its path. Ions of different isotopes converge at different points on the photographic plate. The relative abundance of the isotopes is measured from the densities of the photographic images they produce.

Figure 30–10 shows the mass spectrum of germanium. The numbers shown beside the isotope images are not the atomic masses of the isotopes but the integers nearest the atomic masses. These integers are called *mass numbers*; isotopes are written with the mass number as a superscript to the chemical symbol. Thus the isotopes shown would be written Ge^{70}, Ge^{72}, etc. The mass number is represented by the letter A. As will be discussed in a later chapter, A is equal to the total number of protons and neutrons in the nucleus of the atom; different isotopes of an element have the same number of protons but differing numbers of neutrons in their nuclei.

Masses of atoms are often expressed in *atomic mass units*. By definition, one atomic mass unit (1 u) is $\frac{1}{12}$ the mass of one atom of the most abundant isotope of carbon, C^{12}. Since the mass of an atom in grams is equal to its atomic mass divided by Avogadro's number, it follows that

$$1\,u = \frac{(1/12)(12\,\text{g}\cdot\text{mol}^{-1})}{6.022 \times 10^{23}\,\text{mol}^{-1}}$$

$$= 1.661 \times 10^{-24}\,\text{g} = 1.661 \times 10^{-27}\,\text{kg}.$$

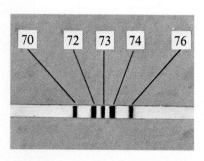

30-10 The mass spectrum of germanium, showing the isotopes of mass numbers 70, 72, 73, 74, 76.

Questions

30-1 The direction of the **E** field is defined to be the direction of the force on a positive charge; why can't we define the direction of **B** similarly?

30-2 Does the earth's magnetic field have a significant effect on the electron beam in a TV picture tube?

30-3 If an electron beam in a cathode-ray tube travels in a straight line, can you be sure there is no magnetic field present?

30-4 A charge moving in a magnetic field experiences a force. But suppose the phenomenon is viewed in a frame of reference moving with the charge, in which the charge is therefore at rest. Then presumably *no* magnetic force would be observed. How is this apparent inconsistency to be reconciled?

30-5 If the magnetic-field force does no work on a charged particle, how can it have any effect on the particle's motion? Are there other examples of forces that do no work but have a significant effect on a particle's motion?

30-6 A permanent magnet can be used to pick up a string of nails, tacks, or paper clips, even though these are not magnets by themselves. How can this be?

30-7 There are regions on earth where the direction of the magnetic field is nearly perpendicular to the earth's surface. What difficulties would this cause in the use of a compass? Where would you expect to find such regions?

30-8 Can one build a compass that does not depend on magnetism for its operation?

30-9 How could a compass be used for a *quantitative* determination of magnitude and direction of magnetic field at a point?

30-10 The direction in which a compass points (magnetic north) is not in general exactly the same as the direction toward the north pole (true north). The difference is called *magnetic declination;* it varies from point to point on the earth and also varies with time. What are some possible explanations for magnetic declination?

30-11 Can a charged particle move through a magnetic field without experiencing any force? How?

30-12 Could the electron beam in an oscilloscope tube (cathode-ray tube) be used as a compass? How? What advantages and disadvantages would it have, compared with a conventional compass?

30-13 Does a magnetic field exert forces on charges moving within conductors? What evidence is there for your answer?

30-14 Does a magnetic field exert forces on the electrons within atoms? What observable effect might such interaction have on the behavior of the atom?

Problems

30-1 Each of the lettered circles at the corners of the cube in Fig. 30-11 represents a positive charge q moving with a velocity of magnitude v in the direction indicated. The region in the figure is a uniform magnetic field **B,** parallel to the x-axis and directed toward the right. Copy the figure, find the magnitude and direction of the force on each charge, and show the force in your diagram.

a) What is the magnetic flux across the surface *abcd* in the figure?

b) What is the magnetic flux across the surface *befc*?

c) What is the magnetic flux across the surface *aefd*?

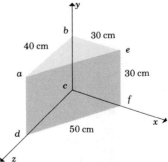

Figure 30-12

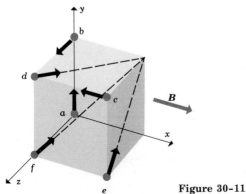

Figure 30-11

30-2 The magnetic field **B** in a certain region is 2 T and its direction is that of the positive x-axis in Fig. 30-12.

30-3 A particle having a mass of 0.5 g carries a charge of 2.5×10^{-8} C. The particle is given an initial horizontal velocity of $6 \times 10^4 \, \text{m·s}^{-1}$. What is the magnitude and direction of the minimum magnetic field that will keep the particle moving in a horizontal direction?

30-4 A deuteron, an isotope of hydrogen whose mass is very nearly 2 u, travels in a circular path of radius 40 cm in a magnetic field of magnitude 1.5 T.

a) Find the speed of the deuteron.

b) Find the time required for it to make one-half a revolution.

c) Through what potential difference would the deuteron have to be accelerated to acquire this velocity?

30-5 An electron at point A in Fig. 30–13 has a speed v_0 of 10^7 m·s^{-1}. Find

a) the magnitude and direction of the magnetic field that will cause the electron to follow the semicircular path from A to B, and

b) the time required for the electron to move from A to B.

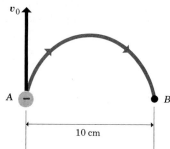

Figure 30-13

30-6 In Problem 30–5, suppose the particle is a proton rather than an electron. Answer the same questions as in that problem.

30-7 In a magnetic field directed vertically upward, a particle initially moving north is deflected toward the west. What is the sign of the charge on the particle?

30-8 In a TV picture tube, an electron in the beam is accelerated by a potential difference of 20,000 V. Then it passes through a region of transverse magnetic field where it moves in a circular arc with radius 12 cm. What is the magnitude of the field?

30-9 Estimate the effect of the earth's magnetic field on the electron beam in a TV picture tube. Suppose the accelerating voltage is 20,000 V; calculate the approximate deflection of the beam over a distance of 0.4 m from the electron gun to the screen, under the action of a transverse field of magnitude 0.5×10^{-4} T (comparable to the magnitude of the earth's field), assuming there are no other deflecting fields. Is this deflection significant?

30-10 A particle carries a charge of 4×10^{-9} C. When it moves with a velocity v_1 of 3×10^4 m·s^{-1} at 45° above the x-axis in the xy-plane, a uniform magnetic field exerts a force F_1 along the negative z-axis. When the particle moves with a velocity v_2 of 2×10^4 m·s^{-1} along the z-axis, there is a force F_2 of 4×10^{-5} N exerted on it along the x-axis. What are the magnitude and direction of the magnetic field? (See Fig. 30–14.)

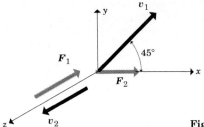

Figure 30-14

30-11 An electron moves in a circular path of radius 1.2 cm perpendicular to a uniform magnetic field. The speed of the electron is 10^6 m·s^{-1}. What is the total magnetic flux encircled by the orbit?

30-12 An electron and an alpha particle (a doubly ionized helium atom) both move in circular paths in a magnetic field with the same tangential speed. Compare the number of revolutions they make per second. The mass of the alpha particle is 6.68×10^{-27} kg.

30-13 In the Bainbridge mass spectrometer (Fig. 30–9), suppose the magnetic field B in the velocity selector is 1.0 T, and ions having a speed of 0.4×10^7 m·s^{-1} pass through undeflected.

a) What should be the electric field between the plates P and P′?

b) If the separation of the plates is 0.5 cm, what is the potential difference between plates?

30-14

a) What is the velocity of a beam of electrons when the simultaneous influence of an electric field of 34×10^4 V·m^{-1} and a magnetic field of 2×10^{-2} T, both fields being normal to the beam and to each other, produces no deflection of the electrons?

b) Show in a diagram the relative orientation of the vectors v, E, and B.

c) What is the radius of the electron orbit when the electric field is removed?

30-15 A singly charged Li^7 ion has a mass of 1.16×10^{-23} g. It is accelerated through a potential difference of 500 V and then enters a magnetic field of 0.4 T, moving perpendicular to the field. What is the radius of its path in the magnetic field?

30-16 Suppose the electric field between the plates P and P′ in Fig. 30–9 is 150 V·cm^{-1}, and the magnetic field is 0.5 T. If the source contains the three isotopes of magnesium, $_{12}$Mg^{24}, $_{12}$Mg^{25}, and $_{12}$Mg^{26}, and the ions are singly charged, find the distance between the lines formed by the three isotopes on the photographic plate. Assume the atomic masses of the isotopes equal to their mass numbers.

30-17 The electric field between the plates of the velocity selector in a Bainbridge mass spectrograph is 1200 V·cm^{-1}, and the magnetic field in both regions is 0.6 T. A stream of singly charged neon moves in a circular path of 7.28-cm radius in the magnetic field. Determine the mass number of the neon isotope.

30-18 A particle of mass m and charge $+q$ starts from rest at the origin in Fig. 30–15. There is a uniform electric field E in the positive y-direction and a uniform magnetic field B directed toward the reader. It is shown in more advanced books that the path is a *cycloid* whose radius of curvature at the top points is twice the y-coordinate at that level.

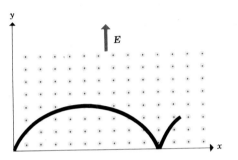

Figure 30-15

a) Explain why the path has this general shape and why it is repetitive.

b) Prove that the speed at any point is equal to $\sqrt{2qEy/m}$. [*Hint.* Use energy conservation.]

c) Applying Newton's second law at the top point, prove that the speed at this point is $2E/B$.

30-19 Two positive ions having the same charge q but different masses, m_1 and m_2, are accelerated horizontally from rest through a potential difference V. They then enter a region where there is uniform magnetic field B normal to the plane of the trajectory.

a) Show that if the beam entered the magnetic field along the x-axis, the value of the y-coordinate for each ion at any time t is approximately

$$y = Bx^2 \left(\frac{q}{8mV}\right)^{1/2},$$

provided x remains much smaller than y.

b) Can this arrangement be used for isotope separation?

MAGNETIC FORCES ON CURRENT-CARRYING CONDUCTORS

31

31-1 Force on a conductor

When a current-carrying conductor lies in a magnetic field, magnetic forces are exerted on the moving charges within the conductor. These forces are transmitted to the material of the conductor, and the conductor as a whole experiences a force distributed along its length. The electric motor and the moving-coil galvanometer both depend for their operation on the magnetic forces on conductors carrying currents.

Figure 31–1 represents a portion of a conducting wire of length l and cross-sectional area A, in which the current density J is from left to right. The wire is in a magnetic field B, perpendicular to the plane of the diagram, and directed *into* the plane. A positive charge q_1 within the wire, moving with drift velocity v_1, is acted on by a force F_1 given by Eq. (30–2), $F_1 = q_1(v_1 \times B)$. As the figure shows, the direction of this force is upward, and since in this case v_1 and B are perpendicular, the magnitude of the force is $F_1 = q_1 v_1 B$. Similarly, a negative charge q_2, with drift velocity v_2 in a direction opposite to that of J, experiences a force $F_2 = q_2 v_2 \times B$. Because q_1 and q_2 have opposite signs and v_1 and v_2 opposite directions, F_2 has the *same* direction as F_1, as shown.

The *total* force on all the moving charges in a length l of conductor can be expressed in terms of the current, using the considerations of Sec. 28–1, Eqs. (28–3), (28–4), and (28–5). Let n_1 and n_2 represent the numbers of positive and negative charge carriers, respectively, per unit volume. The numbers of carriers in the portion are then $n_1 A l$ and $n_2 A l$; the

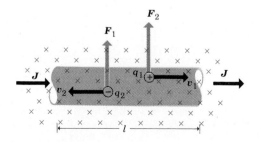

31-1 Forces on the moving charges in a current-carrying conductor. The forces on both positive and negative charges are in the same direction.

total force F on all carriers (and hence the total force on the wire) has magnitude

$$F = (n_1 Al)(q_1 v_1 B) + (n_2 Al)(q_2 v_2 B)$$
$$= (n_1 q_1 v_1 + n_2 q_2 v_2) AlB.$$

But $n_1 q_1 v_1 + n_2 q_2 v_2$ (or more generally, $\Sigma\, nqv$) equals the current density J, and the product JA equals the current I, so finally

$$F = IlB. \tag{31-1}$$

If the B field is not perpendicular to the wire but makes an angle ϕ with it, the situation is just like that discussed in Sec. 30–2 for a single charge. The component of B parallel to the wire (and thus to the drift velocities of the charges) exerts no force; the component perpendicular to the wire is given by $B_\perp = B \sin \phi$ so, in general,

$$F = IlB_\perp = IlB \sin \phi. \tag{31-2}$$

To find the direction of the force on a current-carrying conductor placed in a magnetic field, we may use the same righthand-screw rule that was used in the case of a moving positive charge. Rotate a righthand screw from the direction of I toward B; the direction of advance will be the direction of F. The situation is the same as that shown in Fig. 30–1, but with the direction of v replaced by the direction of I.

Thus this force, like the force on a single moving charge, may be expressed as a vector product. We represent the section of wire by a vector l along the wire and in the direction of the current. Then the force on this section is given by

$$F = Il \times B. \tag{31-3}$$

Example A straight horizontal wire carries a current of 50 A from west to east, in a region where the magnetic field is toward the northeast (i.e., 45° north of east) with magnitude 1.2 T. Find the magnitude and direction of the force on a 1-m section of wire.

Solution The angle ϕ between the directions of current and field is 45°. From Eq. (31–2) we obtain

$$F = (50 \text{ A})(1 \text{ m})(1.2 \text{ T})(\sin 45°) = 42.4 \text{ N}.$$

Consistency of units may be checked by observing, as noted in Sec. 30–2, that $1 \text{ T} = 1 \text{ N} \cdot \text{A}^{-1} \cdot \text{m}^{-1}$. The direction of the force is perpendicular to the plane of the current and the field, both of which lie in the horizontal plane. Thus the force must be along the vertical; the righthand rule shows that it is vertically upward.

31-2 The Hall effect

The reality of the forces on the moving charges in a conductor in a magnetic field is strikingly demonstrated by the *Hall effect*. The conductor in Fig. 31–2 is in the form of a flat strip. The current carriers within it are driven toward the upper edge of the strip by the magnetic force $qv \times B$.

This force is an example of the *non-electrostatic* forces F_n discussed in Chapter 28, and in this case the equivalent *non-electrostatic field* E_n

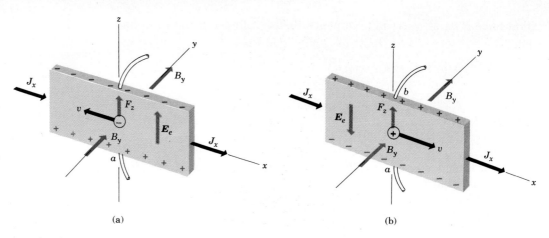

(a)

(b)

31-2 Forces on charge carriers in a conductor in a magnetic field. (a) Negative current carriers (electrons) are pushed toward top of slab, leading to charge distribution as shown. Point a is at higher potential than point b. (b) Positive current carriers; polarity of potential difference is opposite to that of (a).

(the force per unit charge) has magnitude

$$E_n = vB. \tag{31-4}$$

If the charge carriers are electrons, as in Fig. 31-2a, an excess negative charge accumulates at the upper edge of the strip, leaving an excess positive charge at its lower edge, until the transverse electrostatic field E_e within the conductor is equal and opposite to the *non-electrostatic* field E_n. Because the final transverse current is zero, the conductor is on "open circuit" in the transverse direction, and the potential difference between the edges of the strip, which can be measured with a potentiometer, is equal to the *Hall emf* in the strip. The study of this Hall emf has provided much information regarding the conduction process. It is found that, for the metals, the upper edge of the strip in Fig. 31-2a *does* become charged negatively relative to the lower, which justifies our belief that the charge carriers in a metal are negative electrons.

Suppose, however, that the charge carriers are *positive*, as in Fig. 31-2b. Then *positive* charge accumulates at the upper edge, and the potential difference is *opposite* to that resulting from the deflection of negative charges. Soon after the discovery of the Hall effect, in 1879, it was observed that many materials, notably the *semiconductors*, exhibited a Hall emf opposite to that of the metals, *as if* their charge carriers were *positively* charged. We now know that these metals conduct by a process known as *hole conduction*. There are sites within the material that would normally be occupied by an electron but are actually empty, and a *missing negative* charge is equivalent to a *positive* charge. When an electron moves in one direction to fill a hole, it leaves another hole behind it, and the result is that the *hole* (equivalent to a positive charge) migrates in the direction *opposite* to that of the electron.

In terms of the set of coordinate axes in Fig. 31-2a, the electrostatic field E_e is in the z-direction, and we write it as E_z. The magnetic field is in the y-direction, and we write it as B_y. The non-electrostatic field E_n (in the negative z-direction) equals vB_y. The current density, J_x, is in the x-direction.

In the final steady state, when the fields E_e and E_n are equal,

$$E_z = vB_y.$$

The current density J_x is

$$J_x = nqv.$$

When v is eliminated, we have

$$nq = \frac{J_x B_y}{E_z}. \qquad (31\text{-}5)$$

Thus, from measurements of J_x, B_y, and E_z, one can compute the product nq. In both metals and semiconductors, q is equal in magnitude to the electron charge, so the Hall effect permits a direct measurement of n, the density of current-carrying charges in the material.

31-3 Force and torque on a complete circuit

The *total* force and torque on any conductor in a magnetic field may be calculated by dividing it into segments and applying Eq. (31–2) or (31–3) to each segment. In general, this requires an integration, but when the magnetic field is uniform the calculation is much simpler. As we shall see, the total *force* on a complete circuit in a uniform field is always zero, but the *torque* in general is not. Three simple examples will be analyzed in detail, a rectangular loop, a circular loop, and a cylindrical coil or solenoid.

1. *Rectangular loop.*   Figure 31–3 shows a rectangular loop of wire with sides of lengths a and b. The normal to the plane of the loop (a line perpendicular to the plane) makes an angle α with the direction of a uniform magnetic field, and the loop carries a current I. (Provision must be made for leading the current into and out of the loop, or for inserting a seat of emf. This is omitted from the diagram, for simplicity.)

The force F on the right side, of length a, is in the direction of the x-axis, toward the right, as shown. In this side B is perpendicular to the current direction, and the total force on this side (the sum of the forces

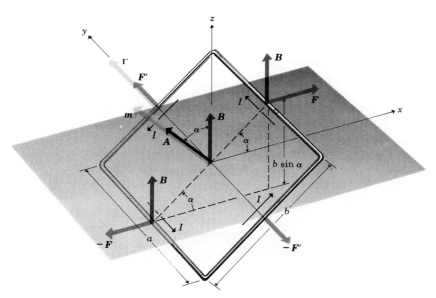

31-3 Forces on the sides of a current-carrying loop in a magnetic field. The resultant of the set of forces is a couple of moment $\Gamma = IAB \sin \alpha$.

distributed along the side) has magnitude

$$F = IaB.$$

A force of the same magnitude but in the opposite direction acts on the opposite side, as shown in the figure.

The forces on the sides of length b, represented by the vectors $\boldsymbol{F}'$ and $-\boldsymbol{F}'$, have magnitude

$$IbB \sin (90° - \alpha), \quad \text{or} \quad IbB \cos \alpha.$$

The lines of action of both lie along the y-axis.

The total force on the loop is zero, since the forces on opposite sides cancel out in pairs. The two forces $\boldsymbol{F}'$ and $-\boldsymbol{F}'$ lie along the same line and so have no torque with respect to any axis, but the two forces $\boldsymbol{F}$ and $-\boldsymbol{F}$ constitute a *couple*, as defined in Sec. 8–4. The torque of a couple has the same value with respect to any point, and is given by the magnitude of either force multiplied by the distance between lines of action of the two forces. From Fig. 31–3, this distance is $b \sin \alpha$, so the torque is

$$\Gamma = (IBa)(b \sin \alpha). \tag{31-6}$$

The torque is greatest when $\alpha = 90°$ (i.e., when the plane of the coil is parallel to the field), and it is zero when α is zero or $180°$ and the plane of the coil is perpendicular to the field. One of the latter positions is a position of *stable* equilibrium, the other of *unstable* equilibrium; which is which?

Since ab is the area A of the coil, Eq. (31–6) may also be written

$$\Gamma = IBA \sin \alpha. \tag{31-7}$$

The product IA is called the *magnetic moment m* of the loop. It is analogous to the dipole moment of an electric dipole, introduced in Sec. 25–1. Thus the torque on a magnetic moment corresponding to a current loop can be expressed simply as

$$\Gamma = mB \sin \alpha. \tag{31-8}$$

Because of the directional relations indicated, the effect of the torque Γ is to tend to rotate the loop in the direction of *decreasing* α, toward its equilibrium position, in which it lies in the xy-plane, perpendicular to the direction of the field $\boldsymbol{B}$.

The interaction of a conducting loop with a magnetic field may be expressed more compactly in terms of vector torque, introduced in Sec. 8–5. We first define a vector area $\boldsymbol{A}$, as in Sec. 25–4, defined as having magnitude A and direction perpendicular to its plane; the sense is determined by the righthand rule applied to the direction of circulation of current around the loop, as shown in Fig. 31–3. Then, as Eq. (31–7) and the figure show, the vector torque Γ is given by

$$\boldsymbol{\Gamma} = I\boldsymbol{A} \times \boldsymbol{B}. \tag{31-9}$$

We may also define a vector magnetic moment

$$\boldsymbol{m} = I\boldsymbol{A}; \tag{31-10}$$

in terms of this the torque is expressed simply as

$$\boldsymbol{\Gamma} = \boldsymbol{m} \times \boldsymbol{B}. \tag{31-11}$$

The effect of the torque Γ is to tend to rotate the loop toward its equilibrium position, in which it lies with its vector magnetic moment m in the same direction as the field B. The torque is greatest when m and B are perpendicular, and zero when they are parallel or antiparallel. Equation (31–11) is the analog of Eq. (25–3) in Sec. 25–1, for the torque on an *electric* dipole in an electric field.

When a magnetic dipole changes its orientation in a magnetic field, the field does work on it, given by $\int \Gamma \, d\theta$, and there is a corresponding potential energy. As the above discussion suggests, the potential energy is least when m and B are parallel and greatest when they are antiparallel. If the potential energy U is taken as zero when $\alpha = \pi/2$, then the potential energy $U(\alpha)$ at any other position is given by

$$U(\alpha) = \int_{\alpha}^{\pi/2} \Gamma \, d\theta.$$

The torque Γ is expressed in terms of the variable angle θ by use of Eq. (31–8):

$$\Gamma = -mB \sin \theta.$$

The extra minus sign is included here because the torque is in the direction of *decreasing* θ. Combining the above relations and evaluating the integral, we find

$$U(\alpha) = -mB \int_{\alpha}^{\pi/2} \sin \theta \, d\theta$$
$$= -mB \cos \theta,$$

or

$$U = -\boldsymbol{m} \cdot \boldsymbol{B}. \tag{31–12}$$

2. *Circular loop.* A circular loop may be approximated by a very large number of small rectangular loops; then the sum of the areas of the rectangular loops may be made to approach the area of the circular one as closely as we please. Furthermore, the boundary of the rectangular loops will approximate the circular loop with any desired accuracy. Currents in the same sense in all the rectangular loops will give rise to forces that will cancel at all points except on the boundary. It can therefore be proved quite rigorously that, *not only for a circular loop, but for a plane loop of **any shape whatever,** carrying a current I in a magnetic field of magnitude B, the torque is given by Eqs. (31–7) through (31–11). In particular, for a circular loop of radius R,

$$\Gamma = \pi I B R^2 \sin \alpha \tag{31–13}$$

and

$$\boldsymbol{\Gamma} = \pi I R^2 \boldsymbol{n} \times \boldsymbol{B}, \tag{31–14}$$

where n is a unit vector perpendicular to the plane of the loop as determined by the sense of circulation of current.

3. *Solenoid.* A helical winding of wire, such as that obtained by winding wire around the surface of a circular cylinder, is called a *solenoid.* If the windings are closely spaced, the solenoid can be approximated by a number of circular loops lying in planes at right angles to its long axis. The total torque on a solenoid in a magnetic field is simply the sum of

the torques on the individual turns. Hence, for a solenoid of N turns in a uniform field B,

$$\Gamma = NIAB \sin \alpha, \tag{31–15}$$

where α is the angle between the axis of the solenoid and the direction of the field. The torque is maximum when the magnetic field is parallel to the planes of the individual turns or perpendicular to the long axis of the solenoid. The effect of this torque, is to tend to rotate the solenoid into a position in which each turn is perpendicular to the field and the axis of the solenoid is parallel to the field.

Although little has been said thus far regarding permanent magnets, the behavior of the solenoid as described above closely resembles that of a bar magnet or compass needle, in that both the solenoid and the magnet will, if free to turn, set themselves with their axes parallel to a magnetic field. The behavior of a bar magnet or compass is sometimes "explained" by ascribing the torque on it to magnetic forces exerted on "poles" at its ends. We see, however, that no such interpretation is demanded in the case of the solenoid. The similarity suggests, in fact, that the orbiting and spinning electrons in a bar of magnetized iron are equivalent to the current in the windings of a solenoid, and that the observed torque arises from the same cause in both instances. We shall return to this matter in Chapter 35.

Example 1   A circular coil of wire 0.05 m in radius, having 30 turns, lies in a horizontal plane, as shown in Fig. 31–4. It carries a current of 5 A, in a counterclockwise sense when viewed from above. The coil is in a magnetic field directed toward the right, with magnitude 1.2 T. Find the magnetic moment and the torque on the coil.

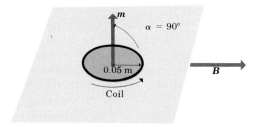

31–4

Solution   The area of the coil is

$$A = \pi r^2 = \pi(0.05 \text{ m})^2 = 7.85 \times 10^{-3} \text{ m}^2.$$

The magnetic moment of each turn of the coil is

$$m = IA = (5 \text{ A})(7.85 \times 10^{-3} \text{ m}^2) = 3.93 \times 10^{-2} \text{ A} \cdot \text{m}^2,$$

and the total magnetic moment of all 30 turns is

$$m = (30)(3.93 \times 10^{-2} \text{ A} \cdot \text{m}^2) = 1.18 \text{ A} \cdot \text{m}^2.$$

The angle α between the direction of B and the normal to the plane of the coil is 90°. From Eq. (31–7) the torque on each turn of the coil is

$$\Gamma = IBA \sin \alpha = (5 \text{ A})(1.2 \text{ T})(7.85 \times 10^{-3} \text{ m}^2)(\sin 90°) = 0.047 \text{ N} \cdot \text{m},$$

and the total torque on the coil is

$$\Gamma = (30)(0.047 \text{ N} \cdot \text{m}) = 1.41 \text{ N} \cdot \text{m}.$$

Alternatively, from Eq. (31–8),

$$\Gamma = mB \sin \alpha = (1.18 \text{ A} \cdot \text{m}^2)(1.2 \text{ T})(\sin 90°) = 1.41 \text{ N} \cdot \text{m}.$$

The direction of the torque is such as to tend to push the right side of the coil down and the left side up and to rotate it into a position where the normal to its plane is parallel to **B.**

Example 2   If the coil in Example 1 rotates from its initial position to a position where its magnetic moment is parallel to **B,** what is the change in potential energy?

Solution   From Eq. (31–12), the initial potential energy U_1 is

$$U_1 = -(1.18 \text{ A} \cdot \text{m}^2)(1.2 \text{ T})(\cos 90°) = 0,$$

and the final potential energy U_2 is

$$U_2 = -(1.18 \text{ A} \cdot \text{m}^2)(1.2 \text{ T})(\cos 0°) = -1.41 \text{ J}.$$

Thus the change in potential energy is -1.41 J.

Example 3   What vertical forces applied to the left and right edges of the coil of Fig. 31–4 would be required to hold it in equilibrium in its initial position?

Solution   An upward force of magnitude F at the right side and a downward force of equal magnitude on the left side would have a total torque (with respect to the coil's center) of

$$\Gamma = (2)(0.05 \text{ m})F.$$

This must be equal to the magnitude of the magnetic-field torque of 1.41 N·m, and we find that the required forces have magnitude 14.1 N.

31-4 The pivoted-coil galvanometer

The pivoted-coil galvanometer, a modified version of an instrument invented by d'Arsonval, has been described in Sec. 29–3. It is widely used in a variety of electrical measurements. A moving coil is pivoted between two bearings, which in high-quality instruments are jeweled bearings. The field of the permanent magnet is shaped by the configuration of its poles and the iron slug in the center of the coil, so that the coil rotates in a field that is everywhere *radial*, as shown in Fig. 29–7. Thus the side thrusts on the coil are always perpendicular to the plane of the coil, and the magnetic-field torque is directly proportional to the current, no matter what the position of the coil.

A restoring torque proportional to the angular displacement of the coil is provided by two hairsprings, which also serve as current leads to the coil. When current is supplied to the coil, it rotates, along with its attached pointer, until the restoring spring torque just balances the magnetic-field torque; hence the deflection is proportional to the current. The frictional torque of the bearings, while small, is not entirely negligible, and it is one of the limitations on the sensitivity of such in-

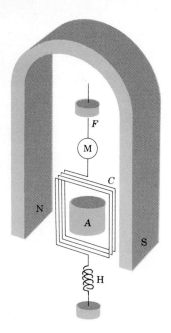

31-5   Principle of the d'Arsonval galvanometer.

struments; the smallest currents that can be measured with these instruments is of the order of 0.1 μA (10^{-7} A).

When greater current sensitivity is needed, an instrument closer to d'Arsonval's original conception can be used, as shown schematically in Fig. 31-5. Instead of being pivoted, the coil is suspended by a fine conducting wire or thin flat strip F, which provides a restoring torque when the coil is deflected from its normal position, and which also serves as one current lead to the coil. The other terminal of the coil is connected to the loosely wound helix H which serves as the second lead, but which exerts a negligible torque on the coil.

The angle of deflection is observed with the aid of a beam of light reflected from a small mirror M cemented to the upper suspension. The light beam serves as a weightless pointer, just as with the Cavendish balance, described in Sec. 4-4.

Because of the nature of its suspension, this type of galvanometer can be made much more sensitive than the pivoted-coil type; currents as small as 10^{-10} A can be detected. However, it is fragile and not readily portable; in fact, such instruments are usually mounted permanently on rigid walls. For measurements involving great sensitivity or great precision, galvanometers have been largely supplanted by electronic instruments using solid-state amplifying devices and often direct digital display for the read-out.

*31-5   The direct-current motor

The direct-current motor is illustrated schematically in Fig. 31-6. The armature or rotor, A, is a cylinder of soft steel mounted on a shaft so that it can rotate about its axis. Embedded in longitudinal slots in the surface of the rotor are a number of copper conductors C. Current is led into and out of these conductors through graphite brushes making contact with a cylinder on the shaft called the *commutator* (not shown in Fig. 31-6). The commutator is an automatic switching arrangement, which maintains the currents in the conductors in the directions shown in the figure, whatever the position of the rotor. The current in the field coils F and F' sets up a magnetic field in the motor frame M and in the gap between the pole pieces P and P' and the rotor. Some of the magnetic field lines are shown as broken lines. With the directions of field and rotor currents shown, the side thrust on each conductor in the rotor is such as to produce a counterclockwise torque on the rotor.

If the rotor and the field windings are connected in series, we have a *series* motor; if they are connected in parallel, we have a *shunt* motor. In some motors the field windings are in two parts, one in series with the rotor and the other in parallel with it; the motor is then *compound*.

A motor converts electrical energy to mechanical energy or work, and thus requires electrical energy input. If the potential difference between terminals is V_{ab} and the current is I, then the power input is $P = V_{ab}I$. Even if the motor coils have no resistance, there must be a potential difference if P is to be different from zero. This potential difference results principally from magnetic forces exerted on the charges in the conductors of the rotor as they rotate through the magnetic field. These forces are an example of non-electrostatic forces, and give rise to a corresponding equivalent non-electrostatic field (concepts introduced in Sec. 28-4). In the present context the resulting electromotive force $\mathcal{E}$ is

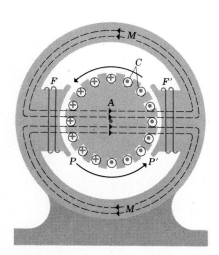

31-6   Schematic diagram of a dc motor.

called an *induced* emf or sometimes a *back* emf, referring to the fact that its sense is opposite to that of the current. Induced emf's resulting from motion of conductors in magnetic fields will be considered in greater detail in Chapter 33.

The situation is analogous to that of a battery being charged, as discussed in Sec. 28–4. In that case, electrical energy is converted to chemical rather than mechanical energy. But in either case, if the device has internal resistance r, then V_{ab} is greater than $\mathcal{E}$, the difference being the drop Ir across the internal resistance of the device. Thus, for either a motor or a battery being charged,

$$V_{ab} = \mathcal{E} + Ir. \qquad (31\text{--}16)$$

For a battery, $\mathcal{E}$ is often constant or nearly so; for a motor, $\mathcal{E}$ is *not* constant but depends on the speed of rotation of the rotor.

Example A dc motor with its rotor and field coils connected in series has an internal resistance of 2.0 Ω. When running at full load on a 120-V line, it draws a current of 4.0 A.

a) What is the emf in the rotor?

$$V_{ab} = \mathcal{E} + Ir,$$
$$120 \text{ V} = \mathcal{E} + (2.0 \ \Omega)(4.0 \text{ A}),$$
$$\mathcal{E} = 112 \text{ V}.$$

b) What is the power delivered to the motor?

$$P = IV_{ab} = (4.0 \text{ A})(120 \text{ V}) = 480 \text{ W}.$$

c) What is the rate of dissipation of energy in the motor?

$$P = I^2 r = (4.0 \text{ A})^2(2.0 \ \Omega) = 32 \text{ W}.$$

d) What is the mechanical power developed?

Mechanical power = total power − rate of dissipation of energy
$$= 480 \text{ W} - 32 \text{ W} = 488 \text{ W}.$$

The mechanical power may also be calculated from the relation

Mechanical power $= \mathcal{E}I = (112 \text{ V})(4.0 \text{ A}) = 448 \text{ W}.$

*31-6 The electromagnetic pump

Magnetic forces acting on conducting fluids provide a convenient means of pumping these fluids. An example is found in nuclear-reactor design; in some types of reactors, heat is transferred from the reactor core to the point where it is to be utilized, by a circulating flow of liquid metal. (Sodium, lithium, bismuth, and sodium-potassium alloys have been used.) The flow is maintained by an *electromagnetic pump*, one form of which is shown schematically in Fig. 31–7.

A current is sent transversely through the liquid metal, in a direction at right angles to a transverse magnetic field. The resulting side thrust on the current-carrying metal drives it along the pipe in which it is contained. The system is completely sealed and the only moving part is the metal itself.

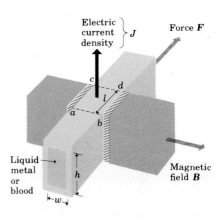

31-7 An electromagnetic pump.

Magnetic pumps are also finding a variety of applications in medical technology, particularly for pumping blood. Ordinary mechanical pumps with moving parts can damage blood cells. Blood contains ions, making it an electrical conductor, and magnetic pumping is possible. The pump is again completely sealed, reducing danger of contamination, and there are no moving parts except the blood itself. Magnetic pumps are now in use in some heart–lung machines and artificial kidney machines.

Questions

31-1 How might a loop of wire carrying a current be used as a compass? Could such a compass distinguish between north and south?

31-2 How could the direction of a magnetic field be determined by making only *qualitative* observations of the magnetic force on a straight wire carrying a current?

31-3 Do the currents in the electrical system of a car have a significant effect on a compass placed in the car?

31-4 A student claimed that, if lightning strikes a metal flagpole, the force exerted by the earth's magnetic field on the current in the pole can be large enough to bend it. Typical lightning currents are of the order of 10^4 to 10^5 A; is the student's opinion justified?

31-5 Is it possible that, in a Hall-effect experiment, *no* transverse potential difference will be observed? Under what circumstances might this happen?

31-6 Can a circular current loop in a uniform magnetic field ever experience a torque tending to rotate it about an axis perpendicular to the plane of the loop?

31-7 A student tried to make an electromagnetic compass by suspending a coil of wire from a thread (with the plane of the coil vertical) and passing a current through it. He expected the coil to align itself perpendicular to the horizontal component of the earth's magnetic field; but instead, the coil went into what appeared to be angular simple harmonic motion (cf. Sec. 11–7), turning back and forth past the expected direction. What was happening? Was the motion truly simple harmonic?

31-8 In the pivoted-coil galvanometer shown in Fig. 29–7, is the magnetic field uniform? Is Eq. (31–7) directly appli-

cable to this situation? If not, how *is* the torque on the coil determined?

31-9 Hall-effect voltages are much *larger* for relatively poor conductors such as germanium than for good conductors such as copper, for comparable currents, fields, and dimensions. Why?

31-10 When the polarity of the voltage applied to a dc motor is reversed, the direction of rotation *does not* reverse. Why not? How *could* the direction of rotation be reversed?

31-11 If an emf is produced in a dc motor, would it be possible to use the motor somehow as a *generator* or *source*, taking power out of it instead of putting power into it? How might this be done?

31-12 In an electromagnetic pump used to pump liquid sodium, as in Fig. 31–7, it was found that there was a difference in pressure between the top and bottom of the pipe passing through the pump. Gravity is one reason for this difference, but there is also an electromagnetic effect. What is it?

31-13 Sodium chloride (common salt) is not a good conductor, yet melted sodium chloride can be pumped by an electromagnetic pump. How is this possible?

31-14 It was noted in Chapter 30 that a magnetic force on a moving charge does no work on the charge. Yet work is required to push a fluid through a pipe, and so a magnetic pump must do work on the pumped fluid. How can this apparent inconsistency be resolved?

Problems

31-1 The cube in Fig. 31–8, 0.5 m on a side, is in a uniform magnetic field of 0.6 T, parallel to the x-axis. The wire *abcdef* carries a current of 4 A in the direction indicated. Determine the magnitude and direction of the force acting on the segments *ab*, *bc*, *cd*, *de*, and *ef*.

31-2 Figure 31–9 shows a portion of a silver ribbon with $z_1 = 2$ cm and $y_1 = 1$ mm, carrying a current of 200 A in

the positive x-direction. The ribbon lies in a uniform magnetic field, in the y-direction, of magnitude 1.5 T. If there are 7.4×10^{28} free electrons per m^3, find

a) the drift velocity of the electrons in the x-direction,

b) the magnitude and direction of the electric field in the z-direction due to the Hall effect, and

c) the Hall emf.

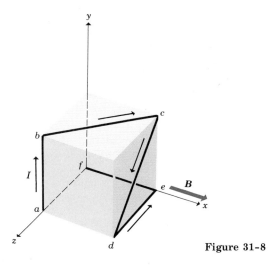

Figure 31-8

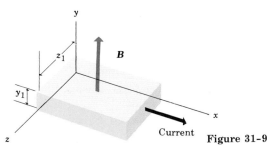

Current

Figure 31-9

31-3 Let Fig. 31-9 represent a strip of copper of the same dimensions as those of the silver ribbon of the preceding problem. When the magnetic field is 5 T and the current is 100 A, the Hall emf is found to be 45.4 μV. What is the density of free electrons in the copper?

31-4 The plane of a rectangular loop of wire 5 cm $\times$ 8 cm is parallel to a magnetic field of magnitude 0.15 T.

a) If the loop carries a current of 10 A, what torque acts on it?

b) What is the magnetic moment of the loop?

c) What is the maximum torque that can be obtained with the same total length of wire carrying the same current in this magnetic field?

31-5 An electromagnet produces a magnetic field of 1.2 T in a cylindrical region of radius 5 cm between its poles. A wire carrying a current of 20 A passes through this region, intersecting the axis of the cylinder and perpendicular to it. What force is exerted on the wire?

31-6 A wire 0.5 m long lies along the y-axis and carries a current of 10 A in the $+y$-direction. The magnetic field is uniform and has components $B_x = 0.3$ T, $B_y = -1.2$ T, and $B_z = 0.5$ T.

a) Find the components of force on the wire.

b) What is the magnitude of the total force on the wire?

31-7 A horizontal rod 0.2 m long is mounted on a balance and carries a current. In the vicinity of the rod is a uni-

form horizontal magnetic field of magnitude 0.05 T, perpendicular to the rod. The force on the rod is measured by the balance and is found to be 0.24 N. What is the current?

31-8 In Problem 31-7, suppose the magnetic field is horizontal but makes an angle of 30° with the rod. What is the current in the rod?

31-9 The rectangular loop in Fig. 31-10 is pivoted about the y-axis and carries a current of 10 A in the direction indicated.

a) If the loop is in a uniform magnetic field of magnitude 0.2 T, parallel to the x-axis, find the force on each side of the loop and the torque required to hold the loop in the position shown.

b) Same as (a) except that the field is parallel to the z-axis.

c) What torque would be required if the loop were pivoted about an axis through its center, parallel to the y-axis?

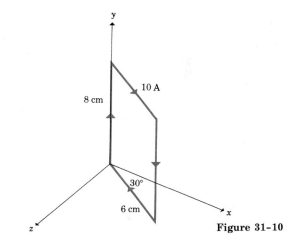

Figure 31-10

31-10 The rectangular loop of wire in Fig. 31-11 has a mass of 0.1 g per centimeter of length, and is pivoted about

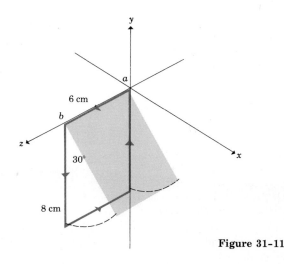

Figure 31-11

side ab as a frictionless axis. The current in the wire is 10 A in the direction shown.

a) Find the magnitude and sense of the magnetic field, parallel to the y-axis, that will cause the loop to swing up until its plane makes an angle of 30° with the yz-plane.

b) Discuss the case where the field is parallel to the x-axis.

31-11 What is the maximum torque on a coil 5×12 cm, of 600 turns, when carrying a current of 10^{-5} A in a uniform field of magnitude 0.1 T?

31-12 The coil of a pivoted-coil galvanometer has 50 turns and encloses an area of 6 cm². The magnetic field in the region in which the coil swings is 0.01 T and is radial. The torsional constant of the hairsprings is

$$k' = 1.0 \times 10^{-8} \text{ N·m·(degree)}^{-1}.$$

Find the angular deflection of the coil for a current of 1 mA.

31-13 A circular coil of wire 8 cm in diameter has 12 turns and carries a current of 5 A. The coil is in a region where the magnetic field is 0.6 T.

a) What is the maximum torque on the coil?

b) In what position would the torque be one-half as great as in (a)?

31-14 In a shunt-wound dc motor (Fig. 31–12), the resistance R_f of the field coils is 150 Ω and the resistance R_f of the rotor is 2 Ω. When a difference of potential of 120 V is applied to the brushes, and the motor is running at full

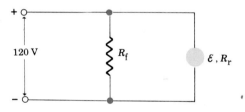

Figure 31-12

speed delivering mechanical power, the current supplied to it is 4.5 A.

a) What is the current in the field coils?

b) What is the current in the rotor?

c) What is the emf developed by the motor?

d) How much mechanical power is developed by this motor?

31-15 Figure 31–12 is a diagram of a shunt-wound dc motor, operating from a 120-V dc power line. The resistance of the field windings. R_f, is 240 Ω. The resistance of the rotor, R_r, is 3 Ω. When the motor is running the rotor develops an emf $\mathcal{E}$. The motor draws a current of 4.5 A from the line. Compute

a) the field current,

b) the rotor current,

c) the emf $\mathcal{E}$,

d) the rate of development of heat in the field windings,

e) the rate of development of heat in the rotor,

f) the power input to the motor, and

g) the efficiency of the motor,

if friction and windage losses amount to 50 W.

31-16 A horizontal tube of rectangular cross section (height h, width w) is placed at right angles to a uniform magnetic field of induction B, so that a length l is in the field. (See Fig. 31–7.) The tube is filled with liquid sodium and an electric current of density J is maintained in the third mutually perpendicular direction.

a) Show that the difference of pressure between a point in the liquid on a vertical plane through ab (Fig. 31–7) and a point in the liquid on another vertical plane through cd, under conditions in which the liquid is prevented from flowing, is

$$\Delta p = JlB.$$

b) What current density would be needed to provide a pressure difference of 1 atm between these two points if $B = 1$ T and $l = 0.1$ m?

MAGNETIC FIELD OF A CURRENT 32

32-1 Sources of magnetic field

In Chapters 30 and 31 we discussed one aspect of the magnetic interaction of moving charges; we took the existence of the magnetic field as a given fact and studied the *forces* exerted by the magnetic field on moving charges and on currents in conductors. We now return to the other aspect of this interaction, the principles governing the *creation* of magnetic fields. As noted in the beginning of Chapter 30, the fundamental source of magnetic field is electric charge in motion; in many cases this takes the form of electric current in a conductor.

The first recorded observations of magnetic fields set up by currents were those of Oersted, who discovered that a pivoted compass needle, beneath a wire in which there was a current, set itself with its long axis perpendicular to the wire. Later experiments by Biot and Savart, and by Ampère, led to a relation which can be used to compute the magnetic field at any point in space around a circuit in which there is a current.

32-2 Magnetic field of a moving charge

We consider first the **B** field produced by a single moving point charge, as in Fig. 32-1. Experiment shows that this field is proportional to the charge q and to its speed v, and inversely proportional to the square of the distance r from the charge. The direction of the magnetic field at a point P is *not* along the line from the charge to P, but rather is perpendicular to the plane containing this line and the velocity vector **v**. Furthermore, the field magnitude is proportional to the sine of the angle θ between these two directions. Thus the magnitude of the magnetic field at point P is given by

$$B = k'\frac{qv\sin\theta}{r^2}, \qquad (32\text{-}1)$$

where k' is a proportionality constant with a numerical value that depends on the system of units being used.

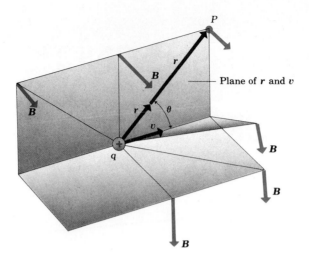

32-1 Magnetic field vectors due to a moving positive point charge q.

The magnitude and direction of $\boldsymbol{B}$ can be incorporated into a single vector equation by use of the vector product. We introduce a *unit vector* $\widehat{r} = \boldsymbol{r}/r$, pointing in the direction from charge q to point P. Then the above description is summarized by the equation

$$B = k'\frac{q\boldsymbol{v} \times \widehat{r}}{r^2}. \qquad (32\text{-}2)$$

Figure 32–1 shows the magnetic field at several points in the vicinity of the charge. The field is zero at all points along a line through the charge in the direction of the velocity $\boldsymbol{v}$, since at all such points $\sin\theta = 0$. At a given distance r from the charge, $\boldsymbol{B}$ has maximum magnitude at points lying in the plane through the charge perpendicular to $\boldsymbol{v}$, since at all such points $\theta = 90°$ and $\sin\theta = 1$. The charge also produces an *electric* field in its vicinity, just as when it is at rest; the electric-field vectors are not shown in the figure.

The field lines describing the *electric* field of a point charge radiate outward from the charge, but the *magnetic* field lines are completely different in character; they are *circles* in planes perpendicular to $\boldsymbol{v}$, with centers along the line of $\boldsymbol{v}$.

As discussed in Sec. 30–2, the unit of B is

$$1\,\text{T} = 1\,\text{N·s·C}^{-1}\text{·m}^{-1} = 1\,\text{N·A}^{-1}\text{·m}^{-1}.$$

Using this with Eq. (32–1) or (32–2) and solving for k', we find that the unit of the magnetic constant k' is

$$1\,\text{N·s}^2\text{·C}^{-2} = 1\,\text{N·A}^{-2} = 1\,\text{Wb·A}^{-1}\text{·m}^{-1}$$
$$= 1\,\text{T·A}^{-1}\text{·m}.$$

In SI units, the numerical value of k' is arbitrarily assigned to be *exactly* 10^{-7}. Thus

$$k' = 10^{-7}\,\text{N·s}^2\text{·C}^{-2} = 10^{-7}\,\text{N·A}^{-2}$$
$$= 10^{-7}\,\text{Wb·A}^{-1}\text{·m}^{-1} = 10^{-7}\,\text{T·A}^{-1}\text{·m} \qquad \text{(exactly).}$$

The reason for this choice of the numerical value of k' will be explained later in the chapter.

It will be recalled that the electrical constant k has the value 8.98755×10^9 N·m²·C⁻². The ratio k/k' is therefore

$$\frac{k}{k'} = \frac{8.98755 \times 10^9 \text{ N·m}^2\text{·C}^{-2}}{10^{-7} \text{ N·s}^2\text{·C}^{-2}}$$

$$= 8.98755 \times 10^{16} \text{ m}^2\text{·s}^{-2},$$

which is equal to the square of the speed of light, c. This suggests that there may be a close relation between electricity, magnetism, and light. We shall return to this question in Chapter 37.

In electrostatics we found it convenient to express electric-field relations not in terms of the constant k in Eq. (24-1) but in terms of ϵ_0, where $k = 1/4\pi\epsilon_0$. Similarly, in magnetic-field relations it is convenient to introduce a constant μ_0, defined by the relation

$$k' = \frac{\mu_0}{4\pi}.$$

Thus

$$\mu_0 = 4\pi \times 10^{-7} \text{ Wb·A}^{-1}\text{·m}^{-1} = 4\pi \times 10^{-7} \text{ T·A}^{-1}\text{·m}.$$

In terms of μ_0, Eqs. (32-1) and (32-2) become

$$B = \frac{\mu_0}{4\pi} \frac{qv\sin\theta}{r^2}, \tag{32-3}$$

$$B = \frac{\mu_0}{4\pi} \frac{qv \times \hat{r}}{r^2}. \tag{32-4}$$

We can now write the expression for the magnetic force between two point charges, both of which are in motion relative to an observer. Thus a charge q', moving with velocity v' in a magnetic field **B**, experiences a force

$$\boldsymbol{F} = q'\boldsymbol{v}' \times \boldsymbol{B,}$$

and if the field **B** is set up by a charge q moving with velocity v,

$$F = \frac{\mu_0}{4\pi} \frac{qq'v' \times (v \times \hat{r})}{r^2}, \tag{32-5}$$

which corresponds to Coulomb's law for the electrical force between the charges.

As noted above, the constants k and k' are related to the speed of light c. The corresponding relation for ϵ_0 and μ_0, as the reader may verify, is

$$\epsilon_0\mu_0 = \frac{1}{c^2}. \tag{32-6}$$

The *principle of superposition,* which has been used to calculate the total electric field from several charges, is found to be valid also for magnetic-field calculations. *The total magnetic field caused by several moving charges is the vector sum of the fields caused by the individual charges.* This fact makes it possible to calculate the field caused by a current in a circuit in terms of fields caused by individual segments of the conductors. This useful relationship is developed in the next section.

32-3 Magnetic field of a current element. The Biot law

The magnetic field set up at any point by the current in a conductor is the resultant (vector sum) of the fields due to all of the moving charges in the conductor. The conductor is to be divided, in imagination, into short segments of length dl, one of which is shown in Fig. 32-2. The volume of a segment equals $A\,dl$, where A is its cross-section area. If there are n charges q per unit volume, the total moving charge dQ in the element is

$$dQ = nqA\,dl.$$

The moving charges are therefore equivalent to a single charge dQ, traveling with a velocity equal to the *drift* velocity v. (Fields due to the *random* velocities of the carriers will, on the average, cancel out at every point.) From Eq. (32-3), the magnitude of dB at any point is

$$dB = \frac{\mu_0}{4\pi}\frac{dQ\,v\sin\theta}{r^2} = \frac{\mu_0}{4\pi}\frac{nqvA\,dl\sin\theta}{r^2}.$$

But $(nqvA)$ equals the current I in the element, so

$$dB = \frac{\mu_0}{4\pi}\frac{I\,dl\sin\theta}{r^2}, \tag{32–7}$$

or in vector form,

$$d\mathbf{B} = \frac{\mu_0}{4\pi}\frac{I\,d\mathbf{l}\times\hat{\mathbf{r}}}{r^2}, \tag{32–8}$$

where $d\mathbf{l}$ is a vector of length dl.

As shown in Fig. 32-2, the field vectors $d\mathbf{B}$ are exactly like those set up by a finite positive charge Q, moving in the direction of the drift velocity v.

The resultant $\mathbf{B}$ at any point in space, due to the current in a complete circuit, is the *vector integral* of the values of $d\mathbf{B}$ due to all elements of the circuit. Thus

$$\mathbf{B} = \frac{\mu_0}{4\pi}\int\frac{I\,d\mathbf{l}\times\hat{\mathbf{r}}}{r^2}. \tag{32–9}$$

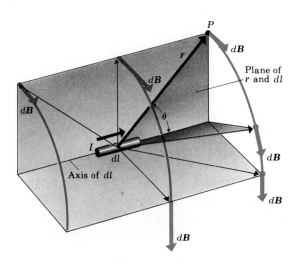

32-2 Magnetic field vectors due to a current element.

It should be pointed out that it is impossible to verify Eq. (32–8) experimentally, since one can never obtain an isolated element of a current-carrying circuit. Only the *resultant* field B, given by the integral in Eq. (32–9), can be measured experimentally. Equation (32–8) was *deduced* originally by the French physicist Jean Biot, in 1820, from experimental studies of the magnetic fields around circuits of various shapes. It is known as the *Biot law* (pronounced "bee-OH").

If there is matter in the space around a circuit, the field at a point will be due not entirely to currents in conductors, but in part to the *magnetization* of this matter. The magnetic properties of matter are discussed more fully in Chapter 35. However, unless iron or some other ferromagnetic material is present, this effect is so small that, while strictly speaking Eq. (32–9) holds only for conductors in vacuum, as a practical matter it can be used without correction for conductors in air, or in the vicinity of any nonferromagnetic material.

32-4 Magnetic field of a long straight conductor

Let us use the Biot law to compute the magnetic field B at point P in Fig. 32–3 due to a long straight conductor carrying a current I. We consider a conductor of total length $2L$, with its midpoint at the origin. The various factors in Eq. (32–9) are:

$$dl = dy\,j,$$
$$\hat{r} = \frac{r}{r} = \frac{-y\,j + z\,k}{\sqrt{y^2 + z^2}},$$
$$r^2 = y^2 + z^2.$$

Thus for our problem Eq. (32–9) takes the form

$$B = \frac{\mu_0}{4\pi} \int_{-L}^{L} \frac{I(dy\,j) \times (-y\,j + z\,k)}{(y^2 + z^2)^{3/2}}$$
$$= \frac{\mu_0 I}{4\pi} \int_{-L}^{L} \frac{z\,dy}{(y^2 + z^2)^{3/2}}\,i. \qquad (32\text{–}10)$$

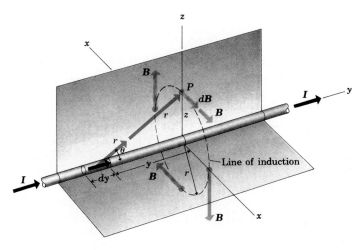

Integrating by trigonometric substitution or use of a table of integrals, we find

$$\boldsymbol{B} = \frac{\mu_0 I}{4\pi} \frac{2L}{z\sqrt{L^2 + z^2}} \boldsymbol{i}. \tag{32-11}$$

Or Eq. (32–7) may be used, with $dl = dy$, $\sin\theta = z/\sqrt{y^2 + z^2}$, and $r^2 = y^2 + z^2$, to obtain

$$B = \frac{\mu_0 I}{4\pi} \int_{-L}^{L} \frac{z\, dy}{(y^2 + z^2)^{3/2}},$$

which is the same integral as Eq. (32–10).

When the length of the conductor is very large compared to its distance from point P, it may be considered to be infinitely long. In Eq. (32–11) the quantity $\sqrt{L^2 + z^2}$ becomes approximately equal to L, and in the limit as $L \to \infty$, Eq. (32–11) becomes

$$\boldsymbol{B} = \frac{\mu_0 I}{2\pi z}\boldsymbol{i} \qquad \text{(long straight wire)}. \tag{32-12}$$

Because the physical situation has axial symmetry about the y-axis, B has the same *magnitude* at all points on a circle centered on the conductor and lying in a plane perpendicular to the conductor, and its *direction* is everywhere tangent to such a circle, as shown in Fig. 32–3. Thus, at all points on a circle of radius r around the conductor (shown as a broken line in the figure), the magnitude B is given by

$$B = \frac{\mu_0 I}{2\pi r} \qquad \text{(long straight wire)}. \tag{32-13}$$

This relation was deduced from experimental observations by Biot and Savart before the differential form, Eq. (32–8), had been discovered. It is called the *Biot–Savart law*, although this name is sometimes applied also to Eqs. (32–8) and (32–9).

Example   A long straight conductor carries a current of 100 A. At what distance from the conductor is the magnetic field caused by the current equal in magnitude to the earth's magnetic field in Pittsburgh (about 0.5×10^{-4} T)?

Solution   We use Eq. (32–13); everything except r is known, so we solve for r and insert the appropriate numbers:

$$r = \frac{\mu_0 I}{2\pi B} = \frac{(4\pi \times 10^{-7}\,\text{T·m·A}^{-1})(100\,\text{A})}{(2\pi)(0.5 \times 10^{-4}\,\text{T})} = 0.4\,\text{m}.$$

At smaller distances the field becomes stronger; when $r = 0.2$ m, $B = 1.0 \times 10^{-4}$ T, and so on.

A portion of the magnetic field around a long straight conductor is shown in Fig. 32–4. The shape of the magnetic-field lines in this situation is completely different from that of the electric-field lines in the analogous electrical situation. Electric-field lines radiate outward from the charges that are their sources (or inward for negative charges). By contrast, magnetic-field lines *encircle* the current that acts as their source.

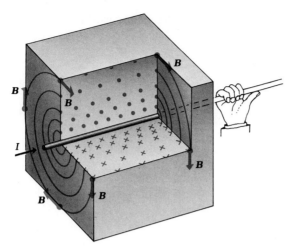

32-4 Magnetic field around a long straight conductor. The field lines are circles, with directions determined by the righthand rule.

Electric-field lines begin and end at charges, while magnetic-field lines *never* have endpoints, irrespective of the shape of the conductor which sets up the field.

If an imaginary closed surface is constructed in a magnetic field, no field line can start or end inside this surface. Thus the number of lines emerging from the surface must equal the number entering it. We have shown that the number of lines crossing a surface is proportional to the flux Φ across the surface. Hence in a magnetic field, the flux across any *closed* surface is zero, or

$$\oint B_\perp \, dA = \oint \boldsymbol{B} \cdot d\boldsymbol{A} = 0. \qquad (32\text{-}14)$$

This should be compared with Gauss's law for electrostatic fields, in which the surface integral of $\boldsymbol{E}$ over a closed surface equals $1/\epsilon_0$ times the enclosed charge. Thus Eq. (32-14) can be interpreted as saying that there is no such thing as "magnetic charge" to act as a source of $\boldsymbol{B}$. The sources of $\boldsymbol{B}$ are moving electric charges, as outlined above.

Finally, we note that the direction of the $\boldsymbol{B}$ lines around a straight conductor is given by a righthand rule: Grasp the conductor, with the thumb extended in the direction of the current; the fingers then curl around the conductor in the direction of the $\boldsymbol{B}$ lines. This rule is illustrated in Fig. 32-4.

32-5 Force between parallel conductors; the ampere

Figure 32-5 shows portions of two long straight parallel conductors separated by a distance a and carrying currents I and I', respectively, in the same direction. Since each conductor lies in the magnetic field set up by the other, each experiences a force. The diagram shows some of the field lines set up by the current in the lower conductor.

From Eq. (32-13), the magnitude of the $\boldsymbol{B}$-vector at the upper conductor is

$$B = \frac{\mu_0 I}{2\pi r}.$$

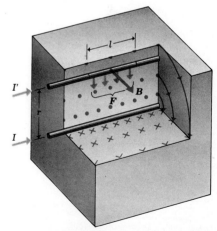

32-5 Parallel conductors carrying currents in the same direction attract each other.

From Eq. (31–1), the force on a length l of the upper conductor is

$$F = IlB = \frac{\mu_0 I I' l}{2\pi r},$$

and the force *per unit length* is therefore

$$\frac{F}{l} = \frac{\mu_0 I I'}{2\pi r}. \tag{32–15}$$

The righthand rule shows that the direction of the force on the upper conductor is downward. There is an equal and opposite force per unit length on the lower conductor, as may be seen by considering the field set up by the upper conductor. Hence, the conductors *attract* one another.

If the direction of either current is reversed, the forces reverse also. Parallel conductors carrying currents in *opposite* directions *repel* one another.

The fact that two straight parallel conductors exert forces of attraction or repulsion on one another is made the basis of the SI definition of the ampere. The ampere is defined as follows:

> *One ampere is that unvarying current which, if present in each of two parallel conductors of infinite length and one meter apart in empty space, causes each conductor to experience a force of exactly* 2×10^{-7} *newtons per meter of length.*

It follows from this and the preceding equation that, by definition, the constant μ_0 is *exactly* $4\pi \times 10^{-7} \, \text{N} \cdot \text{A}^{-2}$, as stated in Sec. 32–2.

From the definition above, the ampere can be established, in principle, with the help of a meter stick and a spring balance. For the practical standardization of the ampere, coils of wire are used instead of straight wires, and their separation is made only a few centimeters. The complete instrument, which is capable of measuring currents with a high degree of precision, is called a *current balance.*

Having defined the ampere, we can now define the coulomb as *the quantity of charge that in one second crosses a section of a circuit in which there is a constant current of one ampere.*

Mutual forces of attraction exist not only between *wires* carrying currents in the same direction, but between each of the longitudinal elements into which a single current-carrying conductor may be subdivided. If the conductor is a liquid or an ionized gas (a plasma), these forces result in a constriction of the conductor as if its surface were acted on by an external, inward, pressure force. The constriction of the conductor is called the *pinch effect,* and attempts are being made to utilize the high temperature produced by the pinch effect in a plasma to bring about nuclear fusion.

32–6 Magnetic field of a circular loop

In many devices in which a current is used to establish a magnetic field, as in an electromagnet or a transformer, the wire carrying the current is wound into a *coil,* often consisting of many circular loops. We therefore consider next the magnetic field set up by a single circular loop of wire carrying a current. Figure 32–6 shows a circular loop of wire of radius a,

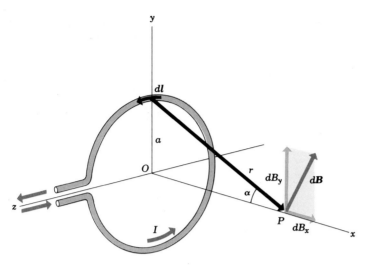

32-6 Magnetic field of a circular loop. The segment dl causes the field $d\boldsymbol{B}$, lying in the xy-plane. Other dl's have different components perpendicular to the x-axis; these add to zero, while the x-components combine to give the total $\boldsymbol{B}$ field at point P.

carrying a current I that is led into and out of the loop through two long straight wires side by side. The currents in the straight wires are in opposite directions and annul each other's magnetic effects.

We may use the law of Biot, Eq. (32–9), to calculate the magnetic field at a point P along a line perpendicular to the plane of the loop, a distance x from its center. As the figure shows, dl and r are perpendicular, and the direction of the field $d\boldsymbol{B}$ caused by element dl lies in the xy-plane. Also, $r = (x^2 + a^2)^{1/2}$. Thus for the element dl we find

$$dB = \frac{\mu_0 I}{4\pi} \frac{dl}{x^2 + a^2}. \tag{32–16}$$

In terms of components,

$$dB_x = dB \sin \alpha = \frac{\mu_0 I}{4\pi} \frac{dl}{x^2 + a^2} \cdot \frac{a}{(x^2 + a^2)^{1/2}}, \tag{32–17}$$

$$dB_y = dB \cos \alpha = \frac{\mu_0 I}{4\pi} \frac{dl}{x^2 + a^2} \cdot \frac{x}{(x^2 + a^2)^{1/2}}. \tag{32–18}$$

Because of the symmetry about the x-axis, the components of $\boldsymbol{B}$ perpendicular to this axis must add to zero. In fact, an element dl on the opposite side of the loop, having direction opposite to that of the dl shown, gives a contribution to the field having the same x-component as given by Eq. (32–17) but an *opposite y-component*. Similarly, the components of $\boldsymbol{B}$ perpendicular to the x-axis set up by pairs of diametrically opposite elements cancel, leaving only the x-components. The total x-component is obtained by integrating Eq. (32–17); everything is constant in this integration except dl, and the integral of dl for the entire loop is simply the circumference of the circle, $2\pi a$. Thus we obtain

$$B_x = \int \frac{\mu_0 I}{4\pi} \cdot \frac{a \, dl}{(x^2 + a^2)^{3/2}} = \frac{\mu_0 I a}{4\pi(x^2 + a^2)^{3/2}} \int dl,$$

or

$$B_x = \frac{\mu_0 I a^2}{2(x^2 + a^2)^{3/2}} \qquad \text{(circular loop)}. \tag{32–19}$$

Equations (32–17) and (32–18) may be obtained somewhat more compactly by use of unit vectors. We write $dl = dl\,\boldsymbol{k}$, $\boldsymbol{r} = x\boldsymbol{i} - a\boldsymbol{j}$, and

$$\frac{dl \times \hat{\boldsymbol{r}}}{r^2} = \frac{dl \times \boldsymbol{r}}{r^3} = \frac{(dl\,\boldsymbol{k}) \times (x\boldsymbol{i} - a\boldsymbol{j})}{(x^2 + a^2)^{3/2}}$$

$$= \frac{x\,dl\,\boldsymbol{j} + a\,dl\,\boldsymbol{i}}{(x^2 + a^2)^{3/2}}.$$

As before, the components perpendicular to the x-axis add to zero, and only the term containing $\boldsymbol{i}$ survives. Using the above expression in Eq. (32–9) and again using the fact that $\int dl = 2\pi a$, we obtain

$$\boldsymbol{B} = \int \frac{\mu_0 I}{4\pi} \frac{a\,dl\,\boldsymbol{i}}{(x^2 + a^2)^{3/2}} = \frac{\mu_0 I a^2 \boldsymbol{i}}{2(x^2 + a^2)^{3/2}}, \tag{32–20}$$

which is equivalent to Eq. (32–19).

At the center of the loop, $x = 0$, and Eq. 32–19 reduces to

$$B_x = \frac{\mu_0 I}{2a} \qquad \text{(center of circular loop)}. \tag{32–21}$$

If, instead of a single loop as in Fig. 32–6, one has a coil of N closely spaced loops, all having the same radius, each loop contributes equally to the field, and Eq. (32–21) becomes

$$B_x = \frac{\mu_0 N I}{2a} \qquad \text{(N circular loops)}. \tag{32–22}$$

Similarly, Eqs. (32–19) and (32–20) may be adapted to the case of N loops by the insertion of a factor N.

Some of the magnetic-field lines surrounding a circular loop and lying in planes through the axis are shown in Fig. 32–7. Again we see that the field lines encircle the conductor, and that their directions are given by the righthand rule, as for the long straight conductor. The field lines for the circular loop are *not* circles, but they are closed curves that link the conductor.

It was pointed out in Chapter 31 that a current-carrying conducting loop of *any* shape is acted on by a *torque* when placed in a magnetic field

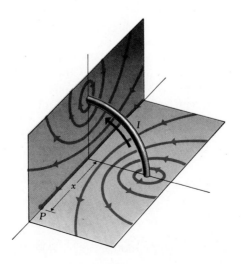

32-7 Field lines surrounding a circular loop.

caused by sources other than the loop. The position of stable equilibrium is one in which the plane of the loop is perpendicular to the externally caused field. Now we see that this position is such that *within the area enclosed by the loop, the **B** field produced by the loop itself is in the same direction as that of the external field.* In other words, a loop, if free to turn, will set itself in such a plane that the flux passing through the area enclosed by it has its *maximum* possible value. This is found to be true in all instances and is a useful general principle. For example, if a current is sent through an irregular loop of flexible wire in an external magnetic field, the loop assumes a circular form with its plane perpendicular to the field and with its own flux adding to that of the field. The same conclusion can, of course, be drawn by analyzing the side thrusts on the elements of the conductor.

32-7 Ampère's law

Ampère's law is a useful relation that is analogous to Gauss's law. The latter, it will be recalled, is a relation between the integral of the normal component of electric field over a closed surface and the net electric charge enclosed by the surface. Ampère's law is a relation between the *tangential* component of *magnetic* field at points on a closed curve and the net current through the *area* bounded by the curve.

Ampère's law is formulated in terms of the *line integral* of **B** around a closed path, denoted by

$$\oint \boldsymbol{B} \cdot d\boldsymbol{l}.$$

This is the same sort of integral used to define work in Chapter 6 and electric potential in Chapter 26. We divide the path into infinitesimal segments $d\boldsymbol{l}$ and for each one calculate the scalar product of **B** and $d\boldsymbol{l}$. In general **B** varies from point to point, and the **B** at the location of each $d\boldsymbol{l}$ must be used. An alternative notation is $\int B_\parallel dl$, where $B_\parallel$ is the component of **B** parallel to $d\boldsymbol{l}$ at each point. The circle on the integral sign indicates that this integral will always be computed for a *closed* path whose beginning and end points are the same. As with Gauss's law, the path need not be an actual physical object; usually it is an imaginary curve constructed to permit the application of Ampère's law to a specific situation.

We consider first a long, straight conductor carrying a current I, passing through the center of a circle of radius r in a plane perpendicular to the conductor. In Fig. 32–3, such a circle is shown as a broken line. From Eq. (32–13), the field has magnitude $\mu_0 I / 2\pi r$ at every point on the circle, and it is tangent to the circle at each point. Thus for every point on the circle $B_\parallel = B = \mu_0 I / 2\pi r$. The line integral of **B** around the circle is

$$\oint \boldsymbol{B} \cdot d\boldsymbol{l} = \oint \frac{\mu_0 I}{2\pi r} dl = \frac{\mu_0 I}{2\pi r} \oint dl,$$

or, since $\oint dl$ is just the circumference ($2\pi r$) of the circle,

$$\oint \boldsymbol{B} \cdot d\boldsymbol{l} = \mu_0 I. \tag{32-23}$$

That is, the line integral of B is equal to μ_0 multiplied by the current passing through the area bounded by the circle.

This result may also be derived for a more general integration path, such as the one in Fig. 32-8. At the position of the line element dl, the angle between dl and B is θ, and

$$\boldsymbol{B}\cdot d\boldsymbol{l} = B\,dl\cos\theta.$$

From the figure, $dl\cos\theta = r\,d\phi$, where $d\phi$ is the angle subtended by dl at the position of the conductor. Thus

$$\oint \boldsymbol{B}\cdot d\boldsymbol{l} = \oint \left(\frac{\mu_0 I}{2\pi r}\right)(r\,d\phi) = \frac{\mu_0 I}{2\pi}\oint d\phi.$$

But $\oint d\phi$ is just the total angle swept out by the radial line from the conductor to dl during a complete trip around the path, that is, 2π. Thus

$$\oint \boldsymbol{B}\cdot d\boldsymbol{l} = \mu_0 I. \tag{32-24}$$

The line integral does not depend on the shape of the path, or on the position of the wire within it. If the current in the wire were opposite to that shown, the integral would have the opposite sign. Hence if any number of long straight conductors pass through the surface bounded by the path, the line integral equals μ_0 times the *algebraic sum* of the currents. The positive direction for current through the surface is determined by the direction in which the path is traversed, in accordance with the usual righthand rule. Thus in Fig. 32-8, a current directed into the plane of the figure is considered positive, and one out of the figure is negative.

If a closed path does not encircle the wire (or if a wire lies outside the path), the line integral of the B field of that wire is zero, because the angle ϕ then has the same value at the start and finish of any round trip. Hence if there are other conductors present that do not pass through a given path, they may contribute to the value of B at every point, but the *line integrals* of their fields are zero.

It follows that if we interpret I in Eq. (32-24) to mean the *algebraic sum* of the currents across the area bounded by a closed path, this equation implies all of the statements above and is the analytic form of Ampère's law:

The line integral of the magnetic field B around any closed path is equal to μ_0 times the net current across the area bounded by the path.

Although derived only for the special case of the field of a number of long straight parallel conductors, the law is true for conductors and paths of any shape. The general derivation is no different in principle from that above, but it is complicated geometrically and will not be given.

Finally, we emphasize that the statement $\oint \boldsymbol{B}\cdot d\boldsymbol{l} = 0$ *does not* necessarily mean that $\boldsymbol{B} = \boldsymbol{0}$ everywhere along the path, but only that no current is linked by the path.

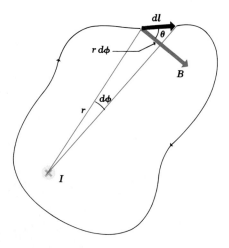

32-8 Line integral of B around a closed path linking a long straight conductor, carrying current I *into* the plane of the figure.

32-8 Applications of Ampère's law

In some cases of practical importance, symmetry considerations make it possible to use Ampère's law to compute the magnetic field caused by a certain current-carrying conductor. A few guiding principles, analogous to those stated for Gauss's law at the end of Sec. 25–4, are the following:

1. If **B** is everywhere tangent to the integration path and has the same magnitude B at every point on the path, then its line integral is equal to B multiplied by the circumference of the path.

2. If **B** is everywhere perpendicular to the path, for all or some portion of the path, that portion of the path makes no contribution to the line integral.

3. In the integral $\oint \boldsymbol{B} \cdot d\boldsymbol{l}$, **B** is always the *total* magnetic field at each point on the path. In general this field is caused partly by currents linked by the path and partly by currents outside. Even when *no* current is linked by the path, the field at points on the path need not be zero. In that case, however, $\oint \boldsymbol{B} \cdot d\boldsymbol{l}$ is always zero.

4. Some judgment is required in choosing an integration path. Two useful guiding principles are that the point or points at which the field is to be determined must lie on the path, and that the path must have enough symmetry so that the integral can be evaluated.

1. Field of a solenoid. A solenoid is constructed by winding wire in a helix around the surface of a cylindrical form, usually of circular cross section. The turns of the winding are ordinarily closely spaced and may consist of one or more layers. For simplicity, we have represented a solenoid in Fig. 32–9 by a relatively small number of circular turns, each carrying a current I. The resultant field at any point is the vector sum of the **B**-vectors due to the individual turns. The diagram shows the field lines in the xy- and yz-planes. Exact calculations show that for a long, closely wound solenoid, half of the lines passing through a cross section at the center emerge from the ends and half "leak out" through the windings between center and end.

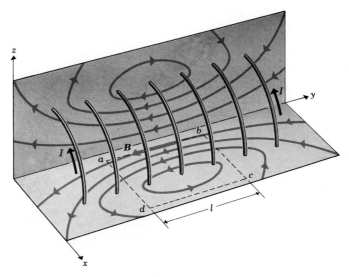

32-9 Magnetic-field lines surrounding a solenoid. The dashed rectangle $abcd$ is used to compute the flux density **B** in the solenoid from Ampère's law.

If the length of the solenoid is large compared with its cross-sectional diameter, the *internal* field near its center is very nearly uniform and parallel to the axis, and the *external* field near the center is very small. The internal field at or near the center can then be found by use of Ampère's law.

We select as a closed path the dashed rectangle $abcd$ in Fig. 32–9. Side ab, of length l, is parallel to the axis of the solenoid. Sides bc and da are to be taken very long so that side cd is far from the solenoid and the field at this side is negligibly small.

By symmetry, the $\boldsymbol{B}$ field along side ab is parallel to this side and is constant, so that for this side $B_{\parallel} = B$ and

$$\int \boldsymbol{B} \cdot d\boldsymbol{l} = Bl.$$

Along sides bc and da, $B_{\parallel} = 0$ since B is perpendicular to these sides; and along side cd, $B_{\parallel} = 0$ also since $B = 0$. The sum around the entire closed path therefore reduces to Bl.

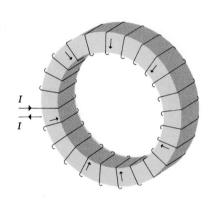

Let n be the number of turns *per unit length* in the windings. The number of turns in length l is then nl. Each of these turns passes once through the rectangle $abcd$ and carries a current I, where I is the current in the windings. The total current through the rectangle is then nlI and, from Ampère's law,

$$Bl = \mu_0 nlI,$$
$$\boldsymbol{B} = \mu_0 n\, I \quad \text{(solenoid).} \tag{32–25}$$

Since side ab need not lie on the axis of the solenoid, the field is uniform over the entire cross section.

2. Field of a toroidal solenoid. Figure 32–10a represents a *toroid,* wound with wire carrying a current I. The black lines in Fig. 32–10b are paths to which we wish to apply Ampère's law. Consider first path 1. By symmetry, if there is any field at all in this region, it will be *tangent* to the path at all points, and $\oint \boldsymbol{B} \cdot d\boldsymbol{l}$ will equal the product of B and the circumference $l = 2\pi r$ of the path. The current through the path, however, is zero, and hence, from Ampère's law (since the circumference is *not* zero), the field $\boldsymbol{B}$ must be zero.

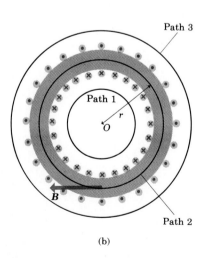

(b)

Similarly, if there is any field at path 3, it will also be tangent to the path at all points. Each turn of the winding passes *twice* through the area bounded by this path, carrying equal currents in opposite directions. The *net* current through the area is therefore zero, and hence $B = 0$ at all points of the path. *The field of the toroidal solenoid is therefore confined wholly to the space enclosed by the windings!* The toroid may be thought of as a solenoid that has been bent into a circle.

32-10 (a) A toroidal winding. (b) Closed paths (black circles) used to compute the magnetic field $\boldsymbol{B}$ set up by a current in a toroidal winding. The field is very nearly zero at all points except those within the space enclosed by the windings.

Finally, we consider path 2, a circle of radius r. Again by symmetry, the $\boldsymbol{B}$ field is tangent to the path, and $\int \boldsymbol{B} \cdot d\boldsymbol{l}$ equals $2\pi rB$. Each turn of the winding passes *once* through the area bounded by path 2, and the total current through the area is NI, where N is the *total* number of turns in the winding. Then, from Ampère's law,

$$2\pi rB = \mu_0 NI,$$

and

$$B = \frac{\mu_0 NI}{2\pi r} \qquad \text{(toroidal solenoid).} \qquad (32\text{–}26)$$

The magnetic field is *not* uniform over a cross section of the core, because the path length l is larger at the outer side of the section than at the inner side. However, if the radial thickness of the core is small compared with the toroid radius r, the field varies only slightly across a section. In that case, considering that $2\pi r$ is the circumferential length of the toroid and that $N/2\pi r$ is the number of turns per unit length n, the field may be written

$$B = \mu_0 nI,$$

just as at the center of a long *straight* solenoid.

The equations derived above for the field in a closely wound straight or toroidal solenoid are strictly correct only for windings in *vacuum*. For most practical purposes, however, they can be used for windings in air, or on a core of any nonferromagnetic material. We shall show in Chapter 35 how they are modified if the core is of iron.

32-9 Magnetic fields and displacement currents

When the Biot and Savart law was discussed in Sec. 32-3, it was assumed that the current I in an element $d\boldsymbol{l}$ was a *conduction* current. Experiment shows, however, that a *displacement* current I_D (Sec. 29-6) contributes to a magnetic field in exactly the same way as a conduction current, I_C, so the general form of the Biot law is

$$d\boldsymbol{B} = \frac{\mu_0}{4\pi} \frac{(I_C + I_D)\, d\boldsymbol{l} \times \hat{\boldsymbol{r}}}{r^2}, \qquad (32\text{–}27)$$

and the general form of Ampère's law is

$$\oint \boldsymbol{B} \cdot d\boldsymbol{s} = \mu_0 (I_C + I_D). \qquad (32\text{–}28)$$

Note that in the Biot law, the symbols I_C and I_D refer to the currents in an element $d\boldsymbol{l}$, while in Ampère's law they refer to the *total* currents across an area bounded by the path of integration.

As an example, Fig. 32–11 is a sectional view of two circular conducting plates of radius R separated by a narrow gap in vacuum (i.e., a parallel-plate capacitor). If the gap is small compared to R, the $\boldsymbol{E}$ field caused by the charge distributions on the plates is confined almost entirely to the region between plates and is nearly uniform, as shown. Suppose we wish to calculate the $\boldsymbol{B}$ field at some point in the midplane, shown in the figure as a broken line. In principle, this could be calculated from the Biot law. It would be necessary to subdivide the whole of space (including the wires and the plates) into elementary filaments $d\boldsymbol{l}$, find the conduction and displacement currents in each filament, and perform a vector integration over all space. Such a calculation would be extremely complicated, if not impossible. The result can be obtained quite easily,

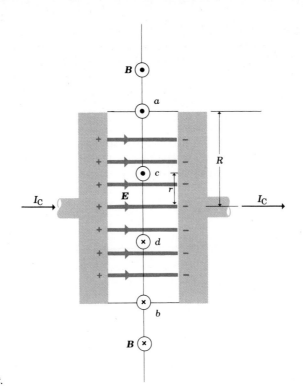

32-11 A capacitor being charged by a current I_C has a displacement current I_D between its plates equal to I_C, with displacement current density $J_D = \epsilon_0 \, dE/dt$.

however, if we use Ampère's law and take advantage of symmetry considerations.

As is evident from the diagram, there is no *conduction* current across the midplane; the current across this plane is a *displacement* current only. Also, since every line crosses the plane at some point, the *total* displacement current across the plane is the *displacement* current *out of* the left plate; from the definition of displacement current in Sec. 29-6, this must equal the *conduction* current *into* the plate. Furthermore, by symmetry, the $\boldsymbol{B}$ field lines are circles with centers on the axis. Thus, at any point on a circle perpendicular to the axis and passing through points such as a and b, the $\boldsymbol{B}$-vector has the same magnitude and is tangent to the circle. At point a, the $\boldsymbol{B}$-vector points toward the reader and at point b it points away from the reader, as indicated by the symbols $\odot$ and $\otimes$.

The displacement current density J_D is given by Eq. (29-20):

$$J_D = \epsilon_0 \frac{dE}{dt}. \tag{32-29}$$

At any instant E is uniform over the region between plates, so J_D is also uniform; the *total* displacement current I_D between plates is just J_D times the area πR^2 of the plates. As remarked above, this must also equal the total *conduction* current I_C. Thus

$$I_D = I_C = \pi R^2 J_D$$

$$J_D = \frac{I_C}{\pi R^2}. \tag{32-30}$$

Now we may find the magnetic field at a point in the region between plates, at a distance r from the axis. We apply Ampère's law to a circle of radius r passing through the point; such a circle passes through points c and d in Fig. 32–11. The total current enclosed by the circle is J_D times its area, or $(I_C/\pi R^2)(\pi r^2)$. The integral $\oint \boldsymbol{B} \cdot d\boldsymbol{l}$ in Ampère's law is just B times the circumference $2\pi r$ of the circle, and Ampère's law becomes

$$\oint \boldsymbol{B} \cdot d\boldsymbol{l} = 2\pi r B = \mu_0 \left(\frac{r^2}{R^2} \right) I_C,$$

or

$$B = \frac{\mu_0}{2\pi} \left(\frac{r}{R^2} \right) I_C. \tag{32–31}$$

This result shows that in the region between the plates $\boldsymbol{B}$ is zero at the axis, increasing linearly with distance from the axis. A similar calculation shows that *outside* the region between the plates, $\boldsymbol{B}$ is the same as though the wire were continuous and the plates not present at all.

The role of displacement current as a source of magnetic field can be expressed in a different way using the concept of *electric flux* Ψ, introduced in Secs. 24–4 and 29–6. As stated in Eq. (29–22), the displacement current may be expressed as

$$I_D = \epsilon_0 \frac{d\Psi}{dt}.$$

Thus in empty space, where there is no conduction current ($I_C = 0$), Ampère's law, Eq. (32–28), may be rewritten

$$\oint \boldsymbol{B} \cdot d\boldsymbol{l} = \mu_0 \epsilon_0 \frac{d\Psi}{dt}. \tag{32–31}$$

A significant feature of this equation is its resemblance to another relation, *Faraday's law*, to be introduced in Chapter 33. Faraday's law describes the *electric* field induced by a changing *magnetic* field, and one form of the relation is

$$\oint \boldsymbol{E} \cdot d\boldsymbol{l} = -\frac{d\Phi}{dt}, \tag{32–32}$$

where Φ is *magnetic* flux. Apart from constants, Eqs. (32–31) and (32–32) have the same form, but the roles of electric and magnetic fields are reversed. Thus the concept of displacement current introduces a kind of symmetry into electric- and magnetic-field relations.

The fact that *displacement* current acts as a source of magnetic field plays an essential role in the understanding of electromagnetic *waves*. A changing *electric* field in a region of space induces a *magnetic* field in neighboring regions, even when no conduction current and no matter are present. This relationship, first proposed by Maxwell in 1865, provides the key to a theoretical understanding of electromagnetic radiation and of light as a particular example of this radiation. We return to this topic in Chapter 37.

Questions

32-1 Streams of charged particles emitted from the sun during unusual sunspot activity create a disturbance in the earth's magnetic field. How does this happen?

32-2 A topic of current interest in physics research is the search (thus far unsuccessful) for an isolated magnetic pole or magnetic *monopole*. If such an entity were found, how could it be recognized? What would its properties be?

32-3 What are the relative advantages and disadvantages of Ampère's Law and the Biot–Savart Law for practical calculations of magnetic fields?

32-4 A student proposed to obtain an isolated magnetic pole by taking a bar magnet (N pole at one end, S at the other) and breaking it in half. Will this work?

32-5 Pairs of conductors carrying current into and out of the power-supply components of electronic equipment are sometimes twisted together to reduce magnetic-field effects. Why does this help?

32-6 The text discusses the magnetic field of an infinitely long straight conductor carrying a current. Of course, there's no such thing as an infinitely long *anything*. How do you decide whether a particular wire is long enough to be considered infinite?

32-7 Suppose one has three long parallel wires, arranged so that in cross section they are at the corners of an equilateral triangle. Is there any way to arrange the currents so that all three wires attract each other? So that all three wires repel each other?

32-8 Two parallel conductors carrying current in the same direction attract each other. If they are permitted to move toward each other, the forces of attraction do work. Where does the energy come from? Does this contradict the assertion in Chapter 30 that magnetic forces on moving charges do no work?

32-9 Considering the magnetic field of a circular loop of wire, would you expect the field to be greatest at the center, or is it greater at some points in the plane of the loop but off-center?

32-10 Two concentric circular loops of wire, of different diameter, carry currents in the same direction. Describe the nature of the forces exerted on the inner loop.

32-11 A current was sent through a helical coil spring. The spring appeared to contract, as though it had been compressed. Why?

32-12 Using the fact that magnetic-field lines never have a beginning or end, explain why it is reasonable for the field of a toroidal solenoid to be confined entirely to its interior, while a straight solenoid *must* have some field outside it.

32-13 In the discussion of the **B** field of a solenoid (Sec. 32-8), suppose the cross section of the solenoid is not circular but some other shape, such as elliptical or square. Is Eq. (32-25) still valid? Why or why not?

32-14 Can one have a displacement current as well as a conduction current within a conductor?

32-15 A character in a popular comic strip has at various times proposed the possibility of "harnessing the earth's magnetic field" as a nearly inexhaustible source of energy. Comment on this concept.

Problems

$$1\,\text{T} = 1\,\text{Wb·m}^{-2} = 1\,\text{N·A}^{-1}\text{·m}^{-1},$$
$$\mu_0 = 4\pi \times 10^{-7}\,\text{N·A}^{-2} = 4\pi \times 10^{-7}\,\text{T·A}^{-1}\text{·m}$$
$$= 4\pi \times 10^{-7}\,\text{Wb·A}^{-1}\text{·m}^{-1}.$$

32-1 An overhead transmission line 5.0 m above the ground carries a current of 400 A in a direction from south to north. Find the magnitude and direction of the magnetic field at a point on the ground directly under the conductor. Would this field cause appreciable error in a magnetic compass located at this point?

32-2 A long straight conductor passes vertically through the center of a laboratory; the direction of the current is upward. The magnetic field at a point 0.20 m from the wire is found to have magnitude $5.0 \times 10^{-4}\,\text{T}$. What is the current in the conductor?

32-3 A long straight wire, carrying a current of 200 A, runs through a cubical wooden box, entering and leaving through holes in the centers of opposite faces, as in Fig. 32–12. The length of each side of the box is 20 cm. Consider

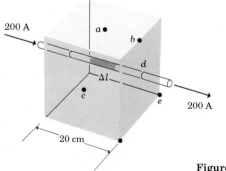

Figure 32–12

an element of the wire 1 cm long at the center of the box. Compute the magnitude of the magnetic field ΔB produced by this element at the points lettered a, b, c, d, and e in Fig. 32–12. Points a, c, and d are at the centers of the faces of the cube, point b is at the midpoint of one edge, and point e

is at a corner. Copy the figure and show by vectors the directions and relative magnitudes of the field vectors.

32–4 A long straight wire carries a current of 10 A directed along the negative y-axis, as shown in Fig. 32–13. A uniform magnetic field $\boldsymbol{B}_0$ of magnitude 10^{-6} T is directed parallel to the x-axis. What is the resultant magnetic field at the following points?

a) $x = 0$, $z = 2$ m;

b) $x = 2$ m, $z = 0$;

c) $x = 0$, $z = -0.5$ m.

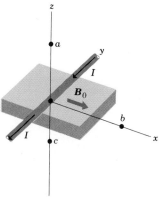

Figure 32-13

32–5 A magnetic field of magnitude 5.0×10^{-4} T is to be produced at a distance of 5 cm from a long straight wire.

a) What current is required to produce this field?

b) With the current found in (a), what is the magnitude of the field at a distance of 10 cm from the wire? At 20 cm?

32–6 A long straight telephone cable contains six wires, each carrying a current of 0.5 A. The distances between wires can be neglected.

a) If the currents in all six wires are in the same direction, what is the magnitude of the magnetic field 10 cm from the cable?

b) If four wires carry currents in one direction and the other two in the opposite direction, what is the field magnitude 10 cm from the cable?

32–7 Figure 32–14 is an end view of two long parallel wires perpendicular to the xy-plane, each carrying a current I, but in opposite directions.

a) Copy the diagram, and show by vectors the $\boldsymbol{B}$ field of each wire, and the resultant $\boldsymbol{B}$ field at point P.

b) Derive the expression for the magnitude of $\boldsymbol{B}$ at any point on the x-axis, in terms of the coordinate x of the point.

c) Construct a graph of the magnitude of $\boldsymbol{B}$ at any point on the x-axis.

d) At what value of x is B a maximum?

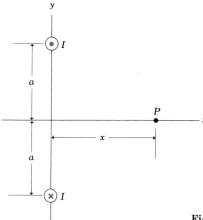

Figure 32-14

32–8 Same as Problem 32–7, except that the current in both wires is *away* from the reader.

32–9 Two long, straight, horizontal parallel wires, one above the other, are separated by a distance $2a$. If the wires carry equal currents in opposite directions, what is the field magnitude in the plane of the wires at a point

a) midway between them, and

b) at a distance a above the upper wire?

If the wires carry equal currents in the same direction, what is the field magnitude in the plane of the wires at a point,

c) midway between them, and

d) at a distance a above the upper wire?

32–10 Two long, straight, parallel wires are 1.0 m apart, as in Fig. 32–15. The upper wire carries a current I_1 of 6 A into the plane of the paper.

a) What must be the magnitude and direction of the current I_2 for the resultant field at point P to be zero?

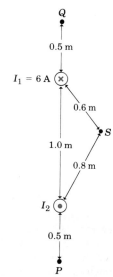

Figure 32-15

b) What is then the resultant field at Q?

c) At S?

32-11 In Fig. 32-14 suppose a third long straight wire, parallel to the other two, passes through point P, and that each wire carries a current $I = 20$ A. Let $a = 30$ cm and $x = 40$ cm. Find the magnitude and direction of the force per unit length on the third wire

a) if the current in it is away from the reader,

b) if the current is *toward* the reader.

32-12 A long straight wire carries a current of 1.5 A. An electron travels with a speed of 5×10^6 cm·s^{-1} parallel to the wire, 10 cm from it, and in the same direction as the current. What force does the magnetic field of the current exert on the moving electron?

32-13 A long horizontal wire AB rests on the surface of a table. (See Fig. 32-16). Another wire CD vertically above the first is 100 cm long and is free to slide up and down on the two vertical metal guides C and D. The two wires are connected through the sliding contacts and carry a current of 50 A. The mass per unit length of the wire CD is 0.05 g·cm^{-1}. To what equilibrium height h will the wire CD rise, assuming the magnetic force on it to be due wholly to the current in the wire AB?

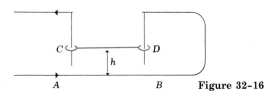

Figure 32-16

32-14 Two long parallel wires are hung by cords of 4 cm length from a common axis, as shown in Fig. 32-17. The wires have a mass per unit length of 50 g·m^{-1} and carry the same current in opposite directions. What is the current if the cords hang at an angle of 6° with the vertical?

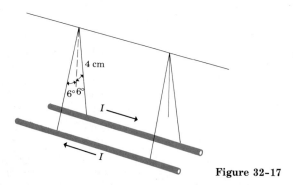

Figure 32-17

32-15 The long straight wire AB in Fig. 32-18 carries a current of 20 A. The rectangular loop whose long edges are parallel to the wire carries a current of 10 A. Find the magnitude and direction of the resultant force exerted on the loop by the magnetic field of the wire.

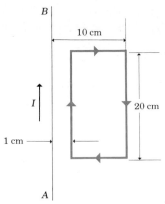

Figure 32-18

32-16 The long straight wire AB in Fig. 32-19 carries a constant current I.

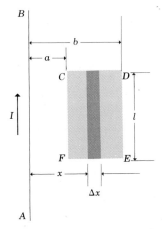

Figure 32-19

a) What is the magnetic field at the shaded area at a perpendicular distance x from the wire?

b) What is the magnetic flux $\Delta\Phi$ through the shaded area?

c) What is the total flux through the rectangle $CDEF$?

32-17 Refer to Fig. 32-7. Sketch a graph of the magnitude of the B field on the axis of the loop, from $x = -3R$ to $x = +3R$.

32-18 Figure 32-20 is a sectional view of two circular coils of radius a, each wound with N turns of wire carrying a current I, circulating in the same direction in both coils. The coils are separated by a distance a equal to their radii.

a) Derive the expression for the magnetic field at point P, midway between the coils.

b) Calculate the magnitude of B if $N = 100$ turns, $I = 5$ A, and $a = 30$ cm?

32-19 Considering the magnetic field along the axis of a circular loop of radius R, at what distance from the center of the loop is the field $\frac{1}{10}$ of its value at the center?

32-20 A circular coil of radius 5 cm has 200 turns and carries a current of 0.2 A. What is the magnetic field

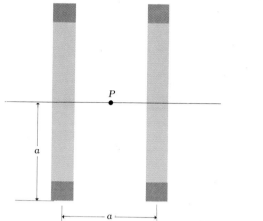

Figure 32-20

a) at the center of the coil?

b) At a point on the axis of the coil, 10 cm from its center?

32-21 A closely wound coil has a diameter of 40 cm and carries a current of 2.5 A. How many turns does it have if the magnetic field at the center of the coil is 1.26×10^{-4} T?

32-22 A thin disk of dielectric material, having a total charge $+Q$ distributed uniformly over its surface, and of radius a, rotates n times per second about an axis perpendicular to the surface of the disk and passing through its center. Find the magnetic field at the center of the disk.

32-23 A solenoid is 30 cm long and is wound with two layers of wire. The inner layer consists of 300 turns, the outer layer of 250 turns. The current is 3 A in the same direction in both layers. What is the magnetic field at a point near the center of the solenoid?

32-24 A wire of circular cross section and radius R carries a current I, uniformly distributed over its cross-sectional area.

a) In terms of I, R, and r_1, what is the current through a circular area of radius r_1, inside the wire?

b) Use Ampère's law to find B inside the wire, at a distance r_1 from the axis.

c) What is B outside the wire, at a distance r_2 from the axis?

d) What would the field be at this distance if the current were concentrated in a very fine wire along the axis?

e) Sketch a graph of the magnitude of $\boldsymbol{B}$ as a function of r, from $r = 0$ to $r = 2R$.

32-25 A coaxial cable consists of a solid conductor of radius R_1, supported by insulating disks on the axis of a tube of inner radius R_2 and outer radius R_3. If the central con-

ductor and the tube carry equal currents in opposite directions, find the magnetic field

a) at points outside the axial conductor but inside the tube, and

b) at points outside the tube.

32-26 A solenoid of length 20 cm and radius 2 cm is closely wound with 200 turns of wire. The current in the winding is 5 A. Compute the magnetic field at a point near the center of the solenoid.

32-27 A wooden ring whose mean diameter is 10 cm is wound with a closely spaced toroidal winding of 500 turns. Compute the field at a point on the mean circumference of the ring when the current in the windings is 0.3 A.

32-28 A conductor is made in the form of a hollow cylinder with inner and outer radii a and b, respectively. It carries a current I, uniformly distributed over the cross section. Derive expressions for the magnetic field in the regions $r < a$, $a < r < b$, and $r > b$.

32-29 A solenoid is to be designed to produce a magnetic field of 0.1 T at its center. The radius is to be 5 cm and the length 50 cm, and the available wire can carry a maximum current of 10 A.

a) How many turns per unit length should the solenoid have?

b) What total length of wire is required?

32-30 A copper wire with circular cross section of area 4 mm² carries a current of 50 A. The resistivity of the material is $2.0 \times 10^{-8} \ \Omega \cdot \mathrm{m}$.

a) What is the electric field in the material?

b) What is the magnetic field 5 cm from the wire?

c) If the current is changing at the rate of 5000 A·s⁻¹, at what rate is the electric field in the material changing?

d) What is the displacement current density in the material in (c)?

e) What is the additional magnetic field 5 cm from the wire, due to the displacement current? Is it significant compared to that due to the conduction current?

32-31 A capacitor has two parallel plates of area A separated by a distance d. The capacitor is given an initial charge Q but, because the damp air between the plates is slightly conductive, the charge slowly leaks through the air. The charge initially changes at a rate dQ/dt.

a) In terms of dQ/dt, what is the initial rate of change of electric field between the plates?

b) Show that the displacement current density has the same magnitude as the conduction current density, but the opposite direction. Hence show that the magnetic field in the material is exactly zero at all times.

INDUCED ELECTROMOTIVE FORCE

33

33-1 Motional electromotive force

The present large-scale production, distribution, and use of electrical energy would not be economically feasible if the only seats of emf available were those of *chemical* nature, such as dry cells. The development of electrical engineering, as we now know it, began with Faraday and Henry who, independently and at nearly the same time, discovered the principles of *magnetically* induced emf's and the methods by which mechanical energy can be converted directly to electrical energy.

Figure 33–1 represents a conductor of length l in a uniform magnetic field perpendicular to the plane of the diagram and directed away from the reader. If the conductor is set in motion toward the right with a velocity v, perpendicular both to its own length and to the magnetic field, a charged particle q within it experiences a force F equal to $qv \times B$ directed along the length of the conductor. The direction of the force on a positive charge is from b toward a in Fig. 33–1, while the force on a negative charge is from a toward b. Because this force is of non-electrostatic origin, we denote it by F_n:

$$F_n = qv \times B. \qquad (33-1)$$

The state of affairs within the conductor is the same as though it had been inserted in an *electric* field equal to $v \times B$, directed from b toward a. Paralleling the discussion of Sec. 28–4, we define the equivalent non-electrostatic field E_n as the non-electrostatic force per unit charge:

$$E_n = v \times B. \qquad (33-2)$$

The free charges in the conductor move in the direction of the force acting on them until the accumulation of excess charges at the ends of the conductor establishes an electrostatic field E_e such that the *resultant* force on every charge within the conductor is zero. The charges are then in equilibrium. The upper end of the wire acquires an excess positive charge, and the lower end an excess negative charge.

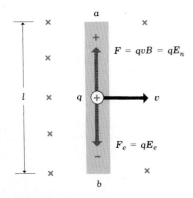

$F = qvB = qE_n$

$F_e = qE_e$

33-1 Conducting rod in uniform magnetic field.

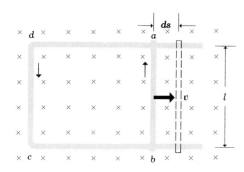

33-2   Current produced by the motion of a conductor in a magnetic field.

Suppose now that the moving conductor slides along a stationary U-shaped conductor, as in Fig. 33-2. There is no magnetic force on the charges within the stationary conductor but, since it lies in the electrostatic field surrounding the moving conductor, a current will be established within it; the direction of this current (defined as usual as the direction of positive-charge motion) is counterclockwise, or from b toward a. As a result of this current, the excess charges at the ends of the moving conductor are reduced, the electrostatic field within the moving conductor is weakened, and the magnetic forces cause a further displacement of the free electrons within the conductor from a toward b. As long as the motion of the conductor is maintained there is, therefore, a continual current in a counterclockwise direction. The moving conductor corresponds to a seat of electromotive force, and is said to have *induced* within it a *motional electromotive force.*

The magnitude of this emf can be found as follows. When a charge q moves from b to a through a distance l, the work of the force F is

$$W = Fl = vBql.$$

The emf $\mathcal{E}$ is the work per unit charge, so

$$\mathcal{E} = \frac{W}{q} = vBl. \tag{33-3}$$

If the velocity of the conductor makes an angle ϕ with the field, we must replace v by $v \sin \phi$, and the induced emf becomes

$$\mathcal{E} = v \sin \phi \, Bl. \tag{33-4}$$

More generally, the electromotive force $\mathcal{E}$ may be defined as in Sec. 28-4, that is, as the line integral of the non-electrostatic field:

$$\mathcal{E} = \int_b^a \boldsymbol{E}_n \cdot d\boldsymbol{l} = \int_b^a \boldsymbol{v} \times \boldsymbol{B} \cdot d\boldsymbol{l}. \tag{33-5}$$

In Fig. 33-2, $\boldsymbol{v}$ and $\boldsymbol{B}$ are perpendicular, and their vector product is parallel to $d\boldsymbol{l}$; thus in this case the right side of Eq. (33-5) reduces to vBl, in agreement with Eq. (33-3) or (33-4).

The emf associated with the moving conductor in Fig. 33-2 is analogous to that of a battery with its positive terminal at a and negative at b. In each the direction of the non-electrostatic field is from b and a, and when the device is connected to an external conductor, the direction of current is from a to b in the external circuit. Electromotive force is not a vector quantity, but the quantity defined by Eq. (33–5) may be positive or negative, depending on the direction of $\boldsymbol{E}_n$.

If v is expresed in m·s⁻¹, B in T, and l in meters, $\mathscr{E}$ is in joules per coulomb or volts, as the reader should verify.

Example 1   Let the length l in Fig. 33–2 be 0.1 m, the velocity v be 0.1 m·s⁻¹, the resistance of the loop be 0.01 Ω, and let $B = 1$ T.

Solution   The emf $\mathscr{E}$ is

$$\mathscr{E} = vBl = 0.01 \text{ V}.$$

The current in the loop is

$$I = \frac{\mathscr{E}}{R} = 1 \text{ A}.$$

Because of this current, there is a force F on the loop, in the opposite direction to its motion, and equal to

$$F = IBl = (1 \text{ A})(1 \text{ T})(0.1 \text{ m})$$
$$= 0.1 \text{ N}.$$

The power necessary to move the loop against this force is

$$P = Fv = (0.1 \text{ N})(0.1 \text{ m·s}^{-1}) = 0.01 \text{ W}.$$

The product $\mathscr{E}I$ is

$$\mathscr{E}I = (0.01 \text{ V})(1 \text{ A}) = 0.01 \text{ W}.$$

Thus, the rate of energy conversion, $\mathscr{E}I$, equals the mechanical power input, Fv, to the system.

Example 2   The rectangular loop in Fig. 33–3, of length a and width b, is rotating with uniform angular velocity ω about the y–axis. The entire loop lies in a uniform, constant $\boldsymbol{B}$ field, parallel to the z-axis. We wish to calculate the induced emf in the loop, from Eq. (33–3).

Solution   The velocity v of either of the sides of the loop of length a has magnitude

$$v = \omega \left(\frac{b}{2} \right).$$

The motional emf in each of the two sides of length a is

$$\mathscr{E} = vB \sin \theta \, \alpha = \tfrac{1}{2}\omega B \, ab \sin \theta.$$

These two emf's add, so the total emf due to these two sides is

$$\mathscr{E} = \omega B \, ab \sin \theta.$$

The magnetic forces on the other two sides of the loop are transverse to these sides and so contribute nothing to the emf. As an alternative approach, the emf may be calculated from Eq. (33–5). The direction of the motional field $\boldsymbol{E}_n = \boldsymbol{v} \times \boldsymbol{B}$ in each of the sides of length a is shown in the diagram. Its magnitude is

$$E_n = vB \sin \theta = \omega \frac{b}{2} B \sin \theta.$$

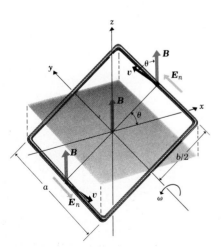

33-3   Rectangular loop rotating with constant angular velocity in a uniform magnetic field.

The motional fields in the other two sides of the loop are transverse to these sides and contribute nothing to the emf. The line integral of $\boldsymbol{E}_n$ around the loop reduces to $2E_n a$, so

$$\mathcal{E} = \oint \boldsymbol{E}_n \cdot d\boldsymbol{l} = 2E_n a = 2\left(\omega \frac{b}{2} B \sin\theta\right)a$$

$$= \omega Bab \sin\theta.$$

The product ab equals the area A of the loop, and if the loop lies in the xy-plane at $t = 0$, then $\theta = \omega t$. Hence

$$\mathcal{E} = \omega AB \sin\omega t. \tag{33–6}$$

The emf therefore varies sinusoidally with time. The *maximum* emf $\mathcal{E}_m$, which occurs when $\sin\omega t = 1$, is

$$\mathcal{E}_m = \omega AB,$$

so we can write Eq. (33–6) as

$$\mathcal{E} = \mathcal{E}_m \sin\omega t. \tag{33–7}$$

The rotating loop is the prototype of one type of *alternating current generator*, or *alternator*; we say it develops a *sinusoidal alternating emf*.

The emf is maximum (in absolute value) when $\theta = 90°$ or $270°$ and the long sides are moving at right angles to the field. The emf is zero when $\theta = 0$ or $180°$ and the long sides are moving parallel or antiparallel to the field. The emf depends only on the *area A* of the loop and not on its shape. This is most easily verified by use of Faraday's law, to be introduced in the next section.

The rotating loop in Fig. 33–3 can be utilized as the source in an external circuit by making connections to *slip rings* S, S, which rotate with the loop, as shown in Fig. 33–4a. Stationary brushes bearing against the rings are connected to the output terminals a and b. The instantaneous terminal voltage v_{ab}, on open circuit, equals the instantaneous emf. Figure 33–4b is a graph of v_{ab} as a function of time.

A similar scheme may be used to obtain an emf which, though varying with time, always has the same sign. The loop is connected to a split ring or commutator, as in Fig. 33–5a. At angular positions where the emf reverses, the connections to the external circuit are interchanged. The resulting emf is shown in Fig. 33–5b.

This device is the prototype of a dc generator. Commercial dc generators have a large number of coils and commutator segments; their terminal voltage is not only unidirectional but also practically constant.

Example 3   A disk of radius R, shown in Fig. 33–6, lies in the xy-plane and rotates with uniform angular velocity ω about the z-axis. The disk is in a uniform, constant $\boldsymbol{B}$ field parallel to the z-axis. Consider a short portion of a narrow radial segment of the disk, of length dr. Its velocity is $v = \omega r$, and since $\boldsymbol{v}$ is at right angles to $\boldsymbol{B}$, the motional field $\boldsymbol{E}_n$ in the segment is

$$E_n = vB = \omega rB.$$

The direction of $\boldsymbol{E}_n$ is radially outward.

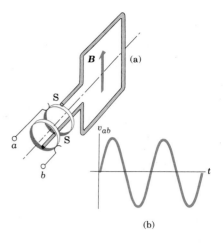

33–4 (a) Schematic diagram of an alternator, using a conducting loop rotating in a magnetic field. Connections to the external circuit are made by means of the slip rings S. (b) The resulting emf at terminals ab.

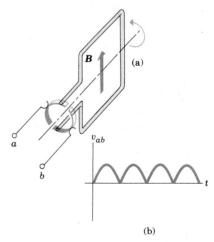

33–5 (a) Schematic diagram of a dc generator, using a split-ring commutator. (b) The resulting induced emf at terminals ab.

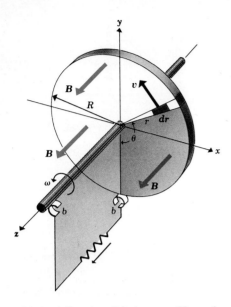

33-6 A Faraday disk dynamo. The emf is induced along radial lines of the rotating disk and is connected to an external circuit through the sliding contacts *bb*.

The motional emf due to this segment is

$$d\mathcal{E} = \boldsymbol{E}_n \cdot d\boldsymbol{l} = E_n dr = \omega r B \, dr.$$

The total emf between center and rim is

$$\mathcal{E} = \int_0^R \boldsymbol{E}_n \cdot d\boldsymbol{l} = \omega B \int_0^R r \, dr = \tfrac{1}{2} \omega B R^2. \qquad (33\text{-}8)$$

All of the radial segments of the disk are in *parallel*, so the emf between center and rim equals that in any radial segment. The entire disk can therefore be considered a *source* for which the emf between center and rim equals $\omega B R^2 / 2$. The source can be included in a closed circuit by completing the circuit through sliding contacts of brushes *b, b*.

The emf in such a disk was studied by Faraday, and the device is called a *Faraday disk dynamo*.

It was pointed out in Sec. 30-4 that the magnetic-field force on a moving charged particle never does any *work* on the particle because $\boldsymbol{F}$ and $\boldsymbol{v}$ are always perpendicular. Yet in defining motional emf we have spoken about the work done by the equivalent non-electrostatic field $\boldsymbol{E}_n = \boldsymbol{v} \times \boldsymbol{B}$. How is this apparent paradox resolved?

The resolution is somewhat subtle and cannot be discussed in detail here. Very briefly, the vertical motion of the charges in the moving conductor in Fig. 33-2 causes a transverse (horizontal) magnetic-field force and thus a transverse displacement of charge, corresponding to the Hall effect discussed in Sec. 31-2. Thus there is a transverse *electrostatic* field in the conductor, from left to right in the example of Fig. 33-2. As the conductor moves to the right, it is this electrostatic force which actually does work on the moving charges. Detailed analysis shows that this work is the same *as though* it had been done by the vertical non-electrostatic or motional force $q\boldsymbol{v} \times \boldsymbol{B}$ during the vertical displacement of the moving charges.

33-2 Faraday's law

The induced emf in the circuit of Fig. 33-2 may be considered from another viewpoint. While the conductor moves toward the right a distance *ds*, the *area* enclosed by the circuit *abcd* increases by

$$dA = l \, ds,$$

and the change in magnetic *flux* through the circuit is

$$d\Phi = B \, dA = Bl \, ds.$$

The *rate of change* of flux is therefore

$$\frac{d\Phi}{dt} = \left(\frac{ds}{dt} \right) Bl = vBl. \qquad (33\text{-}9)$$

But the product vBl equals the induced emf $\mathcal{E}$, so this equation states that *the induced emf in the circuit is numerically equal to the rate of change of the magnetic flux through it*. That is,

$$\mathcal{E} = \frac{d\Phi}{dt}. \qquad (33\text{-}10)$$

This equation does not tell us the *sense* of the emf, that is, which end is at higher potential or the direction of the non-electrostatic force. The sense of the emf can be determined by other considerations, to be discussed later.

Equation (33–10) is known as *Faraday's law*. It may appear to be merely an alternative form of Eq. (33–3) for the emf in a moving conductor. It turns out, however, that the relation has a much broader significance than might be expected from its derivation. That is, it is found to apply to all circuits through which the flux is caused to vary, even when there is no *motion* of any part of the circuit and hence no emf directly attributable to a force on a moving charge.

Suppose, for example, that two loops of wire are located as in Fig. 33–7. A current in circuit 1 sets up a magnetic field whose magnitude at all points is proportional to this current. A part of this flux passes through circuit 2, and if the current in circuit 1 is increased or decreased, the flux through circuit 2 will also vary. Circuit 2 is not moving in a magnetic field and hence no "motional" emf is induced in it, but there is a change in the flux through it, and it is found experimentally that an emf appears in circuit 2 of magnitude $\mathcal{E} = d\Phi/dt$. In such a situation, no one portion of circuit 2 can be considered the seat of emf; the *entire circuit* constitutes the seat.

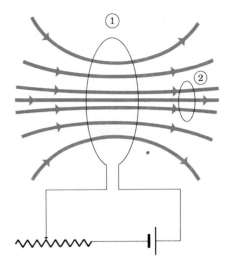

33-7 As the current in circuit 1 is varied, the magnetic flux through circuit 2 changes.

Here is another example. Suppose we set up a magnetic field within the toroidal winding of Fig. 33–8, link the toroid with a conducting ring, and vary the current in the winding of the toroid. We have shown that the flux lines set up by a steady current in a toroidal winding are wholly confined to the space enclosed by the winding; not only is the ring not *moving* in a magnetic field, but if the current were steady it would not even be *in* a magnetic field! However, field lines do pass through the area bounded by the ring, and the flux changes as the current in the windings changes. Equation (33–10) predicts an induced emf in the ring, and we find by experiment that the emf actually exists. In case the reader has not identified the apparatus in Fig. 33–8, it may be pointed out that it is merely a *transformer* with a one-turn secondary, so that the phenomenon we are now discussing is the basis of the operation of every transformer.

To sum up, then, an emf is induced in a circuit whenever the magnetic flux through the circuit varies with time. The flux may be caused to change in various ways: For example, (1) a conductor may move through a stationary magnetic field, as in Figs. 33–2 and 33–3, or (2) the magnetic field through a stationary conducting loop may change with time, as in Figs. 33–7 and 33–8. For case (1), the emf may be computed *either* from

$$\mathcal{E} = vBl$$

or from

$$\mathcal{E} = \frac{d\Phi}{dt}.$$

For case (2), the emf may be computed *only* by

$$\mathcal{E} = \frac{d\Phi}{dt}.$$

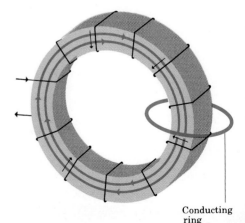

33-8 An emf is induced in the ring when the flux in the toroid varies.

If we have a coil of N turns and the flux varies at the same rate through each, the induced emf's in the turns are in *series*, and the total emf is

$$\mathcal{E} = N \frac{d\Phi}{dt}. \tag{33-11}$$

Example 1 A certain coil of wire consists of 500 circular loops of radius 4 cm. It is placed between the poles of a large electromagnet, where the magnetic field is uniform, perpendicular to the plane of the coil, and increasing at a rate of 0.2 T·s⁻¹. What is the magnitude of the resulting induced emf?

Solution The flux Φ at any time is given by $\Phi = BA$, and the rate of change of flux by $d\Phi/dt = (dB/dt)A$. In our problem, $A = \pi \times (0.04 \text{ m})^2 = 0.00503 \text{ m}^2$ and

$$\frac{d\Phi}{dt} = (0.2 \text{ T·s}^{-1})(0.00503 \text{ m}^2)$$
$$= 0.00101 \text{ T·m}^2\text{·s}^{-1} = 0.00101 \text{ Wb·s}^{-1}.$$

From Eq. (33–11), the magnitude of the induced emf is

$$\mathcal{E} = N \frac{d\Phi}{dt} = (500)(0.00101 \text{ Wb·s}^{-1}) = 0.503 \text{ V}.$$

If the coil is tilted so that a line perpendicular to its plane makes an angle of 30° with $\boldsymbol{B}$, then only the component $B \cos 30°$ contributes to the flux through the coil. In that case the induced emf has magnitude $\mathcal{E} = (0.503 \text{ V})(\cos 30°) = 0.435 \text{ V}$.

Example 2 We consider again the rotating rectangular loop in Fig. 33–3, discussed in Example 2 of Section 33–1; this time we compute the emf from Eq. (33–10). The flux through the loop equals its area A multiplied by the component of B perpendicular to the area, that is, $B \cos \theta$. This is also equal to the magnitude B multiplied by the projected area shaded in Fig. 33–3 on a plane (the xy-plane) perpendicular to $\boldsymbol{B}$.

$$\Phi = \boldsymbol{B} \cdot \boldsymbol{A} = BA \cos \theta = BA \cos \omega t.$$

Then

$$\mathcal{E} = \frac{d\Phi}{dt} = -\omega BA \sin \omega t.$$

The negative sign is associated with the direction or sense of $\mathcal{E}$, to be discussed later. Apart from this sign, the result agrees with Eq. (33–6).

Note that the maximum value of $\mathcal{E}$ occurs when $\theta = 90°$ and the flux through the loop is zero, and that $\mathcal{E} = 0$ when $\theta = 0$ and the flux is maximum. That is, the emf depends not on the flux through the loop, but on its *rate of change*.

Example 3 We consider again the rotating disk in Fig. 33–6, discussed in Example 3 of Section 33–1. To compute the emf from Faraday's law, Eq. (33–10), we consider the circuit to be the periphery of the shaded areas in Fig. 33–6. The rectangular portion in the yz-plane is fixed. The area of the shaded section in the xy-plane is $\frac{1}{2}R^2\theta$, and the flux through it is

$$\Phi = \tfrac{1}{2}BR^2\theta.$$

As the disk rotates, the shaded area increases. In a time dt, the angle θ increases by $d\theta = \omega\, dt$. The flux increases by

$$d\Phi = \tfrac{1}{2}Br^2\, d\theta = \tfrac{1}{2}BR^2\omega\, dt,$$

and the induced emf is

$$\mathcal{E} = \frac{d\Phi}{dt} = \tfrac{1}{2}Br^2\omega,$$

in agreement with the previous result.

Example 4   *The search coil.* A useful experimental method of measuring magnetic field strength uses a small, closely wound coil of N turns called a *search coil* or a *snatch coil*. Assume first, for simplicity, that the search coil is placed with its plane perpendicular to a magnetic field $\boldsymbol{B}$. If the area enclosed by the coil is A, the flux Φ through it is $\Phi = BA$. Now if the coil is quickly given a quarter-turn about one of its diameters so that its plane becomes parallel to the field, or if it is quickly snatched from its position to another where the field is known to be zero, the flux through it decreases rapidly from BA to zero. During the time that the flux is decreasing, an emf of short duration is induced in the coil and there is a momentary induced current in the external circuit to which the coil is connected. As we shall see, the total flux change is proportional to the total *charge* which flows around the circuit.

The current at any instant is

$$i = \frac{\mathcal{E}}{R},$$

where R is the combined resistance of external circuit and search coil, $\mathcal{E}$ is the instantaneous induced emf, and i the instantaneous current.

From Faraday's law,

$$\mathcal{E} = N\frac{d\Phi}{dt}, \qquad i = \frac{N}{R}\frac{d\Phi}{dt}.$$

The total charge Q which flows through the circuit is given by

$$Q = \int_0^t i\, dt = \frac{N}{R}\int d\Phi = \frac{N\Phi}{R}.$$

Thus Φ and B are given by

$$\Phi = \frac{RQ}{N} \quad \text{and} \quad B = \frac{\Phi}{A} = \frac{RQ}{NA}. \tag{33-12}$$

It is not difficult to construct an instrument that measures the total charge which passes through it. Thus if the external circuit contains such an instrument, Q may be measured directly. Equation (33–12) may then be used to compute B.

Strictly speaking, while this method gives correctly the total flux through the coil, it is only the *average* field over the area of the coil which is measured. However, if the area is sufficiently small, this approximates closely the field at, say, the center of the coil.

This discussion has assumed the plane of the coil to be initially perpendicular to the direction of the field. If one is "exploring" a field whose direction is not known in advance, the same apparatus may be used to find the direction by performing a series of experiments in which the coil is placed at a given point in the field in various orientations, and snatched out of the field from each orientation. The charge flow Q will be maximum for the particular orientation in which the plane of the coil was perpendicular to the field. Thus both the magnitude and direction of an unknown field can be found by this method.

33-3 Induced electric fields

The examples of induced emf considered thus far have all involved conductors moving in a magnetic field, but we have also pointed out, in introducing Faraday's law, that induced emf can also occur with *stationary* conductors. An example is shown in Fig. 33–9, a solenoid encircled by a conducting loop of arbitrary shape. A current I in the windings of the solenoid sets up a magnetic field B along the solenoid axis and a magnetic flux $\Phi = BA$ passes through any surface bounded by the loop. Suppose that a small galvanometer G is inserted in the loop. It is found that if the current I is changed (and hence if the flux Φ is changed), the galvanometer indicates an emf $\mathcal{E}$ in the wire *during the time that the flux is changing*, and that this emf is equal to the *time rate of change of flux* through the surface bounded by the wire. That is,

$$\mathcal{E} = \frac{d\Phi}{dt}. \tag{33–13}$$

Equation (33–13) is Faraday's law; the essential difference between this and preceding examples is that here it is applied not to a moving conductor but to a stationary one in which there is a changing flux because the magnetic field in the area bounded by the conductor is changing.

Example Suppose the long solenoid in Fig. 33–9 is wound with 1000 turns per meter, and the current in its windings is increasing at the rate of $100\ \mathrm{A \cdot s^{-1}}$. The cross-sectional area of the solenoid is $4\ \mathrm{cm^2} = 4 \times 10^{-4}\ \mathrm{m^2}$.

The magnetic field magnitude inside the solenoid, at points not too near its ends, is given by Eq. (32–25). The flux Φ in the solenoid is

$$\Phi = BA = \mu_0 nIA.$$

The rate of change of flux is

$$\begin{aligned}
\frac{d\Phi}{dt} &= \mu_0 nA \frac{dI}{dt} \\
&= (4\pi \times 10^{-7}\ \mathrm{Wb \cdot A^{-1} \cdot m^{-1}})(1000\ \mathrm{turns \cdot m^{-1}}) \times \\
&\quad (4 \times 10^{-4}\ \mathrm{m^2})(100\ \mathrm{A \cdot s^{-1}}) \\
&= 16\pi \times 10^{-6}\ \mathrm{Wb \cdot s^{-1}}.
\end{aligned}$$

The magnitude of the induced emf is

$$|\mathcal{E}| = \frac{d\Phi}{dt} = 16\pi \times 10^{-6}\ \mathrm{V} = 16\pi\ \mu V.$$

We have previously associated electromotive force with the line integral of an electric field E_n of nonelectrostatic origin, and the same point

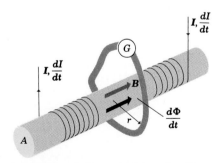

33-9 The windings of a long solenoid carry a current I that is increasing at a rate dI/dt. The magnetic flux in the solenoid is increasing at a rate $d\Phi/dt$, and this changing flux passes through a wire loop of arbitrary size and shape. An emf $\mathcal{E}$ is induced in the loop, given by $\mathcal{E} = d\Phi/dt$.

of view must be taken here. The emf is the line integral of $\boldsymbol{E}_n$ around the loop:

$$\mathcal{E} = \oint \boldsymbol{E}_n \cdot d\boldsymbol{l}$$

Hence the Faraday law states that

$$\oint \boldsymbol{E}_n \cdot d\boldsymbol{l} = \frac{d\Phi}{dt}. \tag{33–14}$$

As an example, suppose the loop in Fig. 33–9 is a circle of radius r. Because of the axial symmetry, the non-electrostatic field $\boldsymbol{E}_n$ has the same magnitude at all points on the circle and is everywhere tangent to this circle. The line integral in Eq. (33–14) becomes simply the magnitude E_n times the circumference $2\pi r$ of the circle. Thus, the induced electric field at a distance r from the axis is given by

$$E_n = \frac{1}{2\pi r} \frac{d\Phi}{dt}, \tag{33–15}$$

where Φ is the flux through a circle of radius r.

If it should happen that at a particular instant the $\boldsymbol{B}$ field is uniform across this circle (as is *not* the case with the solenoid above), then

$$\Phi = \pi r^2 B,$$
$$\frac{d\Phi}{dt} = \pi r^2 \frac{dB}{dt},$$

and

$$E_n = \frac{1}{2\pi r} \pi r^2 \frac{dB}{dt} = \frac{r}{2} \frac{dB}{dt}. \tag{33–16}$$

In summary, Eq. (33–10) is valid for two rather different situations. In the first, an emf is induced by the magnetic forces on charges in a conductor moving through a magnetic field, while in the second a time-varying magnetic field induces an electric field of non-electrostatic nature in a stationary conductor and hence induces an emf. The $\boldsymbol{E}_n$ field in the latter case differs from an electro*static* field in two significant ways. First, its line integral around a closed path is *not* zero, so it is not a *conservative* field. In contrast, an electro*static* field is *always* conservative, as discussed in Section 26–1. Second, the non-electrostatic field is not produced by static charges; it can be shown that the surface integral of $\boldsymbol{E}_n$ over a closed surface (the same integral that appears in Gauss's law) always has the value zero, whether charges are enclosed or not.

Thus, a changing magnetic field acts as a source of electric field, but one that differs from an electrostatic field in two ways: It is nonconservative, and its integral over a closed surface is zero. The reader may also note that this situation is analogous to that of the displacement current discussed in Section 29–6, in which a changing *electric* field acts as a source of *magnetic* field. These relations thus exhibit a kind of symmetry in the behavior of the two fields. We shall return to this relationship in Chapter 37, in connection with the analysis of electromagnetic waves.

33-4 Lenz's law

We return now to the question of determining the sign or direction of an induced emf or current or the direction of the associated non-electrostatic field. A very useful principle is *Lenz's law*. H. F. E. Lenz (1804–1864) was a German scientist who, without knowledge of the work of Faraday and Henry, duplicated many of their discoveries nearly simultaneously. The law states:

The direction of an induced current is such as to oppose the cause producing it.

The "cause" of the current may be the motion of a conductor in a magnetic field, or it may be the change of flux through a stationary circuit. In the first case, the direction of the induced current in the moving conductor is such that the direction of the side-thrust exerted on the conductor by the magnetic field is opposite in direction to its motion. The motion of the conductor is therefore "opposed."

In the second case, the induced current sets up a magnetic field of its own which, within the area bounded by the circuit, is (a) *opposite* to the original field if this is *increasing*, but (b) is in the *same* direction as the original field if the latter is *decreasing*. Thus it is the *change in flux* through the circuit (not the flux itself) which is "opposed" by the induced current.

In order to have an induced current, we must have a closed circuit. If a conductor does not form a closed circuit, then we mentally complete the circuit between the ends of the conductor and use Lenz's law to determine the direction of the current. The polarity of the ends of the open-circuit conductor may then be deduced.

Example 1 In Fig. 33–2, when the conductor moves to the right, a counterclockwise current is induced in the loop. The force exerted by the field on the moving conductor as a result of this current is to the left, *opposing* the conductor's motion.

Example 2 In Fig. 33–3, the direction of the induced current in the loop is the same as that of the non-electrostatic field E_n. The magnetic field exerts forces on the loop as a result of this current; the force on the right side of the loop (of length a) is in the $+x$-direction, and that on the left side is in the $-x$-direction. The resulting torque is opposite in direction to ω and thus opposes the motion.

Example 3 In Fig. 33–9, when the solenoid current is increasing, the induced current in the loop is counterclockwise. The additional field caused by this current, at points inside the loop, is opposite in direction to that of the solenoid; hence the induced current tends to oppose the increase in flux through the loop by causing flux in the opposite sense.

Lenz's law is also directly related to energy conservation. For instance, in Example 1 above, the induced current in the loop dissipates energy at the rate I^2R, and this energy must be supplied by the force that makes the conductor move despite the magnetic force opposing its motion. The work done by this applied force, in fact, must equal the energy dissipated in the circuit resistance. If the induced current were to have the opposite direction, the resulting force on the moving conductor would make it move faster and faster, violating energy conservation.

The content of Lenz's law can be incorporated into the formal statement of Faraday's law by means of a set of sign rules. We first designate one side of the surface bounded by the conducting circuit to be the positive; which side is chosen is immaterial. The positive side may be identified in diagrams by drawing a unit vector n away from the surface on the positive side.

A magnetic flux through this surface is positive if the B field has a component normal to the surface in the same direction as n, and negative if opposite. This defines a sign for $d\Phi/dt$. For example, if B has the same direction as n and is increasing in magnitude, or if the area is increasing, $d\Phi/dt$ is positive. If it has the same direction as n but is decreasing, or if the area is decreasing, $d\Phi/dt$ is negative. If the direction of B is opposite to n but its magnitude is decreasing, $d\Phi/dt$ is positive because Φ is becoming less negative. The reader may check other possibilities.

Finally, we specify the direction used to evaluate $\oint E_n \cdot dl$; we choose the direction around the boundary circuit defined by applying the right-hand rule to n. Thus if the circuit lies in the plane of this page and n points toward the reader, the circuit is to be traversed in the counterclockwise direction. With these conventions, the correct statement of Faraday's law is

$$\mathcal{E} = -\frac{d\Phi}{dt}. \tag{33–17}$$

This differs from Eqs. (33–10) and (33–13) in the negative sign.

Example 4   In Fig. 33–2 let us take n as parallel to B, away from the reader. Then Φ and $d\Phi/dt$ are both positive. The integral $\oint E_n \cdot dl$ is to be taken clockwise around the loop. E_n is different from zero only along the line ab, where its direction is upward. Thus

$$\oint E_n \cdot dl = -E_n l.$$

This quantity is negative because E_n and dl have opposite directions, and it is consistent with Eq. (33–17), in which both sides are positive. Alternatively, we could have taken n as toward the reader. In that case Φ and $d\Phi/dt$ are both negative, the loop is traversed in the counterclockwise sense, and $\oint E_n \cdot dl$ is positive. Either way, consistency with Eq. (33–17) requires that the direction of E_n must be from b to a, and this is the direction of current in the moving conductor.

Example 5   In Fig. 33–9, suppose the current has the direction shown and is decreasing. What is the direction of the induced current in the loop?

Solution   Take the normal to the loop's surface as pointing to the right, parallel to B. Then the flux is positive but decreasing, and $d\Phi/dt$ is negative. Hence, according to Eq. (33–17), when we calculate $\mathcal{E} = \oint E_n \cdot dl$ by traversing the loop clockwise (looking in the direction of B) the result must be positive. Hence E_n points clockwise around the

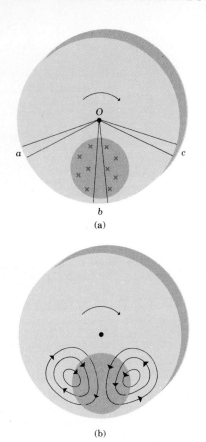

(a)

(b)

33-10 Eddy currents in a rotating disk.

33-11 Reduction of eddy currents by use of a laminated core.

loop, and the induced current must have this direction. Lenz's law gives the same result.

33-5 Eddy currents

Thus far we have considered only instances in which the currents resulting from induced emf's were confined to well-defined paths provided by the wires and apparatus of the external circuit. In many pieces of electrical equipment, however, one finds masses of metal moving in a magnetic field or located in a changing magnetic field, with the result that induced currents circulate throughout the volume of the metal. Because of their general circulatory nature, these are referred to as *eddy currents*.

Consider a disk rotating in a magnetic field perpendicular to the plane of the disk but confined to a limited portion of its area, as in Fig. 33–10a. Element *Ob* is moving across the field and has an emf induced in it. Elements *Oa* and *Oc* are not in the field but, in common with all other elements located outside the field, provide return conducting paths along which charges displaced along *Ob* can return from *b* to *O*. A general eddy-current circulation is therefore set up in the disk, somewhat as sketched in Fig. 33–10b.

The currents in the neighborhood of radius *Ob* experience a side thrust that *opposes* the motion of the disk, while the return currents, since they lie outside the field, do not experience such a thrust. The interaction between the eddy currents and the field therefore results in a braking action on the disk. The apparatus finds some technical applications and is known as an "eddy current brake."

As a second example of eddy currents, consider the core of an alternating current transformer, shown in Fig. 33–11. The alternating current in the primary winding *P* sets up an alternating flux within the core, and an induced emf develops in the secondary winding *S* because of the continual change in flux through it. The iron core, however, is also a conductor, and any section such as that at *AA* can be thought of as a number of closed conducting circuits, one within the other. The flux through each of these circuits is continually changing, so that there is an eddy-current circulation in the entire volume of the core, the lines of flow lying in planes perpendicular to the flux. These eddy currents are very undesirable both because of the energy they dissipate and because of the flux they themselves set up.

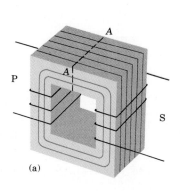

(a)

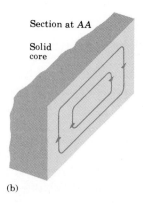

Section at *AA*

Solid core

(b)

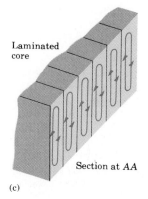

Laminated core

Section at *AA*

(c)

In all actual transformers, the eddy currents are greatly reduced by the use of a *laminated* core, that is, one built up of thin sheets or laminae. The electrical resistance between the surfaces of the laminations (due either to a natural coating of oxide or to an insulating varnish) effectively confines the eddy currents to individual laminae. The resulting length of path is greatly increased, with consequent increase in resistance. Hence, although the induced emf is not altered, the currents and their heating effects are minimized.

In small transformers where eddy-current losses must be kept to an absolute minimum, the cores are sometimes made of *ferrites*, which are complex oxides of iron and other metals. These materials are ferromagnetic but have relatively high resistivity.

Questions

33-1 In most parts of the northern hemisphere the earth's magnetic field has a vertical component directed *into* the earth. An airplane flying east generates an emf between its wingtips. Which wingtip acquires an excess of electrons and which a deficiency?

33-2 A sheet of copper is placed between the poles of an electromagnet, so the magnetic field is perpendicular to the sheet. When it is pulled out, a considerble force is required, and the force required increases with speed. What's happening?

33-3 In Fig. 33-4, if the angular velocity ω of the loop is doubled, the frequency with which the induced current changes direction doubles, and the maximum emf also doubles. Why? Does the torque required to turn the loop change?

33-4 Comparing the conventional dc generator (Fig. 33-5) and the Faraday disk dynamo (Fig. 33-6), what are some advantages and disadvantages of each?

33-5 Some alternating-current generators operate on a principle similar to that shown in Fig. 33-4 but using a rotating permanent magnet and stationary coils. What advantages does this scheme have? What disadvantages?

33-6 When a conductor moves through a magnetic field, the magnetic forces on the charges in the conductor cause an emf. But if this phenomenon is viewed in a frame of reference moving with the conductor, there is no motion, but there is still an emf. How is this paradox resolved?

33-7 Two circular loops lie adjacent to each other. One is connected to a source that supplies an increasing current; the other is a simple closed ring. Is the induced current in the ring in the same direction as that in the ring connected to the source, or opposite? What if the current in the first ring is decreasing?

33-8 A farmer claimed that the high-voltage transmission lines running parallel to his fence induced dangerously large voltages on the fence. Is this within the realm of possibility?

33-9 Small one-cylinder gasoline engines sometimes use a device called a *magneto* to supply current to the spark plug. A permanent magnet is attached to the flywheel, and there is a stationary coil mounted adjacent to it. What happens when the magnet passes the coil?

33-10 A current-carrying conductor passes through the center of a metal ring, perpendicular to its plane. If the current in the conductor increases, is a current induced in the ring?

33-11 A student asserted that, if a permanent magnet is dropped down a vertical copper pipe, it eventually reaches a terminal velocity, even if there is no air resistance. Why should this be? Or should it?

Problems

33-1 In Fig. 33-2, a rod of length $l = 0.40$ m moves in a magnetic field of magnitude $B = 1.2$ T. The emf induced in the moving rod is found to be 2.40 V.

a) What is the speed of the rod?

b) If the total circuit resistance is 1.2 Ω, what is the induced current?

c) What force (magnitude and direction) does the field exert on the rod as a result of this current?

33-2 In Fig. 33-1 a rod of length $l = 0.25$ m moves with constant speed of 6.0 m·s^{-1} in the direction shown. The induced emf is found to be 3.0 V.

a) What is the magnitude of the magnetic field?

b) Which point is at higher potential, a or b?

c) What is the direction of the electrostatic field in the rod? The non-electrostatic field?

d) What is the magnitude of the non-electrostatic field?

33-3 In Fig. 33-1, let $l = 1.5$ m, $B = 0.5$ T, $v = 4$ m·s^{-1}.

a) What is the motional emf in the rod?

b) What is the potential difference between its terminals?

c) Which end is at the higher potential?

33-4 The cube in Fig. 33-12, 1 m on a side, is in a uniform magnetic field of 0.2 T, directed along the positive y-axis. Wires A, C, and D move in the directions indicated, each with a speed of 0.5 m·s^{-1}.

a) What is the motional emf in each wire?

b) What is the potential difference between the terminals of each?

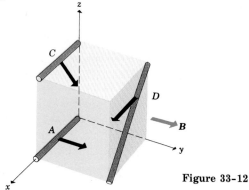

Figure 33-12

33-5 A conducting rod AB in Fig. 33-13 makes contact with the metal rails CA and DB. The apparatus is in a uniform magnetic field 0.5 T, perpendicular to the plane of the diagram.

a) Find the magnitude and direction of the emf induced in the rod when it is moving toward the right with a speed 4 m·s^{-1}.

b) If the resistance of the circuit $ABCD$ is 0.2 Ω (assumed constant), find the force required to maintain the rod in motion. Neglect friction.

c) Compare the rate at which mechanical work is done by the force (Fv) with the rate of development of heat in the circuit (i^2R).

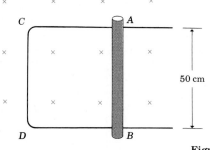

Figure 33-13

33-6 A closely wound rectangular coil of 50 turns has dimensions of 12 cm $\times$ 25 cm. The plane of the coil is rotated from a position where it makes an angle of 45° with a

magnetic field 2 T to a position perpendicular to the field in time $t = 0.1$ s. What is the average emf induced in the coil?

33-7 A coil of 1000 turns enclosing an area of 20 cm^2 is rotated from a position where its plane is perpendicular to the earth's magnetic field to one where its plane is parallel to the field, in 0.02 s. What average emf is induced if the earth's magnetic field is 6×10^{-5} T?

33-8 A square loop of wire with resistance R is moved at constant velocity v across a uniform magnetic field confined to a square region whose sides are twice the length of those of the square loop. (See Fig. 33-14.)

a) Sketch a graph of the external force F needed to move the loop at constant velocity, as a function of the distance x, from $x = -2l$ to $x = +2l$.

b) Sketch a graph of the induced current in the loop as a function of x, plotting clockwise currents upward and counterclockwise currents downward.

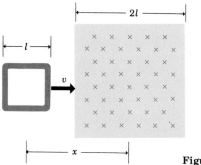

Figure 33-14

33-9 The long rectangular loop in Fig. 33-15, of width l, mass m, and resistance R, starts from rest in the position shown, and is acted on by a constant force $\boldsymbol{F}$. At all points in the colored area there is a uniform magnetic field $\boldsymbol{B}$, perpendicular to the plane of the diagram.

a) Sketch a graph of the velocity of the loop as a function of time.

b) Find the terminal velocity.

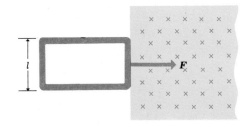

Figure 33-15

33-10 A slender rod 1 m long rotates about an axis through one end and perpendicular to the rod, with an angular velocity of 2 rev·s^{-1}. The plane of rotation of the rod is perpendicular to a uniform magnetic field 0.5 T.

a) What is the induced emf in the rod?

b) What is the potential difference between its terminals?

33–11 A flat square coil of 10 turns has sides of length 12 cm. The coil rotates in a magnetic field 0.025 T.

a) What is the angular velocity of the coil if the maximum emf produced is 20 mV?

b) What is the average emf at this velocity?

33–12 A rectangular coil of wire having 10 turns with dimensions of 20 cm × 30 cm rotates at a constant speed of 600 rpm in a magnetic field 0.10 T. The axis of rotation is perpendicular to the field. Find the maximum emf produced.

33–13 The rectangular loop in Fig. 33–16, of area A and resistance R, rotates at uniform angular velocity ω about the y-axis. The loop lies in a uniform magnetic field B in the direction of the x-axis. Sketch the following graphs:

a) the flux Φ through the loop as a function of time (let $t = 0$ in the position shown in Fig. 33–16);

b) the rate of change of flux $d\Phi/dt$;

c) the induced emf in the loop;

d) the torque Γ needed to keep the loop rotating at constant angular velocity;

e) the induced emf if the angular velocity is doubled. (Neglect the self-inductance of the loop.)

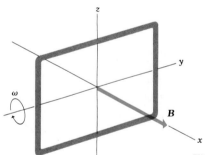

Figure 33–16

33–14 In Problem 33–13 and Fig. 33–16, let $A = 400$ cm^2, $R = 2\,\Omega$, $\omega = 10$ rad·s^{-1}, $B = 0.5$ T. Find

a) the maximum flux through the loop,

b) the maximum induced emf,

c) the maximum torque.

d) Show that the work of the external torque in one revolution is equal to the energy dissipated in the loop.

33–15 Suppose the loop in Fig. 33–16 is

a) rotated about the z-axis;

b) rotated about the x-axis;

c) rotated about an edge parallel to the y-axis.

What is the maximum induced emf in each case, if the angular velocity is the same as in Problem 33–14?

33–16 A flexible circular loop 10 cm in diameter lies in a magnetic field 1.2 T, directed into the plane of the diagram in Fig. 33–17. The loop is pulled at the points indicated by the arrows, forming a loop of zero area in 0.2 s.

a) Find the average induced emf in the circuit.

b) What is the direction of the current in R?

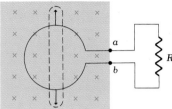

Figure 33–17

33–17 The magnetic field within a long, straight solenoid of circular cross section and radius R is increasing at a rate dB/dt.

a) What is the rate of change of flux through a circle of radius r_1 inside the solenoid, normal to the axis of the solenoid, and with center on the solenoid axis?

b) Find the induced electric field E_n inside the solenoid, at a distance r_1 from its axis. Show the direction of this field in a diagram.

c) What is the induced electric field *outside* the solenoid, at a distance r_2 from the axis?

d) Sketch a graph of the magnitude of E_n as a function of the distance r from the axis, from $r = 0$ to $r = 2R$. Compare with part (e) of Problem 32–24.

e) What is the induced emf in a circular turn of radius $R/2$?

f) Of radius R?

g) Of radius $2R$?

33–18 A long, straight solenoid of cross-sectional area 6 cm^2 is wound with 10 turns of wire per centimeter, and the windings carry a current of 0.25 A. A secondary winding of 2 turns encircles the solenoid. When the primary circuit is opened, the magnetic field of the solenoid becomes zero in 0.05 s. What is the average induced emf in the secondary?

33–19 The magnetic field B at all points within the colored circle of Fig. 33–18 is 0.5 T. It is directed into the plane of the diagram and is decreasing at the rate of 0.1 T·s^{-1}.

a) What is the shape of the field lines of the induced E_n-field in Fig. 33–18, within the colored circle?

b) What are the magnitude and direction of this field at any point of the circular conducting ring of radius 10 cm, and what is the emf in the ring?

c) What is the current in the ring, if its resistance is 2 Ω?

d) What is the potential difference between points a and b of the ring?

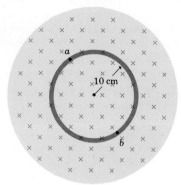

Figure 33-18

e) How do you reconcile your answers to (c) and (d)?

f) If the ring is cut at some point and the ends separated slightly, what will be the potential difference between the ends?

33-20 A square conducting loop, 20 cm on a side, is placed in the same magnetic field as in Problem 33-19. (See Fig. 33-19.)

a) Copy Fig. 33-19, and show by vectors the directions and relative magnitudes of the induced electric field E_n at points a, b, and c.

b) Prove that the component of E_n along the loop has the same value at every point of the loop and is equal to that at the ring of Fig. 33-18.

c) What is the current in the loop, if its resistance is 2 Ω?

d) What is the potential difference between points a and b?

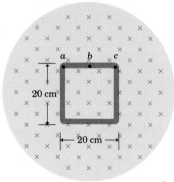

Figure 33-19

33-21 A square conducting loop, 20 cm on a side, is placed in the same magnetic field as in Problem 33-19 with side ac along a diameter and with point b at the center of the field. (See Fig. 33-20.)

a) Copy Fig. 33-20, and show by vectors the directions and relative magnitudes of the induced electric field E_n at the lettered points.

b) What is the induced emf in side ac?

c) What is the induced emf in the loop?

d) What is the current in the loop, if its resistance is 2 Ω?

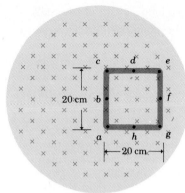

Figure 33-20

e) What is the potential difference between points a and c? Which is the higher potential?

33-22

a) Find the induced field E_n in each side of the square loop in Problem 33-21 and Fig. 33-20.

b) Find the induced emf in each side.

c) Find the *electrostatic* field E_e in each side.

d) Find the potential differences V_{ac}, V_{ce}, V_{eg}, and V_{ga}. What should be the sum of these potential differences?

33-23

a) What is the direction of the drift velocity of an electron at point b in the wire loop of Fig. 33-20?

b) What would be the direction of the drift velocity if the loop were at the left side of the center line instead of at the right side?

c) What would be the direction of the force on an electron if it were at rest at the center of the magnetic field, with no wire loop present?

d) How do you reconcile the answers to (a), (b), and (c)?

33-24 The magnetic flux in a toroid of small cross-sectional area and radius R is increasing at a constant rate $d\Phi/dt$.

a) What are the magnitude and direction of the induced E_n-field at a point on the axis of the toroid at a distance x from its center? (See Sec. 32-6 for the corresponding expression for the B-field at a point on the axis of a circular loop of wire carrying a constant current.)

b) Sketch a graph of the magnitude of E_n as a function of x.

c) Evaluate $\int_{-\infty}^{+\infty} E_n \, dx$ to find the induced emf in a wire that extends along the axis of the toroid from $x = -\infty$ to $x = +\infty$.

d) If the ends of the wire are joined by a conductor very far from the toroid, what is the induced emf in this circuit?

e) What is the induced emf in a ring that links the toroid closely?

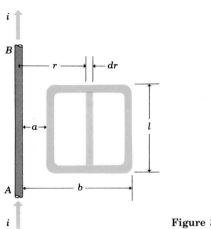

Figure 33-21

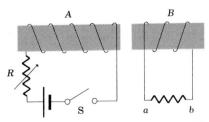

Figure 33-23

33-25 The current in the wire AB of Fig. 33-21 is upward and is increasing steadily at a rate dI/dt.

a) At an instant when the current is i, what are the magnitude and direction of the field $\boldsymbol{B}$ at a distance r from the wire?

b) What is the flux $d\Phi$ through the narrow shaded strip?

c) What is the total flux through the loop?

d) What is the induced emf in the loop?

e) Evaluate the numerical value of the induced emf if $a = 10$ cm, $b = 30$ cm, $l = 20$ cm, and $dI/dt = 2$ A·s^{-1}.

33-26 A cardboard tube is wound with two windings of insulated wire, as in Fig. 33-22. Terminals a and b of winding A may be connected to a battery through a reversing switch. State whether the induced current in the resistor R is from left to right, or from right to left, in the following circumstances:

a) the current in winding A is from a to b and is increasing;

b) the current is from b to a and is decreasing;

c) the current is from b to a and is increasing.

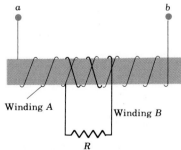

Figure 33-22

33-27 Using Lenz's law, determine the direction of the current in resistor ab of Fig. 33-23 when

a) switch S is opened,

b) coil B is brought closer to coil A,

c) the resistance of R is decreased.

33-28 The cross-section area of a closely wound search coil having 20 turns is 1.5 cm^2 and its resistance is 4 Ω. The coil is connected through leads of negligible resistance to a charge-measuring instrument having internal resistance 16 Ω. Find the quantity of charge displaced when the coil is pulled quickly out of a region where $B = 1.8$ T to a point where the magnetic field is zero. The plane of the coil, when in the field, makes an angle of 90° with the magnetic field.

33-29 A solenoid 50 cm long and 8 cm in diameter is wound with 500 turns. A closely wound coil of 20 turns of insulated wire surrounds the solenoid at its midpoint, and the terminals of the coil are connected to a charge-measuring instrument. The total circuit resistance is 25 Ω.

a) Find the quantity of charge displaced through the instrument when the current in the solenoid is quickly decreased from 3 A to 1 A.

b) Draw a sketch of the apparatus, showing clearly the directions of windings of the solenoid and coil, and of the current in the solenoid. What is the direction of the current in the coil when the solenoid current is decreased?

33-30 A closely wound search coil has an area of 4 cm^2, 160 turns, and a resistance of 50 Ω. It is connected to a charge-measuring instrument whose resistance is 30 Ω. When the coil is rotated quickly from a position parallel to a uniform magnetic field to one perpendicular to the field, the instrument indicates a charge of 4×10^{-5} C. What is the magnitude of the field?

33-31 A Faraday disk dynamo is to be used to supply current to a large electromagnet that requires 20,000 A at 1.0 V. The disk is to be 0.6 m in radius, and it turns in a magnetic field of 1.2 T, supplied by a smaller electromagnet.

a) How many revolutions per second must the disk turn?

b) What torque is required to turn the disk, assuming that all the mechanical energy is dissipated as heat in the electromagnet?

33-32 A search coil used to measure magnetic fields is to be made with a radius of 2 cm. It is to be designed so that flipping it 180° in a field of 0.1 T causes a total charge of 10^{-4} C to flow in a charge-measuring instrument when the total circuit resistance is 50 Ω. How many turns should the coil have?

33-33 A coil 4 cm in radius, containing 500 turns, turns with constant angular velocity about an axis along a diameter perpendicular to the earth's magnetic field, which may be taken as 0.5×10^{-4} T. What angular velocity must it have for the induced emf to have a maximum value of 1.0×10^{-3} V?

33-34 The coil described in Problem 33-33 is placed in a magnetic field which varies with time according to $B = 0.01\, t + (2 \times 10^{-4})t^3$, where B is in teslas and t in seconds. It is connected to a 500-Ω resistor.

a) Find the induced emf in the coil as a function of time.

b) What is the current in the resistor at time $t = 10$ s?

INDUCTANCE 34

34-1 Mutual inductance

An emf is induced in a stationary circuit whenever the magnetic flux through the circuit varies with time. If the variation in flux is brought about by a varying current in a second circuit, it is convenient to express the induced emf in terms of the varying *current*, rather than in terms of the varying *flux*. We shall use the symbol i to represent the instantaneous value of a varying current.

Figure 34–1 is a sectional view of two closely wound coils of wire. A current i_1 in coil 1 sets up a magnetic field, as indicated by the color lines, and some of these lines pass through coil 2. Let the resulting flux through coil 2 be Φ_2. The magnetic field is proportional to i_1, so Φ_2 is also proportional to i_1. When i_1 changes, Φ_2 changes; this changing flux induces an emf $\mathcal{E}_2$ in coil 2, given by

$$\mathcal{E}_2 = N_2 \frac{d\Phi_2}{dt}. \tag{34–1}$$

The proportionality of Φ_2 and i_1 could be represented in the form $\Phi_2 = (\text{constant})\, i_1$, but it is more convenient to include the number of turns N_2 in the relation. Introducing a proportionality constant M, we write

$$N_2 \Phi_2 = M i_1, \tag{34–2}$$

From this,

$$N_2 \frac{d\Phi_2}{dt} = M \frac{di_1}{dt},$$

and Eq. (34–1) may be rewritten

$$\mathcal{E}_2 = M \frac{di_1}{dt}. \tag{34–3}$$

The constant M, which depends only on the geometry of the two coils,

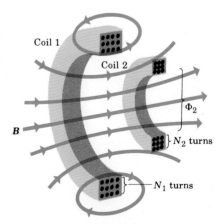

34–1 A portion of the flux set up by a current in coil 1 links with coil 2.

is called their *mutual inductance*. It is defined by Eq. (34–2), which may be rewritten

$$M = \frac{N_2 \Phi_2}{i_1}. \qquad (34\text{–}4)$$

The entire discussion can be repeated for the case where a changing current i_2 in coil 2 causes a changing flux Φ_1 and hence an emf $\mathcal{E}_1$ in coil 1. It might be expected that the constant M would be different in this case, since the two coils are not, in general, symmetric. It turns out, however, that M is always the same in this case as in the case considered above, so a single value of the mutual inductance characterizes completely the induced-emf interaction of two coils.

The SI unit of mutual inductance, from Eq. (34–4), is *one weber per ampere*. An equivalent unit, obtained by reference to Eq. (34–3), is *one volt-second per ampere*. These two equivalent units are called *one henry* (1 H), in honor of Joseph Henry (1797–1878), a pioneer in the development of electromagnetic theory and, incidentally, the first director of the Smithsonian Institution. Thus the unit of mutual inductance is

$$1\,\text{H} = 1\,\text{Wb·A}^{-1} = 1\,\text{V·s·A}^{-1}.$$

Example A long solenoid of length l and cross-sectional area A is closely wound with N_1 turns of wire. A small coil of N_2 turns surrounds it at its center, as in Fig. 34–2. A current i_1 in the solenoid sets up a **B** field at its center, of magnitude

$$B = \mu_0 n i_1 = \frac{\mu_0 N_1 i_1}{l}.$$

The flux through the central section is equal to BA, and since all of this flux links with the small coil, the mutual inductance is

$$M = \frac{N_2 \Phi_2}{i_1} = \frac{N_2}{i_1}\left(\frac{\mu_0 N_1 i_1}{l}\right)A = \frac{\mu_0 A N_1 N_2}{l}.$$

If $l = 0.50\,\text{m}$, $A = 10\,\text{cm}^2 = 10^{-3}\text{m}^2$, $N_1 = 1000$ turns, $N_2 = 10$ turns,

$$M = \frac{(4\pi \times 10^{-7}\,\text{Wb·A}^{-1}\text{·m}^{-1})(10^{-3}\text{m}^2)(1000)(10)}{0.5\,\text{m}}$$

$$= 25.1 \times 10^{-6}\,\text{Wb·A}^{-1} = 25.1 \times 10^{-6}\,\text{H} = 25.1\,\mu\text{H}.$$

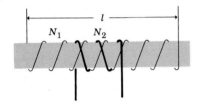

34–2

34-2 Self-inductance

In the preceding section, the circuit acting as the source of the magnetic field linking a circuit was assumed to be independent of the circuit in which the induced emf appeared. But whenever there is a current in any circuit, this current sets up a magnetic field that links with *the same* circuit and varies when the current varies. Hence any circuit in which there is a varying current has induced in it an emf due to the variation in *its own* magnetic field. Such an emf is called a *self-induced electromotive force*.

Suppose that a circuit has N turns of wire and that a flux Φ passes through each turn, as in Fig. 34–3. In analogy to Eq. (34–4), we define

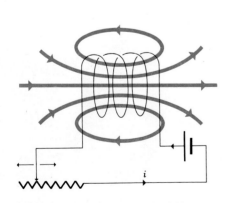

34-3 A flux Φ linking a coil of N turns. When the current in the circuit changes, the flux changes also, and a self-induced emf appears in the circuit.

the *self-inductance L* of the circuit:

$$L = \frac{N\Phi}{i}. \tag{34-5}$$

This can be written as

$$N\Phi = Li.$$

If Φ and i change with time, then

$$N\frac{d\Phi}{dt} = L\frac{di}{dt},$$

and, since the self-induced emf $\mathcal{E}$ is

$$\mathcal{E} = N\frac{d\Phi}{dt},$$

it follows that

$$\mathcal{E} = L\frac{di}{dt}. \tag{34-6}$$

The self-inductance of a circuit is therefore *the self-induced emf per unit rate of change of current*. The SI unit of self-inductance is 1 henry.

A circuit, or part of a circuit, that has inductance is called an *inductor*. An inductor is represented by the symbol

The *direction* of a self-induced (non-electrostatic) field can be found from Lenz's law. We consider the "cause" of this field, and hence of the emf associated with it, to be the *changing current* in the conductor. If the current is *increasing*, the direction of the induced field is *opposite* to that of the current. If the current is *decreasing*, the induced field is in the *same* direction as the current. Thus it is the *change* in current, not the current itself, that is "opposed" by the induced field.

Example 1   An air-core toroidal solenoid of cross-section area A and mean radius r is closely wound with N turns of wire. We neglect the variation of B across the cross section, assuming its average value to be very nearly equal to the value at the center of the cross section, as given by Eq. (32-26). Then the flux in the toroid is

$$\Phi = BA = \frac{\mu_0 NiA}{2\pi r}.$$

Since all of the flux links with each turn, the self-inductance is

$$L = \frac{N\Phi}{i} = \frac{\mu_0 N^2 A}{2\pi r}.$$

Thus, if $N = 100$ turns, $A = 10 \text{ cm}^2 = 10^{-3}\text{m}^2$, $r = 0.10$ m,

$$L = \frac{(4\pi \times 10^{-7}\text{ Wb}\cdot\text{A}^{-1}\cdot\text{m}^{-1})(100)^2(10^{-3}\text{m}^2)}{2\pi(0.10\text{ m})}$$

$$= 20 \times 10^{-6}\text{ H} = 20\mu\text{H}.$$

Example 2 If the current in the coil above increases uniformly from zero to 1 A in 0.1 s, find the magnitude and direction of the self-induced emf.

Solution

$$\mathcal{E} = L\frac{di}{dt} = (20 \times 10^{-6}\,\mathrm{H})\frac{1\,\mathrm{A}}{0.1\,\mathrm{s}}$$

$$= 2.0 \times 10^{-4}\,\mathrm{V}.$$

Since the current is increasing, the direction of this emf is opposite to that of the current. For example, if the inductor terminals are a and b and an increasing current is in the direction a to b in the inductor, then $\mathbf{E}_n$ and the emf in the inductor are in the direction b to a, tending to cause a current *in the external circuit* in the direction a to b.

When Kirchhoff's loop rule (Sec. 28-4) is used with circuits containing inductors, this emf is treated as though it were a potential difference, with a at higher potential than b. That is, if the defined positive direction of the current i is from a to b in the inductor, then

$$V_{ab} = L\frac{di}{dt}. \tag{34-7}$$

The self-inductance of a circuit depends on its size, shape, number of turns, etc. It also depends on the magnetic properties of the material enclosed by the circuit. For example, the self-inductance of a solenoid of given dimensions is much greater if it has an iron core than if it is in vacuum, and when ferromagnetic materials are present, the self-inductance varies in a complicated way as the current varies. For simplicity, we shall consider here only circuits of constant self-inductance.

34-3 Energy in an inductor

Because a changing current in an inductor causes an emf, the source supplying the current must maintain a potential difference between its terminals and hence must supply energy to the inductor. When an inductor carries an instantaneous current i which is changing at the rate di/dt, the induced emf is equal to $L\,di/dt$, and the power P supplied to the inductor is

$$P = \mathcal{E}i = Li\frac{di}{dt}.$$

The energy dW supplied in time dt is $P\,dt$, or

$$dW = Li\,di,$$

and the total energy supplied while the current increases from zero to a final value I is

$$W = L\int_0^I i\,di = \tfrac{1}{2}LI^2. \tag{34-8}$$

After the current has reached its final steady value, $di/dt = 0$, and the power input is zero. The energy that has been supplied to the inductor is used to establish the magnetic field in and around the inductor, where it is stored as a form of potential energy as long as the current is maintained. When the current is reduced to zero, this energy is returned to the circuit which supplied it. If the current is suddenly interrupted by opening a switch, the energy may be dissipated in an arc across the switch contacts.

The energy can be considered as associated with the magnetic field itself, and a relationship can be developed which is analogous to that obtained for electric-field energy in Sec. 27-4, Eq. (27-9). As in that discussion, we consider only one simple case, the toroidal solenoid. As in the preceding example, we assume that the cross-section area A is small enough so the magnetic field may be considered constant over the area, and so the volume in the toroid is approximately equal to the circumferential length $l = 2\pi r$ multiplied by the area A. From the preceding example, the self-inductance of the toroidal solenoid is

$$L = \frac{\mu_0 N^2 A}{l},$$

and the energy stored in the toroid when the current in the windings is I is

$$W = \frac{1}{2}LI^2 = \frac{1}{2}\left(\frac{\mu_0 N^2 A}{l}\right)I^2.$$

We can think of this energy as localized in the volume enclosed by the windings, equal to lA. The energy *per unit volume u* is then

$$u = \frac{W}{lA} = \frac{1}{2}\mu_0\left(\frac{N^2 I^2}{l^2}\right).$$

But $N^2 I^2 / l^2 = B^2 / \mu_0^2$, so

$$u = \frac{B^2}{2\mu_0}, \tag{34-9}$$

which is the analog of the expression for the energy per unit volume in the electric field of an air capacitor, $\frac{1}{2}\epsilon_0 E^2$, discussed in Sec. 27-4.

34-4 The *R-L* circuit

Every inductor necessarily has some resistance (unless its windings are superconducting). To distinguish between the effects of the resistance R and the self-inductance L, we represent the inductor as in Fig. 34-4, replacing it with an ideal resistanceless inductor in series with a noninductive resistor. The same diagram can also represent a resistor in series with an inductor, in which case R is the *total* resistance of the combination. By means of the switch, the *R-L* circuit may be connected to a source having constant terminal voltage V, or it may be shorted by the conductor across the lower switch terminals.

Suppose the switch in the diagram is suddenly closed in the "up" position. Because of the self-induced emf, the current does not immediately rise to its final value at the instant the circuit is closed, but grows at a rate that depends on the inductance and resistance of the circuit.

At some instant after the switch is closed, let i represent the current in the circuit and di/dt its rate of increase. The potential difference

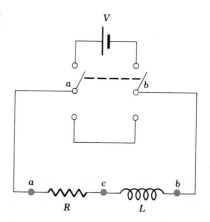

34-4

across the inductor is

$$v_{cb} = L\frac{di}{dt},$$

and that across the resistor is

$$v_{ac} = iR.$$

Since $V = v_{ac} + v_{cb}$, it follows that

$$V = L\frac{di}{dt} + iR. \tag{34-10}$$

The rate of increase of current is therefore

$$\frac{di}{dt} = \frac{V - iR}{L} = \frac{V}{L} - \frac{R}{L}i. \tag{34-11}$$

At the instant the circuit is first closed, $i = 0$ and the current starts to grow at the rate

$$\left(\frac{di}{dt}\right)_{\text{initial}} = \frac{V}{L}.$$

The greater the self-inductance L, the more slowly does the current start to increase.

As the current increases, the term Ri/L increases also, and hence the *rate* of increase of current becomes smaller and smaller, as Eq. (34-11) shows. When the current reaches its final *steady-state* value I, its rate of increase is zero. Then

$$0 = \frac{V}{L} - \left(\frac{R}{L}\right)I$$

and

$$I = \frac{V}{R}.$$

That is, the *final* current does not depend on the self-inductance and is the same as it would be in a pure resistance R connected to a source having emf V.

To obtain an expression for the current as a function of time, we proceed just as we did for the problem of the charging capacitor, discussed in Sec. 29-5. We first rearrange Eq. (34-11):

$$\frac{di}{(V/R) - i} = \frac{R}{L}dt.$$

Integrating both sides, we find

$$-\ln\left(\frac{V}{R} - i\right) = \left(\frac{R}{L}\right)t + \text{constant}.$$

The integration constant is evaluated by noting that the initial current is zero, so $i = 0$ at time $t = 0$.

$$\text{constant} = -\ln\frac{V}{R}.$$

Rearranging again, we obtain

$$\ln\left(\frac{V}{R} - i\right) - \ln\frac{V}{R} = \ln\left(1 - \frac{Ri}{V}\right) = -\left(\frac{R}{L}\right)t,$$

$$i = \frac{V}{R}(1 - e^{-(R/L)t}). \tag{34-12}$$

From this,

$$\frac{di}{dt} = \frac{V}{L}e^{-(R/L)t}. \tag{34-13}$$

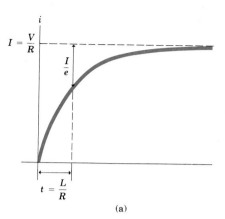

(a)

We note that at time $t = 0$, $i = 0$ and $di/dt = V/L$, and that as $t \to \infty$, $i \to V/R$ and $di/dt \to 0$, as predicted above.

Figure 34-5a is a graph of Eq. (34-12) and shows the variation of current with time. The instantaneous current i first rises rapidly, then increases more slowly and approaches asymptotically the final value $I = V/R$. At a time τ equal to L/R the current has risen to $(1 - 1/e)$ or about 0.63 of its final value. This time is called the *time constant* or the *decay constant* for the circuit.

$$\tau = \frac{L}{R}. \tag{34-14}$$

The time required for the current to reach *half* its final value is obtained by setting

$$e^{-(R/L)t} = \tfrac{1}{2},$$

from which $T_{1/2} = (L/R)\ln 2$. For a given value of R, these times increase when the inductance L increases. Thus, although the graph of i vs. t has the same general shape whatever the inductance, the current rises rapidly to its final value if L is small, and slowly if L is large. For example, if $R = 100\,\Omega$ and $L = 10\,$H,

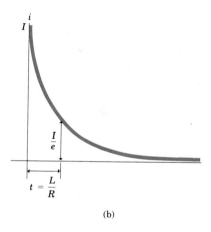

(b)

34-5 (a) Growth of current in a circuit containing inductance and resistance. (b) Decay of current in a circuit containing inductance and resistance.

$$\frac{L}{R} = \frac{10\,\text{H}}{100\,\Omega} = 0.1\,\text{s},$$

and the current increases to about 63% of its final value in 0.1 s. On the other hand, if $L = 0.01\,$H,

$$\frac{L}{R} = \frac{0.01\,\text{H}}{100\,\Omega} = 10^{-4}\,\text{s},$$

and only $10^{-4}\,$s is required for the current to increase to 63% of its final value.

If there is a steady current I in the circuit of Fig. 34-4 and the switch is then quickly thrown to the "down" position, the current decays, as shown in Fig. 34-5(b). The equation of the decaying current is

$$i = Ie^{-Rt/L}, \tag{34-15}$$

and the time constant, L/R, is the time for the current to decrease to $1/e$ or about 37% of its original value. Derivation of Eq. (34-15) is left as a problem. The energy necessary to maintain the current during this decay is provided by the energy stored in the magnetic field of the inductor.

34-5 The L-C circuit

The behavior of an R-C circuit was discussed in Sec. 29-5, and that of an R-L circuit in Sec. 34-4. We now consider the L-C circuit shown in Fig. 34-6, a resistanceless inductor connected between the terminals of a charged capacitor. At the instant connections are made, inFig. 34-6a, the capacitor starts to discharge through the inductor. At a later instant, represented in Fig. 34-6b, the capacitor has completely discharged and the potential difference between its terminals (and those of the inductor) has decreased to zero. The current in the inductor has meanwhile established a magnetic field in the space around it. This magnetic field now decreases, inducing an emf in the inductor in the same direction as the current. The current therefore persists, although with diminishing magnitude, until the magnetic field has disappeared and the capacitor has been charged in the opposite sense to its initial polarity, as in Fig. 34-6c. The process now repeats itself in the reversed direction, and, in the absence of energy losses, the charges on the capacitor surge back and forth indefinitely. This process is called an *electrical oscillation*.

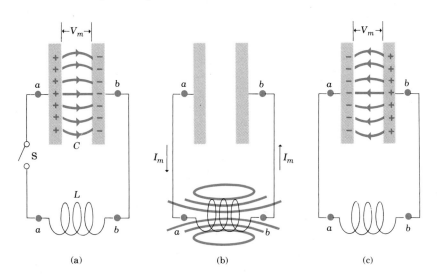

34-6 Energy transfer between electric and magnetic fields in an oscillating L-C circuit.

(a)   (b)   (c)

From the energy standpoint, the oscillations of an electrical circuit consist of a transfer of energy back and forth from the electric field of the capacitor to the magnetic field of the inductor, the *total* energy associated with the circuit remaining constant. This is analogous to the transfer of energy in an oscillating mechanical system from kinetic to potential, and vice versa.

The frequency of the electrical oscillations of a circuit containing inductance and capacitance only (a so-called L-C circuit) may be calculated in exactly the same way as the frequency of oscillation of a body suspended from a spring, discussed in Chapter 11. In the mechanical problem, a body of mass m is attached to a spring of force constant k. Let the body be displaced a distance A from its equilibrium positions and released from rest at time $t = 0$. Then, as shown in the left column of Table 34-1, the kinetic energy of the system at any later time is $\frac{1}{2}mv^2$, and its elastic potential energy is $\frac{1}{2}kx^2$. Because the system is conservative, the sum of these equals the initial energy of the system, $\frac{1}{2}kA^2$. The

Table 34-1   Oscillation of a mass on a spring compared with the electrical oscillation in an *L-C* circuit

Mass on a spring	Circuit containing inductance and capacitance
Kinetic energy $= \frac{1}{2}mv^2$	Magnetic energy $= \frac{1}{2}Li^2$
Potential energy $= \frac{1}{2}kx^2$	Electrical energy $\frac{1}{2}\dfrac{1}{C}q^2$
$\frac{1}{2}mv^2 + \frac{1}{2}kx^2 = \frac{1}{2}kA^2$	$\frac{1}{2}Li^2 + \frac{1}{2}\dfrac{1}{C}q^2 = \frac{1}{2}\dfrac{1}{C}Q^2$
$v = \pm\sqrt{k/m}\,\sqrt{A^2 - x^2}$	$i = \pm\sqrt{1/LC}\,\sqrt{Q^2 - q^2}$
$v = \dfrac{dx}{dt}$	$i = \dfrac{dq}{dt}$
$x = A \cos \sqrt{k/m}\,t = A \cos \omega t$	$q = Q \cos \sqrt{1/LC}\,t = Q \cos \omega t$
$v = -\omega A \sin \omega t = -v_{max}\sin \omega t$	$i = -\omega Q \sin \omega t = -I \sin \omega t$

velocity v at any coordinate x is therefore

$$v = \pm\sqrt{\frac{k}{m}}\,\sqrt{A^2 - x^2}. \qquad (34\text{-}16)$$

The velocity equals dx/dt, and the coordinate x as a function of t has been shown (Chapter 11) to be

$$x = A \cos\left(\sqrt{\frac{k}{m}}\right)t = A \cos \omega t, \qquad (34\text{-}17)$$

where the angular frequency ω is

$$\omega = \sqrt{\frac{k}{m}}.$$

We recall also that the ordinary frequency f, the number of cycles per unit time, is given by $f = \omega/2\pi$.

In the electrical problem, also a conservative system, a capacitor of capacitance C is given an initial charge Q and, at time $t = 0$, is connected to the terminals of an inductor of self-inductance L. The magnetic energy of the inductor at any later time corresponds to the kinetic energy of the vibrating body and is given by $\frac{1}{2}Li^2$. The electrical energy of the capacitor corresponds to the elastic potential energy of the spring and is given by $q^2/2C$, where q is the charge on the capacitor. The sum of these equals the initial energy of the system, $Q^2/2C$. That is,

$$\frac{1}{2}Li^2 + \frac{q^2}{2C} = \frac{Q^2}{2C}.$$

Solving for i, we find that when the charge on the capacitor is q, the current i is

$$i = \sqrt{\frac{1}{LC}}\,\sqrt{Q^2 - q^2}. \qquad (34\text{-}18)$$

Comparing this with Eq. (34-16), we see that the current $i = dq/dt$ varies with time in the same way as the velocity $v = dx/dt$ in the mechanical problem. Continuing the analogy, we conclude that q is given as a function of time by

$$q = Q \cos\left(\sqrt{\frac{1}{LC}}\right)t = Q \cos \omega t. \qquad (34\text{–}19)$$

The angular frequency ω of the electrical oscillations is therefore

$$\omega = \sqrt{\frac{1}{LC}}. \qquad (34\text{–}20)$$

This is called the *natural frequency* of the *L–C* circuit.

These results may also be derived directly from an analysis of the *L–C* circuit. At each instant the capacitor voltage $v_{ab} = q/C$ must equal that of the inductor, $v_{ab} = L\,di/dt$. Also, because of the choice of positive direction for i, we have $i = -dq/dt$. Combining these relations, we find

$$\frac{d^2q}{dt^2} + \frac{1}{LC}q = 0. \qquad (34\text{–}21)$$

The solutions of this differential equation, functions whose second derivative is equal to $-1/LC$ times the original function, are

$$q = Q \cos \omega t, \qquad (34\text{–}22a)$$
$$q = Q \sin \omega t, \qquad (34\text{–}22b)$$
$$q = Q \cos (\omega t + \phi), \qquad (34\text{–}22c)$$

where $\omega = 1/(LC)^{1/2}$, and Q and ϕ are constants. Just as with the harmonic oscillator, the choice of one of these functions is determined by the initial conditions. If at time $t = 0$, the capacitor has maximum charge and $i = 0$, as in the above discussion, then Eq. (34–22a) is to be used. If at $t = 0$, $q = 0$ but i is different from zero, we use Eq. (34–22b), and if both q and i are different from zero at time $t = 0$, the more general form, Eq. (34–22c), must be used.

34–6   The *R–L–C* Circuit

In the above discussion we have assumed that the *L–C* circuit contains no *resistance*. This is an idealization, of course; for every real inductor there is resistance associated with the windings, and there may be resistance in the connecting wires as well. The effect of resistance is to dissipate the electromagnetic energy and convert it to heat; thus, resistance in an electric circuit plays a role analogous to that of friction in a mechanical system.

Suppose an inductor of self-inductance L and a resistor of resistance R are connected in series across the terminals of a capacitor. If the capacitor is initially charged, it starts to discharge at the instant the connections are made but, because of i^2R losses in the resistor, the energy of the inductor when the capacitor is completely discharged is *less* than the original energy of the capacitor. In the same way, the energy of the capacitor when the magnetic field has collapsed is still smaller, and so on.

If the resistance R is relatively small, the circuit oscillates, but with *damped harmonic motion*, as illustrated in Fig. 34–7a. As R is increased, the oscillations die out more rapidly. At a sufficiently large value of R,

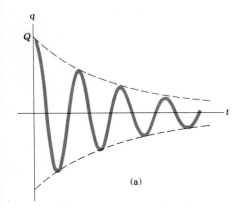

(a)

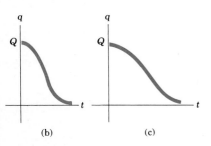

(b)   (c)

34–7 Graphs of q versus t in an *R–L–C* circuit. (a) Small damping. (b) Critically damped. (c) Overdamped.

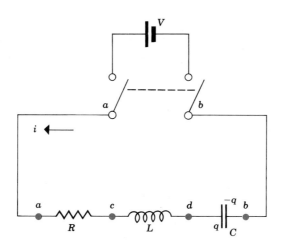

34-8 An *R-L-C* series circuit.

the circuit no longer oscillates and is said to be *critically damped*, as in Fig. 34–7b. For still larger resistances it is *overdamped*, as in Fig. 34–7c.

With the proper electronic circuitry, energy can be fed *into* an *R–L–C* circuit at the same rate as that at which it is dissipated by i^2R or radiation losses. In effect, a *negative resistance* is inserted in the circuit so that its total resistance is zero. the circuit then oscillates with *sustained* oscillations, as does the idealized circuit with no resistance, discussed in Sec. 34–5.

The circuit of Fig. 34–8 may be analyzed in detail to find the charge on the capacitor and the current in the circuit as functions of time; that analysis is sketched below. Suppose the switch is closed in the upward position for a long time, so the capacitor acquires a charge $Q = CV$. Then at time $t = 0$ the switch is flipped to the downward position. Application of Kirchhoff's loop rule to the circuit yields the equation

$$iR + L\frac{di}{dt} + \frac{q}{C} = 0.$$

We replace i with dq/dt and rearrange to obtain

$$\frac{d^2q}{dt^2} + \frac{R}{L}\frac{dq}{dt} + \frac{1}{LC}q = 0. \qquad (34\text{--}23)$$

We note that when $R = 0$ this reduces to Eq. (34–21).

Solutions of Eq. (34–23) can be obtained by general methods of differential equations. The form of the solution depends on whether R is large or small. When R is less than $2(L/C)^{1/2}$, the solution has the form

$$q = VCe^{-(R/2L)t}\cos\sqrt{\frac{1}{LC} - \left(\frac{R}{2L}\right)^2}\,t. \qquad (34\text{--}24)$$

When R is greater than $2(L/C)^{1/2}$, the solution is

$$q = e^{-(R/2L)t}\left[Ae^{\sqrt{\left(\frac{R}{2L}\right)^2 - \frac{1}{LC}}\,t} + Be^{-\sqrt{\left(\frac{R}{2L}\right)^2 - \frac{1}{LC}}\,t}\right], \qquad (34\text{--}25)$$

where A and B are constants determined by V, R, L, and C.

Equation (34–24) corresponds to the *underdamped* behavior shown in Fig. 34–7a; the function represents a sinusoidal oscillation with an exponentially decaying amplitude. We note that the angular frequency of the oscillation is no longer $1/(LC)^{1/2}$ but is *less* than this because of the term containing R. The frequency of the damped oscillations is thus given by

$$\omega = \sqrt{\frac{1}{LC} - \frac{R^2}{4L^2}}. \qquad (34\text{-}26)$$

As R increases, ω becomes smaller and smaller; when $R^2 = 2L/C$ the quantity under the radical becomes zero and the case of *critical damping* has been reached (Fig. 34–7b). For still larger values of R the behavior is no longer oscillatory but is described as the sum of two exponential functions, as in Fig. 34–7c; the circuit is then *overdamped*.

Additional interesting aspects of the behavior of this circuit emerge when a sinusoidally varying emf is included in the circuit. This leads us into the area of alternating-current (ac) circuit analysis, a topic discussed in detail in Chapter 36.

Questions

34-1 A resistor is to be made by winding a wire around a cylindrical form. In order to make the inductance as small as possible, it is proposed that we wind half the wire in one direction and the other half in the opposite direction. Would this achieve the desired result? Why or why not?

34-2 In Fig. 34-1, if coil 2 is turned 90° so that its axis is vertical, does the mutual inductance increase or decrease?

34-3 The toroidal solenoid is one of the few configurations for which it is easy to calculate self-inductance. What features of the toroidal solenoid give it this simplicity?

34-4 Two identical closely wound circular coils, each having self-inductance L, are placed side by side, close together. If they are connected in series, what is the self-inductance of the combination? What if they are connected in parallel? Can they be connected so that the total inductance is zero?

34-5 If two inductors are separated enough so that practically no flux from either links the coils of the other, show that the equivalent inductance of two inductors in series or parallel is obtained by the same rules used for combining resistance.

34-6 Two closely wound circular coils have the same number of turns, but one has twice the radius of the other. How are the self-inductances of the two coils related?

34-7 One of the great problems in the field of energy resources and utilization is the difficulty of storing electri-

cal energy in large quantities economically. Discuss the possibility of storing large amounts of energy by means of currents in large inductors.

34-8 In what regions in a toroidal solenoid is the energy density greatest? Least?

34-9 Suppose there is a steady current in an inductor. If one attempts to reduce the current to zero instantaneously by opening a switch, a big fat arc appears at the switch contacts. Why? What happens to the induced emf in this situation? Is it physically possible to stop the current instantaneously?

34-10 In the R–L circuit of Fig. 34-4, is the current in the resistor always the same as that in the inductor? How do you know?

34-11 In the R–L circuit of Fig. 34-4, when the switch is in the up position and is then thrown to the down position, the potential V_{ab} changes suddenly and discontinuously, but the current does not. Why can the voltage change suddenly but not the current?

34-12 In the R–L–C circuit, what criteria could be used to decide whether the system is overdamped or underdamped? For example, could one compare the maximum energy stored during one cycle to the energy dissipated during one cycle?

Problems

34-1 A solenoid of length 10 cm and radius 2 cm is wound uniformly with 1000 turns. A second coil of 50 turns is wound around the solenoid at its center. What is the mutual inductance of the two coils?

34-2 A toroidal solenoid (cf. Sec. 32–8, Example 2) has a radius of 10 cm and a cross-sectional area of 5 cm^2, and is wound uniformly with 1000 turns. A second coil with 500 turns is wound uniformly on top of the first. What is the mutual inductance?

34-3 Find the self-inductance of the toroidal solenoid in Problem 34-2 if only the 1000-turn coil is used. How would your answer change if the two coils were connected in series?

34-4 A toroidal solenoid has two coils with n_1 and n_2 turns, repectively; it has radius r and cross-sectional area A.

a) Derive an expression for the self-inductance L_1 when only the first coil is used, and that for L_2 when only the second soil is used.

b) Derive an expression for the mutual inductance of the two coils.

c) Show that $M^2 = L_1 L_2$. This result is valid whenever all the flux linked by one coil is also linked by the other.

34-5 Two coils have mutual inductance $M = 0.01$ H. The current i_1 in the first coil increases at a uniform rate of 0.05 A·s^{-1}.

a) What is the induced emf in the second coil? Is it constant?

b) Suppose that the current described is that in the second coil rather than the first; what is the induced emf in the first coil?

34-6 An inductor of inductance 5 H carries a current that decreases at a uniform rate, $di/dt = -0.02$ A·s^{-1}. Find the self-induced emf; what is its polarity?

34-7 An inductor used in a dc power supply has an inductance of 20 H and a resistance of 200 Ω, and carries a current of 0.1 A.

a) What is the energy stored in the magnetic field?

b) At what rate is energy dissipated in the resistor?

34-8

a) Show that the two expressions for self-inductance, namely

$$\frac{N\Phi}{i} \quad \text{and} \quad \frac{\mathcal{E}}{di/dt},$$

have the same units.

b) Show that L/R and RC both have the units of time.

c) Show that 1 Wb·s^{-1} equals 1 V.

34-9 An inductor with $L = 40$ H carries a current i that varies with time according to $i = (0.1 \text{ A}) \sin (120\pi \text{ s}^{-1})t$. Find an expression for the induced emf. What is the phase of $\mathcal{E}$ relative to i?

34-10 A coaxial cable consists of a small solid conductor of radius r_a supported by insulating disks on the axis of a thin-walled tube of inner radius r_b. Show that the self-inductance of a length l of the cable is

$$L = l \frac{\mu_0}{2\pi} \ln \frac{r_b}{r_a}.$$

Assume the inner and outer conductors to carry equal currents in opposite directions. [*Hint:* Use Ampère's law to find the magnetic field at any point in the space between the conductors. Write the expression for the flux $d\Phi$ through a narrow strip of length l parallel to the axis, of width dr, at a distance r from the axis of the cable and lying in a plane containing the axis. Integrate to find the total flux linking a current i in the central conductor.]

34-11 A certain toroidal solenoid has a rectangular cross section, as shown in Fig. 34-9. It has N uniformly spaced turns, with air inside. The magnetic field at a point inside the toroid is given by Eq. (32–26). *Do not* assume the field to be uniform over the cross section.

a) Show that the magnetic flux through a cross section of the toroid is

$$\Phi = \frac{\mu_0 N I h}{2\pi} \ln \frac{b}{a}.$$

b) Show that the self-inductance of the toroidal solenoid is given by

$$L = \frac{\mu_0 N^2 h}{2\pi} \ln \frac{b}{a}.$$

c) The fraction b/a may be written as

$$\frac{b}{a} = \frac{a + b - a}{a} = 1 + \frac{b - a}{a}.$$

The power series expansion for $\ln(1 + x)$ is $\ln(1 + x) = x + x^2/2 + \cdots$. Hence show that when $b - a$ is much less

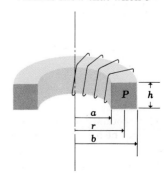

Figure 34-9

than a, the self-inductance is approximately equal to

$$L = \frac{\mu_0 N^2 h(b - a)}{2\pi a}.$$

Compare this result with the result obtained in Sec. 34–1, Example 1.

34-12 For the toroidal solenoid described in Prob. 34–11, calculate the total energy when the current is I. Do not assume the field to be uniform over the cross section. Obtain the total energy in two ways:

a) Use Eq. (34–8), with the expression for L obtained in Prob. 34–11;

b) Integrate Eq. (34–9) through the volume of the toroid.

Compare the results of parts (a) and (b).

34-13 The current in a resistanceless inductor is caused to vary with time as in the graph of Fig. 34–10.

a) Sketch the pattern that would be observed on the screen of an oscilloscope connected to the terminals of the inductor. (The oscilloscope spot sweeps horizontally across the screen at constant speed, and its vertical deflection is proportional to the potential difference between the inductor terminals.)

b) Explain why the inductor can be described as a "differentiating circuit."

Figure 34-10

34-14 The 1000-turn toroidal solenoid described in Problem 34–2 carries a current of 5.0 A.

a) What is the energy density in the magnetic field.

b) What is the total magnetic-field energy? Find using Eq. (34–8) and also by multiplying the energy density from (a) by the volume of the toroid, which is $2\pi rA$; compare the two results.

34-15 A toroidal solenoid has a mean radius of 0.12 m and a cross-sectional area of 20×10^{-4} m²? It is found that when the current is 20 A, the energy stored is 0.1 J. How many turns does the winding have?

34-16 An inductor of inductance 3 H and resistance 6 Ω is connected to the terminals of a battery of emf 12 V and of negligible internal resistance. Find

a) the initial rate of increase of current in the circuit,

b) the rate of increase of current at the instant when the current is 1 A,

c) the current 0.2 s after the circuit is closed,

d) the final steady-state current.

34-17 The resistance of a 10-H inductor is 200 Ω. The inductor is suddenly connected across a potential difference of 10 V.

a) What is the final steady current in the inductor?

b) What is the initial rate of increase of current?

c) At what rate is the current increasing when its value is one-half the final current?

d) At what time after the circuit is closed does the current equal 99% of its final value?

e) Compute the current at the following times after the circuit is closed: 0, 0.025 s, 0.05 s, 0.075 s, 0.10 s. Show the results in a graph.

34-18 An inductor of resistance R and self-inductance L is connected in series with a noninductive resistor of resistance R_0 to a constant potential difference V (Fig. 34–11).

a) Find the expression for the potential difference v_{cb} across the inductor at any time t after switch S_1 is closed.

b) Let $V = 20$ V, $R_0 = 50\,\Omega$, $R = 150\,\Omega$, $L = 5$ H. Compute a few points, and construct graphs of v_{ac} and v_{cb} over a time interval from zero to twice the time constant of the circuit.

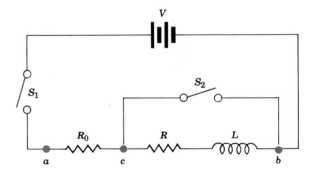

Figure 34-11

34-19 After the current in the circuit of Fig. 34–11 has reached its final steady value the switch S_2 is closed, thus short-circuiting the inductor. What will be the magnitude and direction of the current in S_2, 0.01 s after S_2 is closed?

34-20 Refer to Problem 34–16.

a) What is the power input to the inductor at the instant when the current in it is 0.5 A?

b) What is the rate of dissipation of energy at this instant?

c) What is the rate at which the energy of the magnetic field is increasing?

d) How much energy is stored in the magnetic field when the current has reached its final steady value?

34-21 An inductor having inductance L and resistance R carries a current I. Show that the time constant is equal to *twice the ratio* of the energy stored in the magnetic field to the rate of dissipation of energy in the resistance.

34-22 Show that the quantity $(L/C)^{1/2}$ has units of resistance (ohms).

34-23 The maximum capacitance of a variable air capacitor is 35 pF.

a) What should be the self-inductance of a coil to be connected to this capacitor if the natural frequency of the L–C circuit is to be 550×10^3 Hz, corresponding to one end of the broadcast band?

b) The frequency at the other end of the broadcast band is 1550×10^3 Hz. What must be the minimum capacitance of the capacitor if the natural frequency is to be adjustable over the range of the broadcast band?

34-24 Write an equation corresponding to Eq. (34–11) for the current in Fig. 34–4 just after the switch is thrown to the down position, if the initial current is I. Solve the resulting differential equation and verify Eq. (34–15).

34-25 In Fig. 34–6, equate the voltage across the inductor to that across the capacitor, to show that

$$L \frac{d^2 q}{dt^2} = -\frac{q}{C}.$$

Show that this differential equation is satisfied by the function $q = Q \cos \omega t$, with ω given by $1/(LC)^{1/2}$.

34-26 An inductor having $L = 40$ mH is to be combined with a capacitor to make an L–C circuit with natural frequency 2×10^6 Hz. What value of capacitance should be used?

34-27 An inductor is made with two coils wound close together on a form, so all the flux linking one coil also links the other. The number of turns is the same in each. If the inductance of one coil is L, what is the inductance when the two coils are connected

a) in series;

b) in parallel?

c) If an L–C circuit using this inductor has natural frequency ω using one coil, what is the natural frequency when the two coils are used in series?

34-28 An L–C circuit in an AM radio tuner uses an inductor with inductance 0.1 mH and a variable capacitor. If the natural frequency of the circuit is to be adjustable over the range 0.5 to 1.5 MHz, corresponding to the AM broadcast band, what range of capacitance must the variable capacitor cover?

34-29 The equation preceding Eq. (34–23) may be converted into an energy relation. Multiply both sides of this equation by $i = dq/dt$. Show that the first term on the right is $i^2 R$, the second can be written $d(\frac{1}{2}Li^2)/dt$, and the third can be written $d(q^2/2C)/dt$. Hence, show that the rate at which the battery delivers energy, $V_{ab}i$, is equal to the rate of energy dissipation in the resistor plus the sum of the rates of change of energy stored in the inductor and the capacitor.

MAGNETIC PROPERTIES OF MATTER

35

35-1 Magnetic materials

In Chapter 32 we discussed the magnetic fields set up by moving charges or by currents in conductors, when the charges or conductors are in air (or, strictly speaking, in vacuum). However, pieces of technical equipment, such as transformers, motors, generators, or electromagnets, always incorporate iron or an iron alloy in their structures, both to increase the magnetic flux and to confine it to a desired region. The magnetic field of a galvanometer or loudspeaker is set up by a *permanent* magnet, which produces this field without any apparent circulation of charge. The coating on magnetic tape makes a record of information supplied to it by the extent to which it becomes permanently magnetized.

We therefore turn next to a consideration of the magnetic properties that make iron and a few other ferromagnetic materials so useful. We shall find that magnetic properties are not confined to ferromagnetic materials but are exhibited (to a much smaller extent, to be sure) by *all* substances. A study of the magnetic properties of materials affords another means of gaining an insight into the nature of matter in general.

Magnetic properties of a substance can be demonstrated by supporting a small spherically shaped specimen by a fine thread near the poles of a powerful electromagnet. If the specimen is of iron or one of the *ferromagnetic* substances, it is attracted into the strong part of the magnetic field. Not so familiar is the fact that *every* substance is influenced by the field, although to an extent that is extremely small compared with a substance like iron. Some substances, like iron, are forced into the strong part of the field, while others are pushed toward the weak part of the field. The first type is called *paramagnetic*; the second, *diamagnetic*. All substances, including liquids and gases, fall into one or the other of these classes.

In our discussion of induced emf's in Sec. 33–3, it was assumed that the core of the solenoid in Fig. 33–9 was empty. Suppose, however, that the solenoid is wound on a solid rod. It is found that, for a given change of current in the solenoid windings, the induced emf is not the same as when the core is empty, and hence the change in *flux* is not the same. If the rod is *ferromagnetic*, the change in flux for a given change in current is very much *larger*; if paramagnetic, it will be *slightly* larger; and if diamagnetic, slightly *smaller* than if the core were empty.

35-2 Magnetic permeability

In Chapter 27, our formulation of the properties of a dielectric substance was based on a specimen in the form of a flat slab, inserted in the field between oppositely charged parallel plates. The electric field is almost entirely confined to the region between the plates if their separation is small, and therefore a flat slab between the plates will completely occupy all points of space at which an electric field exists. A specimen of this shape is not as well suited for a study of magnetic effects, however, since magnetic field lines are always *closed* lines and there is no way of producing a magnetic field that is confined to the region between two closely spaced surfaces. The magnetic field within a closely spaced toroidal winding, however, *is* wholly confined to the space enclosed by the winding. We shall accordingly use such a field as a basis for our discussion of magnetic properties of materials, using a specimen in the form of a ring on whose surface the wire is wound. Such a specimen is often called a *Rowland ring*, after J. H. Rowland, who made much use of it in his experimental and theoretical work on electricity and magnetism. The winding of wire around the specimen is called the *magnetizing winding*, and the current in the winding, the *magnetizing current*.

The magnetic field (flux density) within the space enclosed by a toroidal winding *in vacuum* is, from Sec. 32–8,

$$B_0 = \frac{\mu_0 N i}{2\pi r}, \tag{35-1}$$

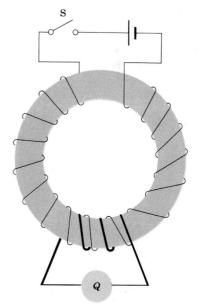

35-1 Magnetic specimen in the form of a ring with a toroidal winding.

where $2\pi r$ is the mean circumference.

Suppose now that the same coil is wound on a Rowland ring, and that a second winding is placed on the ring, as in Fig. 35-1, with its terminals connected to a charge-measuring instrument. The **B** field within the ring may be measured by a procedure essentially the same as that of the search coil, discussed in Sec. 33-2. When switch S is opened and the magnetizing current suddenly drops to zero, the deflection of the charge-measuring instrument is proportional to the total flux change, and the initial flux Φ and magnetic field B can be determined.

The results are found *not* to agree with that computed from Eq. (35-1). If the core is made of a ferromagnetic material, the measured flux density is very much *larger*; if made of a paramagnetic material, very slightly larger; and if made of a diamagnetic material, it is very slightly *smaller* than the calculated value.

Let B represent the magnetic field in a material ring and B_0 the field in a "ring of vacuum." The ratio of B to B_0 is called the *relative permeability* of the material and is represented by K_m:

$$K_\mathrm{m} = \frac{B}{B_0}, \tag{35-2}$$

where

$$K_{\mathrm{m}} \text{ is } \begin{cases} \text{equal to 1 for vacuum,} \\ \text{slightly larger than 1 for paramagnetic materials,} \\ \text{slightly smaller than 1 for diamagnetic materials,} \\ \text{often much larger than 1 for ferromagnetic materials.} \end{cases}$$

Substituting for B_0 its value $\mu_0 Ni/2\pi r$, we get

$$B = K_{\mathrm{m}}\mu_0\left(\frac{Ni}{2\pi r}\right).$$

The product $K_{\mathrm{m}}\mu_0$ is called the *permeability* of the material and is denoted by μ:

$$\mu = K_{\mathrm{m}}\mu_0, \tag{35-3}$$

where

$$\mu \text{ is } \begin{cases} \text{equal to } \mu_0 \text{ for vacuum,} \\ \text{slightly greater than } \mu_0 \text{ for paramagnetic materials,} \\ \text{slightly smaller than } \mu_0 \text{ for diamagnetic materials,} \\ \text{often much larger than } \mu_0 \text{ for ferromagnetic materials.} \end{cases}$$

The unit in which μ is expressed is the same as that for μ_0, namely, webers per ampere·meter or tesla·meters per ampere.

The expression for the magnetic field B in a material substance in the form of a Rowland ring may now be written

$$B = \frac{\mu Ni}{2\pi r}, \tag{35-4}$$

The presence of a magnetic material in a toroidal coil increases the flux for a given current by a factor of K_{m}, the relative permeability of the material. The *self-inductance* of such a coil, defined by Eq. (34-5) in terms of the ratio of flux to current, also increases by this factor. Thus the derivation of the expression for self-inductance of a toroidal coil, Example 1 of Sec. 34-2, must include an additional factor K_{m}. Since $\mu = K_{\mathrm{m}}\mu_0$, the inductance of a toroidal coil with core of permeability μ is given by

$$L = \frac{\mu N^2 A}{2\pi r}. \tag{35-5}$$

The self-inductance of an iron-core coil is therefore very much *larger* than that of the same winding in vacuum.

The expression for the energy density in a magnetic field is modified in the same way. It was shown in Sec. 34-3 that the energy density in a magnetic field in vacuum equals $B^2/2\mu_0$. In a magnetic material of permeability μ, the energy density is

$$u = \frac{1}{2}\frac{B^2}{\mu}. \tag{35-6}$$

35-3 Molecular theory of magnetism

When it was discovered that magnetic effects can be produced by currents as well as by permanent magnets, Ampère proposed the theory that the magnetic properties of a material arise from a large number of tiny closed current loops within the material. The total magnetic field in a material would then be the sum of the field caused by the external current and the additional field caused by these microscopic currents. In diamagnetic materials the direction of the additional field would be *opposite* to that of the field of the external current, *decreasing* the total field, while in paramagnetic materials the two fields would *add* to produce a total field *greater than* that due to the external current alone.

It is now known that this picture is essentially correct. The microscopic current loops are associated with the motion of electrons within atoms. Each current loop has an associated magnetic moment, just as with macroscopic current loops as discussed in Sec. 31-3. In a *paramagnetic* material each atom or molecule has a certain net current distribution and magnetic moment resulting from the combined effect of all its electron motions. In the absence of any externally caused field, these currents and magnetic moments are randomly oriented and cause no net magnetic field; but in the presence of a magnetic field caused by an external current, the magnetic moments tend to become *aligned* with this field, and the additional field caused by the associated currents is in the same direction as the field of the external current, just as for the macroscopic current loop discussed in Sec. 32-6. Thus the total field in the material is *greater* than that of the external current alone.

In a *diamagnetic* material with no external field, the electron currents add to zero for each molecule, and there is no net magnetic moment. But when a field is applied by an external current the electron motion is altered by an effect that is essentially the same as induction of a macroscopic current in a conducting loop by a changing magnetic flux through the loop, as discussed in Sec. 33-2. Lenz's law (Sec. 33-4) requires that the additional magnetic field caused by such a current loop always *oppose* the flux change that caused it. Thus the change in the electron motion in a diamagnetic material in a magnetic field always causes an additional field in the *opposite* direction to that of the applied field. This effect is always very small, typically one part in 10^5, and it acts to decrease the total field in the material.

Diamagnetic effects are in fact present in *all* materials. However, paramagnetic effects, when they occur, are always much *larger* than diamagnetic effects, so a material having diamagnetism and paramagnetism always appears paramagnetic, not diamagnetic.

The alignment of electron currents in a paramagnetic material is shown schematically in Fig. 35-2a. At interior points the currents in adjacent loops are in opposite directions and cancel, so there is no net current in the interior of the material. The parts of the loops adjacent to the outer surface are not canceled however, and the entire collection of loops is equivalent to a current I_S circulating around the outside of the body, as shown in Fig. 35-2b. This equivalent current has the same shape as an additional magnetizing winding around the body, and since its direction is the same as that of the external magnetizing current I_C, the effect is to increase the field within the material. A similar figure could be drawn for a diamagnetic material; in that case, the direction of the current loops is reversed, and the direction of I_S is opposite to that of I_C, leading to a decreased field.

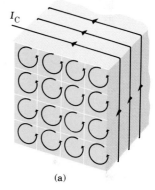

(a)

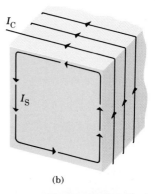

(b)

35-2 (a) Alignment of atomic current loops in a paramagnetic substance. (b) Surface current equivalent to part (a).

*35-4 Magnetization and magnetic intensity

The discussion of the preceding section has shown that the magnetic field in a material is due to the combined effect of the current in external conductors and the microscopic currents in the material, with their associated magnetic moments. For a more complete understanding of magnetic materials, it is useful to develop relationships governing these two separate contributions to the total field. For this purpose we first introduce a quantity called *magnetization*, which is the *density* of magnetic moment in a material.

We consider a small volume V; let a typical molecular magnetic moment in this volume be m. Then the *total* magnetic moment within V is the vector sum Σm. The magnetization, denoted by M, is defined as

$$M = \frac{\Sigma m}{V}. \tag{35-7}$$

Magnetization is therefore *magnetic moment per unit volume*, or *density of magnetic moment*. If the molecular magnetic moments are randomly oriented, or if there are none, as in a diamagnetic material in the absence of an external field, then the vector sum is zero and $M = 0$.

Let us now consider a long rod magnetized by a current I_C in a solenoidal conductor, as shown in Fig. 35–3. At points not too near the ends, the B field is uniform through the material. As shown in the figure, this field is caused jointly by the conduction current I_C and the equivalent surface current I_S due to the microscopic currents. The case shown corresponds to a paramagnetic material, in which I_S has the same direction as I_C.

The magnetization M is directly related to the surface current, and we now need to develop that relationship in detail. The magnetic moment of a current loop is defined as the product of the current and the area of the loop. Let I_S represent the total surface current around the periphery of a portion of the rod of length l. (The surface current is distributed *continuously* along the surface of the rod.) The magnetic moment of this portion is then $I_S A$, where A is the cross-sectional area

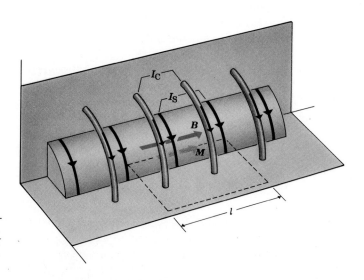

35-3 Magnetic field in a paramagnetic material is caused jointly by the conduction current I_C in the winding and the surface current I_S due to microscopic currents. For a *diamagnetic* material I_C and I_S have opposite directions, and B and M are opposite.

of the rod, and since the volume of the portion is Al, the magnetic moment per unit volume, or the magnetization M, has magnitude

$$M = \frac{I_S A}{Al} = \frac{I_S}{l} = j_S, \qquad (35\text{-}8)$$

where j_S is the *surface current per unit length*. In this special case, the magnetization is numerically *equal* to the surface current per unit length. More generally, the surface current per unit length equals the tangential component of M at the surface.

The broken line in Fig. 35–3 (compare Fig. 32–9) is a closed path of length l. In analogy to Ampère's law, we may consider the integral

$$\oint M \cdot dl$$

around the broken line. Now the magnetization M is uniform within the rod and is zero outside; the sides are perpendicular to M and do not contribute to the integral, which is therefore equal simply to the product Ml. But from Eq. (35–8), $Ml = I_S$, so the integral is also equal to the total surface current I_S through the area bounded by this path. Thus,

$$\oint M \cdot dl = I_S. \qquad (35\text{-}9)$$

Although we have obtained this relation only for a special case, it is found to be true in general; it may be called Ampère's law for the magnetization vector M.

The total B field in the rod is due both to the conduction current in the windings of the solenoid, and to the surface currents. Hence if I_C is the total *conduction current* through the rectangle, Ampère's law for B states that

$$\oint B \cdot dl = \mu_0 (I_C + I_S). \qquad (35\text{-}10)$$

When I_S is eliminated between the preceding equations, we have

$$\oint B \cdot dl = \mu_0 \left(I_C + \oint M \cdot dl \right)$$

or

$$\oint \left(\frac{B}{\mu_0} - M \right) \cdot dl = I_C.$$

Let us define a new quantity called the *magnetic intensity* H (analogous to the displacement D) as the vector difference

$$H = \frac{B}{\mu_0} - M. \qquad (35\text{-}11)$$

Equation (35–10) then takes the simple form,

$$\oint H \cdot dl = I_C. \qquad (35\text{-}12)$$

Thus the magnetic intensity H obeys a relation similar in form to Ampère's law but containing *only* the conduction current I_C, *not* the microscopic currents. In other words, the sources of H are only conduction currents, not magnetization currents. This property is very useful in the analysis of magnetic behavior.

In summary, for any closed path, the line integral of B equals the product of μ_0 and the *total* current across any surface bounded by the path, the line integral of M equals the *surface* current, and the line integral of H equals the *conduction* current. (If displacement currents are also present, they must be added to the conduction current.)

Like the magnetic field B, the magnetic intensity H can be represented by lines, called in this case lines of magnetic intensity. In free space where there is no magnetization, $M = 0$, $B = \mu_0 H$, and the lines of H have the same shape as those of B. In matter, however, the two can be quite different.

Although the symbols H and B are used nearly universally, these fields are given various names by various authors. In some books H is called the *magnetic field* and B the *magnetic induction* or the *magnetic flux density*. Other authors avoid confusion by referring simply to "the H field" and "the B field." In this book we usually call B the *magnetic field* and H the *magnetic intensity*. In the following discussion of the relation of these fields, we shall sometimes call B the *magnetic flux density* to emphasize that the B field is always used in computing magnetic flux for applications of Faraday's law.

As Eq. (35–12) shows, the units of H are amperes per meter ($A \cdot m^{-1}$). Also, from Eq. (35–11), the units of H must be the same as those of M. This may be verified by noting that M is magnetic moment per unit volume, with corresponding units $(A \cdot m^2)/(m^3) = A \cdot m^{-1}$.

In a diamagnetic or paramagnetic material the three vector fields H, M, and B are all proportional, and the proportionality can be represented in various ways. The relation of M to H for any particular material is represented by the *magnetic susceptibility* χ of the material:

$$M = \chi H. \qquad (35\text{–}13)$$

Since H and M have the same units, χ is a dimensionless quantity.

Magnetic susceptibilities for a few materials are shown in Table 35–1. Paramagnetic susceptibilities decrease with increasing temperature because of the disorienting effect of thermal motion; diamagnetic susceptibilities are nearly independent of temperature.

In terms of χ, B is

$$B = \mu_0(H + M) = \mu_0(1 + \chi)H. \qquad (35\text{–}14)$$

The relative permeability K_m, defined in Sec. 35–2, is

$$K_m = 1 + \chi, \qquad (35\text{–}15)$$

so

$$B = \mu_0 K_m H.$$

The product $\mu_0 K_m$ is called the *permeability* μ:

$$\mu = \mu_0 K_m. \qquad (35\text{–}16)$$

and hence

$$B = \mu H.$$

In vacuum, $K_m = 1$ and $\mu = \mu_0$. For this reason, the magnetic constant μ_0 is often referred to as "the permeability of vacuum" or "the

Table 35-1 Magnetic susceptibilities of paramagnetic and diamagnetic materials

Materials	Temperature, °C	$\chi_m = K_m - 1$
Paramagnetic		
Iron ammonium alum	−269	4830×10^{-5}
Iron ammonium alum	−183	213
Oxygen, liquid	−183	152
Iron ammonium alum	+20	66
Uranium	20	40
Platinum	20	26
Aluminum	20	2.2
Sodium	20	0.72
Oxygen gas	20	0.19
Diamagnetic		
Bismuth	20	-16.6×10^{-5}
Mercury	20	−2.9
Silver	20	−2.6
Carbon (diamond)	20	−2.1
Lead	20	−1.8
Rock salt	20	−1.4
Copper	20	−1.0

permeability of free space." The "relative permeability of vacuum," where $\chi = 0$, is equal to 1. The relative permeability of a paramagnetic substance is slightly greater than 1, and that of a diamagnetic substance is slightly less than 1.

*35-5 Ferromagnetism

The behavior of ferromagnetic materials differs from that of diamagnetics and paramagnetics in three important ways. First, the magnetic field in such a material may be several hundred or thousand times as large as that due to the external field alone. Second, the relation between magnetization and magnetic intensity (or magnetic field) is no longer a direct proportion; thus the magnetic susceptibility, defined as the ratio of *M* to *H,* is very large but no longer constant. Third, in some cases ferromagnetic materials exhibit large magnetization even when there is no externally applied field at all; that is, they can become permanently magnetized.

Any substance that exhibits these properties is called *ferromagnetic.* Iron, nickel, cobalt, and gadolinium are the only ferromagnetic *elements* at room temperature, but a number of elements at low temperatures and several alloys whose components are not ferromagnetic also show these effects.

Because of the complicated relation between the flux density B and the magnetic intensity H in a ferromagnetic material, it is not possible to express B as a simple function of H. Instead, the relation between these quantities is either given in tabular form or represented by a graph of B versus H, called the *magnetization curve* of the material.

The initial magnetization curve of a specimen of annealed iron is shown in Fig. 35-4 in the curve labeled B versus H. The permeability μ, equal to the ratio of B to H, can be found at any point of the curve by dividing the flux density B, at the point, by the corresponding magnetic

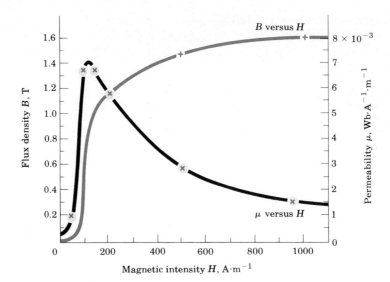

35-4 Magnetization curve and permeability curve of annealed iron.

intensity H. For example, when $H = 150$ A·m^{-1}, $B = 1.01$ T = 1.01 Wb·m^{-2}, and

$$\mu = \frac{B}{H} = \frac{1.01 \text{ T}}{150 \text{ A·m}^{-1}} = 67{,}300 \times 10^{-7} \text{ Wb·A}^{-1}\text{·m}^{-1}.$$

The relative permeability $K_m = \mu/\mu_0$ at this point is

$$K_m = \frac{67{,}300 \times 10^{-7} \text{ Wb·A}^{-1}\text{·m}^{-1}}{4\pi \times 10^{-7} \text{ Wb·A}^{-1}\text{·m}^{-1}} = 5356.$$

For some purposes a more useful quantity is the *differential* permeability, defined as dB/dH and represented by the *slope* of the curve of B versus H in Fig. 35-4. For the point just mentioned, the differential permeability for this material is about $25{,}000 \times 10^{-7}$ Wb·A^{-1}·m^{-1}, and the relative differential permeability is about 2000. Like the permeability, the differential permeability of a particular material varies with B and H.

Figure 35-5 covers a wider range of values for the same specimen, and also shows how the two quantities $\mu_0 H$ and $\mu_0 M$ contribute to B. When H is relatively small, practically all of the flux density B is due to the magnetization M (or to the equivalent surface currents). Beyond the point where H is of the order of 100,000 A·m^{-1} there is very little further increase in the magnetization M, and the iron is said to become *saturated*. Further increases in B are due almost wholly to increases in H resulting from increased current in the windings.

This behavior can be understood qualitatively on the basis of *magnetic domains*, which will be discussed in the next section.

*35-6 Magnetic domains

In most metals, the electrons in the metallic ions that form the crystal lattice structure are paired off, with half spinning one way and half the other way. The ion is thus "magnetically neutral." The ions of the ferromagnetic elements, iron, nickel, cobalt, etc., are exceptions. In particular, that of iron has an excess of two uncompensated electrons; the mag-

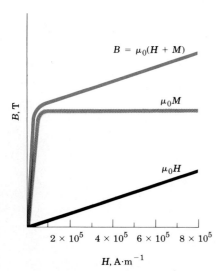

35-5 Graph of B, $\mu_0 H$, and $\mu_0 M$ as functions of H, for annealed iron.

netization of iron is due almost entirely to the alignment of the magnetic moments of these electrons. When a sample of iron is magnetically *saturated*, all of the unpaired electrons are spinning with their axes in the direction of the magnetizing field.

In *small* fields, ferromagnetic materials typically have much larger susceptibility, and therefore larger permeability, than paramagnetic materials. Ferromagnetism results because of a spontaneous, self-aligning, *cooperative* interaction among relatively large numbers of iron atoms in regions called *domains*. As a result of molecular interactions, the molecular magnetic moments in each domain are all aligned parallel to one another. In other words, each domain is *spontaneously magnetized to saturation* even in the absence of any external field. The directions of magnetization in different domains are not necessarily parallel to one another, so that in an unmagnetized specimen the *resultant* magnetization is zero. When the specimen is placed in a magnetic field the resultant magnetization may increase in two different ways, either by an increase in the volume of those domains that are favorably oriented with respect to the field at the expense of unfavorably oriented domains, or by *rotation* of the direction of magnetization toward the direction of the field.

We can now understand the reason for the shape of the graph of $\mu_0 M$ versus H in Fig. 35–5. As H is increased from zero, more and more of the spinning electrons become aligned with the **H** field and M increases steadily. When H is about 100,000 A·m^{-1}, practically *all* of the electrons have lined up with the **H** field. No further increase in M can occur, and the iron is *saturated*.

Saturation can also occur with paramagnetic materials, but in the absence of the cooperative spontaneous alignment that occurs in ferromagnetics, the field required to cause saturation is larger than can be attained with present laboratory equipment. At present, saturation of paramagnetics can be observed only at extremely low temperatures, where the disorienting effect of thermal motion is greatly reduced.

Every ferromagnetic material has a critical temperature called the *Curie temperature*, above which the material is paramagnetic but not ferromagnetic. As the Curie temperature is reached, the energy associated with thermal motion in the material becomes large enough (compared to the interaction energies responsible for the spontaneous alignment of magnetic moments in a domain) so that this alignment disappears. Above the Curie treatment there is no spontaneous magnetization. The transition from ferromagnetic to paramagnetic behavior is a *phase change* analogous to the transitions between solid, liquid, and gaseous phases of matter, discussed in Chapter 15.

*35–7 Hysteresis

A magnetization curve such as that in Fig. 35–4 expresses the relation between the magnetic field B in a ferromagnetic material and the corresponding magnetic intensity H, *provided the sample is initially unmagnetized and the magnetic intensity is steadily increased from zero*. Thus, in Fig. 35–6, if the magnetizing current in the windings of an unmagnetized ring sample, as in Sec. 35–2, is steadily increased, the relation of B to H follows the curve *Oab*. If the current is now *decreased* until point *c* is

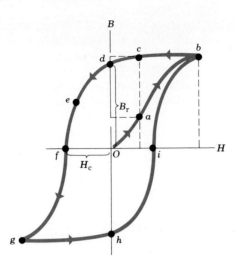

35-6   Hysteresis loop.

reached, B is much larger than at a, even though H is the same. When the current decreases to zero, H is zero, and we reach point d. But at this point B is *not* zero; the material is magnetized even in the absence of a magnetizing current, and thus has become a permanent magnet. In fact, B does not decrease to zero until H has reversed direction and reached point f. As H becomes larger in the reverse direction, point g is reached and the material approaches saturation magnetization in the reverse direction. As H decreases to zero and then increases in the original direction, the path $ghib$ is followed.

The magnetic field B that remains after the material has been magnetized to saturation and then H has been reduced to zero is called the *remanance*; it is denoted by B_r in Fig. 35–6. Because at d, H is zero and so $B = \mu_0 M$, from Eq. (35–14), B_r/μ_0 is also called the *remanent magnetization*. The reverse H needed to reduce B to zero, denoted in the figure by H_c, is called the *coercive field* or *coercivity*.

A significant consequence of hysteresis is the dissipation of *energy* within a ferromagnetic material as it traverses its hysteresis loop. It can be shown that the energy dissipated per unit volume of material in one cycle around the loop is represented by the *area* of the loop.

Different magnetic properties are important for different applications. A permanent magnet material should have large remanent magnetization and, to prevent demagnetization in external fields, large coercivity. Properties of a few permanent-magnet materials are shown in Table 35–2.

For iron cores in inductors, transformers, motors, and other devices, it is usually desirable to have as little hysteresis as possible because of the attendant energy loss and heating when the field undergoes repeated reversals in the presence of an alternating current. In such cases the remanent magnetization and coercivity should be as small as possible. For such materials the curves $bcdefg$ and $bihg$ in Fig. 35–6 are very close together.

A material having small remanence and coercivity is said to be magnetically *soft*, while one having large values is said to be magnetically *hard*. It is interesting to note that soft iron is soft mechanically as well as magnetically, while some (but not all) steels are hard magnetically.

Table 35-2   Remanence and coercivity
of permanent magnet materials

Material	Composition percent	B_r, T	H_c A·m^{-1}
Carbon steel	98 Fe, 0.86 C, 0.9 Mn	0.95	3.6×10^3
Cobalt steel	52 Fe, 36 Co, 7 W, 3.5 Cr, 0.5 Mn, 0.7 C	0.95	18×10^3
Alnico 2	55 Fe, 10 Al, 17 Ni, 12 Co, 6 Cu	0.76	42×10^3
Alnico 5	51 Fe, 8 Al, 14 Ni, 24 Co, 3 Cu	1.25	44×10^3

*35-8 The magnetic field of the earth

The magnetic field of the earth, at distances up to about five earth radii, is approximately the same as that outside a uniformly magnetized sphere. Figure 35–7a represents a section through the earth. The broken line is its axis of rotation, and the *geographic* north and south poles are lettered N_g and S_g. The direction of the (presumed) internal magnetization makes an angle of about 15° with the earth's axis. The dashed line indicates the plane of the magnetic equator, and the letters N_m and S_m represent the so-called *magnetic* north and south poles. Note carefully that *B*-field lines emerge from the earth's surface over the entire southern magnetic hemisphere and enter its surface over the entire northern magnetic hemisphere. Hence, if we wish to attribute the earth's field to magnetic poles, we must assume that magnetic N-poles are distributed over the entire *southern* magnetic hemisphere, and magnetic S-poles over the entire *northern* magnetic hemisphere!

It is interesting to note that the same field at *external* points would result if the earth's magnetism were due to a short bar magnet near its center, as in Fig. 35–7b, with the S-pole of the magnet pointing toward the north magnetic pole. The field within the earth is different in the two cases, but for obvious reasons experimental verification of either hypothesis is impossible.

Except at the magnetic equator, the earth's field is not horizontal. The angle that the field makes with the horizontal is called the *angle of dip* or the *inclination*. At Pittsburgh, Pennsylvania (about 40° N latitude), the magnitude of the earth's field is about 6.1×10^{-5} T and the angle of dip about 72°. Hence, the horizontal component at Pittsburgh is about 1.9×10^{-5} T and the vertical component about 5.8×10^{-5} T. In northern magnetic latitudes, the vertical component is directed downward; in southern magnetic latitudes it is upward. The angle of dip is, of course, 90° at the magnetic poles.

Out to a distance from the earth of about five earth radii, the magnetic field is governed almost entirely by the earth. At larger distances, the motions of ionized particles play an important role. These motions are strongly influenced by the *solar wind*, a thin hot gas expelled by the

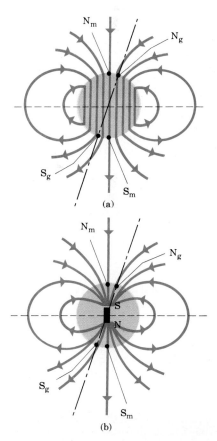

35-7 Simplified diagram of the magnetic field of the earth. The broken lines represent the axis of rotation.

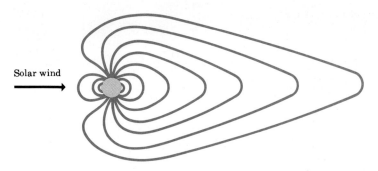

35-8   Probable distortion of the geomagnetic field by the solar wind.

sun. Recent studies using satellites and space probes suggest that at large distances the field is distorted, as suggested in Fig. 35–8.

The angle between the horizontal component of *B* and the true north–south direction is called the *variation* or *declination*. At Pittsburgh, the declination at present is about 5° W, that is, a compass needle points about 5° to the west of true north.

Questions

35-1   Why should the permeability of a paramagnetic material be expected to decrease with increasing temperature?

35-2   In the discussion of magnetic forces on current loops in Sec. 31–3, it was found that no net force is exerted on a complete loop in a uniform magnetic field, only a torque. Yet magnetized materials, which contain atomic current loops, certainly *do* experience net forces in magnetic fields. How is this discrepancy resolved?

35-3   What features of atomic structure determine whether an element is diamagnetic or paramagnetic?

35-4   The magnetic susceptibility of paramagnetic materials is quite strongly temperature-dependent, while that of diamagnetic materials is nearly independent of temperature. Why the difference?

35-5   When a piece of magnetic material is attracted to a permanent magnet, it is moving and hence has kinetic energy just before it strikes the magnet and sticks to it. Where does the energy come from? Where does it go?

35-6   Would a paramagnetic substance have a curved magnetization curve similar to Fig. 35–4 in a sufficiently strong field, or is this behavior limited to ferromagnetic materials?

35-7   You are given two iron bars and told that one is a permanent magnet, the other is not. They look identical. How can you tell which is which without using any equipment other than the bars themselves?

35-8   The magnetic declination at a fixed point on the earth usually varies with time, often by a few degrees per hundred years. What does this tell you about the interior of the earth?

35-9   Are there any points on earth where the magnetic declination is 180°, that is, where a compass points exactly backward?

35-10   You are lost in the woods with only a map, a compass, a fishhook, and a piece of string. Which is more important for you to know, the angle of dip or the declination?

35-11   A student proposed to obtain a single magnetic pole by taking a bar magnet with an N-pole at one end and an S-pole at the other end, and breaking off the N-end. Does this work?

35-12   In a laboratory electromagnet designed to produce very strong fields, what property of the iron core is most important, its permeability, remanence, or coercivity?

Problems

$$\mu_0 = 4\pi \times 10^{-7} \text{ Wb·A}^{-1}\text{·m}^{-1}.$$

35-1   Experimental measurements of the magnetic susceptibility of iron ammonium alum are given in Table 35–3. Make a graph of $1/\chi$ against Kelvin temperature; what conclusion can you draw?

Table 35–3

$T,°C$	χ
−258.15	129×10^{-4}
−173	19.4×10^{-4}
−73	9.7×10^{-4}
27	6.5×10^{-4}

35-2 A toroid having 500 turns of wire and a mean circumferential length of 50 cm carries a current of 0.3 A. The relative permeability of the core is 600.

a) What is the magnetic field in the core?

b) What is the magnetic intensity?

c) What part of the magnetic field is due to surface currents?

35-3 The current in the windings on a toroid is 2.0 A. There are 400 turns and the mean circumferential length is 40 cm. With the aid of a search coil and charge-measuring instrument, the magnetic field is found to be 1.0 T. Calculate

a) the magnetic intensity,

b) the magnetization,

c) the magnetic susceptibility,

d) the equivalent surface current, and

e) the relative permeability.

35-4 Each of the two preceding coils (Problems 35–2 and 35–3) has a cross-sectional area of 8 cm². Calculate the self-inductance L of each coil.

35-5 In 1911, Kamerlingh-Onnes discovered that at low temperatures some metals lose their electrical resistance and become superconductors. Thirty years later, Meissner showed that the magnetic induction inside a superconductor is zero. If the current in the windings of a toroidal superconductor is increased, a critical value H_c may be reached at which the metal suddenly becomes normal with practically zero magnetization.

a) Plot a graph of B/μ_0 against H from $H = 0$ to $H = 2H_c$.

b) Plot a graph of M against H in the same range.

c) Is a superconductor paramagnetic, diamagnetic, or ferromagnetic?

d) If the current in the windings is 10 A, how large are the surface currents on the superconducting ring and in what direction?

e) What happens when you try to place a small permanent magnet on a superconducting plate?

35-6 Table 35–4 lists corresponding values of H and B for a specimen of commercial hot-rolled silicon steel, a material widely used in transformer cores.

a) Construct graphs of B and μ as functions of H, in the range from $H = 0$ to $H = 1000$ A·m⁻¹.

b) What is the maximum permeability?

Table 35-4 Magnetic properties of silicon steel

Magnetic intensity H, A·m⁻¹	Flux density B, T
0	0
10	0.050
20	0.15
40	0.43
50	0.54
60	0.62
80	0.74
100	0.83
150	0.98
200	1.07
500	1.27
1,000	1.34
10,000	1.65
100,000	2.02
800,000	2.92

c) What is the initial permeability ($H = 0$)?

d) What is the permeability when $H = 800,000$ A·m⁻¹?

35-7 Construct graphs of $\mu_0 H$, $\mu_0 M$, and K_m against H, like those in Fig. 35–5 for the specimen of silicon steel in Table 35–4.

35-8 Suppose the ordinate of the point b in Fig. 35–6 corresponds to a flux density of 1.6 T, and the abscissa to a magnetic intensity H of 1000 A·m⁻¹. Approximately what is the relative permeability at points a, b, c, d, e, and f.

35-9 A bar magnet has a coercivity of 4×10^3 A·m⁻¹. It is desired to demagnetize it by inserting it inside a solenoid 12 cm long and having 60 turns. What current should be carried by the solenoid?

35-10 The horizontal component of the earth's magnetic field at Cambridge, Mass., is 1.7×10^{-5} Wb·m⁻². What is the horizontal component of the magnetic intensity?

35-11 A Rowland ring has a cross section of 2 cm², a mean length of 30 cm, and is wound with 400 turns. Find the current in the winding that is required to set up a flux density of 0.1 T in the ring,

a) if the ring is of annealed iron (Fig. 35–4);

b) if the ring is of silicon steel (Table 35–4).

c) Repeat the computations above if a flux density of 1.2 T is desired.

ALTERNATING CURRENTS

36

36-1 Introduction

In Chapters 28 and 29 we have studied the behavior of *direct-current* (dc) circuits, in which all the currents, voltages, and emfs are *constant*, that is, not varying with time. However, many electric circuits of practical importance, including nearly all large-scale electric-power distribution systems and much electronic equipment, use *alternating current* (ac), in which the voltages and currents vary with time, often in a *sinusoidal* manner. In this chapter we study the properties of circuits in which there are sinusoidally varying voltages and currents. Many of the same principles used in Chapters 28 and 29 are applicable here as well, and there are some new ideas related to the circuit behavior of inductors and capacitors.

We have already seen several sources of alternating emf or voltage. A coil of wire, rotating with constant angular velocity in a magnetic field, develops a sinusoidal alternating emf, as explained in Sec. 33–1. This simple device is the prototype of the commercial alternating current generator, or *alternator*. An *L–C* circuit, as discussed in Sec. 34–5, oscillates sinusoidally, and with the proper circuitry provides an alternating potential difference between its terminals having a frequency, depending on the purpose for which it is designed, that may range from a few hertz to many millions of hertz.

We consider first a number of circuits connected to an alternator or oscillator that maintains between its terminals a sinusoidal alternating potential difference

$$v = V \cos \omega t. \tag{36-1}$$

In this expression V is the maximum potential difference or the *voltage amplitude, v* is the *instantaneous* potential difference, and ω the *angular frequency,* equal to 2π times the frequency f. For brevity, the alternator or oscillator will be referred to as an *ac source*. The circuit-diagram symbol for an ac source is $\bigodot$.

Analysis of alternating-current circuits is facilitated by use of vector diagrams similar to those used in the study of harmonic motion (Sec.

11–5). In such diagrams, the instantaneous value of a quantity that varies sinusoidally with time is represented by the *projection* onto a horizontal axis of a vector of length corresponding to the amplitude of the quantity and rotating counterclockwise with angular velocity ω. In the context of ac-circuit analysis, these rotating vectors are often called *phasors*, and diagrams containing them are called *phasor diagrams*.

Alternating currents are of utmost importance in technology and industry. Transmission of power over long distance is very much easier and more economical with alternating than with direct currents. Circuits used in modern communication equipment, including radio and television, make extensive use of alternating currents. Many life processes involve alternating voltages and currents. The beating of the heart induces alternating currents in the surrounding tissues; the detection and study of these currents, called electrocardiography, provide valuable information concerning the health or pathology of the heart. Electroencephalograms, recordings of alternating currents in the brain, provide analogous information regarding brain function. Both electrocardiograms and electroencephalograms are invaluable diagnostic tools in modern medicine.

36–2 Circuits containing resistance, inductance, or capacitance

Let a resistor of resistance R be connected between the terminals of an ac source, as in Fig. 36–1a. The instantaneous potential of point a with respect to point b is $v_{ab} = V \cos \omega t$, and the instantaneous current in the resistor is

$$i = \frac{v_{ab}}{R} = \frac{V}{R} \cos \omega t.$$

The maximum current I, or the *current amplitude*, is evidently

$$I = \frac{V}{R}, \tag{36–2}$$

and we can therefore write

$$i = I \cos \omega t. \tag{36–3}$$

The current and voltage are both proportional to $\cos \omega t$, so the current is *in phase* with the voltage. The current and voltage amplitudes, from Eq. (36–2), are related in the same way as in a dc circuit.

Figure 36–1b shows graphs of i and v as functions of time. The fact that the curve representing the current has the greater amplitude in the diagram is of no significance, because the choice of vertical scales for i and v is arbitrary. The corresponding phasor diagram is given in Fig. 36–1c. Because i and v are *in phase* and have the same frequency, the current and voltage phasors rotate together.

Next, suppose that a capacitor of capacitance C is connected across the source, as in Fig. 36–2a. The instantaneous charge q on the capacitor is

$$q = Cv_{ab} = CV \cos \omega t. \tag{36–4}$$

In this case the instantaneous current is equal to the *rate of change* of

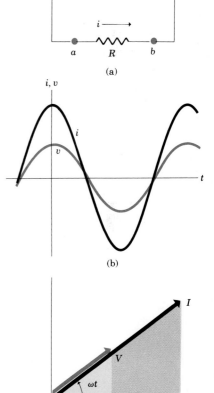

36–1 (a) Resistance R connected across an ac source. (b) Graphs of instantaneous voltage and current. (c) Phasor diagram; current and voltage in phase.

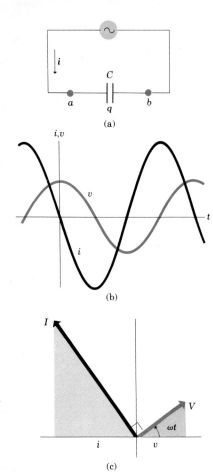

(a)

(b)

(c)

36-2 (a) Capacitor C connected across an ac source. (b) Graphs of instantaneous voltage and current. (c) Phasor diagram; current *leads* voltage by 90°.

the capacitor charge and is thus proportional to the rate of change of voltage. Taking the derivative of Eq. (36–4), we find

$$i = \frac{dq}{dt} = -\omega CV \sin \omega t. \qquad (36\text{–}5)$$

Thus if the voltage is represented by a *cosine* function, the current is represented by a *negative sine* function, as shown in Fig. 36–2b. The current is *not* in phase with the voltage; the current is greatest at times when the v curve is rising or falling most steeply, and is zero at times when the v curve "levels off," that is, when it reaches its maximum and minimum values.

The voltage and current are "out of step" or *out of phase* by a quarter-cycle, with the current a quarter-cycle ahead. The peaks of current occur a quarter-cycle *before* the corresponding voltage peaks. This result is also represented by the vector diagram of Fig. 36–2c, which shows the current vector ahead of the voltage vector by a quarter-cycle, or 90°. This *phase difference* between current and voltage can also be obtained by rewriting Eq. (36–5), using the trigonometric identity $\cos(A + 90°) = -\sin A$:

$$i = \omega CV \cos(\omega t + 90°). \qquad (36\text{–}6)$$

This shows that the expression for i can be viewed as a cosine function with a "head start" of 90° compared with that for the voltage. We say that the current *leads* the voltage by 90° or that the voltage *lags* the current by 90°.

Equations (36–5) and (36–6) show that the maximum current I is given by

$$I = \omega CV. \qquad (36\text{–}7)$$

This expression can be put in the same *form* as that for the maximum current in a resistor ($I = V/R$) if we write Eq. (36–7) as

$$I = \frac{V}{1/\omega C}.$$

and define a quantity X_C, called the *capacitive reactance* of the capacitor, as

$$X_C = \frac{1}{\omega C}. \qquad (36\text{–}8)$$

Then

$$I = \frac{V}{X_C}. \qquad (36\text{–}9)$$

From Eq. (36–9), the unit of capacitive reactance is one *volt per ampere* (1 V·A^{-1}), or one *ohm* (1 Ω).

The reactance of a capacitor is inversely proportional both to the capacitance C and to the angular frequency ω; the greater the capacitance, and the higher the frequency, the *smaller* is the reactance X_C.

Example 1 At an angular frequency of 1000 rad·s^{-1}, the reactance of a 1-μF capacitor is

$$X_C = \frac{1}{\omega C} = \frac{1}{(10^3 \text{ rad·s}^{-1})(10^{-6} \text{ F})} = 1000 \text{ Ω}.$$

At frequency of 10,000 rad·s^{-1}, the reactance of the same capacitor is only 100 Ω, and at a frequency of 100 rad·s^{-1} it is 10,000 Ω.

Finally, suppose a pure inductor having a self-inductance L and zero resistance is connected to an ac source as in Fig. 36–3. Since the potential difference between the terminals of an inductor equals $L\, di/dt$, we have

$$L\frac{di}{dt} = V \cos \omega t \qquad (36\text{--}10)$$

and

$$di = \frac{V}{L} \cos \omega t \, dt.$$

Integration of both sides gives

$$i = \frac{V}{\omega L} \sin \omega t + C.$$

If $i = 0$ at time $t = 0$, then $C = 0$ and

$$i = \frac{V}{\omega L} \sin \omega t. \qquad (36\text{--}11)$$

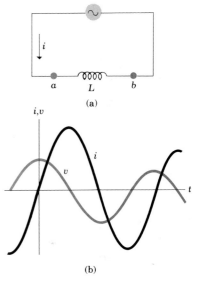

(a)

Again, the voltage and current are a quarter-cycle out of phase, but this time the current *lags* the voltage (or the voltage *leads* the current) by 90°. This may also be seen by rewriting Eq. (36–11) using the trigonometric identity $\cos(A - 90°) = \sin A$:

$$i = \frac{V}{\omega L} \cos(\omega t - 90°). \qquad (36\text{--}12)$$

This result and the vector diagram of Fig. 36–3c show that the current can be viewed as a cosine function with a "late start" of 90°.

In Fig. 36–3b, the points of maximum voltage correspond on the graphs to points of maximum steepness of the current curve, and the points of zero voltage correspond to the points where the current curve levels off at its maximum and minimum values.

The maximum current is

$$I = \frac{V}{\omega L}, \qquad (36\text{--}13)$$

and

$$i = I \sin \omega t. \qquad (36\text{--}14)$$

The *inductive reactance* X_L of an inductor is defined as

$$X_L = \omega L, \qquad (36\text{--}15)$$

and Eq. (36–13) can also be written in the same form as that for a resistor:

$$I = \frac{V}{X_L}. \qquad (36\text{--}16)$$

The unit of inductive reactance is also 1 *ohm*.

The reactance of an inductor is directly proportional both to its inductance L and to the angular frequency ω; the greater the inductance, and the higher the frequency, the *larger* is the reactance.

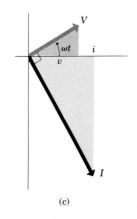

(c)

36–3 (a) Inductance L connected across an ac source. (b) Graphs of instantaneous voltage and current. (c) Phasor diagram; current lags voltage by 90°.

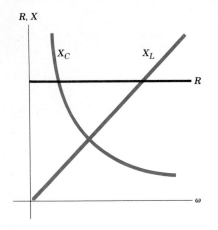

36-4 Graphs of R, X_L, and X_C as functions of frequency.

Example 2 At an angular frequency of 1000 rad·s⁻¹, the reactance of a 1-H inductor is

$$X_L = \omega L = (10^3 \text{ rad·s}^{-1})(1 \text{ H}) = 1000 \, \Omega.$$

At a frequency of 10,000 rad·s⁻¹ the reactance of the same inductor is 10,000 Ω, while at a frequency of 100 rad·s⁻¹ it is only 100 Ω.

The graphs in Fig. 36-4 summarize the variations with frequency of the resistance of a resistor, and of the reactances of an inductor and of a capacitor. As the frequency increases, the reactance of the inductor approaches infinity, and that of the capacitor approaches zero. As the frequency decreases, the inductive reactance approaches zero and the capacitive reactance approaches infinity. The limiting case of zero frequency corresponds to a dc circuit.

36-3 The R–L–C series circuit

In many instances, ac circuits include resistance, inductive reactance, and capacitive reactance. A simple series circuit is shown in Fig. 36-5a. Analysis of this and similar circuits is facilitated by use of a phasor diagram, which includes the voltage and current vectors for the various components. In this instance the instantaneous *total* voltage across all three components is equal to the source voltage at that instant, and its phasor is the *vector sum* of the phasors for the individual voltages. The complete phasor diagram for this circuit is shown in Fig. 36-5b. This may appear complex, so we shall explain it step by step.

The instantaneous current i has the same value at all points of the circuit. Thus, a *single phasor I*, of length proportional to the current amplitude, suffices to represent the current in each circuit element.

Let us use the symbols v_R, v_L, and v_C for the instantaneous voltages across R, L, and C, and V_R, V_L, and V_C for their maximum values. The instantaneous and maximum voltages across the source will be represented by v and V. Then $v = v_{ab}$, $v_R = v_{ac}$, $v_L = v_{cd}$, and $v_C = v_{db}$.

We have shown that the potential difference between the terminals of a resistor is *in phase* with the current in the resistor, and that its maximum value V_R is

$$V_R = IR.$$

Thus the phasor V_R in Fig. 36-5b, in phase with the current vector, represents the voltage across the resistor. Its projection on the horizontal axis, at any instant, gives the instantaneous potential difference v_R.

The voltage across an inductor *leads* the current by 90°. The voltage amplitude is

$$V_L = IX_L.$$

The phasor V_L in Fig. 36-5b represents the voltage across the inductor, and its projection at any instant onto the horizontal axis equals v_L.

The voltage in a capacitor *lags* the current by 90°. The voltage amplitude is

$$V_C = IX_C.$$

The phasor V_C in Fig. 36-5b represents the voltage across the capacitor, and its projection at any instant onto the horizontal axis equals v_C.

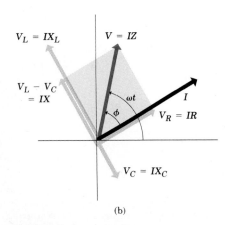

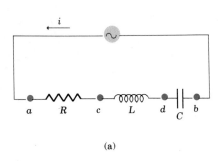

36-5 (a) A series R–L–C circuit. (b) Phasor diagram.

The instantaneous potential difference v between terminals a and b equals at every instant the (algebraic) sum of the potential differences v_R, v_L, and v_C. That is, it equals the sum of the projections of the phasors V_R, V_L, and V_C. But the *projection* of the *vector sum* of these phasors is equal to the *sum* of their *projections*, so this vector sum V must be the phasor that represents the source voltage. To form the vector sum, we first subtract the phasor V_C from the phasor V_L (since these always lie in the same straight line), giving the phasor $V_L - V_C$. Since this is at right angles to the phasor V_R, the magnitude of the phasor V is

$$V = \sqrt{V_R^2 + (V_L - V_C)^2} = \sqrt{(IR)^2 + (IX_L - IX_C)^2}$$
$$= I\sqrt{R^2 + (X_L - X_C)^2}.$$

The quantity $X_L - X_C$ is called the *reactance* of the circuit, denoted by X:

$$X = X_L - X_C. \tag{36-17}$$

Finally, we define the *impedance Z* of the circuit as

$$Z = \sqrt{R^2 + (X_L - X_C)^2} = \sqrt{R^2 + X^2}, \tag{36-18}$$

so we can write

$$V = IZ \qquad \text{or} \qquad I = \frac{V}{Z}. \tag{36-19}$$

Thus, the *form* of the equation relating current and voltage *amplitudes* is the same as that for a dc circuit, the impedance Z playing the same role as the resistance R of the dc circuit. Note, however, that the impedance is actually a function of R, L, and C, as well as of the frequency ω. The complete expression for Z, for a series circuit, is

$$Z = \sqrt{R^2 + X^2}$$
$$= \sqrt{R^2 + (X_L - X_C)^2} \tag{36-20}$$
$$= \sqrt{R^2 + [\omega L - (1/\omega C)]^2}.$$

The unit of impedance, from Eq. (36-19) or (36-20), is one *volt per ampere* (1 V·A⁻¹) or one *ohm*.

The expressions for the impedance Z of (a) an R–L series circuit, (b) an R–C series circuit, and (c) an L–C series circuit can be obtained from Eq. (36-20) by letting (a) $X_C = 0$, (b) $X_L = 0$, and (c) $R = 0$.

Equation (36-20) gives the impedance Z only for a *series* R-L-C circuit. But whatever the nature of R-L-C network, its impedance can be *defined* by Eq. (36-19) as the ratio of the voltage amplitude to the current amplitude.

The angle ϕ, in Fig. 36–5b, is the phase angle of the source voltage V with respect to the current I. From the diagram,

$$\tan \phi = \frac{V_L - V_C}{V_R} = \frac{I(X_L - X_C)}{IR} = \frac{X}{R}. \tag{36-21}$$

Hence, if the source voltage is represented by a cosine function,

$$v = V \cos \omega t,$$

the current *lags* by an angle ϕ between 0 and 90°, and its equation is

$$i = I \cos (\omega t - \phi).$$

Figure 36–5b has been constructed for a circuit in which $X_L > X_C$. If $X_L < X_C$, vector V lies on the opposite side of the current vector I and the current *leads* the voltage. In this case, $X = X_L - X_C$ is a *negative* quantity, tan ϕ is negative, and ϕ is a negative angle between 0 and $-90°$.

To summarize, we can say that the *instantaneous* potential differences in an ac series circuit add *algebraically*, just as in a dc circuit, while the voltage *amplitudes* add *vectorially*.

Example In the series circuit of Fig. 36–5, let $R = 300\,\Omega$, $L = 0.9$ H, $C = 2.0\,\mu$F, and $\omega = 1000$ rad·s^{-1}. Then

$$X_L = \omega L = 900\,\Omega, \qquad X_C = \frac{1}{\omega C} = 500\,\Omega.$$

The reactance X of the circuit is

$$X = X_L - X_C = 400\,\Omega,$$

and the impedance Z is

$$Z = \sqrt{R^2 + X^2} = 500\,\Omega.$$

If the circuit is connected across an ac source of voltage amplitude 50 V, the current amplitude is

$$I = \frac{V}{Z} = 0.10 \text{ A.}$$

The lag angle ϕ is

$$\phi = \tan^{-1}\frac{X}{R} = 53°.$$

The voltage amplitude across the resistor is

$$V_R = IR = 30 \text{ V.}$$

The voltage amplitudes across the inductor and capacitor are, respectively,

$$V_L = IX_L = 90 \text{ V}, \qquad V_C = IX_C = 50 \text{ V.}$$

The above analysis describes the *steady-state* condition of a circuit, the condition that prevails after the circuit has been connected to the source for a long time. When the source is first connected, there may be additional voltages and currents, called *transients*, whose nature depends on the time in the cycle when the circuit is initially completed. A detailed analysis of transients is beyond our scope. In any event, they always die out after a sufficiently long time, and therefore do not affect the steady-state behavior of the circuit.

36-4 Average and root-mean-square values; ac instruments

The *instantaneous* potential difference between two points of an ac circuit can be measured by connecting a calibrated oscilloscope between the points, and the instantaneous current by connecting an oscilloscope across a resistor in the circuit. The usual moving-coil galvanometer, however, has too large a moment of inertia to follow the instantaneous values of an alternating current. It averages out the fluctuating torque on its coil, and its deflection is proportional to the *average* current.

The *average* value f_{av} of any quantity $f(t)$ that varies with time, over a time interval from t_1 to t_2, is defined as

$$f_{av} = \frac{1}{t_2 - t_1} \int_{t_1}^{t_2} f(t)\, dt. \qquad (36\text{-}22)$$

This may be interpreted graphically. The integral represents the *area* bounded by the curve of $f(t)$ and by the two vertical lines t_1 and t_2. The product $f_{av}(t_2 - t_1)$ is the area of a rectangle of height f_{av} and base $(t_2 - t_1)$. Thus f_{av} is the height of a rectangle having the same area as the area under the curve between the same two values of t.

Let us apply this definition to a sinusoidally varying quantity, for example a current given by

$$i = I \sin \omega t.$$

The period τ of this sinusoidal current is given by $\tau = 1/f = 2\pi/\omega$. Let us find the average value of i for a half-period or half-cycle, from $t = 0$ to $t = \pi/\omega$. From Eq. (36-22), this is

$$I_{av} = \frac{\omega}{\pi} \int_0^{\pi/\omega} I \sin \omega t\, dt = \frac{2}{\pi} I. \qquad (36\text{-}23)$$

That is, the average current is $2/\pi$ (about $\frac{2}{3}$) times the maximum current, and the area under the rectangle in Fig. 36-6 equals the area under one loop of the sine curve.

The average current for a *complete cycle* (or any number of complete cycles) is

$$I_{av} = \frac{\omega}{2\pi} \int_0^{2\pi/\omega} I \sin \omega t\, dt = 0,$$

as would be expected, since the *positive* area of the loop between 0 and π/ω is equal to the *negative* area of the loop between π/ω and $2\pi/\omega$. Hence if a sinusoidal current is sent through a moving-coil galvanometer, the meter reads zero!

Such a meter can be used in an ac circuit, however, if it is connected in the *full-wave rectifier circuit* shown in Fig. 36-7a. As explained in Sec. 28-5, an ideal rectifier offers a constant finite resistance to current in the forward direction and an infinite resistance to current in the opposite direction. When the line current is in the direction shown, two of the rectifiers are conducting and two are nonconducting, and the current in the galvanometer is upward. When the line current is in the opposite direction, the rectifiers that are nonconducting in Fig. 36-7a carry the current, and the current in the galvanometer is still upward. Thus, if the line current alternates sinusoidally, the current in the galvanometer has the waveform shown as a solid curve in Fig. 36-7b. Although pulsating, it is always in the same direction and its average value is *not* zero. Then, when provided with the necessary series resistance or shunt, as for a dc voltmeter or ammeter, the galvanometer can serve as an ac voltmeter or ammeter.

The average value of the rectified current, in any number of complete cycles, is the same as the average current in the first half-cycle in

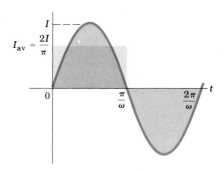

36-6   The average value of a sinusoidal current over a half-cycle is $2I/\pi$. The average over a complete cycle is zero.

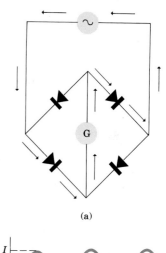

(a)

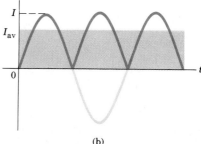

(b)

36-7   (a) A full-wave rectifier. (b) Graph of a full-wave rectified current and its average value.

Fig. 36–6, or $2/\pi$ times the maximum current I. Hence, if the meter deflects full-scale with a steady current I_0 through it, it also deflects full-scale when the *average value* of the rectified current, $2I/\pi$, is equal to I_0. The current amplitude I, when the meter deflects full-scale, is then

$$I = \frac{\pi I_0}{2}.$$

For example, if $I_0 = 1$ A, $I = 1.57$ A.

Most ac meters are calibrated to read not the maximum value of the current or voltage, but the *root-mean-square* value, that is, the *square root* of the *average* value of the *square* of the current or voltage. This special kind of average is usually called simply the rms value. Because i^2 is always positive, the average of i^2 is never zero, even when the average of i itself is zero. Also, rms values of voltage and current are useful in relationships for *power* in ac circuits; we return to this point in Sec. 36–5.

Figure 36–8 shows graphs of a sinusoidally varying current and of its square. If $i = I \sin \omega t$, then

$$i^2 = I^2 \sin^2 \omega t = I^2 \left[\frac{1}{2} (1 - \cos 2\omega t) \right]$$

$$= \frac{1}{2} I^2 - \frac{1}{2} I^2 \cos 2\omega t.$$

The average value of i^2, or the *mean square current*, is equal to the constant term $\frac{1}{2} I^2$, since the average value of $\cos 2\omega t$, over any number of complete cycles, is zero:

$$(I^2)_{\text{av}} = \frac{I^2}{2}.$$

The root-mean-square current is the square root of this, or

$$I_{\text{rms}} = \sqrt{(I^2)_{\text{av}}} = \frac{I}{\sqrt{2}}. \tag{36–24}$$

In the same way, the root-mean-square value of a sinusoidal voltage is

$$V_{\text{rms}} = \frac{V}{\sqrt{2}}. \tag{36–25}$$

Voltages and currents in power distribution systems are always referred to in terms of their rms values. Thus, when we speak of our household power supply as "115-volt ac," this means that the rms voltage is 115 V. The voltage amplitude is

$$V = \sqrt{2} V_{\text{rms}} = 163 \text{ V}.$$

All the voltage-current relations developed in preceding sections have been stated in terms of amplitudes (maximum values) of voltage and current, but they remain valid when rms values are used; in each case the difference is simply a multiplicative factor of $1/\sqrt{2}$.

36–5 Power in ac circuits

The *instantaneous* power input p to an ac circuit is

$$p = vi,$$

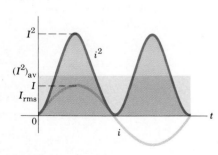

36-8   The average value of the square of a sinusoidally varying current, over any number of half-cycles, is $I^2/2$. The root-mean-square value is $I/\sqrt{2}$.

where v is the instantaneous source potential difference and i is the instantaneous current. We consider first some special cases.

If the circuit consists of a pure resistance R, as in Fig. 36–1, i and v are *in phase*. The graph representing p is obtained by multiplying together at every instant the ordinates of the graphs of v and i in Fig. 36–1b, and it is shown by the full curve in Fig. 36–9a. (The product vi is positive when v and i are both positive or both negative.) That is, energy is supplied *to* the resistor at all instants, whatever the direction of the current, although the *rate* at which it is supplied is not constant.

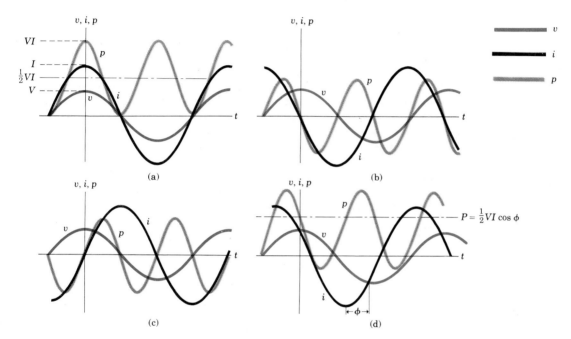

(a)
(b)
(c)
(d)

36-9 Graphs of voltage, current, and power as functions of time, for various circuits. (a) Instantaneous power input to a resistor. The average power is $\frac{1}{2}VI$. (b) Instantaneous power input to a capacitor. The average power is zero. (c) Instantaneous power input to a pure inductor. The average power is zero. (d) Instantaneous power input to an arbitrary ac circuit. The average power is $\frac{1}{2}VI \cos\phi = V_{\text{rms}}I_{\text{rms}}\cos\phi$.

The power curve is symmetrical about a value equal to one-half its maximum ordinate VI, so the *average* power P is

$$P = \frac{1}{2}VI. \qquad (36\text{-}26)$$

The average power can also be written

$$P = \frac{VI}{\sqrt{2}}\left(\frac{I}{\sqrt{2}}\right) = V_{\text{rms}}I_{\text{rms}}. \qquad (36\text{-}27)$$

Furthermore, since $V_{\text{rms}} = I_{\text{rms}}R$, we have

$$P = I_{\text{rms}}^2 R. \qquad (36\text{-}28)$$

Note that Eqs. (36–27) and (36–28) have the same *form* as those for a dc circuit.

Suppose next that the circuit consists of a capacitor, as in Fig. 36–2. The current and voltage are then 90° out of phase. When the curves of v and i are multiplied together (the product vi is *negative* when v and i have *opposite* signs), we get the power curve in Fig. 36–9b, which is symmetrical about the horizontal axis. The average power is therefore zero.

To see why this is so, we recall that positive power means that energy is supplied *to* a device and that negative power means that energy is supplied *by* a device. The process we are considering is merely the charge and discharge of a capacitor. During the intervals when p is positive, energy is supplied to charge the capacitor, and when p is negative the capacitor is discharging and returning energy to the source.

Figure 36–9c is the power curve for a pure inductor. As with a capacitor, the current and voltage are out of phase by 90°, and the average power is zero. Energy is supplied to establish a magnetic field around the inductor and is returned to the source when the field collapses.

In the most general case, current and voltage differ in phase by an angle ϕ and

$$p = [V \sin \omega t][I \sin(\omega t - \phi)]. \qquad (36\text{--}29)$$

The instantaneous power curve has the form shown in Fig. 36–9d. The area under the positive loops is greater than that under the negative loops and the net average power is positive.

The preceding analyses have shown that when v and i are *in phase,* the average power equals $\frac{1}{2}VI$; and when v and i are 90° *out of phase,* the average power is zero. Hence, in the general case, when v and i differ by an angle ϕ, the average power equals $\frac{1}{2}V$, multiplied by $I \cos \phi$, the component of I that is *in phase* with V. That is,

$$P = \frac{1}{2} VI \cos \phi = V_{\text{rms}} I_{\text{rms}} \cos \phi. \qquad (36\text{--}30)$$

This is the general expression for the power input to *any* ac circuit. The factor $\cos \phi$ is called the *power factor* of the circuit. For a pure resistance, $\phi = 0$, $\cos \phi = 1$, and $P = V_{\text{rms}} I_{\text{rms}}$. For a capacitor or inductor, $\phi = 90°$, $\cos \phi = 0$, and $P = 0$.

A low power factor (large angle of lag or lead) is usually undesirable in power circuits because, for a given potential difference, a large current is needed to supply a given amount of power with correspondingly large heat losses in the transmission lines. Since many types of ac machinery draw a lagging current, this situation is likely to arise. It can be corrected by connecting a capacitor in parallel with the load. The leading current drawn by the capacitor compensates for the lagging current in the other branch of the circuit. The capacitor itself takes no net power from the line.

36-6 Series resonance

The impedance of an R–L–C series circuit depends on the frequency, since the inductive reactance is directly proportional to frequency, and the capacitive reactance inversely proportional. This dependence is illustrated in Fig. 36–10a, where a logarithmic frequency scale has been used because of the wide range of frequencies covered. Note that there is one particular frequency at which X_L and X_C are numerically equal. At this frequency, $X = X_L - X_C$ is zero. Hence the impedance Z, equal to $\sqrt{R^2 + X^2}$, is *minimum* at this frequency and is equal to the resistance R.

If an ac source of constant voltage amplitude but variable frequency is connected across the circuit, the current amplitude I varies with frequency, as shown in Fig. 36–10b, and is *maximum* at the frequency at which the impedance Z is *minimum*. The same diagram also shows the phase difference ϕ of the current with respect to the voltage, as a function of frequency. At low frequencies, where capacitive reactance X_C predominates, the current *leads* the voltage. At the frequency where $X = 0$, the current and voltage are *in phase*, and at high frequencies, where X_L predominates, the current *lags* the voltage.

The behavior of the current in an ac series circuit, as the source frequency is varied, is exactly analogous to the response of a spring-mass system having a viscous damping force as the frequency of the driving force is varied. The frequency ω_0 at which the current is maximum is called the *resonant frequency* and is easily computed from the fact that, at this frequency, $X_L = X_C$:

$$X_L = X_C, \qquad \omega_0 L = \frac{1}{\omega_0 C}, \qquad \omega_0 = \frac{1}{\sqrt{LC}}. \tag{36–31}$$

Note that this is equal to the natural frequency of oscillation of an L-C circuit, as derived in Sec. 34–5.

If the inductance L or the capacitance C of a circuit can be varied, the resonant frequency can be varied also. This is the procedure by which a radio or television receiving set may be "tuned" to receive the signal from a desired station.

Example The series circuit in Fig. 36–11 is connected to the terminals of an ac source whose frequency is variable but whose rms terminal voltage is constant and equal to 100 V. The resonant frequency is

$$\omega_0 = \frac{1}{\sqrt{LC}} = \frac{1}{\sqrt{(2\,\text{H})(0.5 \times 10^{-6}\,\text{F})}} = 1000\,\text{rad}\cdot\text{s}^{-1}.$$

At this frequency,

$$X_L = (1000\,\text{rad}\cdot\text{s}^{-1})(2\,\text{H}) = 2000\,\Omega,$$

$$X_C = \frac{1}{(1000\,\text{rad}\cdot\text{s}^{-1})(0.5 \times 10^{-6}\,\text{F})} = 2000\,\Omega,$$

$$X = X_L - X_C = 0.$$

The impedance Z is then equal to the resistance R, and the rms current is

$$I = \frac{V}{Z} = \frac{V}{R} = 0.20\,\text{A}.$$

The rms potential difference across the resistor is

$$V_R = IR = 100\,\text{V}.$$

The rms potential differences across the inductor and capacitor are, respectively,

$$V_L = IX_L = (0.20\,\text{A})(2000\,\Omega) = 400\,\text{V},$$
$$V_C = IX_C = (0.20\,\text{A})(2000\,\Omega) = 400\,\text{V}.$$

The rms potential difference across the inductor–capacitor combination (V_{cb}) is

$$V = IX = I(X_L - X_C) = 0.$$

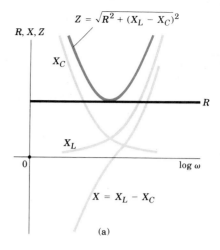

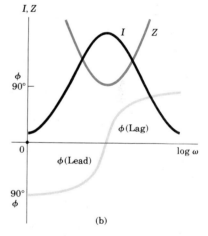

36–10 (a) Reactance, resistance, and impedance as functions of frequency (logarithmic frequency scale). (b) Impedance, current, and phase angle as functions of frequency (logarithmic frequency scale).

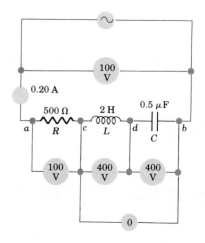

36–11 Series resonant circuit.

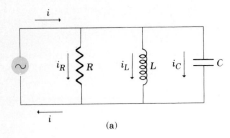

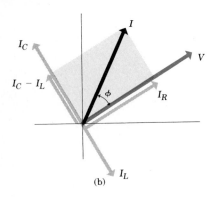

36–12 (a) A parallel R–L–C circuit. (b) Phasor diagram showing current phasors for the three branches. The single voltage phasor V represents the voltage across all three branches.

This is true because the instantaneous potential differences across the inductor and the capacitor have equal amplitudes but are 180° out of phase. Although the rms value of each is substantial, their *resultant* at each instant is zero.

36–7 Parallel circuits

Circuit elements connected in parallel across an ac source can be analyzed by the same procedure as for elements in series. The phasor diagram for the circuit in Fig. 36–12a is given in Fig. 36–12b. In this case, the instantaneous potential difference across each element is the same, and a *single* phasor V represents the common terminal voltage. The phasor I_R, of amplitude V/R and in phase with V, represents the current in the resistor. Phasor I_L, of amplitude V/X_L and lagging V by 90°, represents the current in the inductor, and phasor I_C, of amplitude V/X_C and leading V by 90°, represents the current in the capacitor.

The instantaneous current i, by Kirchhoff's point rule, equals the (algebraic) sum of the instantaneous currents i_R, i_L, and i_C, and is represented by the phasor I, the vector sum of phasors I_R, I_L, and I_C. Angle ϕ is the phase angle of current with respect to source voltage.

From Fig. 36–12,

$$I = \sqrt{I_R{}^2 + (I_C - I_L)^2} = \sqrt{\left(\frac{V}{R}\right)^2 + \left(\omega CV - \frac{V}{\omega L}\right)^2}$$

$$= V\sqrt{\frac{1}{R^2} + \left(\omega C - \frac{1}{\omega L}\right)^2} \qquad (36\text{–}32)$$

The maximum current I is frequency-dependent, as expected. It is *minimum* when the second factor in the radical is zero; this occurs when the two reactances have equal magnitudes, at the resonant frequency ω_0 given by Eq. (36–31).

Thus at resonance the total current in the parallel R–L–C circuit is *minimum*, in contrast to the R–L–C *series* circuit, which has *maximum* current at resonance. This can be understood by noting that, in the parallel circuit, the currents in L and C are *always* exactly a half-cycle out of phase; when they also have equal *magnitudes*, they cancel each other completely, and the total current is simply that through R. Indeed, when $\omega C = 1/\omega L$, Eq. (36–32) becomes simply $I = V/R$. This does *not* mean that there is *no* current in L or C at resonance, but only that the two currents cancel. If R is large, the equivalent impedance of the circuit near resonance is much *larger* than the individual reactances X_L and X_C.

*36–8 The transformer

For reasons of efficiency, it is desirable to transmit electrical power at high voltages and small currents, with consequent reduction of I^2R heating in the transmission line. On the other hand, considerations of safety and of insulation of moving parts require relatively low voltages in generating equipment and in motors and household appliances. One of the most useful features of ac circuits is the ease and efficiency with which voltages (and currents) may be changed from one value to another by *transformers*.

In principle, the transformer consists of two coils electrically insulated from each other and wound on the same iron core. An alternating current in one winding sets up an alternating flux in the core, and the induced electric field produced by this varying flux (see Sec. 33–2) induces an emf in the other winding. Energy is thus transferred from one winding to another via the core flux and its associated induced electric field. The winding to which power is supplied is called the *primary;* that from which power is delivered is called the *secondary.* The circuit symbol for an iron core transformer is

The power output of a transformer is necessarily less than the power input because of unavoidable losses. These losses consist of I^2R losses in the primary and secondary windings, and hysteresis and eddy-current losses in the core. Hysteresis losses are minimized by the use of iron having a narrow hysteresis loop; and eddy currents are minimized by laminating the core. In spite of these losses, transformer efficiencies are usually well over 90%, and in large installations may reach 99%.

For simplicity, we shall consider only an idealized transformer in which there are *no* losses and in which all of the flux is confined to the iron core, so that the same flux links both primary and secondary. The transformer is shown schematically in Fig. 36–13. A primary winding of N_1 turns and a secondary winding of N_2 turns both encircle the core in the same sense. An ac source with voltage amplitude V_1 is connected to the primary and, to begin with, we assume the secondary to be open, so that there is no secondary current. The primary winding then functions merely as an inductor.

The primary current, which is small, lags the primary voltage by 90° and is called the *magnetizing* current. The power input to the transformer is zero. The core flux is in phase with the primary current. Since the same flux links both primary and secondary, the induced emf *per turn* is the same in each. The ratio of primary to secondary induced emf is therefore equal to the ratio of primary to secondary turns, or

$$\frac{\mathcal{E}_2}{\mathcal{E}_1} = \frac{N_2}{N_1}.$$

Since the windings are assumed to have zero resistance, the induced emf's $\mathcal{E}_1$ and $\mathcal{E}_2$ are numerically equal to the corresponding terminal voltages V_1 and V_2, and

$$\frac{V_2}{V_1} = \frac{N_2}{N_1}. \tag{36–33}$$

Hence by properly choosing the turn ratio N_2/N_1, we may obtain any desired secondary voltage from a given primary voltage. If $V_2 > V_1$, we have a *step-up* transformer; if $V_2 < V_1$, a *step-down* transformer.

Consider next the effect of closing the secondary circuit. The secondary current I_2 and its phase angle ϕ_2 will, of course, depend on the nature of the secondary circuit. As soon as the latter is closed, some power must be delivered by the secondary (except when $\phi_2 = 90°$); and, from energy

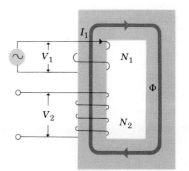

36-13 Schematic diagram of a transformer with secondary open.

considerations, an equal amount of power must be supplied to the primary. The process by which the transformer is enabled to draw the requisite amount of power is as follows. When the secondary circuit is open, the core flux is produced by the primary current only. But when the secondary circuit is closed, *both* primary and secondary currents set up a flux in the core. The secondary current, by Lenz's law, tends to weaken the core flux and therefore to decrease the back-emf in the primary. But (in the absence of losses) the back-emf in the primary must equal the primary terminal voltage, which is assumed to be fixed. The primary current therefore increases until the core flux is restored to its original no-load magnitude.

If the secondary circuit is completed by a resistance R, $I_2 = V_2/R$. From energy considerations, the power delivered to the primary equals that taken out of the secondary (neglecting losses), so

$$V_1 I_1 = V_2 I_2. \tag{36-34}$$

Combining this with the above expression for I_2 and Eq. (36–33) to eliminate V_2 and I_2, we find

$$I_1 = \frac{V_1}{(N_1/N_2)^2 R}. \tag{36-35}$$

Thus, when the secondary circuit is completed through a resistance R, the result is the same as if the *source* had been connected directly to a resistance equal to R multiplied by the reciprocal of the *square* of the turns ratio. In other words, the transformer "transforms" not only voltages and currents, but resistances (more generally, impedances) as well. It can be shown that maximum power is supplied by a source to a resistor when its resistance equals the internal resistance of the source. The same principle applies in ac circuits, with resistance replaced by impedance. When a high-impedance ac source must be connected to a low-impedance circuit, as when an audio amplifier is connected to a loudspeaker, the impedance of the source can be *matched* to that of the circuit by the insertion of a transformer having the correct turns ratio.

Questions

36-1 Some electric-power systems formerly used 25-Hz alternating current instead of the 60-Hz that is now standard. The lights flickered noticeably. Why is this not a problem with 60-Hz?

36-2 Power-distribution systems in airplanes sometimes use 400-Hz ac. What advantages and disadvantages does this have compared to the standard 60-Hz?

36-3 Fluorescent lights often use an inductor, called a "ballast," to limit the current through the tubes. Why is it better to use an inductor than a resistor for this purpose?

36-4 At high frequencies a capacitor becomes a short circuit. Discuss.

36-5 At high frequencies an inductor becomes an open circuit. Discuss.

36-6 Household electric power in most of western Europe is 220 volts, rather than the 110-V which is standard in the United States and Canada. What advantages and disadvantages does each system have?

36-7 The current in an ac power line changes direction 120 times per second, and its average value is zero. So how is it possible for power to be transmitted in such a system?

36-8 Electric-power connecting cords, such as lamp cords, always have two conductors which carry equal currents in opposite directions. How might one determine, using measurements only at the midpoints along the lengths of the wires, the direction of power transmission in the cord?

36-9 Are the equations for the average and rms values of current, Eqs. (36–23) and (36–24), correct when the variation with time is not sinusoidal? Explain.

36-10 Electric power companies like to have their power factors (cf. Sec. 36–5) as close to unity as possible. Why?

36-11 Some electrical appliances operate equally well on ac or dc, while others work only on ac or only on dc. Give examples of each, and explain the differences.

36-12 When a series-resonant circuit is connected across a 110-V ac line, the voltage rating of the capacitor may be exceeded, even if it is rated at 200 or 400 V. How can this be?

36-13 In a parallel-resonant circuit connected across a 110-V line, it is possible for the maximum current rating in the inductor to be exceeded, even if the total current through the circuit is very small. How can this be?

36-14 Can a transformer be used with dc? What happens if a transformer designed for 110-V ac is connected to a 110-V dc line?

36-15 During the last quarter of the nineteenth century there was a great and acrimonious controversy concerning whether ac or dc should be used for power transmission. Edison favored dc, George Westinghouse ac. What arguments might each proponent have used to promote his scheme?

Problems

Angular frequency $= 2\pi$ (frequency),

$$\omega(\text{rad} \cdot \text{s}^{-1}) = 2\pi f(\text{Hz}).$$

36-1

a) At what frequency would a 5-H inductor have a reactance of 4000 Ω?

b) At what frequency would a 5-μF capacitor have the same reactance?

36-2 What is the reactance of a 0.015-μF capacitor at (a) 1 Hz? (b) 5 kHz? (c) 2 MHz?

36-3

a) What is the reactance of a 1-H inductor at a frequency of 60 Hz?

b) What is the inductance of an inductor whose reactance is 1 Ω at 60 Hz?

c) What is the reactance of a 1-μF capacitor at a frequency of 60 Hz?

d) What is the capacitance of a capacitor whose reactance is 1 ohm at 60 Hz?

36-4

a) Compute the reactance of a 10-H inductor at frequencies of 60 Hz and 600 Hz.

b) Compute the reactance of a 10-μF capacitor at the same frequencies.

c) At what frequency is the reactance of a 10-H inductor equal to that of a 10-μF capacitor?

36-5 A 1-μF capacitor is connected across an ac source whose voltage amplitude is kept constant at 50 V, but whose frequency can be varied. Find the current amplitude when the angular frequency is

a) 100 rad $\cdot$ s^{-1},

b) 1000 rad $\cdot$ s^{-1},

c) 10,000 rad $\cdot$ s^{-1}.

d) Construct a log-log plot of current amplitude vs. frequency.

36-6 The voltage amplitude of an ac source is 50 V and its angular frequency is 1000 rad $\cdot$ s^{-1}. Find the current amplitude if the capacitance of a capacitor connected across the source is

a) 0.01 μF,

b) 1.0 μF,

c) 100 μF.

d) Construct a log-log plot of current amplitude vs. capacitance.

36-7 An inductor of self-inductance 10 H and of negligible resistance is connected across the source in Problem 36–5. Find the current amplitude when the angular frequency is

a) 100 rad $\cdot$ s^{-1},

b) 1000 rad $\cdot$ s^{-1},

c) 10,000 rad $\cdot$ s^{-1}.

d) Construct a log-log plot of current amplitude vs. frequency.

36-8 Find the current amplitude if the self-inductance of a resistanceless inductor connected across the source of Problem 36–6 is

a) 0.01 H,

b) 1.0 H,

c) 100 H.

d) Construct a log-log plot of current amplitude versus self-inductance.

36-9 The expression for the impedance Z of an R–L series circuit can be obtained from Eq. (36–20) by setting $X_C = 0$, which corresponds to $C = \infty$. Explain.

36-10 In an R–L–C series circuit, the source has a constant voltage amplitude of 50 V and a frequency of 1000 rad $\cdot$ s^{-1}. $R = 300 \ \Omega$, $L = 0.9$ H, and $C = 2.0 \ \mu$F. Suppose a series circuit contains only the resistor and the inductor in series.

a) What is the impedance of the circuit?

b) What is the current amplitude?

c) What are the voltage amplitudes across the resistor and across the inductor?

d) What is the phase angle ϕ? Does the current lag or lead?

e) Construct the phasor diagram.

36-11 Same as Problem 36-10, except that the circuit consists of the resistor and the capacitor in series.

36-12 Same as Problem 36-10, except that the circuit consists of the inductor and capacitor in series.

36-13

a) Compute the impedance of an R–L–C series circuit at an angular frequency of 500 rad·s^{-1}, using the numbers in Problem 36-10.

b) Describe how the current amplitude varies as the frequency of the source is slowly reduced from 1000 rad·s^{-1} to 500 rad·s^{-1}.

c) What is the phase angle when $\omega = 500$ rad·s^{-1}? Construct the phasor diagram when $\omega = 500$ rad·s^{-1}.

36-14 Five infinite impedance voltmeters, calibrated to read rms values, are connected as shown in Fig. 36-14. What does each voltmeter read? Use the numbers given in Problem 36-10.

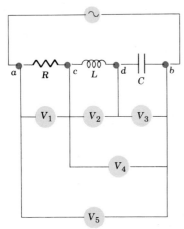

Figure 36-14

36-15 What is the reading of each voltmeter in Fig. 36-14 if the angular frequency $\omega = 500$ rad·s^{-1}? Use the values of V, R, L, and C in Problem 36-10, and assume that the voltmeters measure rms values.

36-16 A 400-Ω resistor is in series with a 0.1-H inductor and a 0.5-μF capacitor. Compute the impedance of the circuit and draw the vector impedance diagram (a) at a frequency of 500 Hz, and (b) at a frequency of 1000 Hz. Compute, in each case, the phase angle between line current and line voltage, and state whether the current lags or leads.

36-17

a) Construct a graph of the current amplitude in the circuit of Fig. 36-5 as the angular frequency of the source is increased from 500 rad·s^{-1} to 1000 rad·s^{-1}. Use the values of V, R, L, and C in Problem 36-10.

b) At what frequency is the circuit in resonance?

c) What is the power factor at resonance?

d) What is the reading of each voltmeter in Fig. 36-14 when the frequency equals the resonant frequency?

e) What would be the resonant frequency if the resistance were reduced to 100 Ω?

f) What would then be the rms current at resonance?

36-18

a) At what frequency is the voltage amplitude across the inductor in an R–L–C series circuit a maximum?

b) At what frequency is the voltage amplitude across the capacitor a maximum?

36-19 In an R–L–C series circuit, $R = 250\,\Omega$, $L = 0.5$ H, and $C = 0.02\,\mu$F.

a) What is the resonant frequency of the circuit?

b) The capacitor can withstand a peak voltage of 350 V. What maximum effective terminal voltage may the generator have at resonant frequency?

36-20 A resistor of 500 Ω and a capacitor of 2 μF are connected in parallel to an ac generator supplying a constant voltage amplitude of 282 V and having an angular frequency of 377 rad·s^{-1}. Find

a) the current amplitude in the resistor,

b) the current amplitude in the capacitor,

c) the phase angle, and

d) the line current amplitude.

36-21 A coil has a resistance of 20 Ω. At a frequency of 100 Hz, the voltage across the coil leads the current in it by 30°. Determine the inductance of the coil.

36-22 An inductor having a reactance of 25 Ω gives off heat at the rate of 2.39 cal·s^{-1} when it carries a current of 0.5 A (rms). What is the impedance of the inductor?

36-23 The circuit of Problem 36-16 carries an rms current of 0.25 A with frequency 100 Hz.

a) What power is consumed in the circuit?

b) In the resistor?

c) In the capacitor?

d) In the inductor?

e) What is the power factor of the circuit?

36-24 A series circuit has a resistance of 75 Ω and an impedance of 150 Ω. What power is consumed in the circuit when a voltage of 120 V (rms) is impressed across it?

36-25 A circuit draws 330 W from a 110-V, 60-Hz ac line. The power factor is 0.6 and the current lags the voltage.

a) Find the capacitance of the series capacitor that will result in a power factor of unity.

b) What power will then be drawn from the supply line?

36-26 A series circuit has an impedance of 50 Ω and a power factor of 0.6 at 60 Hz, the voltage lagging the current.

a) Should an inductor or a capacitor be placed in series with the circuit to raise its power factor?

b) What size element will raise the power factor to unity?

36-27 The internal resistance of an ac source is 10,000 Ω.

a) What should be the turns ratio of a transformer to match the source to a load of resistance 10 Ω?

b) If the voltage amplitude of the source is 100 V, what is the voltage amplitude of the secondary on open circuit?

36-28 A 100-Ω resistor, a 0.1-μF capacitor, and a 0.1-H inductor are connected in parallel to a voltage source with amplitude 100 V.

a) What is the resonant frequency? The resonant angular frequency?

b) What is the maximum total current through the parallel combination at the resonant frequency?

c) What is the maximum current in the resistor at resonance?

d) What is the maximum current in the inductor at resonance?

e) What is the maximum energy stored in the inductor at resonance? In the capacitor?

36-29 The same three components as in Problem 36-28 are connected in *series* to a voltage source with amplitude 100 V.

a) What is the resonant frequency? The resonant angular frequency?

b) What is the maximum current in the resistor at resonance?

c) What is the maximum voltage across the capacitor at resonance?

d) What is the maximum energy stored in the capacitor at resonance?

36-30 A transformer connected to a 120-V ac line is to supply 12 V to a low-voltage lighting system for a model-railroad village. The total equivalent resistance of the system is 2 Ω.

a) What should be the turns ratio of the transformer?

b) What current must the secondary supply?

c) What power is delivered to the load?

d) What resistance connected directly across the 120-V line would draw the same power as the transformer? Show that this is equal to 2 Ω times the square of the turns ratio.

36-31 A step-up transformer connected to a 120-V ac line is to supply 18,000 V for a neon sign. To reduce shock hazard, a fuse is to be inserted in the primary circuit; the fuse is to blow when the current in the secondary circuit exceeds 10 mA.

a) What is the turns ratio of the transformer?

b) What power must be supplied to the transformer when the secondary current is 10 mA?

c) What current rating should the fuse in the primary circuit have?

36-32 The current in a certain circuit varies with time as shown in Fig. 36-15. Find the average current and the rms current, in terms of I_0.

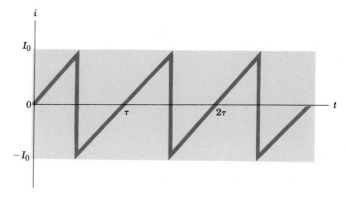

Figure 36-15

ELECTRO-MAGNETIC WAVES

37

37-1 Introduction

Our discussion of electric and magnetic fields in the last several chapters can be classified in two general categories. The first includes fields that do not vary with time. The electrostatic field of a distribution of charges at rest and the magnetic field of a steady current in a conductor are examples of fields which, while they may vary from point to point in space, do not vary with time at any individual point. For such situations we found it possible to treat the electric and magnetic fields independently, without worrying unduly about interactions between the two fields.

The second category includes situations in which the fields *do* vary with time, and in all such cases we have found that it is *not* possible to treat the fields independently. Faraday's law tells us that a time-varying magnetic field acts as a source of electric field. This field is manifested in the induced emf's in inductances and transformers. Similarly, in developing the general formulation of Ampère's law (Sec. 32–9), which is valid for charging capacitors and similar situations as well as for ordinary conductors, we found it necessary to regard a changing electric field as a source of magnetic field.

Thus, when *either* field is changing with time, a field of the other kind is induced in adjacent regions of space. We are thus led naturally to consider the possibility of an electromagnetic disturbance, consisting of time-varying electric and magnetic fields, which can propagate through space from one region to another, even when there is no matter in the intervening region. Such a disturbance, if it exists, will have the properties of a *wave,* and an appropriate descriptive term is *electromagnetic wave*. Such waves do exist, of course, and it is a familiar fact that radio and television transmission, light, x-rays, and many other phenomena are examples of electromagnetic radiation. In the following pages we shall show how the existence of such waves is related to the principles of electromagnetism studied thus far, and shall examine their properties.

As so often happens in the development of science, the theoretical understanding of electromagnetic waves originally took a considerably more devious path than the one outlined above. In the early days of electromagnetic theory (the early nineteenth century), two different units of electric charge were used, one for electrostatics, the other for magnetic phenomena involving currents. The particular system of units in common use at the time had the property that these two units of charge had different physical dimensions; their ratio turned out to have units of velocity. This in itself is not so astounding, but experimental measurements revealed that the ratio had a numerical value precisely equal to the speed of light, $3.00 \times 10^8 \, \text{m} \cdot \text{s}^{-1}$. At the time, physicists regarded this as an extraordinary coincidence, and had no idea how to explain it. It was the search for an explanation of this result that led Maxwell, in 1864, to prove by theoretical reasoning that an electrical disturbance should propagate in free space with a speed equal to that of light, and hence to postulate that light waves were *electromagnetic* waves.

During the course of his analysis, Maxwell also discovered that all the basic principles of electromagnetism can be formulated in terms of four fundamental equations, now called *Maxwell's equations*. We have now studied all four of these principles; they are (1) Gauss's law, (2) the absence of magnetic charge, (3) Ampère's law, including displacement current, and (4) Faraday's law. To summarize, Maxwell's equations, in the integral form used in this text, are:

$$\int \boldsymbol{E} \cdot d\boldsymbol{A} = \frac{Q}{\epsilon_0}, \qquad \text{(Gauss's law)} \qquad (25\text{–}20)$$

$$\int \boldsymbol{B} \cdot d\boldsymbol{A} = 0, \qquad \text{(No magnetic charge)} \qquad (32\text{–}14)$$

$$\oint \boldsymbol{B} \cdot d\boldsymbol{l} = \mu_0 \left(I + \epsilon_0 \frac{d\Psi}{dt} \right), \qquad \text{(Ampère's law)} \qquad (32\text{–}28)$$

$$\oint \boldsymbol{E} \cdot d\boldsymbol{l} = -\frac{d\Phi}{dt}. \qquad \text{(Faraday's law)} \qquad (33\text{–}14)$$

In this form, these equations apply to electric and magnetic fields *in vacuum;* they may also be generalized to include fields in matter. Maxwell's synthesis of electromagnetism in these four equations is one of the great milestones of theoretical physics, comparable in importance to Newton's laws of motion in mechanics.

It remained for Heinrich Hertz, in 1887, actually to *produce* electromagnetic waves, with the aid of oscillating circuits, and to receive and detect these waves with other circuits tuned to the same frequency. Hertz then produced standing electromagnetic waves and measured the distance between adjacent nodes, to measure the wavelength. Knowing the frequency of his resonators, he then found the velocity of the waves from the fundamental wave equation $c = f\lambda$, and verified Maxwell's theoretical value directly.

The possible use of electromagnetic waves for purposes of long-distance communication does not seem to have occurred to Hertz. It remained for the enthusiasm and energy of Marconi and others to make "wireless telegraphy" a familiar household phenomenon.

The unit of frequency, one cycle per second, is named one *hertz* (1 Hz) in honor of Hertz.

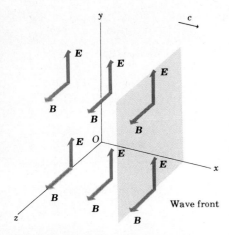

37-1 An electromagnetic wave front. The E and B fields are uniform over the region to the left of the plane, but are zero everywhere to the right of it. The plane representing the wave front moves to the right with speed c.

*37-2 Speed of an electromagnetic wave

In developing the relation of electromagnetic wave propagation to familiar electromagnetic principles, we begin with a particularly simple and somewhat artificial example of an electromagnetic wave. We shall postulate a particular configuration of fields, and then test whether it is consistent with the principles mentioned above, Faraday's law and Ampère's law including displacement current.

Using an x-y-z-coordinate system, as shown in Fig. 37–1, we imagine that all space is divided into two regions by a plane perpendicular to the x-axis (parallel to the yz-plane). At all points to the right of this plane there are no electric or magnetic fields; at all points to the left there is a uniform electric field E in the $+y$-direction and a uniform magnetic field B in the $+z$-direction, as shown. Furthermore, we suppose that the boundary surface, which may also be called the *wave front,* moves to the right with a constant speed c, as yet unknown. Thus the E and B fields travel to the right into previously field-free regions, with a definite speed. The situation, in short, does describe a rudimentary electromagnetic wave, provided it is consistent with the laws of electromagnetism.

We shall not worry about how such a field configuration can be produced. It can be shown that, in principle, an infinitely large sheet of charge in the yz-plane, which is initially at rest and suddenly starts moving with constant velocity in the $-y$-direction, gives such fields, but of course there is no practical way to realize an infinitely large charge sheet. At any rate, we now apply Faraday's law to a rectangle in the xy-plane, as in Fig. 37–2, located so that at some instant the wave front has progressed partway through the rectangle, as shown. In a time dt, the boundary surface moves a distance $c\,dt$ to the right, sweeping out an area $ac\,dt$ of the rectangle, and in this time the magnetic flux increases by $B(ac\,dt)$.

Thus the rate of change of magnetic flux is given by

$$\frac{d\Phi}{dt} = Bac. \tag{37-1}$$

According to Faraday's law, this must equal the line integral $\oint E \cdot dl$ around the boundary of the area. The field at the right end is zero, and the top and bottom sides do not contribute because there the component of E along dl is zero. Thus the integral becomes simply Ea, and Faraday's law gives

$$E = cB. \tag{37-2}$$

Thus the postulated wave is consistent with Faraday's law only if $E, B,$ and c are related as in Eq. (37–2).

Next, we consider a rectangle in the yz-plane, as shown in Fig. 37–3. We wish to apply Ampère's law to this rectangle; the general form of this law, including displacement current, was stated in Chapter 32, Eq. (32–27):

$$\oint B \cdot dl = \mu_0(I_C + I_D). \tag{37-3}$$

In the present situation there is no conduction current, so $I_C = 0$. Previously we have defined the displacement current I_D in terms of a displace-

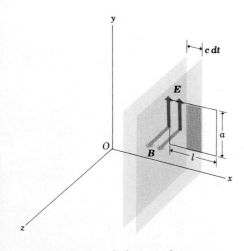

37-2 In time dt the wave front moves to the right a distance $c\,dt$. The magnetic flux through the rectangle in the xy-plane increases by an amount $d\Phi$ equal to the flux through the shaded rectangle of area $ac\,dt$, that is, $d\Phi = Bac\,dt$.

ment current *density,* as in Eq. (29-20):

$$J_{\mathrm{D}} = \epsilon_0 \frac{dE}{dt}. \tag{37-4}$$

Then if the **E** field lines pass through an area A normal to the direction of **E,** we have

$$I_{\mathrm{D}} = J_{\mathrm{D}}A = \epsilon_0 \frac{dE}{dt}A. \tag{37-5}$$

Displacement current may also be expressed in terms of electric flux, as shown in Secs. 29-6 and 32-9. If an area A lies perpendicular to an electric field **E,** the electric flux through the area, denoted by Ψ, is defined as

$$\Psi = EA.$$

In terms of electric flux, Eq. (37-5) can be rewritten as

$$I_{\mathrm{D}} = \epsilon_0 \frac{d\Psi}{dt}, \tag{37-6}$$

and Ampère's law with displacement current but no conduction current is

$$\oint \boldsymbol{B} \cdot d\boldsymbol{l} = \mu_0 \epsilon_0 \frac{d\Psi}{dt}. \tag{37-7}$$

This is the form we shall apply to the rectangle in Fig. 37-3. The change in electric flux $d\Psi$ in time dt is the area $bc\,dt$ swept out by the wave front, multiplied by E. In evaluating the integral $\oint \boldsymbol{B} \cdot d\boldsymbol{l}$, we note that B is zero on the right end and $B_{\parallel} = 0$ on the front and back sides. Thus only the left end contributes, and we find $\oint \boldsymbol{B} \cdot d\boldsymbol{l} = Bb$. Combining these results with Eq. (37-7) and dividing through by b, we obtain

$$B = \mu_0 \epsilon_0 cE. \tag{37-8}$$

Thus Ampère's law is obeyed only if **B,** c, and **E** are related as in Eq. (37-8).

　　Since *both* Ampère's law and Faraday's law must be obeyed simultaneously, Eqs. (37-2) and (37-8) must both be satisfied. This can occur only when $\mu_0 \epsilon_0 c = 1/c$, or

$$c = \frac{1}{\sqrt{\epsilon_0 \mu_0}}. \tag{37-9}$$

Inserting the numerical values of these quantities, we find

$$c = \frac{1}{\sqrt{(8.85 \times 10^{-12})(4\pi \times 10^{-7})}} = 3.00 \times 10^8 \ \mathrm{m \cdot s^{-1}}.$$

The postulated field configuration *is* consistent with the laws of electrodynamics, provided the wave front moves with the speed given above, which of course is recognized as the speed of light.

　　We have chosen a particularly simple wave for study in order to avoid undue mathematical complexity, but this special case nevertheless illustrates several important features of *all* electromagnetic waves:

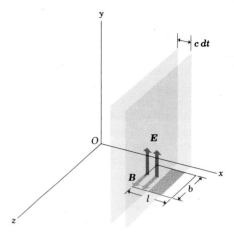

37-3 In time dt the electric flux through the rectangle in the xz-plane increases by an amount equal to E times the area $(bc\,dt)$ of the shaded rectangle; that is, $d\Psi = Ebc\,dt$. Thus $d\Psi/dt = Ebc$.

1. The wave is *transverse;* both E and B are perpendicular to the direction of propagation of the wave, and to each other.

2. There is a definite ratio between E and B.

3. The wave travels in vacuum with a definite and unchanging speed.

It is not difficult to generalize the above discussion to a more realistic situation. Suppose we have several wave fronts in the form of parallel planes perpendicular to the axis and all moving to the right with speed c. Suppose that, within a single region between two planes, the E and B fields are the same at all points in the region, but that they differ from region to region. An extension of the above development shows that such a situation is also consistent with Ampère's and Faraday's laws, provided the wave fronts all move with the speed c given by Eq. (37–9). From this picture it is only a short additional step to a wave picture in which the E and B fields at any instant vary smoothly, rather than in steps, as we move along the x-axis and the entire field pattern moves to the right with speed c. In Sec. 37–5 we consider waves in which the dependence of E and B on position and time is *sinusoidal;* first, however, we consider the *energy* associated with an electromagnetic wave.

37-3 Energy in electromagnetic waves

Analysis of energy needed to charge a capacitor and to establish a current in an inductor has led us (Sections 27–4 and 34–3) to associate an energy density (energy per unit volume) with electric and magnetic fields. Specifically, Eqs. (27–9) and (34–9) show that the total energy density u in a region of space where E and B fields are present is given by

$$u = \frac{1}{2}\epsilon_0 E^2 + \frac{1}{2\mu_0} B^2. \tag{37-10}$$

This and subsequent expressions become somewhat more symmetrical when expressed in terms of the field $H = B/\mu_0$ introduced in Chapter 35. In terms of E and H, the energy density is

$$u = \tfrac{1}{2}\epsilon_0 E^2 + \tfrac{1}{2}\mu_0 H^2. \tag{37-11}$$

Now we have found above that E and B in an electromagnetic wave are not independent but are related by

$$B = \frac{E}{c} = \sqrt{\epsilon_0\mu_0}\,E \quad \text{or} \quad H = \sqrt{\frac{\epsilon_0}{\mu_0}}E. \tag{37-12}$$

Thus the energy density may also be expressed

$$u = \frac{1}{2}\epsilon_0 E^2 + \frac{1}{2}\mu_0\left(\sqrt{\frac{\epsilon_0}{\mu_0}}E\right)^2 = \frac{1}{2}\epsilon_0 E^2 + \frac{1}{2}\epsilon_0 E^2 = \epsilon_0 E^2, \tag{37-13}$$

which shows that, in a wave, the energy density associated with the E field is equal to that of the H field.

Equation (37–12) may also be written

$$\frac{E}{H} = \sqrt{\frac{\mu_0}{\epsilon_0}}. \tag{37-14}$$

The unit of E is $1\ \text{V}\cdot\text{m}^{-1}$ and that of H is $1\ \text{A}\cdot\text{m}^{-1}$, so the unit of the ratio E/H is $1 \text{ V}\cdot\text{A}^{-1}$ or $1\,\Omega$. When numerical values of ϵ_0 and μ_0 are inserted, we find

$$\frac{E}{H} = 377\,\Omega. \qquad (37\text{--}15)$$

For reasons that we cannot discuss in detail, this ratio is called the *impedance of free space.*

Because the $\boldsymbol{E}$ and $\boldsymbol{H}$ fields in the simple wave considered above advance with time into regions where originally there were no fields, it is clear that the wave transports energy from one region to another. This energy transfer is conveniently characterized by considering the energy transferred *per unit time, per unit cross-sectional area* for an area perpendicular to the direction of wave travel. This quantity will be denoted by S. This is analogous to the concept of current density, the charge per unit time transferred across unit area perpendicular to the direction of flow.

To see how the energy flow is related to the fields, we consider a stationary plane perpendicular to the x-axis, which at a certain time coincides with the wave front. In a time dt after this time, the wave front moves a distance $c\,dt$ to the right. Considering an area A on the stationary plane, we note that the energy in the space to the right of this area must have passed through it to reach its new location. The volume dV of the relevant region is the base area A times the length $c\,dt$, and the total energy dU in this region is the energy density u times this volume:

$$dU = \epsilon_0 E^2 A c\,dt. \qquad (37\text{--}16)$$

Since this much energy passed through area A in time dt, the energy flow S per unit time, per unit area is

$$S = \frac{dU}{A\,dt} = \epsilon_0 c E^2.$$

Using Eqs. (37–9) and (37–12), we obtain the alternative forms

$$S = \frac{\epsilon_0}{\sqrt{\epsilon_0 \mu_0}} E^2 = \sqrt{\frac{\epsilon_0}{\mu_0}}\,E^2 = EH. \qquad (37\text{--}17)$$

The unit of S is energy per unit time, per unit area. The SI unit of S is $1\text{ J}\cdot\text{s}^{-1}\cdot\text{m}^{-2}$ or $1\text{ W}\cdot\text{m}^{-2}$.

We can also define a *vector* quantity that describes both the magnitude and direction of the energy flow rate. We define

$$\boldsymbol{S} = \boldsymbol{E} \times \boldsymbol{H}. \qquad (37\text{--}18)$$

$\boldsymbol{S}$ is called the *Poynting vector;* its magnitude is given by Eq. (37–17), and its direction is the direction of propagation of the wave. The magnitude EH gives the flow of energy across a cross section perpendicular to the propagation direction, per unit area and per unit time.

The electric and magnetic fields at any point in a wave vary with time, so the Poynting vector at any point is also a function of time. The *average* value of the magnitude of the Poynting vector at a point is called the *intensity* of the radiation at that point. As mentioned above, the SI unit of intensity is the watt per square meter $(\text{W}\cdot\text{m}^{-2})$. In Sec. 37–5 we consider the relation of the intensity of a sinusoidal wave to the amplitudes of the sinusoidally varying electric and magnetic fields.

Example In the wave discussed above, suppose

$$E = 100 \text{ V} \cdot \text{m}^{-1} = 100 \text{ N} \cdot \text{C}^{-1}.$$

Find the values of B and H, the maximum energy density, and the maximum energy flow S.

Solution From Eq. (37–2),

$$B = \frac{E}{c} = \frac{100 \text{ V} \cdot \text{m}^{-1}}{3.0 \times 10^8 \text{ m} \cdot \text{s}^{-1}} = 3.33 \times 10^{-7} \text{ T};$$

$$H = \frac{B}{\mu_0} = \frac{3.33 \times 10^{-7} \text{ T}}{4\pi \times 10^{-7} \text{ Wb} \cdot \text{A}^{-1} \cdot \text{m}^{-1}} = 0.265 \text{ A} \cdot \text{m}^{-1}.$$

From Eq. (37–13),

$$u = \epsilon_0 E^2 = (8.85 \times 10^{-12} \text{ C}^2 \cdot \text{N}^{-1} \cdot \text{m}^{-2})(100 \text{ N} \cdot \text{C}^{-1})^2$$
$$= 8.85 \times 10^{-8} \text{ N} \cdot \text{m}^{-2} = 8.85 \times 10^{-8} \text{ J} \cdot \text{m}^{-3};$$

$$S = EH = (100 \text{ V} \cdot \text{m}^{-1})(0.265 \text{ A} \cdot \text{m}^{-1})$$
$$= 26.5 \text{ W} \cdot \text{m}^{-2}.$$

The idea that energy can travel through empty space without the aid of any matter in motion may seem strange, yet this is the very mechanism by which energy reaches us from the sun. The conclusion that electromagnetic waves transport energy is as inescapable as the conclusion that energy is required to establish electric and magnetic fields. It is also possible to show that electromagnetic waves carry *momentum,* with a corresponding momentum density of magnitude

$$\frac{EH}{c^2} = \frac{S}{c^2}. \tag{37–19}$$

This momentum is a property of the field alone and is not associated with moving mass. There is also a corresponding momentum flow rate; just as the energy density u corresponds to S, the rate of energy flow per unit area, the momentum density given by Eq. 37–19 corresponds to the momentum flow rate

$$\frac{1}{A}\frac{dp}{dt} = \frac{S}{c} = \frac{EH}{c}, \tag{37–20}$$

which represents the momentum transferred per unit surface area, per unit time.

This momentum is responsible for the phenomenon of *radiation pressure;* when an electromagnetic wave is absorbed by a surface perpendicular to the propagation direction, the time rate of change of momentum is equal to the force on the surface. Thus the force per unit area, or presssure, is equal to S/c. If the wave is totally reflected, the momentum change is twice as great, and the pressure is $2S/c$. For example, the value of S for direct sunlight is about $1.4 \text{ kW} \cdot \text{m}^{-2}$, and the corresponding pressure on a completely absorbing surface is

$$p = \frac{1.4 \times 10^3 \text{ W} \cdot \text{m}^{-2}}{3.0 \times 10^8 \text{ m} \cdot \text{s}^{-1}} = 4.7 \times 10^{-6} \text{ Pa}$$
$$= 4.7 \times 10^{-6} \text{ N} \cdot \text{m}^{-2}.$$

Radiation pressure is important in the structure of stars. Gravitational attractions tend to shrink a star, but this tendency is balanced by radiation pressure in maintaining the size of the star through most stages of its evolution.

*37-4 Electromagnetic waves in matter

The above analysis can be extended to include electromagnetic waves in dielectrics. The wave speed is not the same as in vacuum, and we denote it by w instead of c. Faraday's law is unaltered, but Eq. (37-2) is replaced by $E = wB$. In Ampère's law, the displacement current density is given not by $\epsilon_0 \, dE/dt$ but by $\epsilon \, dE/dt = K\epsilon_0 \, dE/dt$. In addition, the constant μ_0 in Ampère's law must be replaced by $\mu = K_m\mu_0$. Thus, Eq. (37-8) is replaced by

$$B = \mu\epsilon wE,$$

and the wave speed is given by

$$w = \frac{1}{\sqrt{\epsilon\mu}} = \frac{1}{\sqrt{KK_m}} \frac{1}{\sqrt{\epsilon_0\mu_0}}. \tag{37-21}$$

For many dielectrics the relative permeability K_m is very nearly equal to unity; in such cases we can say

$$w \simeq \frac{1}{\sqrt{K}} \frac{1}{\sqrt{\epsilon_0\mu_0}} = \frac{c}{\sqrt{K}}.$$

Because K is always greater than unity, the speed of electromagnetic waves in a dielectric is always *less* than the speed in vacuum, by a factor of $1/\sqrt{K}$. The ratio of the speed in vacuum to the speed in a material is known in optics as the *index of refraction n* of the material; we see that this is given by

$$\frac{c}{w} = n = \sqrt{KK_m} \simeq \sqrt{K}. \tag{37-22}$$

Corresponding modifications are required in the expressions for the energy density and the Poynting vector; these are developed in detail in more advanced texts on electromagnetic theory.

The waves described above cannot propagate in a *conducting* material because the E field leads to currents that provide a mechanism for dissipating the energy of the wave. For an ideal conductor with zero resistivity, E must be zero everywhere inside the material. When an electromagnetic wave strikes such a material, it is totally reflected. Real conductors with finite resistivity permit some penetration of the wave into the material, with partial reflection. A polished metal surface is usually a good *reflector* of electromagnetic waves, but metals are not *transparent* to radiation.

37-5 Sinusoidal waves

Sinusoidal electromagnetic waves are closely analogous to sinusoidal transverse mechanical waves on a stretched string, as discussed in Chapter 21. At any point in space, the E and H fields are sinusoidal functions of time and, at any instant of time, the spatial variation of the fields is also sinusoidal.

The simplest sinusoidal electromagnetic waves share with the wave of Sec. 37–2 the property that at any instant the fields are uniform over any plane perpendicular to the direction of propagation. Such a wave is called a *plane wave*. The entire pattern travels to the right with speed c. Since E and H are at right angles to the direction of propagation, the wave is *transverse*.

The frequency f, the wavelength λ, and the speed of propagation c are related by the equation applicable to any sort of periodic wave motion, namely, $c = f\lambda$. If the frequency f is the power-line frequency of 60 Hz, the wavelength is

$$\lambda = \frac{c}{f} = \frac{3 \times 10^8 \text{ m} \cdot \text{s}^{-1}}{60 \text{ Hz}} = 5 \times 10^6 \text{ m} = 5000 \text{ km},$$

which is of the order of the earth's radius! Hence at this frequency even a distance of many miles includes only a small fraction of a wavelength. On the other hand, if the frequency is 10^8 Hz (100 MHz), typical of commercial FM radio stations, the wavelength is

$$\lambda = \frac{3 \times 10^8 \text{ m} \cdot \text{s}^{-1}}{10^8 \text{ Hz}} = 3 \text{ m},$$

and a moderate distance can include a number of complete waves.

Figure 37–4 represents a sinusoidal electromagnetic wave traveling in the $+x$-direction. The E and H vectors are shown only for a few points on the x-axis; if a plane is constructed perpendicular to the x-axis at a particular point, the fields have the same values at all points in that plane. The values are of course different on different planes. In those planes in which the E vector is in the $+y$-direction, H is in the $+z$-direction; where E is in the $-y$-direction, H is in the $-z$-direction. In both cases the direction of the Poynting vector, given by Eq. (37–18), is along the $+x$-direction.

We recall from Sec. 21–3, Eq. (21–10), that one form of the equation of a transverse wave traveling to the right along a stretched string is $y = A \sin (\omega t - kx)$, where y is the transverse displacement from its equilibrium position, at the time t, of a point of the string whose coordinate is x. The quantity A is the maximum displacement or the *amplitude* of the

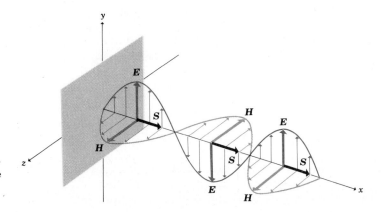

37–4 *E*-, *H*-, and *S*-vectors in a sinusoidal electromagnetic wave traveling in the x-direction. The position of the wave at time $t = 0$ is shown.

wave, ω is its *angular frequency,* equal to 2π times the frequency f, and k is the *propagation constant,* equal to $2\pi/\lambda$, where λ is the wavelength.

Let E and H represent the instantaneous values, and E_{max} and H_{max} the maximum values, or *amplitudes,* of the electric and magnetic fields in Fig. 37–4. The equations of the traveling electromagnetic wave are then

$$E = E_{max} \sin (\omega t - kx), \qquad H = H_{max} \sin (\omega t - kx). \quad (37\text{-}23)$$

The sine curves in Fig. 37–4 represent instantaneous values of E and H, as functions of x at the time $t = 0$. The wave travels to the right with speed c.

The instantaneous value S of the Poynting vector is

$$S = EH = E_{max}H_{max} \sin^2 (\omega t - kx)$$
$$= \tfrac{1}{2}E_{max}H_{max}[1 - \cos 2(\omega t - kx)].$$

The time average value of $\cos 2(\omega t - kx)$ is zero, so the average value S_{av} of the Poynting vector is

$$S_{av} = \tfrac{1}{2}E_{max}H_{max}. \qquad (37\text{-}24)$$

This is the *average power* transmitted per unit area, and, as noted in Sec. 37–3, it is called the *intensity* of the radiation.

Figure 37–5 represents schematically the electric and magnetic fields of a wave traveling in the *negative x*-direction. At points where $\boldsymbol{E}$ is in the positive y-direction, $\boldsymbol{H}$ is in the *negative z*-direction, and where $\boldsymbol{E}$ is in the negative y-direction, $\boldsymbol{H}$ is in the *positive z*-direction. The Poynting vector is in the negative x-direction at all points. (Compare with Fig. 37–4, which represents a wave traveling in the *positive x*-direction.) The equations of the wave are

$$E = -E_{max} \sin (\omega t + kx), \qquad H = H_{max} \sin (\omega t + kx). \quad (37\text{-}25)$$

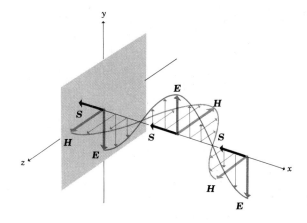

37-5 Electric and magnetic fields of a sinusoidal wave traveling in the negative x-direction. The position of the wave at time $t = 0$ is shown.

37-6 Standing waves

Electromagnetic waves can be reflected, and the superposition of an incident wave and a reflected wave can form a *standing wave* analogous to standing waves on a stretched string, discussed in Sec. 22–2. A conducting surface can serve as a reflector for electromagnetic waves.

Suppose a sheet of an ideal conductor, having zero resistivity, is placed in the yz-plane of Fig. 37–5, and that the wave shown, traveling in the negative x-direction, is incident on it. The essential characteristic of an ideal conductor is that there can never be any electric field in it; any attempt to establish a field is immediately canceled by rearrangement of the mobile charges in the conductor. Thus E must always be zero everywhere in this plane, and the E field of the incident wave induces sinusoidal currents in the conductor so that E is zero inside it.

These induced currents then produce a reflected wave, traveling out from the plane, to the right. From the superposition principle, the total E field at any point to the right of the plane is the vector sum of the E fields of the incident and reflected waves; the same is true for the total H field.

Suppose the incident wave is described by Eqs. (37–25) and the reflected wave by Eqs. (37–23). Then the total fields at any point are given by

$$E = E_{max}[-\sin(\omega t + kx) + \sin(\omega t - kx)],$$
$$H = H_{max}[\sin(\omega t + kx) + \sin(\omega t - kx)].$$

These expressions may be expanded and simplified, using the identity

$$\sin(A \pm B) = \sin A \cos B \pm \cos A \sin B.$$

The results are

$$E = -2E_{max} \cos \omega t \sin kx, \qquad (37\text{–}26a)$$
$$H = 2H_{max} \sin \omega t \cos kx. \qquad (37\text{–}26b)$$

The first of these is analogous to Eq. (22–1) for the stretched string. We see that at $x = 0$, E is *always* zero; this is required by the nature of the ideal conductor, which plays the same role as a fixed point at the end of the string. Furthermore, E is zero at all times in those planes for which $\sin kx = 0$, that is, $kx = 0, \pi, 2\pi, \ldots$, or

$$x = 0, \quad \frac{\lambda}{2}, \quad \lambda, \quad \frac{3\lambda}{2}, \quad \ldots.$$

These are called *nodal planes* of the E field.

The magnetic field is zero at all times in those planes for which $\cos kx = 0$, or at which

$$x = \frac{\lambda}{4}, \quad 3\frac{\lambda}{4}, \quad 5\frac{\lambda}{4}, \quad \text{etc.}$$

These are the nodal planes of the H field. The magnetic field is *not* zero at the conducting surface ($x = 0$), and there is no reason it should be. The nodal planes of one field are midway between those of the other, and the nodal planes of either field are separated by one-half wavelength. Figure 37–6 shows a standing-wave pattern at one instant of time.

The electric field is a *cosine* function of t and the magnetic field is a *sine* function of t. The fields are therefore 90° out of phase. At times when $\cos \omega t = 0$, the electric field is zero *everywhere* and the magnetic field is maximum. When $\sin \omega t = 0$, the magnetic field is zero *everywhere* and the electric field is maximum.

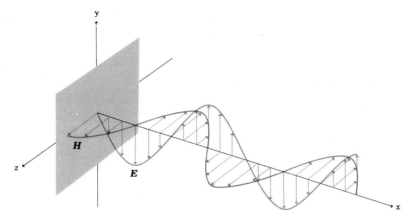

37-6 *E*- and *H*-vectors in a standing wave. The pattern does not move along the *x*-axis, but the *E*- and *H*-vectors grow and diminish with time at each point. At each point *E* is maximum when *H* is minimum, and conversely. The position of the wave at time $t = 0$ is shown.

Pursuing the stretched-string analogy, we may now insert a second conducting plane, parallel to the first and a distance L from it, along the $+x$-axis. This is analogous to the stretched string held at the points $x = 0$ and $x = L$. A standing wave can exist only when L is an integer multiple of $\lambda/2$. Thus the possible wavelengths are

$$\lambda_n = \frac{2L}{n}, \qquad n = 1, 2, 3, \ldots, \qquad (37\text{–}27)$$

and the corresponding frequencies are

$$f_n = \frac{c}{\lambda_n} = n\frac{c}{2L}, \qquad n = 1, 2, 3, \ldots. \qquad (37\text{–}28)$$

Thus there is a set of *normal modes,* each with a characteristic frequency, wave shape, and node pattern. Measurement of the node positions makes it possible to measure the wavelength. If the frequency is known, the wave speed can be determined. This technique was in fact used by Hertz in his pioneering investigations of electromagnetic waves.

Conducting surfaces are not the only reflectors of electromagnetic waves; reflections also occur at an interface between two insulating materials having different dielectric or magnetic properties. The mechanical analog is a junction of two strings with equal tension but different linear mass density. In general, a wave incident on such a boundary surface is partly transmitted into the second material and partly reflected back into the first. The partial transmission and reflection of light at a glass surface is a familiar phenomenon.

*37–7 Radiation from an antenna

The waves discussed above are called *plane waves,* referring to the fact that at any instant the fields are uniform over a plane perpendicular to the direction of propagation of the wave. Although these waves are the simplest to describe and analyze, they are by no means the simplest to produce experimentally. Any charge or current distribution that oscillates sinusoidally with time produces sinusoidal electromagnetic waves. The simplest example of such a source is an oscillating dipole, a pair of electric charges of equal magnitude and opposite sign, with the charge magnitude varying sinusoidally with time. Such an oscillating dipole can

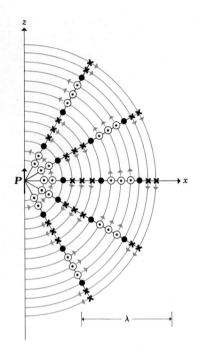

be constructed in a number of ways, the details of which need not concern us.

The radiation from an oscillating dipole is *not* a plane wave, but travels out in all directions from the source. At points far from the source the **E** and **H** fields are perpendicular to the direction from the source and to each other; in this sense the wave is still transverse. The value of S drops off as the square of the distance from the source and also depends on the direction from the source; the radiation is most intense in directions perpendicular to the dipole axis, and $S = 0$ in directions parallel to the axis. The radiation pattern from a dipole source is shown schematically in Fig. 37-7.

37-7 Cross section in the *xz*-plane of radiation from an oscillating electric dipole, showing electric-field vectors at one instant of time. At the points with circles the **B**-field comes out of the plane of the figure; at the points with crosses, it is into the plane.

Questions

37-1 In Ampère's law, is it possible to have both a conduction current and a displacement current at the same time? Is it possible for the effects of the two kinds of current to cancel each other exactly, so that there is *no* magnetic field produced?

37-2 By measuring the electric and magnetic fields at a point in space where there is an electromagnetic wave, can one determine the direction from which the wave came?

37-3 Sometimes neon signs located near a powerful radio station are seen to glow faintly at night, even though they aren't turned on. What is happening?

37-4 Light can be *polarized;* is this a property of all electromagnetic waves, or is it unique to light? What about polarization of sound waves? What fundamental distinction in wave properties is involved?

37-5 How does a microwave oven work? Why does it heat materials that are conductors of electricity, including most foods, but not insulators, such as glass or ceramic dishes?

37-6 Electromagnetic waves can travel through vacuum, where there is no matter. We usually think of vacuum as empty space, but is it *really* empty if electric and magnetic fields are present? What *is* vacuum, anyway?

37-7 Give several examples of electromagnetic waves that are encountered in everyday life. How are they all alike? How do they differ?

37-8 We are surrounded by electromagnetic waves emitted by many radio and television stations. How is a radio or television receiver able to select a single station among all this mishmash of waves? What happens inside a radio receiver when the dial is turned to change stations?

37-9 The metal conducting rods on a television antenna are always in a horizontal plane. Would they work as well if they were vertical?

37-10 If a light beam carries momentum, should a person holding a flashlight feel a recoil analogous to the recoil of a rifle when it is fired? Why is this recoil not actually observed?

37-11 The nineteenth-century inventor Nikolai Tesla proposed to transmit large quantities of electrical energy across space using electromagnetic waves instead of conventional transmission lines. What advantages would this scheme have? What disadvantages?

37-12 Does an electromagnetic *standing wave* have energy? Momentum? What distinction can be drawn between a standing wave and a propagating wave on this basis?

37-13 If light is an electromagnetic wave, what is its frequency? Is this a proper question to ask?

37-14 When an electromagnetic wave is reflected from a moving reflector, the frequency of the reflected wave is different from that of the initial wave. Explain physically how this can happen. (Some radar systems used for highway speed control operate on this principle.)

Problems

37-1 A certain radio station broadcasts at a frequency of 1020 kHz. At a point some distance from the transmitter, the maximum magnetic field of the electromagnetic wave it emits is found to be 1.6×10^{-11} T.

a) What is the speed of propagation of the wave?

b) What is the wavelength?

c) What is the maximum electric field?

37-2 A certain plane electromagnetic wave emitted by a microwave antenna has a wavelength of 3.0 cm and a maximum magnitude of electric field of 2.0×10^{-4} V·cm^{-1}.

a) What is the frequency of the wave?

b) What is the maximum magnetic field?

c) What is the intensity (power per unit area) if the *average* power is half the power computed from the *maximum* values of E and B?

37-3 An electromagnetic wave propagates in a ferrite material having $K = 10$ and $K_M = 1000$. Find

a) the speed of propagation;

b) the wavelength

of a wave having a frequency of 100 MHz.

37-4 The maximum electric field in the vicinity of a certain radio transmitter is 1.0×10^{-3} V·m^{-1}. What is the maximum magnitude of the **B** field? How does this compare in magnitude with the earth's field?

37-5 The energy flow to the earth associated with sunlight is about 1.4 kW·m^{-2}.

a) Find the maximum values of E and B for a wave of this intensity.

b) The distance from the earth to the sun is about 1.5×10^{11} m. Find the total power radiated by the sun.

37-6 For a 50,000-W radio station, find the maximum magnitudes of **E** and **B** at a distance of 100 km from the antenna, assuming that the antenna radiates equally in all directions (which is probably not actually the case).

37-7 The nineteenth-century inventor Nikolai Tesla proposed to transmit electric power via electromagnetic waves. Suppose power is to be transmitted in a beam of cross-sectional area 100 m^2; what **E** and **B** strengths are required to transmit an amount of power comparable to that handled by modern transmission lines (of the order of 500 kV and 1000 A)?

37-8 A bar magnet is mounted on an insulating support, as in the side and end views of Fig. 37–8(a) and (b), and is given a positive electric charge. Copy the diagram, sketch the lines of the **E** and **H** fields around the magnet, and draw the Poynting vector at a number of points.

a) What is the general nature of the energy flow in the electromagnetic field?

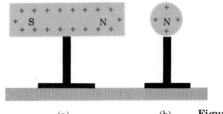

(a)   (b)   **Figure 37-8**

b) What can you say about the "hidden momentum" in this case?

37-9 A cylindrical conductor of circular cross section has a radius a and a resistivity ρ and carries a constant current I.

a) What are the magnitude and direction of the **E**-vector at a point inside the wire, at a distance r from the axis?

b) What are the magnitude and direction of the **H**-vector at the same point?

c) What are the magnitude and direction of the Poynting vector **S** at the same point?

d) Compare your answer to (c) with the rate of dissipation of energy within a volume of the conductor of length l and radius r.

37-10 A wire of radius 1 mm whose resistance per unit length is 3×10^{-3} Ω·m^{-1} carries a current of 25.1 A. At a point very near the surface of the wire, calculate

a) the magnitude of **H**,

b) the component of **E** parallel to the wire,

c) the component of **S** perpendicular to the wire.

37-11 Assume that 10% of the power input to a 100-W lamp is radiated uniformly as light of wavelength 500 nm (1 nm = 10^{-9} m). At a distance of 2 m from the source, the electric and magnetic intensities vary sinusoidally according to the equations $E = E_{max} \sin (\omega t + \phi)$ and $H = H_{max} \sin (\omega t + \phi)$. Calculate E_{max} and H_{max}.

37-12 A capacitor consists of two circular plates of radius r separated by a distance l. Neglecting fringing, show that while the capacitor is being charged, the rate at which energy flows into the space between the plates is equal to the rate at which the electrostatic energy increases.

37-13 A very long solenoid of n turns per unit length and radius a carries an increasing current i.

a) Calculate the induced electric field at a point inside the solenoid at a distance r from the solenoid axis.

b) Compute the magnitude and direction of the Poynting vector at this point.

37-14 An FM radio station antenna radiates a power of 10 kW at a wavelength of 3 m. Assume for simplicity that the radiated power is confined to, and is uniform over, a hemisphere with the antenna at its center. What are the amplitudes of E_{max} and H_{max} in the radiation field at a distance of 10 km from the antenna?

37-15 In a TV picture, ghost images are formed when the signal from the transmitter travels directly to the receiver and also indirectly after reflection from a building or other large metallic mass. In a 25-inch set the ghost is about 1 cm to the right of the principal image if the reflected signal arrives 1 μs after the principal signal. In this case, what is the difference in path length for the two signals?

37-16 For a sinusoidal electromagnetic wave in vacuum, such as that described by Eqs. (37–23), show that the average density of energy in the electric field is the same as that in the magnetic field.

37-17 A plane electromagnetic wave has a wavelength of 3.0 cm and an E-field amplitude of 30 V·m^{-1}.

a) What is the frequency?

b) What is the B-field amplitude?

c) What is the intensity?

d) What average force does the radiation pressure exert on a totally absorbing surface of area 0.5 m^2 perpendicular to the direction of propagation?

37-18 If the intensity of direct sunlight is 1.4 kW·m^{-2}, find the radiation pressure (in pascals) on

a) a totally absorbing surface;

b) a totally reflecting surface.

Also express your results in atmospheres.

THE NATURE AND PROPAGATION OF LIGHT

38

38-1 Nature of light

Until the middle of the seventeenth century, light was generally thought to consist of a stream of some sort of particles or *corpuscles* emanating from light sources. Newton and many other scientists of the day supported the corpuscular theory of light. But about this time the idea was proposed by Huygens and others that light might be a *wave* phenomenon. Indeed, diffraction effects that are now known to be associated with the wave nature of light were observed by Grimaldi as early as 1665, but the significance of his observations was not understood at this time.

Early in the nineteenth century, evidence for the wave theory of light grew more persuasive. The experiments of Fresnel and Thomas Young on interference and diffraction showed conclusively that there are many optical phenomena that can be understood on the basis of the wave theory but for which the corpuscular theory is inadequate. These phenomena will be discussed further in Chapter 41. Young's experiments enabled him to measure the wavelength of the waves, and Fresnel showed that the rectilinear propagation of light, as well as the diffraction effects observed by Grimaldi and others, could be accounted for by the behavior of waves of short wavelength.

The next great forward step in the theory of light was the work of Maxwell, discussed in the preceding chapter. In 1873, Maxwell showed that an oscillating electrical circuit should radiate electromagnetic waves. The speed of propagation of the waves could be computed from purely electrical and magnetic measurements. It turned out to be equal, within the limits of experimental error, to the measured speed of propagation of light. The evidence seemed inescapable that light consisted of electromagnetic waves of extremely short wavelength. In 1887 Heinrich Hertz, using an oscillating circuit of small dimensions, succeeded in producing short-wavelength electromagnetic waves and showed that they possessed all the properties of light waves. They could be reflected, refracted, focused by a lens, polarized, and so on, just as could waves of light. Maxwell's electromagnetic theory of light and its experimental justification by Hertz constituted one of the triumphs of physical science.

By the end of the nineteenth century researchers believed that little, if anything, would be added in the future to our knowledge of the nature of light. How wrong they were!

The classical electromagnetic theory failed to account for several phenomena associated with *emission* and *absorption* of light. One example is the phenomenon of photoelectric emission, that is, the ejection of electrons from a conductor by light incident on its surface. In 1905, Einstein extended an idea proposed five years earlier by Planck, and postulated that the energy in a light beam was concentrated in packets or *photons*. The wave picture was retained, in that a photon was still considered to have a frequency and the energy of a photon was proportional to its frequency. Experiments by Millikan confirmed Einstein's predictions. These experiments will be discussed in more detail in Chapter 44.

Another striking confirmation of the photon nature of light is the Compton effect. A. H. Compton, in 1921 succeeded in determining the motion of a photon and a single electron, both before and after a "collision" between them, and found that they behaved like material bodies having kinetic energy and momentum, both of which were conserved in the collision. The photoelectric effect and the Compton effect, then, both seem to demand a return to a corpuscular theory of light.

The reconciliation of these apparently contradictory experiments has come only since 1930, with the development of quantum electrodynamics, a comprehensive theory that includes *both* wave and particle properties. The phenomena of light *propagation* may best be described by the electromagnetic *wave* theory, while the interaction of light with matter, in the processes of emission and absorption, is a *corpuscular* phenomenon.

38-2 Sources of light

All bodies emit electromagnetic radiation as a result of thermal motion of their molecules; this radiation, called *thermal radiation,* is a mixture of different wavelengths. At a temperature of 300°C the most intense of these waves has a wavelength of 5000×10^{-9} m or 5000 nm, which is the *infrared* region. At a temperature of 800°C a body emits enough visible radiant energy to be self-luminous and appears "red hot." By far the larger part of the energy emitted, however, is still carried by infrared waves. At 3000°C, which is about the temperature of an incandescent lamp filament, the radiant energy contains enough of the "visible" wavelengths, between 400 nm and 700 nm, that the body appears "white hot." In modern incandescent lamps, the filament is a coil of fine tungsten wire. An inert gas such as argon is introduced to reduce evaporation of the filament. Incandescent lamps vary in size from one no larger than a grain of wheat to one with a power input of 5000 W, used for illuminating airfields.

One of the brightest sources of light is the *carbon arc*. Two carbon rods, typically 10 to 20 cm long and 1 cm in diameter, are connected to a 110-V or 220-V dc source. They are touched together momentarily and then are pulled apart a few millimeters. Intense electron bombardment of the positive rod causes an extremely hot crater to form at its end; this crater, whose temperature is typically 4000°C, is the source of light. Carbon-arc lights are used in most theater motion-picture projectors and in large searchlights and lighthouses.

Some light sources use an arc discharge in a metal vapor, such as mercury or sodium, in a sealed bulb containing two electrodes, which are connected to a power source. Some argon is sometimes added to permit a glow discharge that helps vaporize and ionize the metal. The bluish light of mercury-arc lamps and the bright orange-yellow of sodium-vapor lamps are familiar in highway and other outdoor lighting.

An important variation of the mercury-arc lamp is the *fluorescent* lamp. This consists of a glass tube containing argon and a droplet of mercury. The electrodes are of tungsten. When an electric discharge takes place in the mercury–argon mixture, only a small amount of visible light is emitted by the mercury and argon atoms. There is, however, considerable *ultraviolet* light (light of wavelength shorter than that of visible violet). This ultraviolet light is absorbed in a thin layer of material, called a *phosphor,* which is the white coating on the interior walls of the glass tube. The phosphor has the property of *fluorescence,* which means that it emits visible light when illuminated by light of shorter wavelength. Lamps may be obtained that will fluoresce with any desired color, depending on the nature of the phosphor.

A special light source that has attained prominence in the last twenty years is the *laser,* which can produce a very narrow beam of enormously intense radiation. High-intensity lasers have been used to cut through steel, fuse high-melting-point materials, and bring about many other effects that are important in physics, chemistry, biology, and engineering. An equally significant characteristic of laser light is that it is much more nearly *monochromatic,* or single-frequency, than any other light source. The operation of one type of laser will be discussed in Chapter 44.

38-3 The speed of light

The speed of light in free space is one of the fundamental constants of nature. Its magnitude is so great (about $186{,}000 \text{ mi·s}^{-1}$ or $3 \times 10^8 \text{ m·s}^{-1}$) that it evaded experimental measurement until 1676. Up to that time it was generally believed that light traveled with an infinite speed.

The first recorded attempt to measure the speed of light involved a method proposed by Galileo. Two experimenters were stationed on the tops of two hills about a mile apart. Each was provided with a lantern, the experiment being performed at night. One man was first to uncover his lantern and, observing the light from this lantern, the second was to uncover his. The velocity of light could then be computed from the known distance between the lanterns and the time elapsing between the instant when the first observer uncovered his lantern and when he observed the light from the second. While the experiment was correct in principle, we know now that the speed is much too great for the time interval to be measured in this way.

In 1676, the Danish astronomer Olaf Roemer, from astronomical observations made on one of the satellites of the planet Jupiter, obtained the first definite evidence that light travels with finite speed. Jupiter has twelve small satellites or moons, four of which are sufficiently bright to be seen with a moderately good telescope or a pair of field glasses. The satellites appear as tiny bright points at one side or the other of the planet. These satellites revolve about Jupiter just as does our moon about the earth and, since the plane of their orbits is nearly the same as that in which the earth and Jupiter revolve, each is *eclipsed* by the planet during a part of every revolution.

Roemer measured the time of revolution of one of the satellites by taking the time interval between consecutive eclipses (about 42 hr). He found, by a comparison of results over a long period of time, that while the earth was *receding* from Jupiter the periodic times were all somewhat *longer* than the average, and that while it was *approaching* Jupiter the times were all somewhat *shorter*. He concluded rightly that the cause of these variations was the varying distance between Jupiter and the earth.

Roemer concluded from his observations that a time of about 22 min was required for light to travel a distance equal to the diameter of the earth's orbit. The best figure for this distance, in Roemer's time, was about 172,000,000 mi. Although there is no record that Roemer actually made the computation, if he *had* used the data above he would have found a speed of about 130,000 mi·s^{-1} or 2.1×10^8 m·s^{-1}.

The first successful determination of the speed of light from purely *terrestrial* measurements was made by the French scientist Fizeau in 1849. A schematic diagram of his apparatus is given in Fig. 38-1. Lens L_1 forms an image of the light source S at a point near the rim of a toothed wheel T, which can be set into rapid rotation. G is an inclined plate of clear glass. Suppose first that the wheel is stationary and the light passes through one of the openings between the teeth. Lenses L_2 and L_3, which are separated by about 8.6 km, form a second image on the mirror M. The light is reflected from M, retraces its path, and is in part reflected from the glass plate G through the lens L_4 into the eye of an observer at E.

If the wheel T is set in rotation, the light from S is "chopped up" into a succession of wave trains of limited length. If the speed of rotation is such that by the time the front of one wave train has traveled to the mirror and returned, an opaque segment of the wheel has moved into the position formerly occupied by an open portion, *no* reflected light will reach the observer E. At twice this angular velocity, the light transmitted through any one opening will return through the next and an image

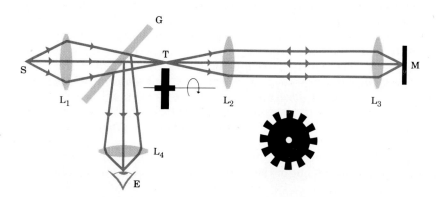

38-1 Fizeau's toothed-wheel method for measuring the velocity of light. S is a light source, L_1, L_2, L_3, and L_4 are lenses. T is the toothed wheel, M is a mirror, and G is a glass plate.

of S will again be observed. From a knowledge of the angular velocity and radius of the wheel, the distance between openings, and the distance from wheel to mirror, the speed of light may be computed. Fizeau's measurements were not of high precision. He obtained a value of

$$3.15 \times 10^8 \, \text{m} \cdot \text{s}^{-1}.$$

Fizeau's apparatus was modified by Foucault, who replaced the toothed wheel with a rotating mirror. The most precise measurements by the Foucault method were made by the American physicist Albert A. Michelson (1852-1931). His first experiments were performed in 1878 while he was on the staff of the Naval Academy at Annapolis. The last, under way at the time of his death, were completed in 1935 by Pease and Pearson.

From analysis of all measurements up to 1976, the most probable value for the speed of light is

$$c = 2.9979246 \times 10^8 \, \text{m} \cdot \text{s}^{-1},$$

which is believed to be correct within $\pm 2 \, \text{m} \cdot \text{s}^{-1}$.

As shown in Chapter 37, the speed of any electromagnetic wave in free space is given by

$$c = \sqrt{\frac{1}{\epsilon_0 \mu_0}},$$

and since, in SI units, μ_0 is assigned a value of *exactly* $4\pi \times 10^{-7} \, \text{N} \cdot \text{s}^2 \cdot \text{C}^{-2}$, the preceding equation provides the most precise means for finding the value of the electrical constant ϵ_0:

$$\epsilon_0 = 1/\mu_0 c^2,$$

since the speed of light can be measured with much higher precision than the force between two charged bodies.

38-4 The electromagnetic spectrum

As indicated in preceding sections, it is now well established that light consists of electromagnetic waves. In this respect it is a small part of a very broad class of electromagnetic radiations, all having the general characteristics described in Chapter 37 (including the common speed of propagation $c = 3.00 \times 10^8 \, \text{m} \cdot \text{s}^{-1}$) but differing in frequency (f) and wavelength (λ). The general wave relation $c = \lambda f$ holds for each.

The wavelengths of visible light (i.e., of electromagnetic waves that are perceived by the sense of sight) can be measured by methods to be discussed in Chapter 41, and are found to lie in the range 4 to 7×10^{-7} m. The corresponding range of frequencies is about 7.5 to 4.3×10^{14} Hz. The position of visible light in relation to the entire electromagnetic spectrum is shown in Fig. 38-2.

Because of the very small magnitudes of light wavelengths, it is convenient to measure them in small units of length. Three such units are commonly used: the micrometer (1 μm), the nanometer (1 nm) (both accented on the *first* syllable), and the angstrom (1 Å):

$$1 \, \mu\text{m} = 10^{-6} \, \text{m} = 10^{-4} \, \text{cm},$$
$$1 \, \text{nm} = 10^{-9} \, \text{m} = 10^{-7} \, \text{cm},$$
$$1 \, \text{Å} = 10^{-10} \, \text{m} = 10^{-8} \, \text{cm} = 0.1 \, \text{nm}.$$

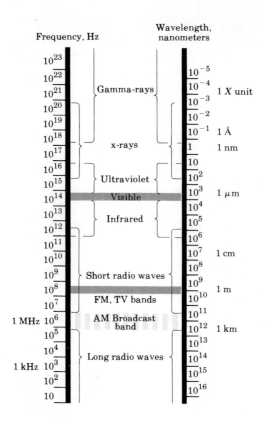

38-2 A chart of the electromagnetic spectrum.

In older literature the micrometer is sometimes called the *micron,* and the nanometer is sometimes called the *millimicron;* these terms are now obsolete. Most workers in the fields of optical instrument design, color, and physiological optics express wavelengths in *nanometers.* For example, the wavelength of the yellow light from a sodium-vapor lamp is 589 nm, but many spectroscopists would identify this same wavelength as 5890 Å.

Different parts of the visible spectrum evoke the sensations of different colors. Wavelengths for colors in the visible spectrum are (very approximately) as follows:

400 to 450 nm	Violet
450 to 500 nm	Blue
500 to 550 nm	Green
550 to 600 nm	Yellow
600 to 650 nm	Orange
650 to 700 nm	Red

By the use of special sources or special filters, it is possible to limit the wavelength spread to a small band, say from 1 to 10 nm. Such light is called roughly *monochromatic light,* meaning light of a single color. Light consisting of only one wavelength is an idealization that is useful in theoretical calculations but represents an experimental impossibility. When the expression "monochromatic light of wavelength 550 nm" is used in theoretical discussions, it refers to one wavelength, but in descriptions of laboratory experiments it means a small band of wavelengths *around* 550 nm. One distinguishing characteristic of light from a *laser* is that it is much more nearly monochromatic than light obtainable in any other way.

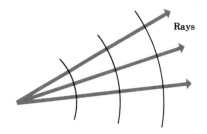

38-5 Waves, wave fronts, and rays

It is convenient to represent a wave of any sort by means of *wave fronts. A wave front is defined as the locus of all points at which the phase of vibration of a physical quantity is the same.* Thus, in the case of sound waves spreading out in all directions from a point source, any spherical surface concentric with the source is a possible wave front. Some of these spherical surfaces are the loci of points at which the pressure is a *maximum,* others where it is a *minimum,* and so on; but the *phase* of the pressure variations is the same over any spherical surface. It is customary to draw only a few wave fronts, usually those that pass through the maxima and minima of the disturbance. Such wave fronts are separated from one another by one-half wavelength.

If the wave is a light wave, the quantity corresponding to the pressure in a sound wave is the electric or magnetic field. It is usually unnecessary to indicate in a diagram either the magnitude or direction of the field, but simply to show the *shape* of the wave by drawing the wave fronts or their intersections with some reference plane. For example, the electromagnetic waves radiated by a small light source may be represented by *spherical* surfaces concentric with the source or, as in Fig. 38–3a, by the intersections of these surfaces with the plane of the diagram. At a sufficiently great distance from the source, where the radii of the spheres have become very large, the spherical surfaces can be considered planes and we have a *plane* wave, as in Fig. 38–3b.

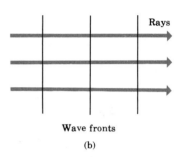

38-3   Wave fronts and rays.

For some purposes, especially in that branch of optics called *geometrical* optics, it is convenient to represent a light wave by *rays* rather than by wave fronts. Indeed, rays were used to describe light long before its wave nature was firmly established; in a corpuscular theory of light, rays are merely the paths of the corpuscles. From the wave viewpoint, *a ray is an imaginary line drawn in the direction in which the wave is traveling.* Thus, in Fig. 38–3a the rays are the radii of the spherical wave fronts, and in Fig. 38–3b they are straight lines perpendicular to the wave fronts. In fact, in every case where waves travel in a homogeneous isotropic medium, the rays are straight lines normal to the wave fronts. At a boundary surface between two media, such as the surface between a glass plate and the air outside it, the direction of a ray may change suddenly, but it is a straight line both in the air and in the glass. If the medium is *not* homogeneous, for instance, if one is considering the passage of light through the earth's atmosphere, where the density and hence the velocity vary with elevation, the rays are curved but are still normal to the wave fronts.

38-6 Reflection and refraction

Many familiar optical phenomena involve the behavior of a wave that strikes an interface between two optical materials, such as air and glass or water and glass. When the interface is smooth, i.e., when its irregulari-

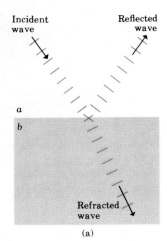

(a)

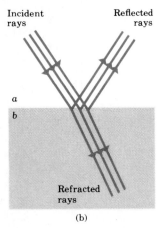

(b)

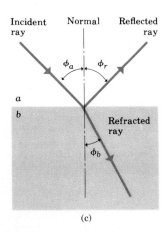

(c)

38-4 (a) A plane wave is in part reflected and in part refracted at the boundary between two media. (b) The waves in (a) are represented by rays. (c) For simplicity, only one example each of incident, reflected, and refracted rays is drawn.

ties are small compared with the wavelength, the wave is in general partly reflected and partly transmitted into the second medium, as shown in Fig. 38-4a. As an example of this behavior, when one looks into a store window from the street, one sees light coming from inside by transmission through the glass and also a reflection of the street scene caused by the air–glass interface.

The segments of plane waves shown in Fig. 38-4a can be represented by bundles of rays forming *beams* of light, as in Fig. 38-4b; and for simplicity in discussing the various angles we often consider only one ray in each beam, as in Fig. 38-4c. Representing these waves in terms of rays is the basis of that branch of optics called *geometrical optics*. This chapter and the two following chapters are devoted entirely to geometrical optics and are concerned with optical phenomena that can be understood on the basis of rays, without having to use the wave nature of light explicitly. It must be understood that geometrical optics, like any idealized model, has its limitations, and that there are optical phenomena that require a more detailed model embodying the *wave* properties of light for their understanding. We return to such phenomena in Chapters 41 and 42.

The directions of the incident, reflected, and refracted beams of light are specified in terms of the angles they make with the *normal* to the surface at the point of incidence. For this purpose it is sufficient to indicate one ray, as in Fig. 38-4c, although a single ray of light is a geometrical abstraction. A careful experimental study of the directions of the incident, reflected, and refracted beams leads to the following results:

1. *The incident, reflected, and refracted rays, and the normal to the surface, all lie in the same plane.* Thus, if the incident ray is in the plane of the diagram, and the surface of separation is perpendicular to this plane, the reflected and refracted rays are in the plane of the diagram.

2. *The angle of reflection ϕ_r is equal to the angle of incidence ϕ_a for all colors and any pair of substances.* Thus

$$\phi_r = \phi_a. \qquad (38-1)$$

The experimental result that $\phi_r = \phi_a$, and that the incident and reflected rays and the normal all lie in the same plane, is known as the *law of reflection*.

3. For monochromatic light and for a given pair of substances, a and b, on opposite sides of the surface of separation, *the ratio of the sine of the angle ϕ_a* (between the ray in substance a and the normal) *and the sine of the angle ϕ_b* (between the ray in substance b and the normal) *is a constant.* Thus

$$\frac{\sin \phi_a}{\sin \phi_b} = \text{constant.} \qquad (38-2)$$

This experimental result, together with the fact that the incident and refracted rays and the normal to the surface all lie in the same plane, is known as the *law of refraction*. The discovery of this law is usually credited to Willebrord Snell (1591–1626), although there is some doubt that it was actually original with him. It is called Snell's law.

In the above discussion, the angles of the various rays are by convention always measured with respect to the *normal* to the surface, not the surface itself. It is essential to keep this in mind when applying Eq. (38-2).

The laws of reflection and refraction relate only to the *directions* of the corresponding rays but say nothing about an equally important question, namely, the *intensities* of the reflected and refracted rays. These depend on the angle of incidence; for the present we simply state that the fraction reflected is smallest at *normal* incidence, where it is a few percent, and that it increases with increasing angle of incidence to almost 100% at grazing incidence or when $\phi_a = 90°$.

When a ray of light is directed from *below* the surface in Fig. 38-4, there are again reflected and refracted rays; these, together with the incident ray and the normal, all lie in the same plane. The same law of reflection applies as when the ray is originally traveling in air, and the same law of refraction, Eq. (38-2). *The passage of a ray of light in going from one medium to another is reversible.* It follows the same path in going from *b* to *a* as when going from *a* to *b*.

Let us now consider a beam of monochromatic light traveling *in vacuum,* making an angle of incidence ϕ_0 with the normal to the surface of a substance *a*, and let ϕ_a be the angle of refraction in the substance. The constant in Snell's law is then called the *index of refraction* of substance *a* and is designated by n_a:

$$\frac{\sin \phi_0}{\sin \phi_a} = n_a. \qquad (38\text{-}3)$$

The index of refraction (also called *refractive index*) is always greater than unity; it depends not only on the substance but on the wavelength of the light. If no wavelength is stated, the index is usually assumed to be that corresponding to the yellow light from a sodium lamp, of wavelength 589 nm. This wavelength is near the middle of the visible spectrum.

The index of refraction of most of the common glasses used in optical instruments lies between 1.46 and 1.96. There are very few substances having indices larger than 1.96, diamond being one, with an index of 2.42, and rutile (crystalline titanium dioxide) another, with an index of 2.7. The values for a number of solids and liquids are given in Table 38-1.

The index of refraction of *air* at standard conditions is about 1.0003; for most purposes the *index of refraction of air can be assumed to be unity*. The index of refraction of a gas increases uniformly as the density of the gas increases.

It follows from Eq. (38-3) that the angle of refraction ϕ_a is always *less than* the angle of incidence ϕ_0 for a ray passing from a vacuum into one of the materials listed in Table 38-1. In such a case, the ray is bent *toward* the normal. If the light is traveling in the opposite direction, the reverse is true and the ray is bent *away from* the normal.

To relate the index of refraction to the constant in Eq. (38-2), we now consider two parallel-sided plates of substances *a* and *b* placed parallel to each other with an arbitrary space between them, as shown in Fig. 38-5a. Let the medium surrounding both plates be vacuum, although the behavior of light would be practically the same if the plates

Table 38-1 Index of refraction for yellow sodium light ($\lambda = 589$ nm)

Substance	Index of refraction
Solids	
Ice (H_2O)	1.309
Fluorite (CaF_2)	1.434
Rock salt (NaCl)	1.544
Quartz (SiO_2)	1.544
Zircon ($ZrO_2 \cdot SiO_2$)	1.923
Diamond (C)	2.417
Fabulite ($SrTiO_3$)	2.409
Glasses (typical values)	
Crown	1.52
Light flint	1.58
Medium flint	1.62
Dense flint	1.66
Lanthanum flint	1.80
Liquids at 20°C	
Methyl alcohol (CH_3OH)	1.329
Water (H_2O)	1.333
Ethyl alcohol (C_2H_5OH)	1.36
Carbon tetrachloride (CCl_4)	1.460
Turpentine	1.472
Glycerine	1.473
Benzene	1.501
Carbon disulfide (CS_2)	1.628

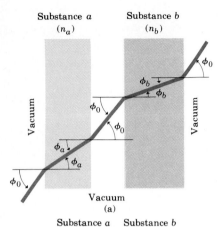

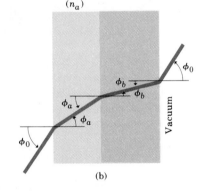

38-5 The transmission of light through parallel plates of different substances. The incident and emerging rays are parallel, regardless of direction and regardless of the thickness of the space between adjacent slabs.

were surrounded by air. If a ray of monochromatic light starts at the lower left with an angle of incidence ϕ_0, the angle between ray and normal in substance a is ϕ_a, and the light emerges from substance a at an angle ϕ_0 equal to its incident angle. The light ray therefore enters plate b with an angle of incidence ϕ_0, makes an angle ϕ_b in substance b, and emerges again at an angle ϕ_0. Exactly the same path would be traversed if the same light ray were to start at the upper right and enter substance b at an angle ϕ_0. Moreover, *the angles are independent of the thickness of the space between the two plates,* and are the same when the space shrinks to nothing, as in Fig. 38-5b.

Applying Snell's law to the refractions at the surface between vacuum and substance a, and at the surface between vacuum and substance b, we have

$$\frac{\sin \phi_0}{\sin \phi_a} = n_a, \qquad \frac{\sin \phi_0}{\sin \phi_b} = n_b.$$

Dividing the second equation by the first, we obtain

$$\frac{\sin \phi_a}{\sin \phi_b} = \frac{n_b}{n_a}, \tag{38-4}$$

which shows that *the constant in Snell's law for the refraction between substances a and b is the ratio of the indices of refraction.* From Eq. (38-4) we see that the simplest and most symmetrical way of writing Snell's law for any two substances a and b, and for any direction, is

$$n_a \sin \phi_a = n_b \sin \phi_b. \tag{38-5}$$

Example   In Fig. 38-4, let material a be water and material b a glass with index of refraction 1.50. If the incident ray makes an angle of 60° with the normal, find the direction of the reflected and refracted rays.

Solution   The reflected-ray angle with the normal is the same as that of the incident ray, according to Eq. (38-1). Hence $\phi_r = \phi_a = 60°$. To find the direction of the refracted ray we use Eq. (38-5), with $n_a = 1.33$, $n_b = 1.50$, and $\phi_a = 60°$. We find

$$(1.33)(\sin 60°) = (1.50)(\sin \phi_b),$$
$$\phi_b = 50.2°.$$

The second material has larger refractive index than the first, and the refracted ray is bent toward the normal.

Two final comments about reflection and refraction need to be added. First, reflection occurs at a highly polished surface of an *opaque* material such as a metal. There is then no refracted ray, but the reflected ray still behaves according to Eq. (38-1). Second, if the reflecting surface of either a transparent or an opaque material is rough, with irregularities on a scale comparable to or larger than the wavelength of light, reflection occurs not in a single direction but in all directions; such reflection is called *diffuse* reflection. Conversely, reflections in a single direction from smooth surfaces are called *regular* reflections or *specular* reflections.

38–7 Total internal reflection

Figure 38–6 shows a number of rays diverging from a point source P in medium a of index n_a and striking the surface of a second medium b of index n_b, where $n_a > n_b$. From Snell's law,

$$\sin \phi_b = \frac{n_a}{n_b} \sin \phi_a.$$

Since n_a/n_b is greater than unity, $\sin \phi_b$ is larger than $\sin \phi_a$. Thus there must be some value of ϕ_a less than 90° for which $\sin \phi_b = 1$ and $\phi_b = 90°$. This is illustrated by ray 3 in the diagram, which emerges just grazing the surface at an angle of refraction of 90°. The angle of incidence for which the refracted ray emerges tangent to the surface is called the *critical angle* and is designated by ϕ_{crit} in the diagram. If the angle of incidence is greater than the critical angle, the sine of the angle of refraction, as computed by Snell's law, would have to be greater than unity. This may be interpreted to mean that beyond the critical angle the ray *does not pass* into the upper medium but is *totally internally reflected* at the boundary surface. Total internal reflection can occur only when a ray is incident on the surface of a medium whose index is *smaller* than that of the medium in which the ray is traveling.

The critical angle for two given substances may be found by setting $\phi_b = 90°$ or $\sin \phi_b = 1$ in Snell's law. We then have

$$\sin \phi_{\text{crit}} = \frac{n_b}{n_a}. \tag{38–6}$$

The critical angle of a glass–air surface, taking 1.50 as a typical index of refraction of glass, is

$$\sin \phi_{\text{crit}} = \frac{1}{1.50} = 0.67, \qquad \phi_{\text{crit}} = 42°.$$

The fact that this angle is slightly less than 45° makes it possible to use a prism of angles 45°–45°–90° as a totally reflecting surface. The advantages of totally reflecting prisms over metallic surfaces as reflec-

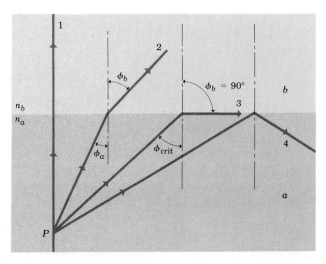

38–6 Total internal reflection. The angle of incidence ϕ_a for which the angle of refraction is 90°, is called the critical angle.

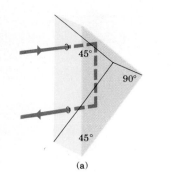

(a)

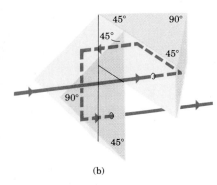

(b)

38-7 (a) A Porro prism. (b) A combination of two Porro prisms.

tors are, first, that the light is *totally* reflected, while no metallic surface reflects 100% of the light incident on it, and second, the reflecting properties are permanent and not affected by tarnishing. Offsetting these is the fact that there is some loss of light by reflection at the surfaces where light enters and leaves the prism, although coating the surfaces with so-called "nonreflecting" films can reduce this loss considerably.

A 45°–45°–90° prism, used as in Fig. 38-7a, is called a *Porro* prism. Light enters and leaves at right angles to the hypotenuse and is totally reflected at each of the shorter faces. The deviation is 180°. Two Porro prisms are sometimes combined, as in Fig. 38-7b, an arrangement often found in binoculars.

If a beam of light enters one end of a transparent rod, as in Fig. 38-8, the light is totally reflected internally and is "trapped" within the rod even if it is curved, provided the curvature is not too great. The rod is sometimes referred to as a *light pipe*. A bundle of fine fibers will behave in the same way and has the advantage of being flexible. The study of the properties of such a bundle is an active field of research known as *fiber optics.*

A bundle may consist of thousands of individual fibers, each the order of 0.002 to 0.01 mm in diameter. If the fibers can be assembled in the bundle so that the relative positions of the ends are the same at both ends, the bundle can transmit an image, as shown in Fig. 38-9. Bundles several feet in length have been made. Devices using fiber optics are finding a wide range of applications in medical science. The interior of lungs and other passages in the human body can be viewed by insertion of a fiber bundle. A bundle can be enclosed in a hypodermic needle for the study of tissues and blood vessels far beneath the skin. Despite the enormous technical difficulties of manufacturing fiber-optic components, such systems promise to become an extremely important class of optical systems.

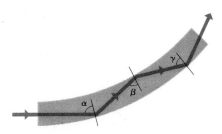

38-8 A light ray "trapped" by internal reflections.

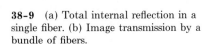

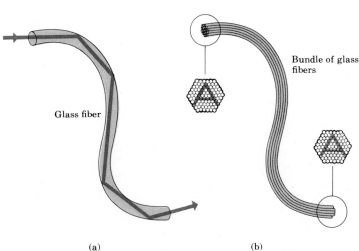

Glass fiber

Bundle of glass fibers

(a)   (b)

38-9 (a) Total internal reflection in a single fiber. (b) Image transmission by a bundle of fibers.

38-8 Huygens' principle

The principles governing reflection and refraction of light rays, discussed in the preceding section, were discovered experimentally long before the wave nature of light was firmly established. These principles may however be derived from *wave* considerations and thus shown to be consistent with the wave nature of light. To establish this connection we use a principle called *Huygens' principle*; as stated originally by Christian Huygens in 1678, this principle is a geometrical method for finding, from the known shape of a wave front at some instant, the shape of the wave front at some later time. Huygens assumed that *every point of a wave front may be considered the source of secondary wavelets, which spread out in all directions with a speed equal to the speed of propagation of the waves.* The new wave front is then found by constructing a surface *tangent* to the secondary wavelets or, as it is called, the *envelope* of the wavelets.

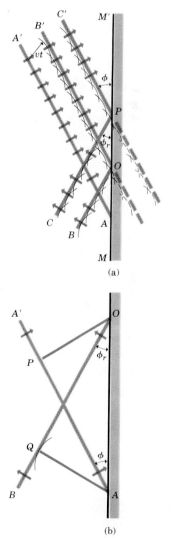

38-10   Huygens' principle.

Huygens' principle is illustrated in Fig. 38-10. The original wave front, AA', is traveling as indicated by the small arrows. We wish to find the shape of the wave front after a time interval t. Let v represent the speed of propagation. We construct a number of circles (traces of spherical wavelets) of radius $r = vt$, with centers along AA'. The trace of the envelope of these wavelets, which is the new wave front, is the curve BB'. The speed v has been assumed the same at all points in all directions.

To derive the law of reflection from Huygens' principle, we consider a plane wave approaching a plane reflecting surface. In Fig. 38-11a, the lines AA', BB', and CC' represent successive positions of a wave front approaching the surface MM'. The actual planes are perpendicular to the plane of the figure. Point A on the wave front AA' has just arrived at the reflecting surface. The position of the wave front after a time interval t may be found by applying Huygens' principle. With points on AA' as centers, draw a number of secondary wavelets of radius vt, where v is the speed of propagation of the wave. Those wavelets originating near the upper end of AA' spread out unhindered, and their envelope gives that portion of the new wave surface OB'. The wavelets originating near the lower end of AA', however, strike the reflecting surface. If the latter had not been there, they would have occupied the positions shown by the dotted circular arcs. The effect of the reflecting surface is to *reverse the direction* of travel of those wavelets that strike it, so that that part of a wavelet that would have penetrated the surface actually lies to the left of it, as shown by the full lines. The envelope of these reflected wavelets is then that portion of the wave front OB. The trace of the entire wave front at this instant is the broken line BOB'. A similar construction gives the line CPC' for the wave front after another interval t.

The angle ϕ between the incident *wave front* and the *surface* is the same as that between the incident *ray* and the *normal* to the surface, and is therefore the angle of incidence. Similarly, ϕ_r is the angle of reflection. To find the relation between these angles, we consider Fig. 38-11b. From O, draw $OP = vt$, perpendicular to AA'. Now OB, by construction, is tangent to a circle of radius vt with center at A. Hence, if AQ is drawn from A to the point of tangency, the triangles APO and OQA are equal (right triangles with the side AO in common and with $AQ = OP$). The angle ϕ therefore equals the angle ϕ_r, and we have the law of reflection.

The law of refraction is obtained by a similar procedure. In Fig. 38-12a, we consider a wave front, represented by line AA', for which point A has just arrived at the boundary surface MM' between two transparent media a and b, of indices of refraction n_a and n_b. (The *reflected* waves are not shown in the figure; they proceed exactly as in Fig.

38-11   (a) Successive positions of a plane wave AA' as it is reflected from a plane surface. (b) A portion of (a).

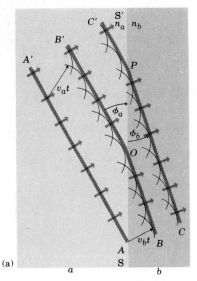

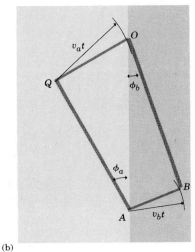

(a)

(b)

38-12 (a) Successive positions of a plane wave front AA' as it is refracted by a plane surface. (b) A portion of (a). The case $v_b < v_a$ is shown.

38-11.) Let us apply Huygens' principle to find the position of the refracted wave front after a time t.

With points on AA' as centers, we draw a number of secondary wavelets. Those originating near the upper end of AA' travel with speed v_a and, after a time interval t, are spherical surfaces of radius $v_a t$. The wavelet originating at point A, however, is traveling in the second medium b with speed v_b and at time t is a spherical surface of radius $v_b t$. The envelope of the wavelets from the original wave front is the plane whose trace is the broken line BOB'. A similar construction leads to the trace CPC' after a second interval t.

The angles ϕ_a and ϕ_b between the surface and the incident and refracted wave fronts are, respectively, the angle of incidence and the angle of refraction. To find the relation between these angles, refer to Fig. 38-12b. Draw $OQ = v_a t$, perpendicular to AQ, and draw $AB = v_b t$, perpendicular to BO. From the right triangle AOQ,

$$\sin \phi_a = \frac{v_a t}{AO},$$

and from the right triangle AOB,

$$\sin \phi_b = \frac{v_b t}{AO}.$$

Hence

$$\frac{\sin \phi_a}{\sin \phi_b} = \frac{v_a}{v_b} \tag{38-7}$$

Since v_a/v_b is a constant, Eq. (38-7) expresses Snell's law, and we have derived Snell's law from a wave theory.

The most general form of Snell's law is given by Eq. (38-4), namely

$$\frac{\sin \phi_a}{\sin \phi_b} = \frac{n_b}{n_a}.$$

Comparing this with Eq. (38-7), we see that

$$\frac{v_a}{v_b} = \frac{n_b}{n_a},$$

and

$$n_a v_a = n_b v_b.$$

When either medium is vacuum, $n = 1$ and the speed is c. Hence

$$n_a = \frac{c}{v_a}, \qquad n_b = \frac{c}{v_b}, \tag{38-8}$$

showing that *the index of refraction of any medium is the ratio of the speed of light in vacuum to the speed in the medium.* For any material, n is always greater than unity; hence the speed of light in a material is always *less* than in vacuum.

In Fig. 38-12, if t is chosen to be the period τ of the wave, the spacing is $v\tau$, which is the wavelength λ. The figure shows that when v_b is less than v_a, the wavelength in the second medium is smaller than that in the first. When a light wave proceeds from one medium to another, where the speed is different, the wavelength changes *but not the frequency*. Since

$$v_a = f\lambda_a \quad \text{and} \quad v_b = f\lambda_b,$$

$$\frac{\lambda_a}{v_a} = \frac{\lambda_b}{v_b} \quad \text{and} \quad \lambda_a\frac{c}{v_a} = \lambda_b\frac{c}{v_b}.$$

Therefore,

$$\lambda_a n_a = \lambda_b n_b.$$

If either medium is vacuum, the index is 1 and the wavelength in vacuum is represented by λ_0. Hence

$$\lambda_a = \frac{\lambda_0}{n_a}, \qquad \lambda_b = \frac{\lambda_0}{n_b}, \qquad (38\text{-}9)$$

showing that *the wavelength in any medium is the wavelength in vacuum divided by the index refraction of the medium.*

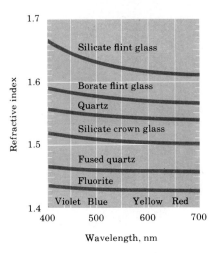

38-13 Variation of index with wavelength.

38-9 Dispersion

Most light beams are a mixture of waves whose wavelengths extend throughout the visible spectrum. While the speed of light waves in a vacuum is the same for all wavelengths, the speed in a material substance is different for different wavelengths. Hence the index of refraction of a substance is a function of wavelength. The dependence of wave speed on wavelength was mentioned in connection with water waves, in Sec. 21-7. A substance in which the speed of a wave varies with wavelength is said to exhibit *dispersion*. Figure 38-13 is a diagram showing the variation of index of refraction with wavelength for a number of the more common optical materials.

Consider a ray of white light, a mixture of all visible wavelengths, incident on a prism, as in Fig. 38-14. Since the deviation produced by the prism increases with increasing index of refraction, violet light is deviated most and red least, with other colors occupying intermediate positions. On emerging from the prism, the light is spread out into a fan-shaped beam, as shown. The light is said to be *dispersed* into a spectrum. A simple measure of the dispersion is provided by the angular separation of the red and violet rays. Dispersion depends on the *difference* between the refractive index for violet light and that for red light. From Fig. 38-13 it can be seen that, for a substance such as fluorite, whose index for yellow light is small, the difference between the indices for red and violet is also small. On the other hand, in the case of silicate flint glass, both the index for yellow light and the difference between extreme indices are large. In other words, for most transparent materials the greater the deviation, the greater the dispersion.

The brilliance of diamond is due in part to its large dispersion. In recent years synthetic crystals of titanium dioxide and of strontium titanate, with about eight times the dispersion of diamond, have been produced.

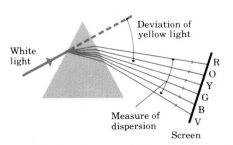

38-14 Dispersion by a prism. The band of colors on the screen is called a spectrum.

*38-10   Absorption

No material is perfectly transparent; as light passes through any optical medium (except vacuum) its energy is partially absorbed, increasing the internal energy in the material, and the intensity (power per unit area) is correspondingly attenuated.

When a beam of light passes through a thin sheet of material, of thickness dx, the decrease dI in its intensity I is found to be proportional to the initial intensity I and to the thickness dx. Thus

$$dI = -\alpha I\,dx. \tag{38-10}$$

The proportionality constant α, which depends on the material, is called the *absorption coefficient*. The intensity after passage through a slab of finite thickness x can be obtained by integrating Eq. (38-10):

$$I = I_0 e^{-\alpha x}, \tag{38-11}$$

where I_0 is the intensity at $x = 0$. Equation (38-11) is called *Lambert's law*.

Example   A certain glass has $\alpha = 10\ \mathrm{m}^{-1}$. Light passing through a flat slab of this glass at normal incidence is decreased in intensity by 5 percent. What is the thickness of the slab?

Solution   We divide both sides of Eq. (38-11) by I_0 and then take natural logs of both sides to obtain

$$\ln \frac{I}{I_0} = -\alpha x, \qquad \text{or} \qquad x = -\frac{1}{\alpha}\ln \frac{I}{I_0}.$$

In our example, $I/I_0 = 0.95$, and we find

$$x = -\frac{1}{10\ \mathrm{m}^{-1}}\ln 0.95 = 0.00513\ \mathrm{m}$$

$$= 0.513\ \mathrm{cm}.$$

The absorption coefficient is often strongly wavelength-dependent. A clear optical glass for which $\alpha = 4\ \mathrm{m}^{-1}$ (a typical value in the middle of the visible spectrum) may have $\alpha = 1000\ \mathrm{m}^{-1}$ at $\lambda = 250\ \mathrm{nm}$ (the near ultraviolet) and hence be essentially opaque to ultraviolet radiation. If α varies substantially *within* the visible spectrum, white light incident on the material appears colored when it emerges. For example, if α is larger at 500 nm than at 600 nm, proportionally more green and blue light is absorbed, and the emergent light appears red.

In some materials, absorption depends on the *polarization* of the incident light, that is, the plane in which the electric-field vector oscillates. A familiar example is the Polaroid filter, discussed in Sec. 42-5, used for sunglasses and many other applications. This material is transparent for light with one plane of polarization but absorbent for light in the perpendicular plane; the two absorption coefficients may differ by as much as a factor of 100.

A beam of light passing through an optical medium may also be attenuated by *scattering*. In contrast to absorption, in which the energy is ordinarily converted to internal energy, scattering simply redirects some of the radiation into directions other than that of the beam. The visibility from the side of a searchlight beam at night results from scattering of light out of the beam by dust particles and water droplets in the air and, to a small extent, by the air molecules themselves.

Scattering of visible light is usually wavelength-dependent, usually increasing with increasing frequency (decreasing wavelength). The sky is blue because the shorter-wavelength (blue) visible light from the sun is scattered more strongly in the earth's atmosphere than is the longer-wavelength (red) light. Similarly, at sunset the sun appears red because proportionally more of the blue light has been scattered out of the beam.

* 38-11 Illumination

We have defined the *intensity* of light and other electromagnetic radiation as power per unit area, measured in watts per square meter. Similarly, the *total* rate of radiation of energy from any of the sources of light discussed in Sec. 38-2 is called the *radiant power* or *radiant flux,* measured in watts. These quantities are not adequate to measure the visual sensation of *brightness,* however, for two reasons: First, not all the radiation from a source lies in the visible spectrum; an ordinary incandescent light bulb radiates more energy in the infrared than in the visible spectrum. Second, the eye is not equally sensitive to all wavelengths; a bulb emitting 1 watt of yellow light appears brighter than one emitting one watt of blue light.

The quantity analogous to radiant power, but compensated to include the above effects, is called *luminous flux,* denoted by F. The unit of luminous flux is the *lumen,* abbreviated lm, defined as that quantity of light emitted by $\frac{1}{60}$ cm^2 surface area of pure platinum at its melting temperature (about 1770°C), within a solid angle of 1 steradian (1 sr). As an example, the total light output (luminous flux) of a 40-watt incandescent light bulb is about 500 lm, while that of a 40-watt fluorescent tube is about 2300 lm.

When luminous flux strikes a surface, the surface is said to be *illuminated.* The intensity of illumination, analogous to the intensity of electromagnetic radiation (which is power per unit area) is the *luminous flux per unit area,* called the *illuminance,* denoted by E. The unit of illuminance is the lumen per square meter, also called the *lux:*

$$1 \text{ lux} = 1 \text{ lm} \cdot \text{m}^{-2}.$$

An older unit, the lumen per square foot, or foot-candle, has become obsolete. If luminous flux F falls at normal incidence on an area A, the illuminance E is given by

$$E = \frac{F}{A}. \tag{38-12}$$

Most light sources do not radiate equally in all directions; it is useful to have a quantity that describes the intensity of a source in a specific direction, without using any specific distance from the source. We place the source at the center of an imaginary sphere of radius R. A small area A of the sphere subtends a solid angle Ω given by $\Omega = A/R^2$. If the luminous flux passing through this area is F, we define the *luminous intensity* I in the direction of the area as

$$I = \frac{F}{\Omega}. \tag{38-13}$$

The unit of luminous intensity is one lumen per steradian, also called one *candela,* abbreviated cd:

$$1 \text{ cd} = 1 \text{ lm} \cdot \text{sr}^{-1}.$$

The term "luminous intensity" is somewhat misleading. The usual usage of *intensity* connotes power per unit area, and the intensity of radiation from a point source decreases as the square of the distance. Luminous intensity, however, is flux per unit *solid angle,* not per unit *area,* and the luminous intensity of a source in a particular direction *does not* decrease with increasing distance.

Example A certain 100-watt lightbulb emits a total luminous flux of 1200 lm, distributed uniformly over a hemisphere. Find the illuminance and the luminous intensity at a distance of 1 m, and at 5 m.

Solution The area of a half-sphere of radius 1 m is

$$(2\pi)(1 \text{ m})^2 = 6.28 \text{ m}^2.$$

The illuminance at 1 m is

$$E = \frac{1200 \text{ lm}}{6.28 \text{ m}^2} = 191 \text{ lm} \cdot \text{m}^{-2} = 191 \text{ lux}.$$

Similarly, the area of a half-sphere of radius 5 m is

$$(2\pi)(5 \text{ m})^2 = 157 \text{ m}^2,$$

and the illuminance at 5 m is

$$E = \frac{1200 \text{ lm}}{157 \text{ m}^2} = 7.64 \text{ lm} \cdot \text{m}^{-2} = 7.64 \text{ lux}.$$

This is smaller by a factor of 5^2 than the illuminance at 1 m, and illustrates the inverse-square law for illuminance from a point source.

The solid angle subtended by a hemisphere is 2π sr. The luminous intensity is

$$I = \frac{1200 \text{ lm}}{2\pi \text{ sr}} = 191 \text{ lm} \cdot \text{sr}^{-1} = 191 \text{ cd}.$$

The luminous intensity does not depend on distance.

Questions

38-1 During a thunderstorm one always sees the flash of lightning before hearing the accompanying thunder. Discuss this in terms of the various wave speeds. Can this phenomenon be used to determine how far away the storm is?

38-2 When hot air rises around a radiator or from a heating duct, objects behind it appear to shimmer or waver. What is happening?

38-3 Light requires about 8 minutes to travel from the sun to the earth. Is it delayed appreciably by the earth's atmosphere?

38-4 Sometimes when looking at a window one sees two reflected images, slightly displaced from each other. What causes this?

38-5 An object submerged in water appears to be closer to the surface than it actually is. Why? Swimming pools are always deeper than they look; is this the same phenomenon?

38-6 Can radio or television waves be reflected? Refracted? Give examples of situations where this might occur.

38-7 A ray of light in air strikes a glass surface. Is there a range of angles for which total reflection occurs?

38-8 As shown in Table 38–1, diamond has a much larger refractive index than glass. Is there a larger or smaller range of angles for which total internal reflection occurs for diamond, than for glass? Does this have anything to do with the fact that a real diamond has more sparkle than a glass imitation?

38-9 Light is usually observed to travel in straight lines, while radio waves seem to be able to bend around obstacles. If both are electromagnetic waves, why the difference?

38-10 Sunlight or starlight passing through the earth's atmosphere is always bent toward the vertical. Why? Does this mean that a star isn't really where it appears to be?

38-11 The sun or moon usually appears flattened just before it sets. Is this related to refraction in the earth's atmosphere, mentioned in Question 38–10?

38-12 A student claimed that, because of atmospheric refraction (cf. Question 38–10), the sun can be seen after it has set, and that the day is therefore longer than it would be if the earth had no atmosphere. First, what does he mean by saying the sun can be seen after it has set? Second, comment on the validity of his conclusion.

38-13 Can sound waves be reflected? Refracted? Give examples. Does Huygens' principle apply to sound waves?

38-14 Why should the wavelength of light change, but not its frequency, in passing from one material to another.

38-15 When light is incident on an interface between two materials, the angle of the refracted ray depends on the wavelength, but the angle of the reflected ray does not. Why should this be?

38-16 A room is completely lined with mirrors. A light source inside the room is turned on and then off. Does the light continue to reflect forever? If not, what happens to it?

38-17 When light slows down as it enters glass from vacuum, and speeds up again as it emerges, does it gain and lose energy in the process? Momentum?

Problems

38-1 What is the wavelength in meters, microns, nanometers, and angstrom units of (a) soft x-rays of frequency 2×10^{17} Hz? (b) green light of frequency 5.6×10^{14} Hz?

38-2 The visible spectrum includes a wavelength range from about 400 nm to about 700 nm. Express these wavelengths in inches.

38-3 Assuming the radius of the earth's orbit to be 92,900,000 mi, and taking the best value of the speed of light, compute the time required for light to travel a distance equal to the diameter of the earth's orbit. Compare with Roemer's value of 22 min.

38-4 Fizeau's measurements of the speed of light were continued by Cornu, using Fizeau's apparatus but with the distance between mirrors increased to 22.9 km. One of the toothed wheels used was 40 mm in diameter and had 180 teeth. Find the angular velocity at which it should rotate so that light transmitted through one opening will return through the next.

38-5 A ray of light traveling with speed c leaves point 1 of Fig. 38–15 and is reflected to point 2. Show that the time required for the light to travel from 1 to 2 is $(y_1 \sec \theta_1 + y_2 \sec \theta_2)/c$.

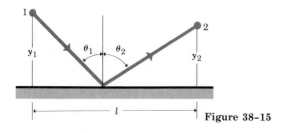

Figure 38-15

38-6 Prove that a ray of light reflected from a plane mirror rotates through an angle 2θ when the mirror rotates through an angle θ about an axis perpendicular to the plane of incidence.

38-7 A parallel beam of light is incident on a prism, as shown in Fig. 38–16. Part of the light is reflected from one face and part from another. Show that the angle θ between the two reflected beams is twice the angle A between the two reflecting surfaces.

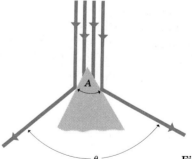

Figure 38-16

38-8 A parallel beam of light makes an angle of 30° with the surface of a glass plate having a refractive index of 1.50.

a) What is the angle between the refracted beam and the surface of the glass?

b) What should be the angle of incidence ϕ with this plate for the angle of refraction to be $\phi/2$?

38-9 Light strikes a glass plate at an angle of incidence of 60°, part of the beam being reflected and part refracted. It is observed that the reflected and refracted portions make an angle of 90° with each other. What is the index of refraction of the glass?

38-10 A ray of light is incident on a plane surface separating two transparent substances of indices 1.60 and 1.40. The angle of incidence is 30° and the ray originates in the medium of higher index. Compute the angle of refraction.

38-11 A parallel-sided plate of glass having a refractive index of 1.60 is held on the surface of water in a tank. A ray coming from above makes an angle of incidence of 45° with the top surface of the glass.

a) What angle does the ray make with the normal in the water?

b) How does this angle vary with the refractive index of the glass?

38-12 An inside corner of a cube is lined with mirrors. A ray of light is reflected successively from each of three mutually perpendicular mirrors; show that its final direction is always exactly opposite to its initial direction. This principle is used in tail-light lenses and reflecting highway signs.

38-13

a) What is the speed of light of wavelength 500 nm (in vacuum) in glass whose index at this wavelength is 1.50?

b) What is the wavelength of these waves in the glass?

38-14 A glass plate 3 mm thick, of index 1.50, is placed between a point source of light of wavelength 600 nm (in vacuum) and a screen. The distance from source to screen is 3 cm. How many waves are there between source and screen?

38-15 The speed of light of wavelength 656 nm in heavy flint glass is 1.60×10^8 m·s^{-1}. What is the index of refraction of this glass?

38-16 Light of a certain frequency has a wavelength in water of 442 nm. What is the wavelength of this light when it passes into carbon disulfide?

38-17 A ray of light goes from point A in a medium where the velocity of light is v_1 to point B in a medium where the velocity is v_2, as in Fig. 38-17.

a) Show that the time required for the light to go from A to B is

$$t = \frac{h_1 \sec \theta_1}{v_1} + \frac{h_2 \sec \theta_2}{v_2}.$$

b) Take the derivative of t with respect to θ_1, noting that θ_1 and θ_2 are functionally related by the fact that $l = h_1 \tan \theta_1 + h_2 \tan \theta_2$. Set this derivative equal to zero to show that this time reaches its *minimum* value when $n_1 \sin \theta_1 = n_2 \sin \theta_2$. This is an example of Fermat's *principle of least time,* which states that

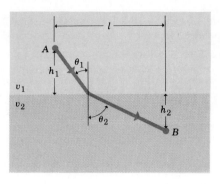

Figure 38-17

among all possible paths between two points, the one actually taken by a ray of light is that for which the time of travel is a *minimum.* (In fact, there are some cases where the time is a maximum rather than a minimum.)

38-18 A point light source is 5 cm below a water–air surface. Compute the angles of refraction of rays from the source making angles with the normal of 10°, 20°, 30°, and 40°, and show these rays in a carefully drawn full-size diagram.

38-19 A glass cube in air has a refractive index of 1.50. Parallel rays of light enter the top obliquely and then strike a side of the cube. Is it possible for the rays to emerge from this side?

38-20 A point source of light is 20 cm below the surface of a body of water. Find the diameter of the largest circle at the surface through which light can emerge from the water.

38-21 The index of refraction of the prism shown in Fig. 38-18 is 1.56. A ray of light enters the prism at point a and follows in the prism the path ab, which is parallel to the line cd.

a) Sketch carefully the path of the ray from a point outside the prism at the left, through the glass, and out some distance into the air again.

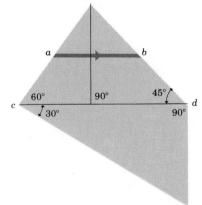

Figure 38-18

b) Compute the angle between the original and final directions in air.

(Black lines are construction lines only.)

38-22 Light is incident normally on the short face of a 30°–60°–90° prism, as in Fig. 38–19. A drop of liquid is placed on the hypotenuse of the prism. If the index of the prism is 1.50, find the maximum index the liquid may have if the light is to be totally reflected.

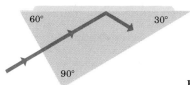

60°   30°

90°

Figure 38–19

38-23 A 45°–45°–90° prism is immersed in water. What is the minimum index of refraction the prism may have if it is to totally reflect a ray incident normally on one of its shorter faces?

38-24 The velocity of a sound wave is $330 \text{ m} \cdot \text{s}^{-1}$ in air and $1320 \text{ m} \cdot \text{s}^{-1}$ in water.

a) What is the critical angle for a sound wave incident on the surface between air and water?

b) Which medium has the higher "index of refraction" for sound?

38-25 Light is incident at an angle ϕ_1 (as in Fig. 38–20) on the upper surface of a transparent plate, the surfaces of the plate being plane and parallel to each other.

a) Prove that $\phi_1 = \phi_2'$.

b) Show that this is true for any number of different parallel plates.

c) Prove that the lateral displacement d of the emergent beam is given by the relation

$$d = t \, \frac{\sin(\phi_1 - \phi_1')}{\cos \phi_1'},$$

where t is the thickness of the plate.

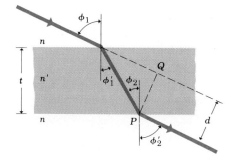

Figure 38–20

38-26 A parallel beam of light containing wavelengths A and B is incident on the face of a triangular glass prism having a refracting angle of 60°. The indices of refraction are $n_A = 1.40$ and $n_B = 1.60$. If beam A goes through the prism at minimum deviation, find (a) the angle of emergence of each beam and (b) the angle of deviation of each.

38-27 A ray of light is incident at an angle of 60° on one surface of a glass plate 2 cm thick, of index 1.50. The medium on either side of the plate is air. Find the transverse displacement between the incident and emergent rays.

38-28 The prism of Fig. 38–21 has a refractive index of 1.414, and the angles A are 30°. Two light rays m and n are parallel as they enter the prism. What is the angle between them after they emerge?

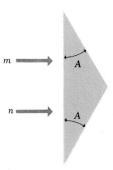

Figure 38–21

38-29 Light passes symmetrically through a prism having apex angle A, as shown in Fig. 38–22. Show that the angle of deviation δ (the angle between the initial and final directions of the ray) is given by

$$\sin \frac{A + \delta}{2} = n \sin \frac{A}{2}.$$

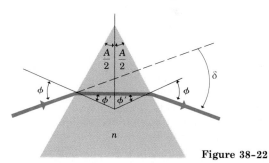

Figure 38–22

38-30 Use the result of Problem 38–29 to find the angle of deviation for a ray of light passing symmetrically through a prism having three equal angles ($A = 60°$) and $n = 1.50$.

38-31 A certain glass has a refractive index of 1.50 for red light (700 nm) and 1.52 for violet light (400 nm). If both colors pass through symmetrically, as described in Problem 38–30, and if $A = 60°$, find the difference between the angles of deviation for the two colors, using the result of Problem 38–29.

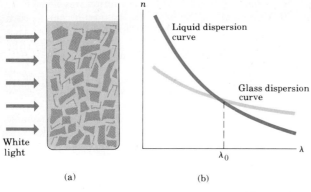

(a)

(b)

Figure 38-23

38-32 The glass vessel shown in Fig. 38-23a contains a large number of small, irregular pieces of glass and a liquid. The dispersion curves of the glass and of the liquid are shown in Fig. 38-23b. Explain the behavior of a parallel beam of white light as it traverses the vessel. (This is known as a *Christiansen filter.*)

38-33 The density of the earth's atmosphere increases as the surface of the earth is approached.

a) Draw a diagram showing how the light from a star or planet is bent as it goes through the atmosphere.

b) Indicate the apparent position of the light source.

c) Explain how one can see the sun after it has set.

d) Explain why the setting sun appears flattened.

38-34 A lens for sunglasses is to be made from a gray-tinted glass having an absorption coefficient of 500 m^{-1}. If the intensity of light passing through the lens is to be de-creased by a factor of $\frac{1}{4}$, what should be the thickness of the lens?

38-35 If a swimming pool is 5 m deep and the water has an absorption coefficient of 2 m^{-1}, by what factor is the intensity of light attenuated as it travels from the surface to the bottom of the pool?

38-36 If the absorption coefficient of sea water is 2 m^{-1}, and if the eye can perceive light of intensity smaller than that of sunlight by a factor of 10^{-18}, what is the greatest depth below the ocean surface at which light can be seen?

38-37 The illuminance of direct sunlight is about 10^5 lux. If a photoflash lamp has an intensity in a certain direction of 5×10^6 cd, at what distance from a surface should it be placed to produce illuminance equal to that of sunlight?

38-38 The *luminous efficiency* of a lamp is defined as the ratio of luminous flux to electric power input. A certain lamp mounted 3 m above a desk top has a luminous efficiency of 20 lm·W^{-1}. What is the power input to the lamp if the illuminance on the desk is equal to that of sunlight, about 10^5 lux? Assume that the lamp radiates uniformly over its lower half-sphere.

38-39 A certain baseball field in the shape of a square 140 m on a side is to be illuminated for night games by six towers supporting banks of 1000-watt incandescent lamps with luminous efficiency (see Problem 38-38) of 30 lm·W^{-1}. The illuminance required on the playing field is 200 lux. Assume that 50% of the luminous flux from the lamps reaches the field.

a) How many lamps are required in each tower?

b) What is the electric power input to each tower?

c) If power for all six towers is supplied by a generator driven by a gasoline engine, what must be the power capacity of the engine?

IMAGES FORMED BY A SINGLE SURFACE 39

39-1 Introduction

In the preceding chapter we discussed reflection and refraction of a ray of light, or of a bundle of parallel rays, at a surface separating two materials having different refractive indices. We now consider the analysis of the paths of several rays that diverge from a common point and strike a reflecting or refracting surface. The point may be self-luminous, or it may be part of a rough surface that reflects light in various directions.

Such a situation is shown in Fig. 39–1. Rays diverge from point P and are reflected or refracted (or both) at the interface. The direction of each *reflected* ray is given by the law of reflection, and that of each transmitted or *refracted* ray by Snell's law. Here and in the following discussion we denote the two refractive indexes as n and n', without using the more elaborate subscript notation of Chapter 38.

A central concept in this discussion is that of *image*. As we shall see, the rays after reflection or refraction emerge with directions characteristic of having passed through some common point, which we call the *image point*. In some cases the emerging rays really *do* meet at a common point and then diverge again after passing it; such a point is called a *real* image point. In other cases the rays diverge *as though* they had passed through such a point, which is then called a *virtual* image point. In many situations the image point exists only in an approximate sense, that is, when certain approximations are used in the calculations. This chapter and the next are devoted primarily to a study of the formation and properties of images.

39-2 Reflection at a plane surface

We begin our analysis by considering reflected rays. The reflection can occur at an interface between two transparent materials or at a highly polished surface of an opaque material such as a metal, in which case the surface is usually called a *mirror*. We shall often use the term mirror to include all these possibilities.

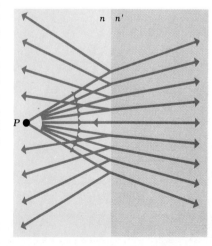

39–1 Reflection and refraction of rays at a plane interface between two transparent materials.

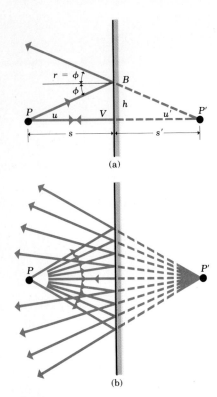

(a)

(b)

39-2 After reflection at a plane surface, all rays originally diverging from the object point P now diverge from the point P', although they do not *originate* at P'. Point P' is called the *virtual image* of point P.

Figure 39–2a shows two rays diverging from a point P at a distance s to the left of a plane mirror. The ray PV, incident normally on the mirror, returns along its original path. The ray PB, making an angle u with PV, strikes the mirror at an angle of incidence $\phi = u$, and is reflected at an angle $r = \phi = u$. When extended backward, the reflected ray intersects the normal to the surface at point P'. The angle u' is equal to r and hence to u.

Figure 39–2b shows several rays diverging from P. The construction of Fig. 39–2a can be repeated for each of them, and we see that the directions of the outgoing rays are as though they had originated at point P', which is therefore the *image* of P. The rays do not, of course, actually pass through this point; in fact, if the mirror is opaque there is no light at all on the right side. Thus P' is a *virtual* image. Nevertheless P' is a very real point in the sense that it describes the final directions of all the rays that originally diverged from P.

From the symmetry of the figure, we see that P' lies on a line perpendicular to the mirror passing through P, and that P and P' are equidistant from the mirror, on opposite sides. Thus *for a plane mirror the image of an object point lies on the extension of the normal from the object point to the mirror, and the object and image points are equidistant from the mirror.*

Before proceeding further, we pause here to introduce some conventions that anticipate later situations in which the object and image may be on either side of a reflecting or refracting surface. We adopt the following:

1. *When the object is on the same side of the reflecting surface as the incoming light, the object distance s is positive; otherwise it is negative.*

2. *When the image is on the same side of the reflecting surface as the outgoing light, the image distance s' is positive; otherwise it is negative.*

For a mirror the incoming and outgoing sides are always the same; in Fig. 39–2 they are both the left side. The above rules have been stated in this form so they may be applied also to *refracting* surfaces, for which light comes in one side and goes out the opposite side.

In Fig. 39–2 the object distance s is *positive* because the object point P is on the incoming side, that is, the left side, of the reflecting surface. The image distance s' is *negative* because the image point P' is *not* on the outgoing side of the surface. Thus object and image distances are related simply by

$$s = -s'. \tag{39–1}$$

Next we consider an object of finite size, parallel to the mirror, represented by the arrow PQ in Fig. 39–3. Point P', the image of P, is found as in Fig. 39–2. Two of the rays from Q are shown, and *all* rays from Q diverge from its image Q' after reflection. Other points of the object PQ have image points between P' and Q'.

Let y and y' represent the lengths of object and image, respectively. The ratio y'/y is called the *lateral magnification m*:

$$m = \frac{y'}{y}.$$

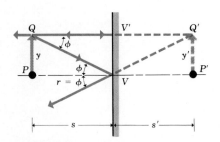

39-3 Construction for determining the height of an image formed by reflection at a plane surface.

From the triangles PQV and $P'Q'V$,

$$\tan \phi = \frac{y}{s} = \frac{y'}{-s'}$$

and since $s' = -s$,

$$m = \frac{y'}{y} = -\frac{s'}{s} = +1. \qquad (39\text{-}2)$$

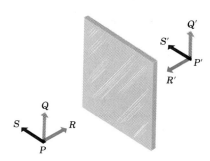

39-4 The image formed by a plane mirror is virtual, erect, and perverted, and is the same size as the object.

The lateral magnification for a plane mirror is unity.

In general, if a transverse object is represented by an arrow, its image may point in the same direction as the object, or in the opposite direction. If the directions are the same, the image is called *erect;* if they are opposite, the image is *inverted.* When the lateral magnification m is positive, the object and image arrows point in the same direction and the image is erect.

The three-dimensional virtual image of a three-dimensional object, formed by a plane mirror, is shown in Fig. 39–4. The image of every object point lies on the normal from that point to the mirror, and the distances from object and image to the mirror are equal. The images $P'Q'$ and $P'S'$ are parallel to their objects, but $P'R'$ is reversed relative to PR. We can think of the arrow PR as generated by the displacement of an object point P from P to R, through a distance Δs. Since

$$s' = -s,$$

it follows that the corresponding displacement $\Delta s'$ of the image, from P' to R', is equal to $-\Delta s$. The ratio $\Delta s'/\Delta s$ (or more generally ds'/ds) is called the *longitudinal magnification m'.* In this case,

$$m' = \frac{\Delta s'}{\Delta s} = -1. \qquad (39\text{-}3)$$

The negative sign means that the arrows PR and $P'R'$ point in opposite directions. Hence the image of a three-dimensional object formed by a plane mirror is the same size as the object in both its lateral and transverse dimensions. However, the image and object are not identical in all respects but are related in the same way as are a right hand and a left hand. To verify this, point your thumbs along PR and $P'R'$, your forefingers along PQ and $P'Q'$, and your middle fingers along PS and $P'S'$. When an object and its image are related in this way the image is said to be *perverted.* When the transverse dimensions of object and image are in the same direction, the image is erect. Thus a plane mirror forms an erect but perverted image.

39-3 Reflection at a spherical mirror

Next we consider the formation of an image by a *spherical* mirror. Figure 39–5a shows a spherical mirror of radius of curvature R, with its concave side facing the incident light. The center of curvature of the surface is at C. Point P is an object point; for the moment we assume that P is farther from V than is the center of curvature. The ray PV, passing through C, strikes the mirror normally and is reflected back on itself. Point V is called the *vertex* of the mirror, and the line PCV the *optic axis.*

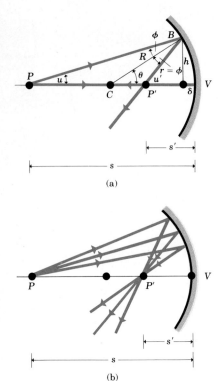

(a)

(b)

39-5   (a) Construction for finding the position of the image P' of a point object P, formed by a concave spherical mirror. (b) If the angle u is small, *all* rays from P intersect at P'.

Ray PB, at an angle u with the axis, strikes the mirror at B, where the angle of incidence is ϕ and the angle of reflection is $r = \phi$. The reflected ray intersects the axis at point P'. We shall show shortly that if the angle u is small, *all* rays from P intersect the axis at the *same* point P', as in Fig. 39–5b, no matter what u is. Point P' is therefore the *image* of the object point P. The object distance, measured from the vertex V, is s, and the image distance is s'.

The object point P is on the same side as the incident light, so the object distance s is positive. The image point P' is on the same side as the reflected light, so the image distance s' is also positive.

Unlike the reflected rays in Fig. 39–2, the reflected rays in Fig. 39–5b actually intersect at point P', and then diverge from P' *as if* they had originated at this point. The image P' is called *real,* and a real image corresponds to a *positive* image distance.

Making use of the fact that an exterior angle of a triangle equals the sum of the two opposite interior angles, and considering the triangles PBC and $P'BC$ in Fig. 39–5a, we have

$$\theta = u + \phi, \qquad u' = \theta + \phi.$$

Eliminating ϕ between these equations gives

$$u + u' = 2\theta. \qquad (39\text{–}4)$$

We may now introduce a sign convention for radii of curvature.

When the center of curvature C is on the same side as the outgoing (reflected) light, the radius of curvature is positive; otherwise, it is negative.

In Fig. 39–5, R is positive, because the center of curvature C is on the same side of the mirror as the reflected light. This is always the case for reflection from the concave side of a surface; for a convex surface, the center of curvature is on the opposite side from the reflected light, and R is negative.

We may now compute the image position s'. Let h represent the height of point B above the axis, and δ the short distance from V to the foot of this vertical line. Now write expressions for the tangents of u, u', and θ, remembering that s, s', and R are all positive quantities:

$$\tan u = \frac{h}{s - \delta}, \qquad \tan u' = \frac{h}{s' - \delta}, \qquad \tan \theta = \frac{h}{R - \delta}.$$

These *trigonometric* equations cannot be solved as simply as the corresponding *algebraic* equations for a plane mirror. However, *if the angle u is small,* the angles u' and θ will be small also. Since the tangent of a small angle is nearly equal to the angle itself (in radians), we can replace $\tan u'$ by u', etc., in the equations above. Also if u is small, the distance δ can be neglected compared with s', s, and R. Hence, *approximately, for small angles,*

$$u = \frac{h}{s}, \qquad u' = \frac{h}{s'}, \qquad \theta = \frac{h}{R}.$$

Substituting in Eq. (39–4) and canceling h, we obtain

$$\frac{1}{s} + \frac{1}{s'} = \frac{2}{R} \qquad (39\text{–}5)$$

as a general relation among the three quantities s, s', and R. Since the equation above does not contain the angle u, *all* rays from P making sufficiently small angles with the axis will, after reflection, intersect at P'. Such rays, nearly parallel to the axis, are called *paraxial* rays.

It must be understood that Eq. (39–5), as well as many similar relations to be derived later, is the result of a calculation containing approximations and is valid only for paraxial rays. (The term *paraxial approximation* is also commonly used.) As the angle increases, the point P' moves closer to the vertex; a spherical mirror, unlike a plane mirror, does not form precisely a point image of a point object. This property of a spherical mirror is called *spherical aberration*.

If $R = \infty$, the mirror becomes *plane* and Eq. (39–5) reduces to Eq. (39–1), previously derived for this special case.

Now suppose we have an object of finite size, represented by the arrow PQ in Fig. 39-6, perpendicular to the axis PV. The image of P formed by paraxial rays is at P'. Since the object distance for point Q is very nearly equal to that for the point P, the image $P'Q'$ is nearly straight and is perpendicular to the axis. Herein lies another approximation; if the height PQ is not sufficiently small compared to the object distance s, the image $P'Q'$ is not a straight line but is curved; this is another aberration of spherical surfaces, called *curvature of field*. We shall assume height PQ is small enough that this effect is negligible.

We now compute the lateral magnification m. In Fig. 39-6 the ray QCQ' passes through the center of curvature; hence it strikes the spherical reflecting surface at normal incidence and is reflected back along the same radial line on which it approached the mirror. The ray QV makes an angle of incidence ϕ and an angle of reflection $r = \phi$. From the triangles PQV and $P'Q'V$,

$$\tan \phi = \frac{y}{s} = \frac{-y'}{s'}.$$

The negative sign arises because object and image are on opposite sides of the optic axis; if y is positive, y' is negative. Hence,

$$m = \frac{y'}{y} = -\frac{s'}{s}. \tag{39–6}$$

A negative value of m indicates that the image is *inverted* relative to the object, as the figure shows. In cases to be considered later, where m may

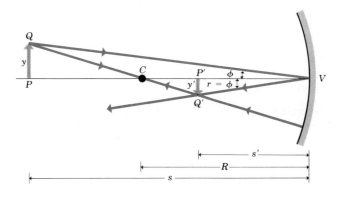

39-6 Construction for determining the height of an image formed by a concave spherical mirror.

be either positive or negative, a positive value always corresponds to an erect image, a negative value to an inverted one. For a plane mirror, $s = -s'$ and hence $y' = y$, as we have already shown.

The longitudinal magnification m' is defined as before, as ds'/ds. Taking the derivative of Eq. (39-5) with respect to s, we find

$$-\frac{1}{s^2} - \frac{1}{s'^2}\frac{ds'}{ds} = 0,$$

$$m' = \frac{ds'}{ds} = -\frac{s'^2}{s^2} = -m^2, \tag{39-7}$$

since the lateral magnification is $m = -s'/s$. Thus the longitudinal magnification equals the *square* of the lateral magnification and is always *negative*, regardless of the sign of m, since m^2 is always positive. The negative sign means that when the object is displaced slightly along the axis, image and object always move in *opposite* directions.

Although the ratio of image size to object size is referred to as the *magnification*, the image formed by a mirror or lens may be *smaller* than the object. The magnification is then a small fraction and might more appropriately be called the *reduction*. The image formed by an astronomical telescope mirror, or by a camera lens, is much smaller than the object. Since the longitudinal magnification is the square of the lateral magnification, the two are not equal except in the special case of *unit* magnification when $m = 1$. In particular, if m is a small fraction, m^2 is very small and the three-dimensional image of a three-dimensional object is reduced *longitudinally* much more than it is reduced *transversely*. Figure 39-7 illustrates this effect. Also, the image formed by a spherical mirror, like that of a plane mirror, is always perverted.

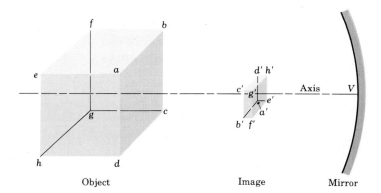

39-7 Schematic diagram of an object and its real, inverted, reduced image formed by a concave mirror.

Example 1 (a) What type of mirror is required to form an image, on a wall 3 m from the mirror, of the filament of a headlight lamp 10 cm in front of the mirror? (b) What is the height of the image if the height of the object is 5 mm?

Solution

a)
$$s = 10\ \text{cm}, \qquad s' = 300\ \text{cm},$$

$$\frac{1}{10\ \text{cm}} + \frac{1}{300\ \text{cm}} = \frac{2}{R},$$

$$R = 19.4\ \text{cm}.$$

Since the radius is positive, a concave mirror is required.

b)
$$m = \frac{y'}{y} = -\frac{s'}{s} = -\frac{300 \text{ cm}}{10 \text{ cm}} = -30.$$

The image is therefore inverted (m is negative) and is 30 times the height of the object, or $(30)(5 \text{ mm}) = 150 \text{ mm}$.

Example 2   A wire frame in the form of a small cube 3 cm on a side is placed with its center on the axis of a concave mirror of radius of curvature 30 cm. The sides of the cube are parallel or perpendicular to the axis. The face toward the mirror is 60 cm to the left of the vertex. Find (a) the position of the image of the cube, (b) the lateral magnification, (c) the longitudinal magnification.

Solution
a) Let us calculate the position of the image of a point at the center of the right face of the cube. From the given data, $s = +60 \text{ cm}$, $R = +30 \text{ cm}$:

$$\frac{1}{s} + \frac{1}{s'} = \frac{2}{R}, \qquad \frac{1}{60} = \frac{1}{s'} = \frac{2}{30}, \qquad s' = +20 \text{ cm}.$$

The image of this point is therefore 20 cm to the left of the vertex (s' is positive) and is real.

b)
$$m = \frac{y'}{y} = -\frac{s'}{s} = -\frac{1}{3}, \qquad y' = -\frac{1}{3}y.$$

The image of the right face is therefore inverted (m is negative) and is a square 1 cm on a side.

c) The object distance for the left face of the cube is 63 cm. To find the position and magnification of the left face of the cube, we could repeat the calculations above with $s = 63 \text{ cm}$, but since the dimensions of object and image are small compared with the object distance, let us instead find the longitudinal magnification.

$$m' = \frac{\Delta s'}{\Delta s} = -m^2 = -\frac{1}{9}, \qquad \Delta s' = -\frac{1}{9}\Delta s.$$

The longitudinal magnification is negative, so the image of the left face is nearer the mirror than the image of the right face, and the images of the cube sides parallel to the axis are $\frac{1}{3}$ cm in length.

Figure 39–7 (not to scale) represents schematically the object and image, with corresponding points designated by unprimed and primed letters. It has been assumed that the lateral magnification is the same for both transverse faces.

In Fig. 39–8a, the *convex* side of a spherical mirror faces the incident light so that R is negative. Ray PB is reflected with the angle of reflection r equal to the angle of incidence ϕ, and the reflected ray, projected backward, intersects the axis at P'. As in the case of a concave mirror, *all* rays from P will, after reflection, diverge from the same point P', provided that the angle u is small; so P' is the image of P. The object distance s is positive, the image distance s' is negative, and the radius of curvature R is negative.

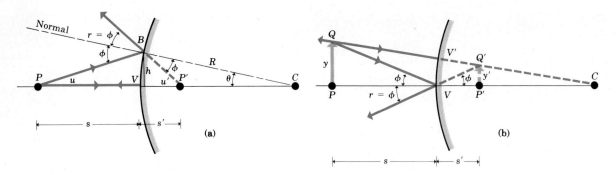

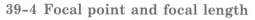

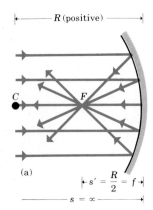

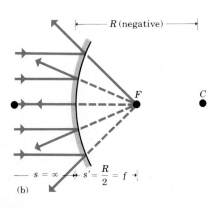

39-8 Construction for finding (a) the position and (b) the magnification of the image formed by a convex mirror.

Figure 39-8b shows two rays diverging from the head of the arrow PQ, and the virtual image $P'Q'$ of this arrow. It is left to the reader to show, by the same procedure as that used for a concave mirror, that:

$$\frac{1}{s} + \frac{1}{s'} = \frac{2}{R}.$$

The lateral magnification is

$$m = \frac{y'}{y} = -\frac{s'}{s};$$

and the longitudinal magnification again equals the negative square of the lateral magnification. These expressions are exactly the same as those for a concave mirror, as they should be when a consistent sign convention is used.

39-4 Focal point and focal length

When an object point is a very large distance from a mirror, all rays from that point that strike the mirror are parallel to one another. The object distance is $s = \infty$ and, from Eq. (39-5),

$$\frac{1}{\infty} + \frac{1}{s'} = \frac{2}{R}, \qquad s' = \frac{R}{2}.$$

The image distance s' then equals one-half the radius of curvature and has the same sign as R. This means that, if R is positive, as in Fig. 39-9a, the image point F lies to the left of the mirror and is real, while if R is negative, as in Fig. 39-9b, the image point F lies to the right of the mirror and is virtual.

Conversely, if the image distance s' is very large, the object distance s is

$$\frac{1}{s} + \frac{1}{\infty} = \frac{2}{R}, \qquad s = \frac{R}{2}.$$

The object distance then equals half the radius of curvature. If R is positive, as in Fig. 39-10a, s is positive. If R is negative, as in Fig. 39-10b, s is negative and the *object is behind* the mirror. That is, rays previously made converging by some other surface are converging toward F.

The point F in Figs. 39-9 and 39-10 is called the *focal point* of the mirror, or simply the *focus*. It may be considered either as the image point of an infinitely distant object point on the mirror axis, or as the

39-9 Incident rays parallel to the axis (a) converge to the focal point F of a concave mirror, (b) diverge as though coming from the focal point F of a convex mirror.

object point of an infinitely distant image point. Thus the mirror of an astronomical telescope forms at its focal point an image of a star on the axis of the mirror.

The distance between the vertex of a mirror and the focal point is called the *focal length* of the mirror and is represented by f. As seen from the preceding discussion, the magnitude of the focal length equals one-half the radius of curvature:

$$f = \frac{R}{2}. \tag{39-8}$$

The relation between object and image distances, Eq. (39-5), for a mirror may now be written as

$$\frac{1}{s} + \frac{1}{s'} = \frac{1}{f}. \tag{39-9}$$

39-5 Graphical methods

The position and size of the image formed by a mirror may be found by a simple graphical method. This method consists of finding the point of intersection, after reflection from the mirror, of a few particular rays diverging from some point of the object *not* on the mirror axis, such as point Q in Fig. 39-11. Then (neglecting aberrations) *all* rays from this point that strike the mirror will intersect at the same point. Four rays whose paths may readily be traced are shown in Fig. 39-11. These are often called *principal rays*.

1. *A ray parallel to the axis,* after reflection, passes through the focal point of a concave mirror or appears to come from the focal point of a convex mirror.

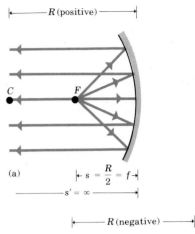
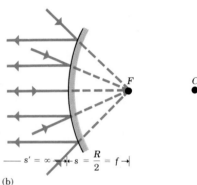

39-10 Rays from a point object at the focal point of a spherical mirror are parallel to the axis after reflection. The object in part (b) is virtual.

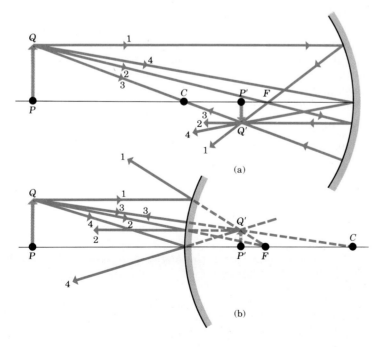

39-11 Rays used in the graphical method of locating an image.

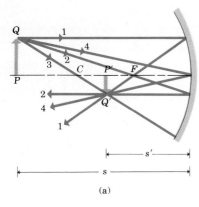

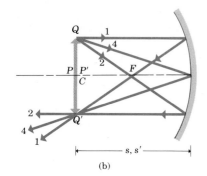

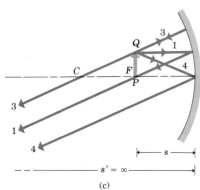

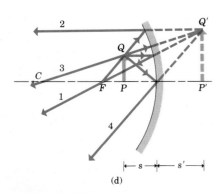

39–12 Image of an object at various distances from a concave mirror, showing principal-ray construction.

2. *A ray from* (*or proceeding toward*) *the focal point* is reflected parallel to the axis.

3. *A ray along the radius* (extended if necessary) intersects the surface normally and is reflected back along its original path.

4. *A ray to the center* is reflected with equal angles with the optic axis.

Once the position of the image point has been found by means of the intersection of any two of these principal rays (1, 2, 3, 4) the paths of all other rays from the same object point may be drawn.

Example A concave mirror has a radius of curvature of magnitude 20 in. Find graphically the image of an object in the form of an arrow perpendicular to the axis of the mirror at each of the following object distances: 30 in., 20 in., 10 in., and 5 in. Check the construction by computing the size and magnification of the image.

The graphical construction is indicated in the four parts of Fig. 39–12. (Note that in (b) and (c) only three of the four rays can be used.) The calculations are left as a problem.

In every problem involving formation of an image by a reflecting surface or a lens, the student should *always* draw a principal-ray diagram. This is useful not only as a graphical check on numerical calculations but also as an aid to understanding the basic concepts, especially the concept of *image*.

39–6 Refraction at a plane surface

The method of finding the image of a point object formed by rays *refracted* at a plane or spherical surface is essentially the same as for reflection; the only difference is that Snell's law replaces the law of reflection. We let n represent the refractive index of the medium on the "incoming" side of the surface and n' that of the medium on the "outgoing" side. The same convention of signs is used as in reflection.

Consider first a plane surface, shown in Fig. 39–13, and assume $n' > n$. A ray from the object point P toward the vertex V is incident normally and passes into the second medium without deviation. A ray making an angle u with the axis is incident at B with an angle of incidence $\phi = u$. The angle of refraction, ϕ', is found from Snell's law,

$$n \sin \phi = n' \sin \phi'.$$

The two rays both appear to come from the image point P' after refraction. From the triangles PVB and $P'VB$,

$$\tan \phi = \frac{h}{s}, \qquad \tan \phi' = \frac{h}{-s'}. \qquad (39\text{--}10)$$

We must write $-s'$, since the image point is on the side *opposite* to that of the refracted (outgoing) light.

If the angle u is small, the angles ϕ, u', and ϕ' are small also, and therefore, approximately,

$$\tan \phi = \sin \phi, \qquad \tan \phi' = \sin \phi'.$$

Then Snell's law can be written as

$$n \tan \phi = n' \tan \phi',$$

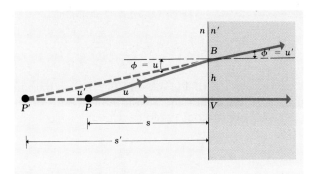

39-13 Construction for finding the position of the image P' of a point object P, formed by refraction at a plane surface.

and from Eq. (39-10), after cancelling h, we have

$$\frac{n}{s} = \frac{n'}{s'},$$

or

$$\frac{s'}{s} = -\frac{n'}{n}. \qquad (39\text{--}11)$$

This is an *approximate* relation, *valid for paraxial rays only.* That is, a plane refracting surface does *not* image all rays from a point object at the same image point.

Consider next the image of a finite object, as in Fig. 39-14. The two rays diverging from point Q appear to diverge from its image Q' after refraction. From the triangles PQV and $P'Q'V$,

$$\tan \phi = \frac{y}{s}, \qquad \tan \phi' = \frac{y'}{-s'}.$$

Combining with Snell's law and using the small-angle approximation,

$$\sin \phi = \tan \phi, \qquad \sin \phi' = \tan \phi',$$

we obtain

$$\frac{ny}{s} = -\frac{n'y'}{s'},$$

and

$$m = \frac{y'}{y} = -\frac{ns'}{n's}. \qquad (39\text{--}12)$$

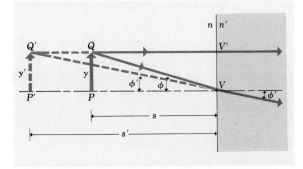

39-14 Construction for determining the height of an image formed by refraction at a plane surface.

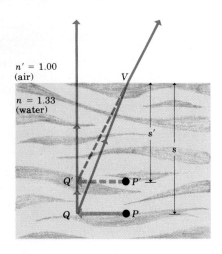

39-15 Arrow $P'Q'$ is the image of the underwater object PQ.

However, from Eq. (39–11), $ns' = -n's$, so

$$m = \frac{y'}{y} = 1, \tag{39-13}$$

in agreement with Fig. 39–14. The image distance is greater than the object distance, but image and object are the same height.

The longitudinal magnification m' is easily found from Eq. (39–11):

$$m' = \frac{ds'}{ds} = -\frac{n'}{n}. \tag{39-14}$$

A common example of refraction at a plane surface is afforded by looking vertically downward into the quiet water of a pond or a swimming pool; the apparent depth is less than the actual depth. Figure 39–15 illustrates this situation. Two rays are shown diverging from a point Q at a distance s below the surface. Here, n' (air) is less than n (water) and the ray through V is deviated *away from* the normal. The rays after refraction appear to diverge from Q', and the arrow PQ, to an observer looking vertically downward, appears lifted to the position $P'Q'$. From Eq. (39–11).

$$s' = -\frac{n'}{n}s = -\frac{1.00}{1.33}s = -0.75\,s.$$

The apparent depth s' is therefore only three-fourths of the actual depth s. The same phenomenon accounts for the apparent sharp bend in an oar when a portion of it extends below a water surface. The submerged portion appears lifted above its actual position.

39-7 Refraction at a spherical surface

Finally, we consider refraction at a spherical surface. In Fig. 39–16, P is an object point at a distance s to the left of a spherical surface of radius R, with center of curvature C. The refractive indexes at the left and right of the surface are n and n', respectively. Ray PV, incident normally, passes into the second medium without deviation. Ray PB, making an angle u with the axis, is incident at an angle ϕ with the normal and is refracted at an angle ϕ'. These rays intersect at P' at a distance s' to the right of the vertex.

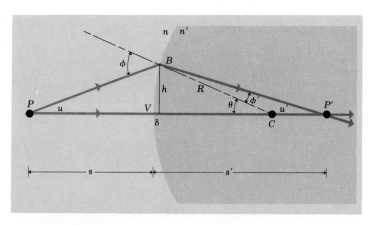

39-16 Construction for finding the position of the image P' of a point object P, formed by refraction at a spherical surface.

We shall show that if the angle u is small, all rays from P intersect at the same point P', so P' is the *real image* of P. The object and image distances are both positive. The radius of curvature is positive also, because the center of curvature is on the side of the outgoing or refracted light.

When considering rays from P that are *reflected* at the surface, as in Fig. 39-8, the radius of curvature is negative. Thus the radius of curvature of a given surface has one sign for reflected light and the opposite sign for refracted light. This apparent inconsistency is resolved by noting that in both cases R is positive when the center of curvature is on the "outgoing" side of the reflecting or refracting surface and negative when it is on the "incoming" side.

From the triangles PBC and $P'BC$, we have

$$\phi = \theta + u, \qquad \theta = u' + \phi'. \tag{39-15}$$

From Snell's law,

$$n \sin \phi = n' \sin \phi'.$$

Also, the tangents of u, u', and θ are

$$\tan u = \frac{h}{s + \delta}, \qquad \tan u' = \frac{h}{s' - \delta}, \qquad \tan \theta = \frac{h}{R - \delta}.$$

For paraxial rays we may approximate both the sine and tangent of an angle by the angle itself, and neglect the small distance δ. Snell's law then becomes

$$n\phi = n'\phi',$$

and, combining with the first of Eqs. (39-15), we get

$$\phi' = \frac{n}{n'}(u + \theta).$$

Substituting this in the second of Eqs. (39-15) gives

$$nu + n'u' = (n' - n)\theta.$$

Using the small-angle approximations and cancelling h, we obtain

$$\frac{n}{s} + \frac{n'}{s'} = \frac{n' - n}{R}. \tag{39-16}$$

This equation does not contain the angle u, so the image distance is the same for all paraxial rays from P.

If the surface is plane, $R = \infty$ and this equation reduces to Eq. (39-11), already derived for the special case of a plane surface.

The magnification is found from the construction in Fig. 39-17. From point Q draw two rays, one through the center of curvature C and the other incident at the vertex V. From the triangles PQV and $P'Q'V$,

$$\tan \phi = \frac{y}{s}, \qquad \tan \phi' = \frac{-y'}{s'},$$

and from Snell's law,

$$n \sin \phi = n' \sin \phi'.$$

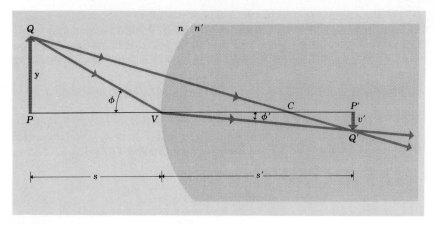

39-17 Construction for determining the height of an image formed by refraction at a spherical surface.

For small angles,

$$\tan \phi = \sin \phi, \qquad \tan \phi' = \sin \phi',$$

and hence

$$\frac{ny}{s} = -\frac{n'y'}{s'},$$

or

$$m = \frac{y'}{y} = -\frac{ns'}{n's}, \qquad (39\text{--}17)$$

which agrees with the relation previously derived for a plane surface. The longitudinal magnification is

$$m' = -\frac{n'}{n}m^2. \qquad (39\text{--}18)$$

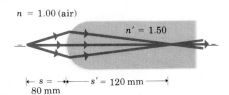

$n = 1.00$ (air)

$n' = 1.50$

$s = 80$ mm $\qquad s' = 120$ mm

39-18

Example 1 One end of a cylindrical glass rod (Fig. 39-18) is ground to a hemispherical surface of radius $R = 20$ mm. Find the image distance of a point object on the axis of the rod, 80 mm to the left of the vertex. The rod is in air.

Solution

$$n = 1, \qquad n' = 1.5,$$
$$R = +20 \text{ mm}, \qquad s = +80 \text{ mm}.$$
$$\frac{1}{80 \text{ mm}} + \frac{1.5}{s'} = \frac{1.5 - 1}{+20 \text{ mm}},$$
$$s' = +120 \text{ mm}.$$

The image is therefore formed at the right of the vertex (s' is positive) and at a distance of 120 mm from it. Suppose that the object is an arrow 1 mm high, perpendicular to the axis. Then

$$m = -\frac{ns'}{n's} = -\frac{(1)(120 \text{ mm})}{(1.5)(80 \text{ mm})} = -1.$$

That is, the image is the same height as the object, but is inverted.

Example 2 Let the same rod be immersed in water of index 1.33, the other quantities having the same values as before. Find the image distance (Fig. 39-19).

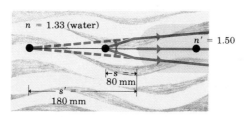

39-19

Solution

$$\frac{1.33}{80 \text{ mm}} + \frac{1.5}{s'} = \frac{1.5 - 1.33}{+20 \text{ mm}}, \qquad s' = -180 \text{ mm}.$$

The fact that s' is negative means that the rays, after refraction by the surface, are not converging but *appear* to diverge from a point 180 mm to the *left* of the vertex. We have met a similar case before in the refraction of spherical waves by a plane surface, and have called the point a *virtual image*. In this example, then, the surface forms a virtual image 180 mm to the left of the vertex.

Equations (39–16), (39–17), and (39–18) can be applied to both convex and concave refracting surfaces when a consistent sign convention is used, and they apply whether n' is greater or less than n. The reader should construct diagrams like Figs. 39–16 and 39–17, when R is negative and $n' < n$, and use them to derive Eqs. (39–16) and (39–17).

The concepts of focal point and focal length can also be applied to a refracting surface. Such a surface is found to have *two* focal points. The first is the *object* point when the image is at infinity, the second is the *image* point of an infinitely distant object. These points lie on opposite sides of the surface and at different distances from it, so that a single refracting surface has two focal lengths. The positions of the focal points can readily be found from Eq. (39–16) by setting s or s' equal to infinity.

39-8 Summary

The results of this chapter are summarized in Table 39–1. Note that if we let $R = \infty$, the equation for a plane surface follows immediately from the appropriate equation for a curved surface.

Table 39-1

	Plane mirror	Spherical mirror	Plane refracting surface	Spherical refracting surface
Object and image distances	$\dfrac{1}{s} + \dfrac{1}{s'} = 0$	$\dfrac{1}{s} + \dfrac{1}{s'} = \dfrac{2}{R} = \dfrac{1}{f}$	$\dfrac{n}{s} + \dfrac{n'}{s'} = 0$	$\dfrac{n}{s} + \dfrac{n'}{s'} = \dfrac{n' - n}{R}$
Lateral magnification	$m = -\dfrac{s'}{s} = 1$	$m = -\dfrac{s'}{s}$	$m = -\dfrac{ns'}{n's} = 1$	$m = -\dfrac{ns'}{n's}$
Longitudinal magnification	$m' = \dfrac{\Delta s'}{\Delta s} = -1$	$m' = -\dfrac{s'^2}{s^2}$	$m' = -\dfrac{n'}{n}$	$m' = -\dfrac{ns'^2}{n's^2}$

Questions

39-1 Can a person see a real image by looking backward along the direction from which the rays come? A virtual image? Can you tell by looking whether an image is real or virtual? How *can* the two be distinguished?

39-2 Why does a plane mirror reverse left and right but not top and bottom?

39-3 For a spherical mirror, if $s = f$, then $s' = \infty$, and the lateral magnification m is infinite. Does this make sense? If so, what does it mean?

39-4 According to the discussion of the preceding chapter, light rays are reversible. Are the formulas in Table 39–1 still valid if object and image are interchanged? What does reversibility imply with respect to the *forms* of the various formulas?

39-5 If a spherical mirror is immersed in water, does its focal length change?

39-6 For what range of object positions does a concave spherical mirror form a real image. What about a convex spherical mirror?

39-7 If a piece of photographic film is placed at the location of a real image, the film will record the image. Can this be done with a virtual image? How might one record a virtual image?

39-8 When a room has mirrors on two opposing walls, an infinite series of reflections can be seen. Discuss this phenomenon in terms of images. Why do the distant images appear darker?

39-9 When observing fish in an aquarium filled with water, one can see clearly only when looking nearly perpendicularly to the glass wall; objects viewed at an oblique angle always appear blurred. Why? Do the fish have the same problem when looking at you?

39-10 Can an image formed by one reflecting or refracting surface serve as an object for a second reflection or refraction? Does it matter whether the first image is real or virtual?

39-11 A concave mirror (sometimes surrounded by lights) is often used as an aid for applying cosmetics to the face. Why is such a mirror always concave rather than convex? What considerations determine its radius of curvature?

39-12 A student claimed that one can start a fire on a sunny day by use of the sun's rays and a concave mirrror. How is this done? Is the concept of image relevant? Could one do the same thing with a convex mirror?

39-13 A person looks at his reflection in the concave side of a shiny spoon. Is it right side up or inverted? What if he looks in the convex side?

Problems

39-1 What is the size of the smallest vertical plane mirror in which an observer standing erect can see his full-length image?

39-2 The image of a tree just covers the length of a 5-cm plane mirror when the mirror is held 30 cm from the eye. The tree is 100 m from the mirror. What is its height?

39-3 An object is placed between two mirrors arranged at right angles to each other.

a) Locate all of the images of the object.

b) Draw the paths of rays from the object to the eye of an observer.

39-4 An object 1 cm high is 20 cm from the vertex of a concave spherical mirror whose radius of curvature is 50 cm. Compute the position and size of the image. Is it real or virtual? Erect or inverted?

39-5 A concave mirror is to form an image of the filament of a headlight lamp on a screen 4 m from the mirror. The filament is 5 mm high, and the image is to be 40 cm high.

a) What should be the radius of curvature of the mirror?

b) How far in front of the vertex of the mirror should the filament be placed?

39-6 The diameter of the moon is 3480 km and its distance from the earth is 386,000 km. Find the diameter of the image of the moon formed by a spherical concave telescope mirror of focal length 4 m.

39-7 A spherical concave shaving mirror has a radius of curvature of 30 cm.

a) What is the magnification when the face is 10 cm from the vertex of the mirror?

b) Where is the image?

39-8 A concave spherical mirror has a radius of curvature of 10 cm. Make a diagram of the mirror to scale, and show rays incident on it parallel to the axis and at distances of 1, 2, 3, 4, and 5 cm from the axis. Using a protractor, construct the reflected rays and indicate the points at which they cross the axis.

39-9 An object is 16 cm from the center of a silvered spherical glass Christmas tree ornament 8 cm in diameter. What are the position and magnification of its image?

39-10 A concave mirror of radius 5 cm has a radius of curvature of 20 cm.

a) What is its focal length?

b) If the mirror is immersed in water (refractive index 1.33), what is its focal length?

39-11 An object 2 cm high is placed 5 cm away from a concave spherical mirror having radius of curvature of 20 cm.

a) Draw a principal-ray diagram showing formation of the image.

b) Determine the position, size, orientation, and nature of the image.

39-12 Prove that the image formed of a real object by a convex mirror is always virtual, no matter what the object position.

39-13 If light striking a convex mirror does not diverge from an object point but instead is converging toward a point at a (negative) distance s to the right of the mirror, this point is called a *virtual object*.

a) For a convex mirror having radius of curvature 10 cm, for what range of virtual-object positions is a real image formed?

b) What is the orientation of the image?

c) Draw a principal-ray diagram showing formation of such an image.

39-14 A tank whose bottom is a mirror is filled with water to a depth of 20 cm. A small object hangs motionless 8 cm under the surface of the water. What is the apparent depth of its image when viewed at normal incidence?

39-15 A ray of light in air makes an angle of incidence of 45° at the surface of a sheet of ice. The ray is refracted within the ice at an angle of 30°.

a) What is the critical angle for the ice?

b) A speck of dirt is embedded 2 cm below the surface of the ice. What is its apparent depth when viewed at normal incidence?

39-16 A microscope is focused on the upper surface of a glass plate. A second plate is then placed over the first. In order to focus on the bottom surface of the second plate, the microscope must be raised 1 mm. In order to focus on the upper surface it must be raised 2 mm *farther*. Find the index of refraction of the second plate. (This problem illustrates one method of measuring index of refraction.)

39-17 A layer of ether ($n = 1.36$) 2 cm deep floats on water ($n = 1.33$) 4 cm deep. What is the apparent distance from the ether surface to the bottom of the water layer, when viewed at normal incidence?

39-18 The end of a long glass rod 8 cm in diameter has a hemispherical surface 4 cm in radius. The refractive index of the glass is 1.50. Determine each position of the image if an object is placed on the axis of the rod at the following distances from its end:

a) infinitely far,   b) 16 cm,   c) 4 cm.

39-19 The rod of Problem 39–18 is immersed in a liquid. An object 60 cm from the end of the rod and on its axis is imaged at a point 100 cm inside the rod. What is the refractive index of the liquid?

39-20 What should be the index of refraction of a trans-parent sphere in order that paraxial rays from an infinitely distant object will be brought to a focus at the vertex of the surface opposite the point of incidence?

39-21 The left end of a long glass rod 10 cm in diameter, of index 1.50, is ground and polished to a convex hemispherical surface of radius 5 cm. An object in the form of an arrow 1 mm long, at right angles to the axis of the rod, is located on the axis 20 cm to the left of the vertex of the convex surface. Find the position and magnification of the image of the arrow formed by paraxial rays incident on the convex surface.

39-22 A transparent rod 40 cm long is cut flat at one end and rounded to a hemispherical surface of 12 cm radius at the other end. A small object is embedded within the rod along its axis and halfway between its ends. When viewed from the flat end of the rod the apparent depth of the object is 12.5 cm. What is its apparent depth when viewed from the curved end?

39-23 A solid glass hemisphere having a radius of 10 cm and a refractive index of 1.50 is placed with its flat face downward on a table. A parallel beam of light of circular cross section 1 cm in diameter travels directly downward and enters the hemisphere along its diameter. What is the diameter of the circle of light formed on the table?

39-24 A small tropical fish is at the center of a spherical fish bowl 30 cm in diameter. Find its apparent position and magnification to an observer outside the bowl. The effect of the thin walls of the bowl may be neglected.

39-25 In Fig. 39–20a the first focal length f is seen to be the value of s corresponding to $s' = \infty$; in (b) the second focal length f' is the value of s' when $s = \infty$.

a) Prove that $n/n' = f/f'$.

b) Prove that the general relation between object and image distances is

$$\frac{f}{s} + \frac{f'}{s'} = 1.$$

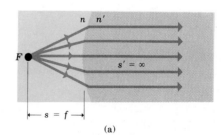

(a)

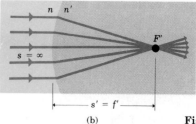

(b)   **Figure 39-20**

LENSES AND OPTICAL INSTRUMENTS

40

40-1 Images as objects

Most optical systems include more than one reflecting or refracting surface. The image formed by the first surface serves as the object for the second; the image formed by the second surface serves as the object for the third; etc. Figure 40–1 illustrates the various situations that may arise, and the following discussion of them should be studied carefully.

In Fig. 40–1, the arrow at point O represents a small object at right angles to the axis. A narrow cone of rays diverging from the head of the arrow is traced through the system. Surface 1 forms a real image of the arrow at point P. Distance OV_1 is the object distance for the first surface and distance V_1P is the image distance. Both of these are positive.

The image at P, formed by surface 1, serves as the object for surface 2. The object distance is PV_2 and is positive, since the direction from P to V_2 is the same as that of the incident light. The second surface forms a virtual image at point Q. The image distance is V_2Q and is negative because the direction from V_2 to Q is opposite to that of the refracted light.

The image at Q, formed by surface 2, serves as the object for surface 3. The object distance is QV_3 and is positive. The image at Q, although virtual, constitutes a *real object* so far as surface 3 is concerned. The rays incident on surface 3 are rendered converging and, except for the interposition of surface 4, would converge to a real image at point R. Even though this image is never formed, distance V_3R is the image distance for surface 3 and is positive.

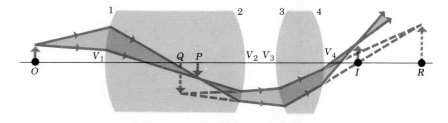

40-1 The object for each surface, after the first, is the image formed by the preceding surface.

750

The rays incident on surfaces 1, 2, and 3 have all been diverging, and the object distance has been the distance from the surface to the point from which the rays were actually or apparently diverging. The rays incident on surface 4, however, are *converging* and there is no point at the left of the vertex from which they diverge or appear to diverge. The *image at R, toward which the rays are converging*, is the object for surface 4, and since the direction from R to V_4 is opposite to that of the incident light, the object distance RV_4 is negative. The image at R is called a *virtual object* for surface 4. In general, whenever a *converging* cone of rays is incident on a surface, the point toward which the rays are converging serves as the object, the object distance is negative, and the point is called a virtual object.

Finally, surface 4 forms a real image at I, the image distance being V_4I and positive.

The meanings of virtual object and of virtual image are further exemplified by Fig. 40-2.

40-2 The thin lens

A lens is an optical system including two refracting surfaces. The general problem of refraction by a lens is solved by applying the methods of Sec. 39-7 to each surface in turn, the object for the second surface being the image formed by the first. Figure 40-3 shows a pencil of rays diverging from point Q of an object PQ. The first surface of lens L forms a virtual image of Q at Q'. This virtual image serves as a real object for the second surface of the lens, which forms a real image of Q' at Q''. Distance s_1 is the object distance for the first surface; s_1' is the corresponding image distance. The object distance for the second surface is s_2, equal to the sum of s_1' and the lens thickness t, and s_2' is the image distance for the second surface.

If, as is often the case, the lens is so thin that its thickness t is negligible in comparison with the distances s_1, s_1', s_2, s_2', we may assume that s_1' equals $-s_2$, and measure object and image distances from either vertex of the lens. We shall also assume the medium on both sides of the lens to be air, with index of refraction 1.00. For the first refraction, Eq. (39-16) becomes

$$\frac{1}{s_1} + \frac{n}{s_1'} = \frac{n-1}{R_1}.$$

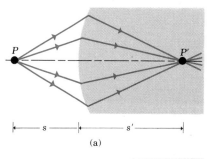

(a)

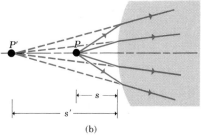

(b)

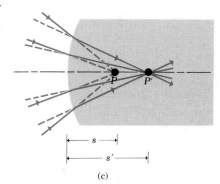

(c)

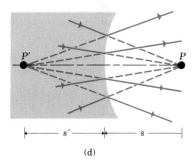

(d)

40-2 (a) A real image of a real object. (b) A virtual image of a real object. (c) A real image of a virtual object. (d) A virtual image of a virtual object.

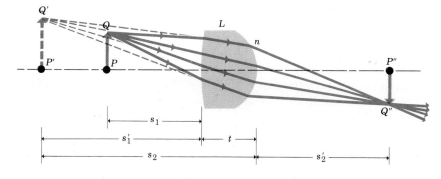

40-3 The image formed by the first surface of a lens serves as the object for the second surface.

Refraction at the second surface yields the equation

$$\frac{n}{s_2} + \frac{1}{s_2{'}} = \frac{1 - n}{R_2}.$$

Adding these two equations, and remembering that the lens is so thin that $s_2 = -s_1{'}$, we get

$$\frac{1}{s_1} + \frac{1}{s_2{'}} = (n - 1)\left(\frac{1}{R_1} - \frac{1}{R_2}\right).$$

Since s_1 is the object distance for the thin lens and $s_2{'}$ is the image distance, the subscripts may be omitted, and we get finally

$$\frac{1}{s} + \frac{1}{s'} = (n - 1)\left(\frac{1}{R_1} - \frac{1}{R_2}\right). \tag{40–1}$$

The usual sign conventions apply to this equation. Thus, in Fig. 40–4, s, s', and R_1 are positive quantities, but R_2 is negative.

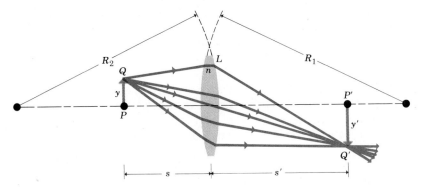

40-4   A thin lens.

The *focal length* f of a thin lens may be defined either as (a) the object distance of a point object on the lens axis whose image is at infinity, or (b) the image distance of a point object on the lens axis at an infinite distance from the lens. When we set either s or s' equal to infinity in Eq. (40–1) we find, for the focal length,

$$\frac{1}{f} = (n - 1)\left(\frac{1}{R_1} - \frac{1}{R_2}\right), \tag{40–2}$$

which is known as the *lensmaker's equation*.

Example   In Fig. 40–4, let the absolute magnitudes of the radii of curvature of the lens surfaces be respectively 20 cm and 5 cm. Since the center of curvature of the first surface is on the side of the outgoing light, $R_1 = +20$ cm, and since that of the second surface is not, $R_2 = -5$ cm. Let $n = 1.50$. Then

$$\frac{1}{f} = (1.50 - 1)\left(\frac{1}{20\text{ cm}} - \frac{1}{-5\text{ cm}}\right),$$
$$f = +8\text{ cm}.$$

Substituting Eq. (40–2) in Eq. (40–1), we obtain the thin-lens equation

$$\frac{1}{s} + \frac{1}{s'} = \frac{1}{f}. \tag{40-3}$$

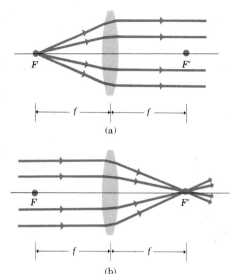

(a)

This is known as the *gaussian* form of the thin-lens equation, after Karl F. Gauss, the same mathematician responsible for the law in electrostatics bearing his name. Note that Eq. (40–3) has exactly the same form as the equation for a spherical mirror.

The object point for which the image is at infinity is called the *first focal point* or *focus* of the lens and is lettered F in Fig. 40–5a. The image point for an infinitely distant object is called the *second focal point* or *focus* and is lettered F' in Fig. 40–5b. The focal points (foci) of a thin lens lie on opposite sides of the lens at distances from it equal to its focal length f.

Figure 40–6 corresponds to Fig. 40–5, except that it is drawn for object and image points not on the axis of the lens. Planes through the first and second focal points of a lens, perpendicular to the axis, are called the first and second *focal planes*. Paraxial rays from point Q in Fig. 40–6a, in the first focal plane of the lens, are parallel to one another after refraction. In other words, they converge to an infinitely distant image of point Q. In Fig. 40–6b a bundle of parallel rays from an infinitely distant point, not on the lens axis, converges to an image Q' lying in the second focal plane of the lens.

The lateral magnification produced by a lens may be obtained by inspection of Fig. 40–4; the object PQ, the image $P'Q'$, and the lines PP' and QQ' form two similar triangles. Hence

$$\frac{-y'}{y'} = \frac{s'}{s},$$

and, since m, as usual, is y'/y, we obtain

$$m = -\frac{s'}{s}. \tag{40-4}$$

Taking the derivative of Eq. (40–3) with respect to s, we find that the *longitudinal* magnification is

$$m' = \frac{ds'}{ds} = -\left(\frac{s'}{s}\right)^2 = -m^2. \tag{40-5}$$

Although Eqs. (40–3) and (40–4) were derived for the special case of rays making small angles with the axis and, in general, do not apply to rays making large angles, they may be used for any lens that has been corrected so that all rays are imaged at the same point. These are therefore two of the most important equations in geometrical optics.

The three-dimensional image of a three-dimensional object, formed by a lens, is shown in Fig. 40–7. Since point R is nearer the lens than point P, its image, from Eq. (40–3), is farther from the lens than is point P', and the image $P'R'$ points in the same direction as the object PR. Arrows $P'S'$ and $P'Q'$ are reversed in space, relative to PS and PQ. Although we speak of the image as "inverted," only its transverse dimensions are reversed.

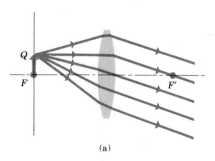

40-5 First and second focal points of a thin lens.

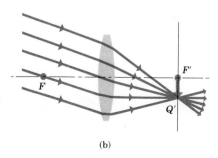

40-6 Planes through the focal points of a lens are called focal planes.

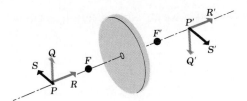

40-7   A lens forms a three-dimensional image of a three-dimensional object.

Figure 40–7 should be compared with Fig. 39–4, showing the image formed by a plane mirror. Note that the image formed by a lens, although it is inverted, is *not* perverted. That is, if the object is a left hand, its image is a left hand also. This may be verified by pointing the left thumb along *PR*, the left forefinger along *PQ*, and the left middle finger along *PS*. A rotation of 180° about the thumb as an axis then brings the fingers into coincidence with *P'Q'* and *P'S'*. In other words, *inversion* of an image is equivalent to a rotation of 180° about the lens axis.

40-3 Diverging lenses

A bundle of parallel rays incident on the lens shown in Figs. 40–5 and 40–6 converges to a real image after passing through the lens. The lens is called a *converging lens*. Its focal length, as computed from Eq. (40–2), is a positive quantity and therefore the lens is also called a *positive lens*.

A bundle of parallel rays incident on the lens in Fig. 40–8 becomes diverging after refraction, and the lens is called a *diverging lens*. Its focal length, computed by Eq. (40–2), is a negative quantity and therefore the lens is also called a *negative lens*. The focal points of a negative lens are reversed, relative to those of a positive lens. The second focal point, *F'*, of a negative lens is the point from which rays, originally parallel to the axis, appear to diverge after refraction, as in Fig. 40–8a. Incident rays converging toward the first focal point *F*, as in Fig. 40–8b, emerge from the lens parallel to its axis. That is, just as for a positive lens, the second focal point is the (virtual) image of an infinitely distant object on the axis of the lens, while the first focal point is the object point (a virtual

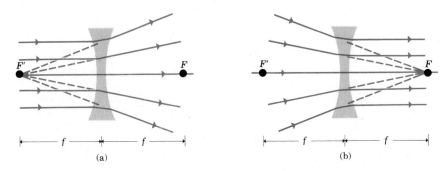

40-8   Focal points of a diverging lens.

object if the lens is diverging) for which an image is formed at infinity. Equations (40–2) through (40–5) apply both to negative and to positive lenses. Various types of lenses, both converging and diverging, are illustrated in Fig. 40–9.

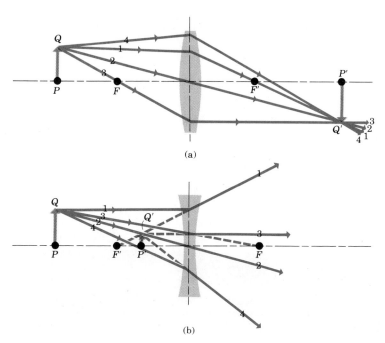

40-9 (a) Meniscus, plano-convex, and double-convex converging lenses. (b) Meniscus, plano-concave, and double concave diverging lenses.

(a)   (b)

40-4 Graphical methods

The position and size of the image of an object formed by a thin lens may be found by a simple graphical method. This method consists of finding the point of intersection, after passing through the lens, of a few rays (called *principal rays*) diverging from some chosen point of the object *not* on the lens axis, such as point Q in Fig. 40–10. Then (neglecting lens aberrations) all rays from this point that pass through the lens will intersect at the same point. In using the graphical method, the entire deviation of any ray is assumed to take place at a plane through the center of the lens. Three principal rays whose paths may readily be traced are shown in Fig. 40–10.

1. *A ray parallel to the axis,* after refraction by the lens, passes through the second focal point of a converging lens, or appears to come from the second focal point of a diverging lens.

2. *A ray through the center of the lens* is not appreciably deviated, since the two lens surfaces through which the central ray passes are very nearly parallel if the lens is thin. We have seen that a ray passing through a plate with parallel faces is not deviated, but only dis-

(a)

(b)

40-10 Principal-ray diagram, showing graphical method of locating an image.

placed. For a thin lens, the displacement may be neglected.

3. *A ray through* (*or proceeding toward*) *the first focal point* emerges parallel to the axis.

Once the position of the image point has been found by means of the intersection of any two rays 1, 2, and 3, the paths of all other rays from the same point, such as ray 4 in Fig. 40–10, may be drawn. A few examples of this procedure are given in Fig. 40–11. Not all the rays in Fig. 40–11 are principal rays.

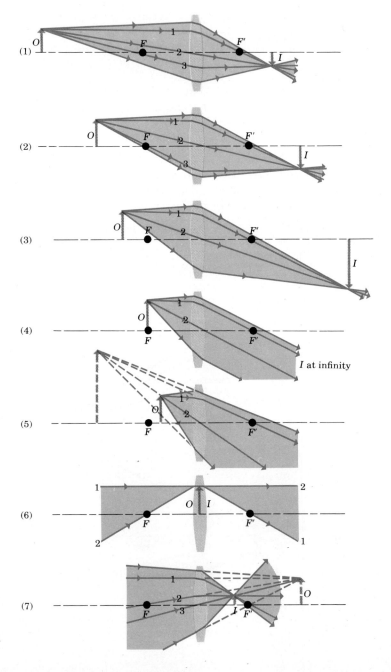

40-11 Formation of an image by a thin lens. The principal rays are labeled.

In every problem involving formation of an image by a lens, the student should *always* draw a principal-ray diagram; this serves not only as a graphical check on numerical calculations but also as an aid to the understanding of basic concepts, especially the concept of *image*.

40-5 Images as objects for lenses

It was shown in Sec. 40-1 and Fig. 40-1 that the image formed by any one *surface* in an optical system serves as the object for the next surface. This principle was used to derive the thin-lens relations, Eqs. (40-1) through (40-3), with the assumption that the distance between the two surfaces is negligible. A similar analysis can be carried out for a thick lens, where this distance (which we may call d) is *not* negligible compared to the other distances. In this case the object distance s_2 for the second surface is not equal to the negative of the image distance s_1' for the first, but instead we have

$$s_2 = -(s_1' - d), \qquad (40\text{-}6)$$

and the resulting formulas are not so simple. Examples of thick-lens analysis are found in the problems for this chapter.

The same principle can be used in systems containing several *lenses*. Many optical systems, such as camera lenses, microscopes, and telescopes, use more than one lens. In each case the image formed by any one lens serves as the object for the next lens. Figure 40-12 shows various possibilities. Lens 1 forms a real image at P of a real object at O. This real image serves as a real object for lens 2. The virtual image at Q formed by lens 2 is a real object for lens 3. If lens 4 were not present, lens 3 would form a real image at R. Although this image is never formed, it serves as a *virtual object* for lens 4, which forms a final real image at I.

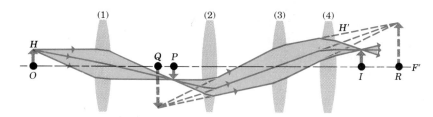

40-12   The object for each lens, after the first, is the image formed by the preceding lens.

*40-6 Lens aberrations

The relatively simple equations we have derived relating object and image distances, focal lengths, radii of curvature, etc., were based upon the *paraxial approximation;* all rays were assumed to be paraxial, that is, to make small angles with the axis. In general, however, a lens must image not only points on its axis, but points that lie off the axis as well. Furthermore, because of the finite size of the lens, the cone of rays that forms the image of any point is of finite size. Nonparaxial rays proceeding from a given object point *do not*, in general, all intersect at the same point after refraction by a lens. Consequently, the image formed by these rays is not a sharp one. Furthermore, the focal length of a lens depends upon its index of refraction, which varies with wavelength. Therefore, if the light proceeding from an object is not monochromatic, a lens forms a number of colored images, which lie in different positions and are of different sizes, even if formed by paraxial rays.

The departures of an actual image from the predictions of simple theory are called *aberrations*. Those caused by the variation of index with wavelength are the *chromatic aberrations*. The others, which would arise even if the light were monochromatic, are the *monochromatic aberrations*. Lens aberrations are not caused by faulty construction of the lens, such as the failure of its surfaces to conform to a truly spherical shape, but are simply consequences of the laws of refraction at spherical surfaces.

The monochromatic aberrations are all related to the failure of the paraxial-ray approximation for lenses of finite aperture, but it is customary to distinguish various aspects of this difficulty, each with its characteristic effect. *Spherical aberration* is the failure of rays from a point object on the optic axis to converge to a point image; instead, the rays converge to a circle of minimum radius, called *the circle of least confusion*, and then diverge again, as shown in Fig. 40–13. The corresponding effect for points off the axis produces images that are comet-shaped figures rather than circles; this is called *coma*.

Astigmatism is the imaging of a point off the axis as a *line;* in this aberration the rays from a point object converge at some distance from the lens to a line in the place defined by the optic axis and the object point and, at a somewhat different distance from the lens, to a line *perpendicular* to this plane. The circle of least confusion appears between these two positions, at a location that depends on the object point's distance from the axis as well as its distance from the lens. As a result, object points lying in a plane are, in general, imaged not in a plane but in some curved surface; this effect is called *curvature of field*. Finally, the image of a straight line that does not pass through the axis may be curved; as a result the image of a square with the axis through its center may resemble a barrel (sides bent outward) or a pincushion (sides bent inward). This effect, called *distortion*, is not related to lack of sharpness of the image but results from a change in lateral magnification with distance from the axis.

Chromatic aberrations result directly from the variation of index of refraction with wavelength. Even in the absence of all monochromatic aberration, different wavelengths are imaged at different points, and when an object is illuminated with white light containing a mixture of wavelengths, there is no single point at which a point object is imaged. The magnification of a lens also varies with wavelength; this effect is responsible for the rainbow-fringed images seen with inexpensive binoculars or telescopes.

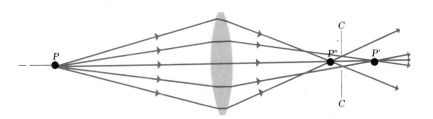

40-13   Spherical aberration. The circle of least confusion is shown by *C–C*.

It is impossible to eliminate these aberrations from a single lens, but in a compound lens of several elements, the aberrations of one element may partially cancel those of another element. Design of such lenses is an extremely complex problem, which has been aided greatly in recent years by the use of high-speed computers. It is still impossible to eliminate all aberrations, but it *is* possible to decide which ones are most troublesome for a particular application and to design accordingly.

40-7 The eye

Since the purpose of most optical instruments is to enable us to see better, the logical place to begin a discussion of such instruments is with the eye. The essential parts of the eye, considered as an optical system, are shown in Fig. 40-14.

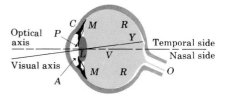

40-14   The eye.

The eye is very nearly spherical in shape, and about an inch in diameter. The front portion is somewhat more sharply curved, and is covered by a tough, transparent membrane C, called the *cornea*. The region behind the cornea contains a liquid A called the *aqueous humor*. Next comes the *crystalline lens, L,* a capsule containing a fibrous jelly, hard at the center and progressively softer at the outer portions. The crystalline lens is held in place by ligaments that attach it to the ciliary muscle M. Behind the lens, the eye is filled with a thin jelly V consisting largely of water, called the *vitreous humor*. The indices of refraction of both the aqueous humor and the vitreous humor are nearly equal to that of water, about 1.336. The crystalline lens, while not homogeneous, has an "average" index of 1.437. This is not very different from the indices of the aqueous and vitreous humors, so that most of the refraction of light entering the eye occurs at the cornea.

A large part of the inner surface of the eye is covered with a delicate film of nerve fibers, R, called the *retina*. A cross section of the retina is shown in Fig. 40-15a. Nerve fibers branching out from the *optic nerve O* terminate in minute structures called rods and cones. The rods and cones, together with a bluish liquid called the visual purple, which circulates among them, receive the optical image and transmit it along the optic nerve to the brain. There is a slight depression in the retina at Y called the yellow spot or macula. At its center is a minute region, about 0.25 mm in diameter, called the *fovea centralis,* which contains cones exclusively. Vision is much more acute at the fovea than at other portions of the retina, and the muscles controlling the eye always rotate the eyeball until the image of the object toward which attention is directed

40-15   (a) Section of the human retina (500X). Light is incident from the left. (b) Figure for demonstrating the blind spot.

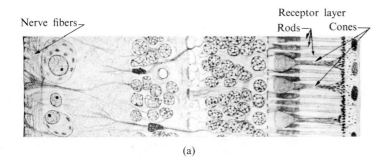

(a)

(b)

falls on the fovea. The outer portion of the retina merely serves to give a general picture of the field of view. The fovea is so small that motion of the eye is necessary to focus distinctly two points as close together as the dots in a colon (:).

There are no rods or cones at the point where the optic nerve enters the eye, and an image formed at this point cannot be seen. This region is called the *blind spot.* The existence of the blind spot can be demonstrated by closing the left eye and looking with the right eye at the cross in Fig. 40–15b. When the diagram is about 25 cm from the eye, the image of the square falls on the blind spot and the square disappears. At a smaller distance, the square reappears while the circle disappears. At a still smaller distance, the circle again appears.

In front of the crystalline lens is the iris, at the center of which is an opening *P* called the *pupil,* which regulates the quantity of light entering the eye, dilating if the brightness of the field is low, and contracting if the brightness is increased. This process is known as *adaptation.* However, the range of pupillary diameter is only about fourfold (hence, the range in area is about sixteenfold) over a range of brightness that is 100,000-fold. The receptive mechanism of the retina can also adapt itself to large differences in quantity of light.

To see an object distinctly, a sharp image of it must be formed on the retina. If all the elements of the eye were rigidly fixed in position, there would be but one object distance for which a sharp retinal image would be formed, while in fact the normal eye can focus sharply on an object at any distance from infinity up to about 25 cm in front of the eye. This is made possible by the action of the crystalline lens and the ciliary muscle to which it is attached. When relaxed, the normal eye is focused on objects at infinity, i.e., the second focal point is at the retina. When it is desired to view an object nearer than infinity, the ciliary muscle tenses and the crystalline lens assumes a more nearly spherical shape. This process is called *accommodation.*

The extremes of the range over which distinct vision is possible are known as the *far point* and the *near point* of the eye. The far point of a normal eye is at infinity. The position of the near point evidently depends on the extent to which the curvature of the crystalline lens may be increased in accommodation. The range of accommodation gradually diminishes with age as the crystalline lens loses its flexibility. For this reason the near point gradually recedes as one grows older. This recession of the near point with age is called *presbyopia,* and should not be considered a defect of vision, since it proceeds at about the same rate in all normal eyes. The following is a table of the approximate average position of the near point at various ages:

Age, years	Near point, cm
10	7
20	10
30	14
40	22
50	40
60	200

*40-8 Defects of vision

Several common defects of vision result from an incorrect relation between the parts of the optical system of the eye. A normal eye forms an
image on the retina of an object at infinity when the eye is relaxed, as in
Fig. 40-16a. In the *myopic* (nearsighted) eye, the eyeball is too long from
front to back in comparison with the radius of curvature of the cornea,
and rays from an object at infinity are focused in front of the retina. The
most distant object for which an image can be formed on the retina is
then nearer than infinity. In the *hyperopic* (farsighted) eye, the eyeball
is too short and the image of an infinitely distant object would be formed
behind the retina. By accommodation, these parallel rays may be made
to converge on the retina, but evidently, if the range of accommodation
is normal, the near point will be more distant than that of a normal eye.
The myopic eye produces too much convergence in a parallel bundle of
rays for an image to be formed on the retina; the hyperopic eye, not
enough convergence.

Astigmatism refers to a defect in which the surface of the cornea is
not spherical, but is more sharply curved in one plane than another. (It
should not be confused with the lens aberration of the same name, which
applies to the behavior, after passing through a lens having spherical
surfaces, of rays making a large angle with the axis.) Astigmatism makes
it impossible, for example, to focus clearly on the horizontal and vertical
bars of a window at the same time.

These defects can be corrected by the use of corrective lenses
("glasses"). The near point of either a presbyopic or a hyperopic eye is
farther from the eye than normal. To see clearly an object at normal
reading distance (usually assumed to be 25 cm) we must place in front of
the eye a lens of such focal length that it forms an image of the object at
or beyond the near point. Thus the function of the lens is not to make
the object appear larger, but in effect to move the object farther away
from the eye to a point where a sharp retinal image can be formed.

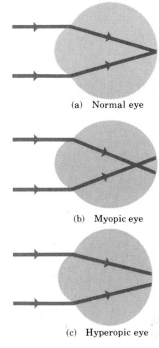

(a)   Normal eye

(b)   Myopic eye

(c)   Hyperopic eye

40-16

Example 1   The near point of a certain eye is 100 cm in front of the eye.
What lens should be used to see clearly an object 25 cm in front of the
eye?

Solution   We have

$$s = +25 \text{ cm}, \qquad s' = -100 \text{ cm},$$

$$\frac{1}{f} = \frac{1}{s} + \frac{1}{s'} = \frac{1}{+25 \text{ cm}} + \frac{1}{-100 \text{ cm}},$$

$$f = +33 \text{ cm}.$$

That is, a converging lens of focal length 33 cm is required.

The far point of a *myopic* eye is nearer than infinity. To see clearly
objects beyond the far point, a lens must be used that will form an image
of such objects, not farther from the eye than the far point.

Example 2   The far point of a certain eye is 1 m in front of the eye.
What lens should be used to see clearly an object at infinity?

Solution Assume the image to be formed at the far point. Then

$$s = \infty, \qquad s' = -100 \text{ cm},$$

$$\frac{1}{f} = \frac{1}{s} + \frac{1}{s'} = \frac{1}{\infty} + \frac{1}{-100 \text{ cm}},$$

$$f = -100 \text{ cm}.$$

A *diverging* lens of focal length 100 cm is required.

Astigmatism is corrected by means of a *cylindrical* lens. The curvature of the cornea in a horizontal plane may have the proper value such that rays from infinity are focused on the retina. In the vertical plane, however, the curvature may not be sufficient to form a sharp retinal image. When a cylindrical lens with axis horizontal is placed before the eye, the rays in a horizontal plane are unaffected, while the additional convergence of the rays in a vertical plane now causes these to be sharply imaged on the retina.

The optometrist describes the converging or diverging effect of lenses in terms, not of the focal length, but of its *reciprocal.* The reciprocal of the focal length of a lens is called its *power,* and if the focal length is in meters the power is in *diopters.* Thus the power of a positive lens whose focal length is 1 m is 1 diopter; if the focal length is 2 m the power is 0.5 diopter, and so on. If the focal length is negative, the power is negative also. Thus in the two examples above, the required powers are $+3.0$ diopter and -1.0 diopter, respectively.

40-9 The magnifier

The apparent size of an object is determined by the size of its retinal image; if the eye is unaided, this depends upon the *angle* subtended by the object at the eye. When one wishes to examine a small object in detail one brings it close to the eye, in order that the angle subtended and the retinal image may be as large as possible. Since the eye cannot focus sharply on objects closer than the near point, a given object subtends the maximum possible angle at an unaided eye when placed at this point. (We shall assume hereafter that the near point is 25 cm from the eye.) By placing a converging lens in front of the eye, the accommodation may, in effect, be increased. The object may then be brought closer to the eye than the near point and will subtend a correspondingly larger angle. A lens used for this purpose is called a *magnifying glass,* a *simple microscope,* or a *magnifier.* The magnifier forms a virtual image of the object and the eye "looks at" this virtual image. Since a (normal) eye can focus sharply on an object anywhere between the near point and infinity, the image can be seen equally clearly if it is formed anywhere within this range. We shall assume that the image is formed at infinity.

The magnifier is illustrated in Fig. 40–17. In (a), the object is at the near point, where it subtends an angle u at the eye. In (b), a magnifier in front of the eye forms an image at infinity, and the angle subtended at the magnifier is u'. The *angular magnification M* (not to be confused with the *lateral magnification m*) is defined as the ratio of the angle u' to the angle u. The value of M may be found as follows.

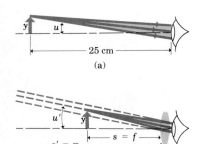

40-17 A simple magnifier.

From Fig. 40–17, u and u' are given (in radians) by

$$u = \frac{y}{25 \text{ cm}} \quad \text{(approximately)},$$

$$u' = \frac{y}{f} \quad \text{(approximately)}.$$

Hence,

$$M = \frac{u'}{u} = \frac{y/f}{y/25} = \frac{25}{f} \qquad (f \text{ in centimeters}). \qquad (40\text{–}7)$$

While it appears at first that the angular magnification may be made as large as desired by decreasing the focal length f, the aberrations of a simple double convex lens set a limit to M of about 2X or 3X. If these aberrations are corrected, the magnification may be carried as high as 20X.

40–10 The camera

The essential elements of a camera are a lens equipped with a shutter, a light-tight enclosure, and a light-sensitive film to record an image. An example is shown in Fig. 40–18. The lens forms a real image in the plane of the film of the object being photographed. The lens may be moved closer to or farther from the film to provide proper image distances for various object distances. All but the most inexpensive lenses have several elements, to permit partial correction of various aberrations. Many modern camera lenses are variations of the Zeiss "Tessar" design, shown in Fig. 40–19.

In order for the image to be recorded properly on the film, the total light energy per unit area reaching the film (the "exposure") must fall within certain limits; this is controlled by the shutter and the lens aperture. The shutter controls the time during which light enters the lens, typically adjustable in steps corresponding to factors of about two, from 1 s to $\frac{1}{1000}$ s or thereabouts. The light-gathering capacity of the lens is proportional to its effective area; this may be varied by means of an adjustable aperture or *diaphragm,* which is a nearly circular hole of variable diameter. The aperture size is usually described in terms of its "f-number," which is the focal length of the lens divided by the diameter of the aperture. Thus a lens having $f = 50$ mm and an aperture diameter of 25 mm would be said to have an aperture of $f/2$.

Because the light-gathering capacity of the lens is proportional to its area and thus to the *square* of its diameter, changing the diameter by a factor of $\sqrt{2}$ corresponds to a factor of two in exposure. Thus, adjustable apertures usually have scales labeled with successive numbers related by factors of $\sqrt{2}$, such as

$$f/2, \quad f/2.8, \quad f/4, \quad f/5.6, \quad f/8, \quad f/11, \quad f/16,$$

and so on, with the larger numbers representing smaller apertures and exposures.

The choice of focal length for a camera lens depends on the film size and the desired angle of view, or *field.* For the popular 35-mm cameras,

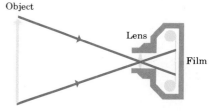

40–18   Essential elements of a camera.

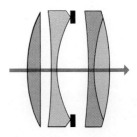

40–19   Zeiss "Tessar" lens design.

with image size of 24 × 36 mm, the normal lens is usually about 50 mm in focal length and permits an angle of about 45°. A longer focal-length lens, used with the same film size, provides a smaller angle of view and a larger image of part of the object, compared with a normal lens; this gives the impression that the camera is closer than it really is, and such a lens is called a *telephoto* lens. At the other extreme, a lens of shorter focal length, such as 35 mm or 28 mm, permits a wider angle of view and is called a *wide-angle* lens.

For a given position of the photographic film only those points lying in the corresponding object plane are sharply focused; objects at greater or lesser distance appear somewhat blurred. However, because of lens aberrations, a point of a given object will be imaged as a small circle, called the *circle of confusion,* even with the best focusing. The circles of confusion of points at other distances will be larger. If extremely sharp definition of the image is not essential, there is evidently a certain range of object distances, called the *depth of field,* such that all objects within this range are simultaneously "in focus" on the plate. That is, the circles of confusion of points within this range are not so large that the image is unsatisfactory. The diameter of the circle of confusion of a particular point is proportional to the aperture diameter; thus, depth of field is increased by "stopping down" the lens to a smaller aperture and larger *f*-number.

40-11 The projector

The optical system of a projector for slides or motion pictures is shown in Fig. 40–20. The arrow at the left represents the light source, for example, the filament of a projection lamp. For simplicity, the slide to be projected is represented as opaque except for a single transparent aperture.

The diagram traces the course of three pencils of rays originating at the ends and at the midpoint of the source. The function of the condensing lens is to deviate the light from the source inward, so that it can pass through the projecting lens. If the condensing lens were omitted, light passing through the outer portions of the slide would not strike the projecting lens and only a small portion of the slide near its center would be imaged on the screen.

A study of the figure shows that, for the three selected points of the source, only those rays within the shaded pencils can pass through the aperture; all others striking the condensing lens are intercepted by the opaque portions of the slide. Similar pencils of rays could be drawn from all other points of the source.

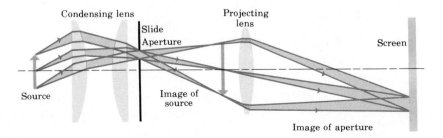

40-20 Essential elements of a projector for slides or motion pictures.

Each of these pencils converges, after passing through the aperture, to form an image of its point of origin just to the left of the projecting lens. In practice, this image would be formed *at* the projecting lens, but for clarity in the diagram the image and the lens have been displaced slightly. The focal length of the condensing lens should be such that the image of the source just fills the projecting lens. If the image of the source is larger than the projecting lens, some of the light passing through the slide is wasted. If it is smaller, the area of the projecting lens is not being fully utilized. Thus, in the diagram, the outer portions of the projecting lens serve no useful purpose.

Three rays tangent to the upper edge of the aperture have been emphasized in the figure. These rays originate at *different* points of the source. Hence, although they intersect at the edge of the aperture, this point of intersection does not constitute an image of any point of the source; but these three rays diverge from a common point of the *slide,* and therefore this point of the slide is imaged, as shown on the screen. Similarly, rays tangent to any point of the edge of the aperture are imaged at a corresponding point on the screen. Thus, if the aperture is circular, a circular spot of light appears on the screen. Note that light from *all* points of the source illuminates *every* point of the image of the aperture, and would do the same were the aperture at any other point of the slide.

The preceding discussion has explained the conditions that determine the *focal length* of the condensing lens and the *diameter* of the projecting lens (the image of the source formed by the condensing lens should just fill the projecting lens). The *diameter* of the condensing lens must evidently be at least as great as the diagonal of the largest slide to be projected, while the *focal length* of the projecting lens is determined by the magnification desired and the distance of the projector from the screen.

40-12 The compound microscope

When an angular magnification larger than that attainable with a simple magnifier is desired, it is necessary to use a *compound microscope,* usually called merely a *microscope.* The essential elements of a microscope are illustrated in Fig. 40-21. The object O to be examined is placed just beyond the first focal point F_1 of the *objective* lens, which forms a real and enlarged image I. This image lies just within the first focal point F_2 of the *eyepiece,* which forms a virtual image of I at I'. As was stated earlier, the position of I' may be anywhere between the near and far points of the eye. Although both the objective and eyepiece of an actual microscope are highly corrected compound lenses, they are shown as simple thin lenses for simplicity.

Since the objective lens merely forms an enlarged real image that is viewed by the eyepiece, the overall angular magnification M of the compound microscope is the product of the lateral magnification m_1 of the objective and the angular magnification M_2 of the eyepiece. The former is given by

$$m_1 = -\frac{s_1{}'}{s_1},$$

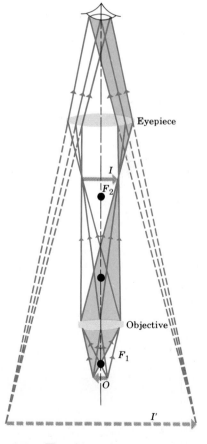

40-21 The microscope.

where s_1 and s_1' are the object and image distances for the objective lens. Ordinarily the object is very close to the focus, resulting in an image whose distance from the objective is much larger than its focal length f_1. Thus s_1 is approximately equal to f_1, and $m_1 = -s_1'/f_1$, approximately. The angular magnification of the eyepiece, from Eq. (40-7), is $M_2 = (25 \text{ cm})/f_2$, where f_2 is the focal length of the eyepiece, considered as a simple lens. Hence the overall magnification M of the compound microscope is, apart from a negative sign, which is customarily ignored,

$$M = m_1 M_2 = \frac{(25 \text{ cm})s_1'}{f_1 f_2}, \qquad (40\text{-}8)$$

where s_1', f_1, and f_2 are measured in centimeters. Microscope manufacturers customarily specify the values of m_1 and M_2 for microscope components rather than the focal lengths of the objective and eyepiece.

40-13 Telescopes

The optical system of a refracting telescope is essentially the same as that of a compound microscope. In both instruments, the image formed by an objective is viewed through an eyepiece. The difference is that the telescope is used to examine large objects at large distances and the microscope to examine small objects close at hand.

The *astronomical* telescope is illustrated in Fig. 40-22. The objective lens forms a real, reduced image I of the object, and a virtual image of I is formed by the eyepiece. As with the microscope, the image I' may be formed anywhere between the near and far points of the eye. In practice, the objects examined by a telescope are at such large distances from the instrument that the image I is formed very nearly at the second focal point of the objective. Furthermore, if the image I' is at infinity, the image I is at the first focal point of the eyepiece. The distance between objective and eyepiece, or the length of the telescope, is therefore the *sum* of the focal lengths of objective and eyepiece, $f_1 + f_2$.

The angular magnification M of a telescope is defined as the ratio of the angle subtended at the eye by the final image I', to the angle subtended at the (unaided) eye by the object. This ratio may be expressed in terms of the focal lengths of objective and eyepiece as follows. In Fig. 40-22, the ray passing through F_1, the first focal point of the objective, and through F_2', the second focal point of the eyepiece, has been emphasized. The object (not shown) subtends an angle u at the objective and would subtend essentially the same angle at the unaided eye. Also, since the observer's eye is placed just to the right of the focal point F_2', the

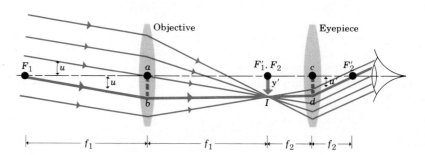

40-22 Telescope; final image at infinity.

angle subtended at the eye by the final image is very nearly equal to the angle u'. The distances ab and cd are evidently equal to each other and to the height y' of the image I. Since u and u' are small, they may be approximated by their tangents. From the right triangle F_1ab and $F_2'cd$,

$$u = \frac{-y'}{f_1}, \qquad u' = \frac{y'}{f_2},$$

Hence,

$$M = \frac{u'}{u} = -\frac{y'/f_2}{y'/f_1} = -\frac{f_1}{f_2}. \qquad (40\text{-}9)$$

The angular magnification M of a telescope is therefore equal to the ratio of the focal length of the objective to that of the eyepiece. The negative sign denotes an inverted image.

An inverted image is not a disadvantage if the instrument is to be used for astronomical observations, but it is desirable that a terrestrial telescope form an erect image. This is accomplished in the *prism binocular,* of which Fig. 40–23 is a cutaway view, by a pair of 45°–45°–90° totally reflecting prisms inserted between objective and eyepiece. The image is inverted by the four reflections from the inclined faces of the prisms. It is customary to stamp on a flat metal surface of a binocular two numbers separated by a multiplication sign, thus, 7×50. The first number is the magnification and the second is the diameter of the objective lenses in millimeters.

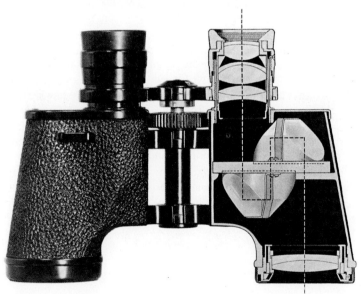

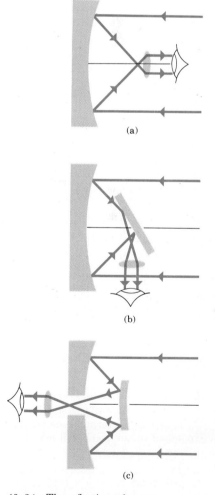

(a)

(b)

(c)

40-23   The prism binocular. (Courtesy of Bausch & Lomb Optical Co.)

In the *reflecting telescope* the objective lens is replaced by a concave mirror, as shown in Fig. 40–24. In large telescopes this scheme has many advantages, both theoretical and practical. The mirror is intrinsically free of chromatic aberrations, and spherical aberrations are much easier to correct than with a lens. The material need not be transparent, and the reflector can be made more rigid than a lens, which has to be sup-

40-24   The reflecting telescope.

ported only at its edges. The largest reflecting telescope in the world has a mirror over 5 m in diameter.

Because the image is formed in a region traversed by incoming rays, this image can be observed directly with an ocular only by blocking off part of the incoming beam; this is practical only for the very largest telescopes. Alternative schemes use a mirror to reflect the image out the side or through a hole in the mirror, as shown in Figs. 40–24b and 40–24c.

Questions

40–1 Sometimes a wine glass filled with white wine forms an image of an overhead light on a white tablecloth. Would the same image be formed with an empty glass? With a glass of gin? Gasoline?

40–2 How could one very quickly make an approximate measurement of the focal length of a converging lens? Could the same method be applied if we wished to use a diverging lens?

40–3 If you look closely at a shiny Christmas-tree ball, you can see nearly the entire room. Does the room appear right-side-up or upside-down? Discuss your observations in terms of images.

40–4 A student asserted that any lens with spherical surfaces has a positive focal length if it is thicker at the center than at the edge, and negative if it is thicker at the edge. Do you agree?

40–5 The focal length of a simple lens depends on the color (wavelength) of light passing through it. Why? Is it possible for a lens to have a positive focal length for some colors and negative for others?

40–6 The human eye is often compared to a camera. In what ways is it similar to a camera? In what ways does it differ?

40–7 How could one make a lens for sound waves?

40–8 A student proposed to use a plastic bag full of air, immersed in water, as an underwater lens. Is this possible? If the lens is to be a converging lens, what shape should the air pocket have?

40–9 When a converging lens is immersed in water, does its focal length increase or decrease, compared with the value in air?

40–10 You are marooned on a desert island and want to use your eyeglasses to start a fire. Can this be done if you are nearsighted? If you are farsighted?

40–11 While lost in the mountains a person who was nearsighted in one eye and farsighted in the other made a crude emergency telescope from the two lenses of his eyeglasses. How did he do this?

40–12 In using a magnifying glass, is the magnification greater when the glass is close to the object or when it is close to the eye?

40–13 When a slide projector is turned on without a slide in it, and the focus adjustment is moved far enough in one direction, a gigantic image of the lightbulb filament can be seen on the screen. Explain how this happens.

40–14 A spherical air bubble in water can function as a lens. Is it a converging or diverging lens? How is its focal length related to its radius?

40–15 As discussed in the text, some binoculars use prisms to invert the final image. Why are prisms better than ordinary mirrors for this purpose?

40–16 There have been reports of round fishbowls starting fires by focusing the sun's rays coming in a window. Is this possible?

40–17 How does a person judge distance? Can a person with vision in only one eye judge distance? What is meant by "binocular vision?"

40–18 Zoom lenses are widely used in television cameras and conventional photography. Such a lens has, effectively, a variable focal length; changes in focal length are accomplished by moving some lens elements relative to others. Try to devise a scheme to accomplish this effect.

Problems

40–1 A thin-walled glass sphere of radius R is filled with water. An object is placed a distance $3R$ from the surface of the sphere. Determine the position of the final image. The effect of the glass wall may be neglected.

40–2 A transparent rod 40 cm long is cut flat at one end and rounded to a hemispherical surface of 12 cm radius at the other end. An object is placed on the axis of the rod, 10 cm from the hemispherical end.

a) What is the position of the final image?

b) What is its magnification? Assume the refractive index to be 1.50.

40–3 Both ends of a glass rod 10 cm in diameter, of index 1.50, are ground and polished to convex hemispherical surfaces of radius 5 cm at the left end and radius 10 cm at the right end. The length of the rod between vertices is 60 cm.

An arrow 1 mm long, at right angles to the axis and 20 cm to the left of the first vertex, constitutes the object for the first surface.

a) What constitutes the object for the second surface?

b) What is the object distance for the second surface?

c) Is the object real or virtual?

d) What is the position of the image formed by the second surface?

e) What is the height of the final image?

40-4 The same rod as in Problem 40-3 is now shortened to a distance of 10 cm between its vertices, the curvatures of its ends remaining the same.

a) What is the object distance for the second surface?

b) Is the object real or virtual?

c) What is the position of the image formed by the second surface?

d Is the image real or virtual, erect or inverted, with respect to the original object?

e) What is the height of the final image?

40-5 A glass rod of refractive index 1.50 is ground and polished at both ends to hemispherical surfaces of 5 cm radius. When an object is placed on the axis of the rod, 20 cm from one end, the final image is formed 40 cm from the opposite end. What is the length of the rod?

40-6 A solid glass sphere of radius R and index 1.50 is silvered over one hemisphere, as in Fig. 40-25. A small object is located on the axis of the sphere at a distance $2R$ from the pole of the unsilvered hemisphere. Find the position of the final image after all refractions and reflections have taken place.

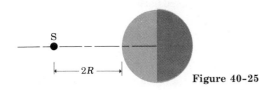

Figure 40-25

40-7 A narrow beam of parallel rays enters a solid glass sphere in a radial direction. At what point outside the sphere are these rays brought to a focus? The radius of the sphere is 3 cm and its index is 1.50.

40-8 A glass plate 2 cm thick, of index 1.50, having plane parallel faces, is held with its faces horizontal and its lower face 8 cm above a printed page. Find the position of the image of the page, formed by rays making a small angle with the normal to the plate.

40-9

a) Show that the equation

$$\frac{1}{s} + \frac{1}{s'} = \frac{1}{f}$$

is that of an equilateral hyperbola having as asymptotes the lines $x = f$ and $y = f$.

b) Construct a graph with object distance s as abscissa, and image distance s' as ordinate for a lens of focal length f, and for object distances from 0 to ∞.

c) On the same set of axes, construct a graph of magnification (ordinate) vs. object distance.

40-10 A converging lens has a focal length of 10 cm. For object distances of 30 cm, 20 cm, 15 cm, and 5 cm determine

a) image position,

b) magnification,

c) whether the image is real or virtual,

d) whether the image is erect or inverted.

40-11 Sketch the various possible thin lenses obtainable by combining two surfaces whose radii of curvature are, in absolute magnitude, 10 cm and 20 cm. Which are converging and which are diverging? Find the focal length of each lens if made of glass of index 1.50.

40-12 The radii of curvature of the surfaces of a thin lens are +10 cm and +30 cm. The index is 1.50.

a) Compute the position and size of the image of an object in the form of an arrow 1 cm high, perpendicular to the lens axis, 40 cm to the left of the lens.

b) A second similar lens is placed 160 cm to the right of the first. Find the position of the final image.

c) Same as (b), except the second lens is 40 cm to the right of the first.

d) Same as (c), except the second lens is diverging, of focal length −40 cm.

40-13 An object is placed 18 cm from a screen.

a) At what points between object and screen may a lens of 4 cm focal length be placed to obtain an image on the screen?

b) What is the magnification of the image for these positions of the lens?

40-14 An object is imaged by a lens on a screen placed 12 cm from the lens. When the lens is moved 2 cm farther from the object, the screen must be moved 2 cm closer to the object to refocus it. What is the focal length of the lens?

40-15 Three thin lenses, each of focal length 20 cm, are aligned on a common axis and adjacent lenses are separated by 30 cm. Find the position of the image of a small object on the axis, 60 cm to the left of the first lens.

40-16 An equiconvex thin lens made of glass of index 1.50 has a focal length in air of 30 cm. The lens is sealed into an opening in one end of a tank filled with water (index = 1.33). At the end of the tank opposite the lens is a plane mirror, 80 cm distant from the lens. Find the position of the image formed by the lens-water-mirror system, of a small object outside the tank on the lens axis and 90 cm to the left of the lens. Is the image real or virtual, erect or inverted?

40-17 A diverging meniscus lens of 1.48 refractive index has concave spherical surfaces whose radii are 2.5 and 4 cm. What would be the position of the image if an object were placed 15 cm in front of the lens?

40-18 Figure 40-26 represents a compound lens consisting of two thin lenses L_1 and L_2, each of focal length 6 cm, separated by a distance of 3 cm.

a) Find the position of the image formed by lens L_1 of a point on the axis at an infinite distance to the left of the lens.

b) Find the position of the image of this image formed by lens L_2. This locates the second focal point F' of the compound lens and, by symmetry, the first focal point F lies at the same distance to the left of L_1.

c) Find the position of the image formed by L_1 of a point on the axis at a distance $x = 2$ cm to the left of the first focal point F.

d) Find the distance x' from the second focal point F' to the image of this image formed by lens L_2.

e) What is the focal length f of the compound lens?

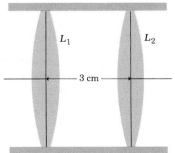

L_1 L_2

3 cm

Figure 40-26

40-19 A lens having one convex and one plane surface is 2 cm thick along its axis. The refractive index is 1.50 and the radius of curvature of the convex surface is 10 cm. The convex surface faces toward the left.

a) Find the distance from the first focal point to the vertex of the convex surface.

b) Find the distance from the vertex on the plane surface to the second focal point.

40-20 When an object is placed at the proper distance in front of a converging lens, the image falls on a screen 20 cm from the lens. A diverging lens is now placed halfway between the converging lens and the screen, and it is found that the screen must be moved 20 cm farther away from the lens to obtain a sharp image. What is the focal length of the diverging lens?

40-21

a) Prove that when two thin lenses of focal lengths f_1 and

f_2 are placed *in contact,* the focal length f of the combination is given by the relation

$$\frac{1}{f} = \frac{1}{f_1} + \frac{1}{f_2}.$$

b) A converging meniscus lens has an index of refraction of 1.50, and the radii of its surfaces are 5 and 10 cm. The concave surface is placed upward and filled with water. What is the focal length of the water–glass combination?

40-22 Two thin lenses, both of 10 cm focal length, the first converging, the second diverging, are placed 5 cm apart. An object is placed 20 cm in front of the first (converging) lens.

a) How far from this lens will the image be formed?

b) Is the image real or virtual?

40-23 Rays from a lens are converging toward a point image P, as in Fig. 40-27. What thickness t of glass of index 1.50 must be interposed, as in the figure, in order that the image shall be formed at P'?

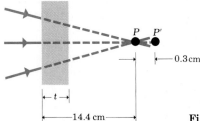

P P'

0.3 cm

t

14.4 cm   **Figure 40-27**

40-24 An object 2.4 m in front of a camera lens is sharply imaged on a photographic film 12 cm behind the lens. A glass plate 1 cm thick, of index 1.50, having plane parallel faces, is interposed between lens and plate, as shown in Fig. 40-28.

a) Find the new position of the image.

b) At what distance in front of the lens will an object be in sharp focus on the film with the plate in place, the distance from lens to film remaining 12 cm? Consider the lens as a simple thin lens.

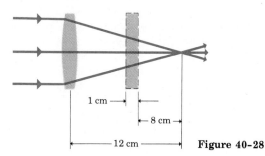

1 cm

8 cm

12 cm   **Figure 40-28**

40-25 The picture size on ordinary 35-mm camera film is 24×36 mm. Focal lengths of lenses available for 35-mm cameras typically include 28 mm, 35 mm, 50 mm (the "standard" lens), 85 mm, 100 mm, 135 mm, 200 mm, and 300 mm, among others. Which of these lenses should be used to photograph the following objects, assuming the object is to fill most of the picture area?

a) A cathedral 100 m high and 150 m long, at a distance of 150 m?

b) An eagle with a wingspan 2.0 m, at a distance of 15 m?

40-26 During a lunar eclipse, a picture of the moon (diameter 3.48×10^6 m, distance from earth 3.8×10^8 m) is taken with a camera whose lens has focal length 50 mm. What is the diameter of the image on the film?

40-27 The *resolution* of a camera lens can be defined as the maximum number of lines per millimeter in the image that can barely be distinguished as separate lines. A certain lens has a focal length of 50 mm and resolution of 100 lines mm^{-1}. What is the minimum separation of two lines in an object 100 m away if they are to be visible in the image as separate lines?

40-28 Show that when two thin lenses are placed in contact, the *power* of the combination in diopters, as defined in Sec. 40-8, is the sum of the powers of the separate lenses. Is this relation valid even when one lens has positive power and the other negative?

40-29 An eyepiece consists of two similar positive thin lenses having focal lengths of 6 cm, separated by a distance of 3 cm. Where are the focal points of the eyepiece?

40-30 When two thin lenses are closely spaced, the power of the combination is the sum of the powers of the individual lenses. Two thin lenses of 25 cm and 40 cm focal lengths are in contact. What is the power of the combination?

40-31 What is the power of the spectacles required (a) by a hyperopic eye whose near point is at 125 cm? (b) by a myopic eye whose far point is at 50 cm?

40-32

a) What spectacles are required for reading purposes by a person whose near point is at 200 cm?

b) The far point of a myopic eye is at 30 cm. What spectacles are required for distant vision?

40-33

a) Where is the near point of an eye for which a spectacle lens of power +2 diopters is prescribed?

b) Where is the far point of an eye for which a spectacle lens of power −0.5 diopter is prescribed for distant vision?

40-34 A thin lens of focal length 10 cm is used as a simple magnifier.

a) What angular magnification is obtainable with the lens?

b) When an object is examined through the lens, how close may it be brought to the eye?

40-35 The focal length of a simple magnifier is 10 cm.

a) How far in front of the magnifier should an object to be examined be placed if the image is formed at the observer's near point, 25 cm in front of his eye?

b) If the object is 1 mm high, what is the height of its image formed by the magnifier? Assume the magnifier to be a thin lens.

40-36 A camera lens is focused on a distant point source of light, the image forming on a screen at a (Fig. 40-29). When the screen is moved backward a distance of 2 cm to b, the circle of light on the screen has a diameter of 4 mm. What is the f/number of the lens?

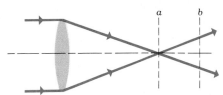

Figure 40-29

40-37 Camera A, having an $f/8$ lens 2.5 cm in diameter, photographs an object using the correct exposure of $\frac{1}{100}$ s. What exposure should camera B use in photographing the same object if it has an $f/4$ lens 5 cm in diameter?

40-38 The focal length of an $f/2.8$ camera lens is 8 cm.

a) What is the diameter of the lens?

b) If the correct exposure of a certain scene is $\frac{1}{200}$ s at $f/2.8$, what would be the correct exposure at $f/5.6$?

40-39 The dimensions of the picture on a 35-mm color slide are 24 mm × 36 mm. It is desired to project an image of the slide, enlarged to 2 m × 3 m, on a screen 10 m from the projection lens.

a) What should be the focal length of the projection lens?

b) Where should the slide be placed?

40-40 The image formed by a microscope objective of focal length 4 mm is 180 mm from its second focal point. The eyepiece has a focal length of 31.25 mm.

a) What is the magnification of the microscope?

b) The unaided eye can distinguish two points as separate if they are about 0.1 mm apart. What is the minimum separation, using this microscope?

40-41 A certain microscope is provided with objectives of focal lengths 16 mm, 4 mm, and 1.9 mm, and with eyepieces of angular magnification 5X and 10X. What is (a) the largest, and (b) the least overall magnification obtainable? Each objective forms an image 160 mm beyond its second focal point.

40-42 The focal length of the eyepiece of a certain microscope is 2.5 cm. The focal length of the objective is 16 mm. The distance between objective and eyepiece is 22.1 cm. The final image formed by the eyepiece is at infinity. Treat all lenses as thin.

a) What should be the distance from the objective to the object viewed?

b) What is the linear magnification produced by the objective?

c) What is the overall magnification of the microscope?

40-43 A microscope with an objective of focal length 9 mm and an eyepiece of focal length 5 cm is used to project an image on a screen 1 m from the eyepiece. What is the lateral magnification of the image? Let the image distance of the objective be 18 cm.

40-44 The moon subtends an angle at the earth of approximately $\frac{1}{2}°$. What is the diameter of the image of the moon produced by the objective of the Lick Observatory telescope, a refractor having a focal length of 18 m?

40-45 The eyepiece of a telescope has a focal length of 10 cm. The distance between objective and eyepiece is 2.1 m. What is the angular magnification of the telescope?

40-46 A crude telescope is constructed of two spectacle lenses of focal lengths 100 cm and 20 cm, respectively.

a) Find its angular magnification.

b) Find the height of the image formed by the objective of a building 80 m high and distant 2 km.

40-47 Figure 40–30 is a diagram of a *Galilean telescope,* or *opera glass,* with both the object and its final image at infinity. The image I serves as a virtual object for the eyepiece. The final image is virtual and erect. Prove that the angular magnification $M = -f_1/f_2$.

40-48 A Galilean telescope is to be constructed, using the same objective as in Problem 40–46.

a) What type lens should be used as an eyepiece and what focal length should it have, if the telescopes are to have the same magnification?

b) Compare the lengths of the telescopes.

40-49 A reflecting telescope is made using a mirror of radius of curvature 0.50 m and an eyepiece of focal length 1.0 cm. What is the angular magnification? What should be the position of the eyepiece if both the object and the final image are at infinity?

40-50 A certain reflecting telescope has a mirror 10 cm in diameter, with radius of curvature 1.0 m, and an eyepiece of focal length 1.0 cm. If the angular magnification is 48 and the object is at infinity, find the position of the lens and the position and nature of the final image.

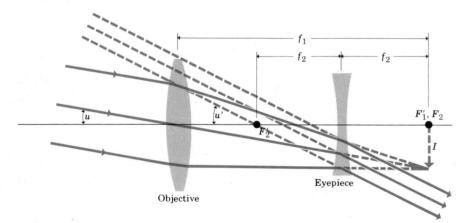

Figure 40–30

INTERFERENCE AND DIFFRACTION 41

41-1 Interference and coherent sources

In analyzing the formation of images by lenses and mirrors, we have represented light as *rays* which travel in straight lines in a homogeneous medium and which are deviated in accordance with simple laws at a reflecting surface or an interface between two optical media. This simple model forms the basis of *geometrical optics,* which, as we have seen, is adequate for understanding a wide variety of phenomena involving lenses and mirrors.

In this chapter we shall discuss the phenomena of *interference* and *diffraction,* for whose understanding the principles of geometrical optics *do not* suffice. Instead, we must return to the more fundamental point of view that light is a *wave motion,* and that the total effect of a number of waves arriving at one point depends on the *phases* of the waves as well as upon their amplitudes. This part of the subject is called *physical optics.*

In our discussions of mechanical waves in Chapter 21 and electromagnetic waves in Chapter 37, we have often considered *sinusoidal* waves having a single frequency and a single wavelength. Such a wave is called a *monochromatic* (single-color) wave. Common sources of light, such as an incandescent lightbulb or a flame, *do not* emit monochromatic light but rather a continuous distribution of wavelengths. Indeed, a strictly monochromatic light wave is an unattainable idealization, like a frictionless plane, a point mass, an inductor with zero resistance, and many other idealizations in physics. Nevertheless, it remains a useful concept for the analysis of phenomena of light interference.

Monochromatic light can be *approximated* in the laboratory. Continuous-spectrum light can be passed through a filter that blocks all but a narrow range of wavelengths. Gas-discharge lamps, such as the mercury-arc lamp, emit line spectra in which the light consists of a discrete set of colors, each having a narrow band of wavelengths called a *spectrum line.* For example, the bright green line in the mercury spectrum has an average wavelength of 546.1 nm, with a spread of wavelength of the order of ±0.001 nm, depending on the pressure and temperature of the mercury vapor in the lamp. By far the most nearly monochromatic source available at present is the *laser,* to be discussed in Chapter 44. The familiar helium–neon laser, inexpensive and readily available, emits visible light at 632.8 nm with a line width (wavelength range) of the order of ± 0.000001 nm, or about one part in 10^9. Laser light also has much greater *coherence* (to be discussed later) than ordinary light.

The term *interference* refers to any situation in which two or more waves overlap in space. This term was introduced in Sec. 22–2 in connection with standing waves on a stretched string, formed by the superposition of two sinusoidal waves traveling in opposite directions. In such cases, the total displacement at any point at any instant of time is governed by the *principle of linear superposition,* introduced in Sec. 22–2. This principle, the most important in all of physical optics, states that *when two or more waves overlap, the resultant displacement at any point and at any instant may be found by adding the instantaneous displacements that would be produced at the point by the individual waves if each were present alone.* The term "displacement," as used here, is a general one. If one is considering surface ripples on a liquid, the displacement means the actual displacement of the surface above or below its normal level. If the waves are sound waves, the term refers to the excess or deficiency of pressure. If the waves are electromagnetic, the displacement means the magnitude of the electric or magnetic field. When light of extremely high intensity passes through matter, the principle of linear superposition is *not* precisely obeyed, and the resulting phenomena are classified under the heading *nonlinear optics.* (These effects are beyond the scope of this book.)

To introduce the essential ideas of interference, we consider first the problem of two identical sources of monochromatic waves, S_1 and S_2, separated in space by a certain distance. The two sources are permanently *in phase,* so that at every point in space there is a definite and unchanging phase relation between waves from the two sources. They might be, for example, two loudspeakers driven by the same amplifier, or two radio antennas powered by the same transmitter, or two small apertures in an opaque screen, illuminated by the same monochromatic light source.

We locate the sources S_1 and S_2 along the y-axis and equidistant from the origin, as shown in Fig. 41–1. Let P_0 be any point on the x-axis. From symmetry, the two distances S_1P_0 and S_2P_0 are equal; waves from the two sources thus require equal times to travel to P_0, and having left S_1 and S_2 in phase, they arrive at P_0 in phase. The total amplitude at P_0 is thus twice the amplitude of each individual wave. Next, we consider a point P_1, located so that its distance from S_2 is exactly one wavelength greater than its distance from S_1. That is,

$$S_2P_1 - S_1P_1 = \lambda.$$

Then any given wave crest from S_1 arrives at P_1 exactly one cycle earlier than the crest emitted at the same time from S_2, and again the two waves arrive in phase. Similarly, waves arrive in phase at all points P_2 for which the path difference is two wavelengths $(S_2P_2 - S_1P_2 = 2\lambda)$, or indeed for *any* positive or negative integer number of wavelengths.

The addition of amplitudes that results when waves from two or more sources arrive at a point in a phase is often called *constructive interference* or *reinforcement,* and the above discussion shows that con-

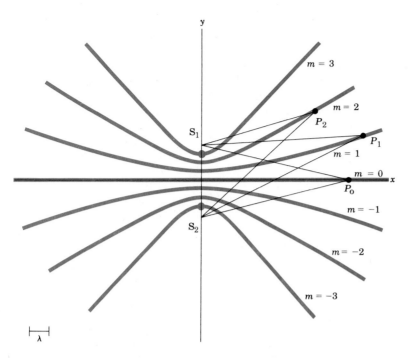

41-1 Curves of maximum intensity in the interference pattern of two monochromatic point sources. In this example the distance between sources is four times the wavelength.

structive interference occurs whenever the path difference for the two sources is an integer multiple of the wavelength:

$$S_2P - S_1P = m\lambda \qquad (m = 0, \pm 1, \pm 2, \pm 3, \ldots). \qquad (41\text{-}1)$$

In our example, the locus of points satisfying this condition is a set of hyperbolas, as shown in Fig. 41–1. Intermediate between these lines is a set of other lines for which the path difference for the two sources is a half-integer number of wavelengths. Waves from the two sources arrive at a point on one of these lines exactly a half-cycle out of phase, and the resultant amplitude is the *difference* of the two individual amplitudes. If the amplitudes are equal, which is approximately the case when the distance from either source to P is much greater than the distance between sources, then the total amplitude at such a point is zero! This condition is called *destructive interference* or *cancellation*. In our example, the condition for destructive interference is

$$S_2P - S_1P = (m + \tfrac{1}{2})\lambda \qquad (m = 0, \pm 1, \pm 2, \pm 3, \ldots). \qquad (41\text{-}2)$$

Thus far we have considered only points in the *xy*-plane, but it should be clear that the discussion can be extended to include other points in the space surrounding the two sources. If the lines in Fig. 41–1 are rotated about the *y*-axis, they trace out surfaces called *hyperboloids of revolution,* and every point on such a surface is a point of maximum intensity of radiation. Similarly, the intermediate lines of zero intensity generate surfaces on which the intensity is zero everywhere.

In the above discussion the constant *phase* relationship between the sources is an essential requirement. If the relative phase of the sources changes, the positions of the maxima and minima in the resulting interference pattern also change. When the radiation is light, it is possible for the two sources to have a definite and constant phase relation *only when they both emit light coming from a single primary source;* it is *not* possible with two separate sources. The reason is a fundamental one associated with the mechanism of light emission.

In ordinary light sources, atoms of the material of the source are given excess energy by thermal agitation or impact with accelerated electrons. An atom thus "excited" begins to radiate and continues until it has lost all the energy it can, typically in a time of the order of 10^{-8} s. A source ordinarily contains a very large number of atoms, which radiate in an unsynchronized and random phase relationship. Thus emission from two such sources has a rapidly varying phase relation; the result is a constantly changing interference pattern which, with ordinary observations, does not reveal a visible interference pattern at all.

However, if the light from a single source is split so that parts of it emerge from two or more regions of space, forming two or more *secondary sources*, any random phase change in the source affects these secondary sources equally and does not change their *relative* phase. Two such sources derived from a single source and having a definite phase relation are said to be *coherent*.

The distinguishing feature of light from a *laser* is that the emission of light from many atoms is *synchronized* in frequency and phase, by mechanisms to be discussed in Chapter 44. As a result, the random phase changes mentioned above occur *much* less frequently. Definite phase relations are preserved over correspondingly much greater lengths in the beam. Accordingly, laser light is said to be much more *coherent* than ordinary light.

41-2 Young's experiment and Pohl's experiment

One of the earliest demonstrations of the fact that light can produce interference effects was performed in 1800 by the English scientist Thomas Young. The experiment was a crucial one at the time, since it added further evidence to the growing belief in the wave nature of light. A corpuscular theory was quite inadequate to account for the effects observed.

Young's apparatus is shown in Fig. 41–2a. Monochromatic light issuing from a narrow slit S_0 is divided into two parts by falling upon a screen in which are cut two other narrow slits S_1 and S_2, very close together. The dimensions in this figure are distorted for clarity. The distance from the source slit S_0 to the screen containing S_1 and S_2 is 20 cm to 100 cm. The distance from the double-slit screen to the final screen is usually from 1 m to 5 m. The slits are 0.1 mm to 0.2 mm wide

41-2 (a) Interference of light waves passing through two slits. (b) Young's experiment.

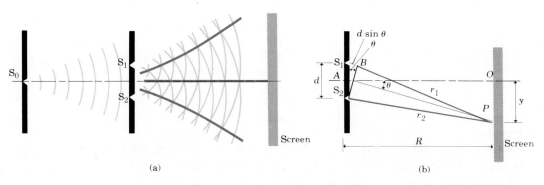

(a)   (b)

and the separation of the slits S_1 and S_2 is less than 1 mm. In short, all slit widths and slit separations are fractions of millimeters; all other distances are hundreds or thousands of millimeters.

According to Huygens' principle (Sec. 38–8), cylindrical wavelets spread out from slit S_0 and reach slits S_1 and S_2 in phase, because they travel equal distances from S_0. A succession of Huygens wavelets diverges from each slit; and the two sets of wavelets leave in phase, and therefore act as *coherent* sources. But they do not necessarily arrive at point P in phase because of the path difference $(r_1 - r_2)$ for the two waves.

When the distance R is very large compared to the distance d between slits, as is usually the case, the path difference can be expressed simply in terms of the angle θ. With P as a center and PS_2 as radius, we draw an arc which intersects line PS_1 at point B. The path difference is then the distance S_1B. When R is much larger than d, the arc S_2B is very nearly a straight line, and it is perpendicular to PS_2, PA, and PS_1. Then triangle BS_1S_2 is a right triangle, similar to POA, and the distance S_1B is equal to $d \sin \theta$. Thus the path difference is given by

$$r_1 - r_2 = d \sin \theta. \tag{41-3}$$

According to the principles discussed in the preceding section, complete reinforcement occurs at point P (that is, P lies at the center of a bright fringe) only when the path difference $d \sin \theta$ is some integral number of wavelengths, say $m\lambda$ $(m = 0, 1, 2, 3,$ etc.). Thus,

$$d \sin \theta = m\lambda \qquad \text{or} \qquad \sin \theta = \frac{m\lambda}{d}. \tag{41-4}$$

Now λ is of the order of 5×10^{-5} cm, while d cannot be made much smaller than about 10^{-2} cm. As a rule, only the first five to ten fringes are bright enough to be seen, so that m is at most, say, 10. Therefore the very largest value of $\sin \theta$ is

$$\sin \theta \ (\text{maximum}) = \frac{(10)(5 \times 10^{-5} \text{ cm})}{10^{-2} \text{ cm}} = 0.05,$$

which corresponds to an angle of only 3°. Figure 41–3 shows a typical pattern.

The central bright fringe at point O, or zeroth fringe $(m = 0)$, corresponds to zero path difference, or $\sin \theta = 0$. If point P is at the center of the mth fringe, the distance y_m from the zeroth to the mth fringe is, from Fig. 41–2b,

$$y_m = R \tan \theta_m.$$

Zeroth fringe

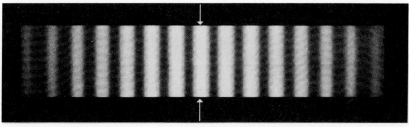

41-3 Interference fringes produced by Young's double-slit interferometer.

If, as is often the case in optical interference experiments, y is much smaller than R, then the angle θ_m for all values of m is extremely small. In that case, $\tan \theta_m \approx \sin \theta_m$ and

$$y_m = R \sin \theta_m.$$

Therefore,

$$y_m = R \frac{m\lambda}{d},$$

and

$$\lambda = \frac{y_m d}{mR}. \tag{41-5}$$

Hence, by measuring the distance d between the slits, the distance R to the screen, and the distance y_m from the center of the zeroth fringe to the center of the mth fringe on either side, one may compute the wavelength of the light producing the interference pattern. Such an experiment, first performed by Thomas Young in 1800, provided the first direct measurement of wavelengths of light, and of course also gave very strong confirmation of the wave nature of light.

Example With two slits spaced 0.2 mm apart, and a screen at a distance of 1 m, the third bright fringe is found to be displaced 7.5 mm from the central fringe. Find the wavelength of the light used.

Solution Let λ be the unknown wavelength. Then

$$\lambda = \frac{y_m d}{mR} = \frac{(0.75 \text{ cm})(0.02 \text{ cm})}{(3)(100 \text{ cm})} = 5 \times 10^{-5} \text{ cm}$$
$$= 500 \times 10^{-9} \text{ m} = 500 \text{ nm}.$$

Circular interference fringes may be produced very easily with the aid of a simple apparatus suggested by Robert Pohl and shown in Fig. 41-4. A small arc lamp S_0 is placed a few centimeters away from a sheet of mica of thickness about 0.05 mm. Some light is reflected from the first surface, as though it were issuing from the virtual image S_1. An approximately equal amount of light is reflected from the back surface, as though it were coming from the virtual image S_2. The circular interference fringes formed by the light issuing from these two mutually coherent sources may be shown on the entire wall of a room, as shown in Fig. 41-4b.

*41-3 Intensity distribution in interference fringes

In Sec. 41-2 we computed the positions of the maxima (bright fringes) in the two-slit interference pattern; we may also compute the intensity at *any* point in the pattern. In Fig. 41-2b, each source produces a sinusoidally varying disturbance at point P; if the sources are in phase, the waves arriving at P differ in phase by an amount proportional to the path difference $(r_1 - r_2)$. If the path difference is one wavelength, the phase difference is $\delta = 2\pi$ (or 360°). If $(r_1 - r_2)/\lambda = \frac{1}{2}$, the phase difference is π, and so on. The general relation is that a path difference $r_1 - r_2$

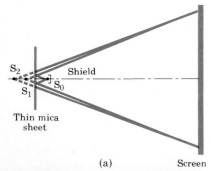

41-4 (a) Pohl's mica-sheet interferometer. (b) Circular interference fringes produced by Pohl's interferometer. The dark rectangle is the shadow of the mercury arc housing.

yields a phase difference δ given by

$$\delta = \frac{2\pi}{\lambda}(r_1 - r_2) = k(r_1 - r_2), \qquad (41\text{-}6)$$

where $k = 2\pi/\lambda$ is the *wave number* introduced in Sec. 21-3.

If the medium in the space between the sources and P is other than vacuum, the wavelength *in the medium* must be used in Eq. (41-5). If the medium has refractive index n, then

$$\lambda = \frac{\lambda_0}{n} \qquad \text{and} \qquad k = nk_0, \qquad (41\text{-}7)$$

where λ_0 and k_0 are the values of λ and k, respectively, in vacuum.

The total wave arriving at P is the superposition of two waves from slits S_1 and S_2, having amplitudes E_1 and E_2, respectively, and with a phase difference δ resulting from their path difference. To find the amplitude of the resulting wave, we represent each wave at P as a rotating vector or *phasor,* just as in Chapters 11 and 36. The appropriate phasor diagram is shown in Fig. 41-5, where the amplitude of the resultant is labeled E_P. Applying the law of cosines, we find

$$E_P{}^2 = E_1{}^2 + E_2{}^2 + 2E_1 E_2 \cos \delta.$$

When the two coherent sources S_1 and S_2 are equally intense, $E_1 = E_2 = E$, and

$$E_P{}^2 = 2E^2 + 2E^2 \cos \delta = 2E^2(1 + \cos \delta)$$

$$= 4E^2 \cos^2 \frac{\delta}{2}. \qquad (41\text{-}8)$$

When the two waves are in phase, $\delta = 0$ and $E_P = 2E$. When they are exactly a half-cycle out of phase, $\delta = \pi$ (180°), $\delta/2 = \pi/2$, $\cos^2(\delta/2) = 0$, and $E = 0$. Thus the superposition of two sinusoidal waves with the same amplitude and frequency but a phase difference yields a sinusoidal wave with amplitude between zero and twice the individual amplitudes, depending on the phase difference.

Now the analysis of electromagnetic waves in Chapter 37, particularly Sec. 37-3, has shown that the *intensity* of an electromagnetic wave is proportional to the *square* of the amplitude of the wave. Thus the intensity is proportional to the final expression in Eq. (41-8). This shows, incidentally, that at points of maximum constructive interference the intensity is *four times* (not twice) as great as it would be from either source alone.

The phase angle δ may be expressed in terms of geometrical quantities in Fig. 41-2 by use of Eqs. (41-3) and (41-6). The path difference is given by

$$r_1 - r_2 = d \sin \theta,$$

provided R is much larger than d. From Eq. (41-6), the phase difference is

$$\delta = k(r_1 - r_2) = kd \sin \theta = \frac{2\pi d}{\lambda}\sin \theta.$$

When this is substituted into Eq. (41-8), we obtain

$$E_P{}^2 = 4E^2 \cos^2(\tfrac{1}{2}kd \sin \theta) = 4E^2 \cos^2\left(\frac{\pi d}{\lambda}\sin \theta\right).$$

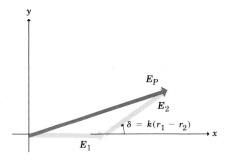

41-5 Phasor diagram of the variations of the electric vector at a point P where two waves meet.

Thus the ratio of intensity I at any angle θ to the intensity I_0 at $\theta = 0$ (where $\delta = 0$) is given by

$$\frac{I}{I_0} = \frac{4E^2 \cos^2(\tfrac{1}{2}kd \sin \theta)}{4E^2 \cos^2(\tfrac{1}{2}kd \sin 0)} = \cos^2(\tfrac{1}{2}kd \sin \theta),$$

and finally

$$I = I_0 \cos^2(\tfrac{1}{2}kd \sin \theta) = I_0 \cos^2\left(\frac{\pi d}{\lambda} \sin \theta\right). \qquad (41\text{-}9)$$

In cases where the further approximation $y/R = \sin \theta$ is justified, i.e., when $y \ll R$ and θ is small, we obtain the simpler expressions

$$I = I_0 \cos^2\left(\frac{kdy}{2R}\right) = I_0 \cos^2\left(\frac{\pi dy}{\lambda R}\right), \qquad (41\text{-}10)$$

which enable us to calculate the intensity of light at any point and to reproduce the pattern of Fig. 41-3.

Example In Fig. 41-1, suppose the two sources are identical radio antennas 10 m apart, radiating waves in all directions with a frequency of $f = 30$ MHz. If the intensity in the $+x$-direction (corresponding to $\theta = 0$ in Fig. 41-2) is $I_0 = 0.02$ W·m^{-2}, what is the intensity in the direction $\theta = 45°$? In what direction is the intensity zero?

Solution We want to use Eq. (41-9); the approximate relation of Eq. (41-10) cannot be used in this case because θ is not small. First we must find the wavelength, using the relation $c = \lambda f$. We find

$$\lambda = \frac{c}{f} = \frac{3.0 \times 10^8 \text{ m·s}^{-1}}{30 \times 10^6 \text{ s}^{-1}} = 10 \text{ m}.$$

The spacing between sources is $d = 10$ m, and Eq. (41-9) becomes

$$I = (0.02 \text{ W·m}^{-2}) \cos^2\left[\frac{\pi(10 \text{ m})}{(10 \text{ m})} \sin \theta\right]$$

$$= (0.02 \text{ W·m}^{-2}) \cos^2(\pi \sin \theta).$$

When $\theta = 45°$,

$$I = (0.02 \text{ W·m}^{-2}) \cos^2(\pi \sin 45°) = 0.0073 \text{ W·m}^{-2}.$$

This is about 37% of the intensity at $\theta = 0$.

The intensity is zero when $\cos(\pi \sin \theta) = 0$; this occurs when $\pi \sin \theta = \pi/2$, $\sin \theta = \tfrac{1}{2}$, and $\theta = 30°$. By symmetry, the intensity is also zero when $\theta = 150°$, $210°$, and $330°$ (or $-150°$, $-30°$).

41-4 Interference in thin films. Newton's rings

The brilliant colors that are often seen when light is reflected from a soap bubble or from a thin layer of oil floating on water are produced by interference effects between the two light waves reflected at opposite surfaces of the thin films of soap solution or of oil. In Fig. 41-6, the line ab is one ray in a beam of monochromatic light incident on the upper surface of a thin film. A part of the incident light is reflected at the first surface, as indicated by ray bc, and a part, represented by bd, is transmitted. At the second surface a part is again reflected, and, of this, a part

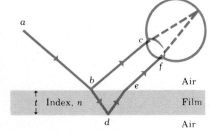

41-6 Interference between rays reflected from the upper and lower surface of a thin film.

emerges as represented by ray *ef*. The rays *bc* and *ef* come together at a point on the retina of the eye. Depending on the phase relationship, they may interfere constructively or destructively. Because different colors have different wavelengths, the interference may be constructive for some colors and destructive for others; hence the appearance of colored rings or fringes.

To discuss these phenomena in as simple a context as possible, let us consider interference of *monochromatic* light reflected from two nearly parallel surfaces. Figure 41–7 shows two plates of glass separated by a wedge of air; we want to consider interference between the two light waves reflected from the surfaces adjacent to the air wedge, as shown. If the observer is at a great distance compared to the other dimensions of the experiment and the rays are nearly perpendicular to the reflecting surfaces, then the path difference between the two waves (corresponding to $r_1 - r_2$ in the preceding section) is just twice the thickness d of the air wedge at each point. At points for which this path difference is an integer number of wavelengths, we expect to see constructive interference and a bright area, and where it is a half-integer number of wavelengths, destructive interference and a dark area. Along the line where the plates are in contact there is *no* path difference and we expect a bright area.

It is easy enough to carry out this experiment; the bright and dark fringes appear as expected, but they are interchanged! Along the line of contact a *dark* fringe, not a bright one, is found. What happened? The inescapable conclusion is that one of the waves has undergone a half-cycle phase shift during its reflection, so that the two reflected waves are a half-cycle out of phase even though they have the same path length.

This phase change can be demonstrated directly with an arrangement called Lloyd's mirror, shown in Fig. 41–8. In this arrangement, the two coherent sources are the actual source slit S_0 and its virtual image S_1. The fringes formed by interference between the light waves from these coherent sources may be viewed on a ground-glass screen placed anywhere beyond the mirror. If, instead, an eyepiece of high magnification is used to view the fringes that form in space in a plane passing through the edge B, the fringe nearest this edge is seen to be black. This is the fringe corresponding to zero path difference, and if the two waves giving rise to this fringe had both traveled in air or had both been reflected from glass, this fringe would have been bright. The zeroth fringe, however, was formed by two waves, of which one had undergone reflection from the glass and one had proceeded directly from S_1. The fact that the zeroth fringe is black indicates that *the waves reflected from glass have undergone a phase shift of* 180°. In other words, the wave has gained (or lost) half a cycle in phase during the process of reflection.

Further experiments show that a half-cycle phase change occurs whenever the material in which the wave is initially traveling before reflection has a smaller refractive index than the second material forming the interface, such as the glass in Fig. 41–8. But when the first material has *greater* refractive index than the second, such as a wave in glass

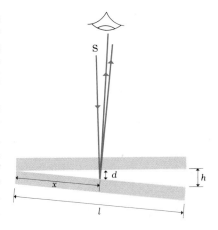

41-7 Interference between two light waves reflected from the two sides of an air wedge separating two glass plates. The path difference is 2*d*.

41-8 Lloyd's mirror. When the eyepiece is focused on the edge *B* of the glass block, the fringe nearest the edge is black.

reflected internally at a glass–air interface, there is found to be *no* phase change. Thus, in Fig. 41–6 the waves reflected at points *b* and *e* undergo the half-cycle phase shift, while that reflected at *d* does not.

Example   Suppose the two glass plates in Fig. 41–7 are two microscope slides 10 cm long; at one end they are in contact, while at the other end they are separated by a thin piece of tissue paper 0.02 mm thick. What is the spacing of the resulting interference fringes? Is the fringe adjacent to the line of contact bright or dark? Assume monochromatic light with $\lambda = 500$ nm.

Solution   To answer the second question first, the fringe at the line of contact is dark because the wave reflected from the lower surface of the air wedge has undergone a half-cycle phase shift, while that from the upper surface has not. For this reason, the condition for *destructive* interference (a dark fringe) is that the path difference ($2d$) be an integer number of wavelengths:

$$2d = m\lambda \quad (m = 0, 1, 2, 3, \ldots). \tag{41–11}$$

From similar triangles, d is proportional to the distance x from the line of contact:

$$\frac{d}{x} = \frac{h}{l}.$$

Combining this with Eq. (41–11), we find

$$\frac{2xh}{l} = m\lambda,$$

or

$$x = m\frac{l\lambda}{2h} = m\frac{(0.1 \text{ m})(500 \times 10^{-9} \text{ m})}{(2)(0.02 \times 10^{-3} \text{ m})} = m(1.25 \text{ mm}).$$

Thus successive dark fringes, corresponding to successive integer values of m, are spaced 1.25 mm apart.

In this example, if the space between plates contains water ($n = 1.33$) instead of air, the phase changes are the same but the wavelength is $\lambda = \lambda_0/n = 376$ nm, and the fringe spacing is 0.94 mm. But now suppose the top plate is a glass with $n = 1.4$, the wedge is filled with a silicone grease having $n = 1.5$, and the bottom plate has $n = 1.6$. In this case there are half-cycle phase shifts at *both* surfaces bounding the wedge, and the line of contact corresponds to a *bright* fringe, not a dark one. The fringe spacing is again obtained using the wavelength in the wedge, $\lambda = 500$ nm$/1.5 = 333$ nm, and is found to be 0.83 nm.

If the arrangement in Fig. 41–7 is illuminated first by blue, then by red light, the spacing of the red fringes is greater than that of the blue, as is to be expected from the greater wavelength of the red light. The fringes produced by intermediate wavelengths occupy intermediate positions. If it is illuminated by white light, the color at any point is that due to the mixture of those colors that may be reflected at that point, while the colors for which the thickness is such as to result in destructive interference are absent. Just those colors that are absent in the reflected light, however, are found to predominate in the transmitted light. At any point, the color of the wedge by reflected light is *complementary* to its color by transmitted light!

If the convex surface of a lens is placed in contact with a plane glass plate, as in Fig. 41-9, a thin film of air is formed between the two surfaces. The thickness of this film is very small at the point of contact, gradually increasing as one proceeds outward. The loci of points of equal thickness are circles concentric with the point of contact. Such a film is found to exhibit interference colors, produced in the same way as the colors in a thin soap film. The interference bands are circular, concentric with the point of contact. When viewed by reflected light, the center of the pattern is black, as is a thin soap film. Note that in this case there is no phase reversal of the light reflected from the upper surface of the film (which here is of smaller index than that of the medium in which the light is traveling before reflection), but the phase of the wave reflected from the lower surface is reversed. When viewed by transmitted light, the center of the pattern is bright. If white light is used, the color of the light reflected from the film at any point is complementary to the color transmitted.

These interference fringes were studied by Newton, and are called *Newton's rings*. Figure 41-10 is a photograph of Newton's rings formed by the air film between a convex and a plane surface.

The surface of an optical part that is being ground to some desired curvature may be compared with that of another surface, known to be correct, by bringing the two in contact and observing the interference fringes. Figure 41-11 is a photograph made at one stage of the process of manufacturing a telescope objective. The lower, larger-diameter, thicker disk is the master. The smaller upper disk is the objective under test. The "contour lines" are Newton's interference fringes, and each one indicates an additional departure of the specimen from the master of $\frac{1}{2}$ wavelength of light. That is, at 10 lines from the center spot the space between the specimen and master is 5 wavelengths, or about 0.0001 inch. This specimen is very poor; high-quality lenses are routinely ground with a precision of less than a wavelength.

41-5 Thin coatings on glass

The phenomenon of interference is utilized in nonreflective coatings for glass. A thin layer or film of hard transparent material with an index of refraction smaller than that of the glass is deposited on the surface of the glass, as in Fig. 41-12. If the coating has the proper index of refraction, equal quantities of light will be reflected from its outer surface and from the boundary surface between it and the glass. Furthermore, since in both reflections the light is reflected from a medium of greater index than that in which it is traveling, the same phase change occurs in each reflection. It follows that if the film thickness is $\frac{1}{4}$ wavelength *in the film* (normal incidence is assumed), the light reflected from the first surface will be 180° out of phase with that reflected from the second, and complete destructive interference will result.

The thickness can, of course, be $\frac{1}{4}$ wavelength for only one particular wavelength. This is usually chosen in the yellow-green portion of the spectrum (about 550 mm) where the eye is most sensitive. Some reflec-

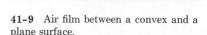

41-9 Air film between a convex and a plane surface.

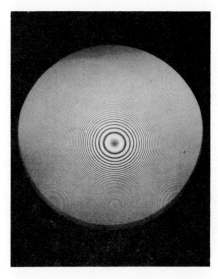

41-10 Newton's rings formed by interference in the air film between a convex and a plane surface. (Courtesy of Bausch & Lomb Optical Co.)

41-11 The surface of a telescope objective under inspection during manufacture. (Courtesy of Bausch & Lomb Optical Co.)

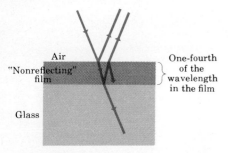

41-12 Destructive interference results when the film thickness is one-quarter of the wavelength in the film.

tion then takes place at both longer and shorter wavelengths, and the reflected light has a purple hue. The overall reflection from a lens or prism surface can be reduced in this way from 4 or 5% to a fraction of 1%. The treatment is extremely effective in eliminating stray reflected light and increasing the contrast in an image formed by highly corrected lenses having a large number of air–glass surfaces. A commonly used material is magnesium fluoride, MgF_2, with an index of 1.38. With this coating, the wavelength of green light in the coating is

$$\lambda = \frac{\lambda_0}{n} = \frac{550 \times 10^{-9} \text{ m}}{1.38} = 4 \times 10^{-5} \text{ cm},$$

and the thickness of a "nonreflecting" film of MgF_2 is 10^{-5} cm.

If a material whose index of refraction is *greater* than that of glass is deposited on glass to a thickness of $\frac{1}{4}$ wavelength, then the reflectivity is *increased*. For example, a coating of index 2.5 will allow 38% of the incident energy to be reflected, instead of the usual 4% when there is no coating. With the aid of multiple coatings it is possible to achieve reflectivity for a particular wavelength of almost 100%.

*41-6 The Michelson interferometer

Young's double slit and Lloyd's mirror are examples of optical interferometers in which a wavefront from a very narrow source slit is subdivided into two wavefronts by reflecting or transmitting regions of the interferometer. Some interferometers can be used in conjunction with a large extended source. Of these, the Michelson interferometer has been most important in the past and is still of some significance.

Figure 41–13 is a diagram of the principal features of the Michelson interferometer. The figure shows the path of one ray from a point A of an extended source. Light from the source strikes a glass plate C, the right side of which has a thin coating of silver. Part of the light is reflected from the silvered surface at point P to the mirror M_2 and back through C to the observer's eye. The remainder of the light passes through the silvered surface and the compensator plate D, and is re-

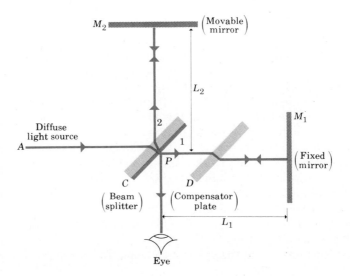

41-13 The Michelson interferometer.

flected from mirror M_1. It then returns through D and is reflected from the silvered surface of C to the observer. The compensator plate D is cut from the same piece of glass as plate C, so that its thickness will not differ from that of C by more than a fraction of a wavelength. Its purpose is to ensure that rays 1 and 2 pass through the same thickness of glass. Plate C is called a *beam splitter*.

The whole apparatus is mounted on a heavy rigid frame and a fine, very accurate screw thread is used to move the mirror M_2. A common commercial model of the interferometer is shown in Fig. 41–14. The source is placed to the left, and the observer is directly in front of the handle that turns the screw.

If the distances L_1 and L_2 in Fig. 41–13 are exactly equal, and the mirrors M_1 and M_2 are exactly at right angles, the virtual image of M_1 formed by reflection at the silvered surface of plate C will coincide with mirror M_2. If L_1 and L_2 are not exactly equal, the image of M_1 will be displaced slightly from M_2; and if the angle between the mirrors is not exactly a right angle, the image of M_1 will make a slight angle with M_2. Then the mirror M_2, and the virtual image of M_1, play the same roles as the two surfaces of a thin film, discussed in Sec. 41–4, and the same sort of interference fringes result from the light that is reflected from these surfaces.

Suppose that the extended source in Fig. 41–13 is monochromatic with wavelength λ, and that the angle between mirror M_2 and the virtual image of M_1 is such that five or six vertical fringes are present in the field of view. If the mirror M_2 is now moved slowly either backward or forward by a distance $\lambda/2$, the effective film thickness will change by λ and each of the fringes will move either to the right or to the left through a distance equal to the spacing of the fringes. If the fringes are observed through a telescope whose eyepiece is equipped with a crosshair, and m fringes cross the crosshair when the mirror is moved a distance x, then

$$x = m\frac{\lambda}{2} \qquad \text{or} \qquad \lambda = \frac{2x}{m}. \tag{41-12}$$

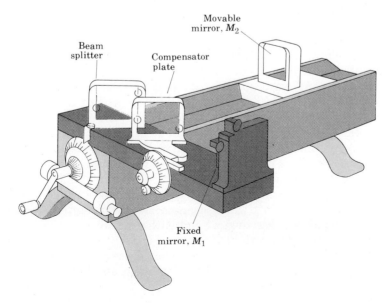

41-14 A common type of Michelson interferometer.

If m is as large as several thousand, the distance x is sufficiently great so that it can be measured with good precision and hence a precise value of the wavelength λ can be obtained.

It was stated in Chapter 1 that the meter is defined as a length equal to a specified number of wavelengths of the orange-red light of krypton-86. Before this standard could be established, it was necessary to measure as accurately as possible the number of these wavelengths in the *former* standard meter, defined as the distance between two scratches on a bar of platinum-iridium. The measurement was made with a modified Michelson interferometer, many times and under very carefully controlled conditions. The number of wavelengths in a distance equal to the old standard meter was found to be 1,650,763.73 wavelengths. The meter is now defined as *exactly* this number of wavelengths.

Another application of the Michelson interferometer of considerable historical interest is the Michelson–Morley experiment. To understand the purpose of this experiment, we must recall that, before the electromagnetic theory of light and Einstein's special theory of relativity became established, physicists believed that light waves were propagated in a medium called *the ether*.

Although the ether was assumed to be very rigid, in order to propagate waves with the enormous speed of light, it was assumed also to be very tenuous, so the planets could move freely through it. A light wave was considered to travel with speed c relative to the ether, just as sound waves in a medium travel with the speed of sound relative to the medium. If a medium is in motion relative to the earth, the velocity of a *sound* wave relative to the earth, v_{WE}, is the vector sum of its velocity relative to the medium , v_{WM}, and the velocity of the medium relative to the earth, v_{ME}:

$$v_{WE} = v_{WM} + v_{ME}. \tag{41–13}$$

This point of view has been substantiated so many times with sound waves and water waves that it seemed obvious that the same results should hold for light waves.

In 1887, Michelson and Morley utilized the Michelson interferometer in an attempt to detect the relative motion of the earth and the ether. Suppose the interferometer in Fig. 41–13 is moving from left to right relative to the ether. According to nineteenth-century theory, this would lead to changes in the speed of light in the horizontal portions of the path, and corresponding fringe shifts relative to the positions the fringes would have if the instrument were at rest in the ether. Then, when the entire instrument was rotated 90°, the vertical paths would be similarly affected, giving a fringe shift in the opposite direction.

A reasonable guess as to the velocity of the earth relative to the ether is that it is of the same order as the orbital velocity of the earth around the sun, about $3 \times 10^4\,\mathrm{m \cdot s^{-1}}$. Michelson and Morley calculated that, for green light, this velocity should cause a fringe shift of about four-tenths of a fringe when the instrument was rotated. The shift actually observed was less than a hundredth of a fringe, and within the limits of experimental uncertainty appeared in fact to be zero! Despite its orbital velocity the earth appeared to be at rest relative to the ether. This negative result baffled physicists of the time, and to this day the Michelson–Morley experiment is the most significant "negative-result" experiment ever performed.

Understanding of this result had to wait for Einstein's special theory of relativity, published in 1905. Einstein realized that the classical equation for combining relative velocities, Eq. (41–13), is only the limiting case of a more general equation, and that this general equation leads to the result that the velocity of a light wave has the same magnitude c relative to *all* reference frames, whatever their velocity may be relative to other frames. Even if there is an ether breeze past an interferometer, this breeze has no effect on the velocity of light relative to the interferometer. The result is that rays 1 and 2 in Fig. 41–13 travel, relative to the interferometer, with the same speed c, regardless of the orientation of the interferometer. There is no path difference, and no shift when the interferometer is rotated. The presumed ether then plays no role, and the very concept of an ether has been given up. The underlying principle of Einstein's theory of special relativity may be stated as follows: *There is no preferred frame of reference for the measurement of the speed of light; the speed is the same for every observer, without regard to the magnitude or direction of his velocity relative to other observers.* This theory, a well-established cornerstone of modern physics, is discussed in detail in Chapter 43.

41-7 Fresnel diffraction

According to *geometrical* optics, if an opaque object is placed between a point light source and a screen, as in Fig. 41–15, the edges of the object cast a sharp shadow on the screen. No light reaches the screen at points within the geometrical shadow, while outside the shadow the screen is uniformly illuminated. Geometrical optics is, however, an idealized model of the behavior of light. There are situations in which the representation in terms of straight-line or ray propagation is inadequate; we now proceed to examine some of these situations. An important class of phenomena in which the ray model of geometrical optics is *not* adequate is grouped under the heading *diffraction*.

The photograph in Fig. 41–16 was made by placing a razor blade halfway between a pinhole illuminated by monochromatic light and a photographic film, so that the film made a record of the shadow cast by the blade. Figure 41–17 is an enlargement of a region near the shadow of an edge of the blade. The boundary of the *geometrical* shadow is indicated by the short arrows. Note that a small amount of light has "bent" around the edge, into the geometrical shadow, which is bordered by alternating bright and dark bands. Note also that in the first bright band, just outside the geometrical shadow, the illumination is actually *greater* than in the region of uniform illumination to the extreme left. This simple experimental setup serves to give some idea of the true complexity of what is often considered the most elementary of optical phenomena, the shadow cast by a small source of light. The distribution of light and dark on any screen after a beam of light has been partially blocked by a perforated diaphragm is called a *diffraction pattern*.

Diffraction patterns such as that in Fig. 41–16 are not commonly observed in everyday life because most ordinary light sources are not

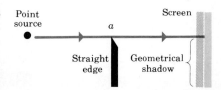

41-15 Geometrical shadow of a straight edge.

41-16 Shadow of a razor blade.

point sources of monochromatic light. If the shadow of a razor blade is cast by a frosted-bulb incandescent lamp, for example, the light from every point of the surface of the lamp forms its own diffraction pattern, but these overlap to such an extent that no individual pattern can be observed.

The term *diffraction* is applied to problems in which one is concerned with *the resultant effect produced by a limited portion of a wave front.* Since in most diffraction problems some light is found within the region of geometrical shadow, diffraction is sometimes defined as "the bending of light around an obstacle." It should be emphasized, however, that the process by which diffraction effects are produced is going on continuously in the propagation of *every* wave. Only if a part of the wave is cut off by some obstacle are the effects commonly called "diffraction effects" observed. But since every optical instrument does, in fact, make use of only a limited portion of a wave (a telescope, for example, utilizes only that portion of a wave admitted by the objective lens), it is evident that a clear comprehension of the nature of diffraction is essential for a complete understanding of practically all optical phenomena.

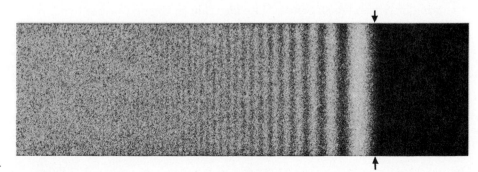

41-17 Shadow of a straight edge.

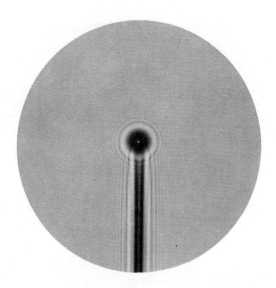

41-18 Fresnel diffraction pattern of a
small circular obstacle. A bright spot is at
the center of the shadow.

A circular *opening* in an opaque screen transmits only a small circular patch of a wave surface; the remainder of the wave is obscured. An
interesting effect is observed if we reverse this procedure and insert a
small circular *obstacle* in the light from a distant point source. A small
circular patch of the wave is then obscured, while the remainder is allowed to proceed. Figure 41–18 is a photograph of the shadow of a small
steel ball, supported from the tip of a magnetized sewing needle. Constructive interference of the wavelets from the unobstructed portion of
the incident wave results in the small bright spot at the center of the
geometrical shadow.

The essential features observed in diffraction effects can be predicted
with the help of Huygens' principle, according to which every point of a
wave surface can be considered the source of a secondary wavelet that
spreads out in all directions. However, instead of finding the new wave
surface by the simple process of constructing the envelope of all the
secondary wavelets, we must combine these wavelets according to the
principles of interference. That is, at every point we must combine the
displacements that would be produced by the secondary wavelets, taking into account their amplitudes and relative phases. The mathematical operations are often quite complicated.

In Fig. 41–15 both the point source and the screen are at large but
finite distances (say several meters) from the obstacle forming the diffraction pattern, and no lenses are used. This situation is described as
Fresnel diffraction (after Augustin Jean Fresnel, 1788–1827), and the
resulting pattern on the screen is called a *Fresnel diffraction pattern*. If
the source is far enough away so that the diffraction pattern appears on
a screen in the second focal plane of the lens, the phenomenon is called
Fraunhofer diffraction (after Joseph von Fraunhofer, 1787–1826.) The
latter situation is simpler to treat in detail, and we shall consider it first.

41-8 Fraunhofer diffraction from a single slit

Suppose a beam of parallel monochromatic light is incident from the left
on an opaque plate having a narrow vertical slit. According to geometri-

(a)   (b)

41-19 (a) Geometrical "shadow" of a slit. (b) Diffraction pattern of a slit. The slit width has been greatly exaggerated.

cal optics, the transmitted beam should have been the same cross section as the slit, and a screen in the path of the beam would be illuminated uniformly over an area of the same size and shape as the slit. What is actually observed is the pattern shown in Fig. 41–19b. The beam spreads out horizontally after passing through the slit, and the diffraction pattern consists of a central bright band, which may be much wider than the slit width, bordered by alternating dark bands and bright bands of decreasing intensity. A diffraction pattern of this nature can readily be observed by looking at a point source such as a distant street light through a narrow slit formed between two fingers in front of the eye. The retina of the eye then corresponds to the screen.

Let us now apply Huygens' principle to compute the distribution of light on the screen. We consider a plane wave front at the moment it reaches the space between the edges of a slit such as that shown in section in Fig. 41–20a. Small elements of area are obtained by subdividing the wave front into narrow strips, parallel to the long edges of the slit, perpendicular to the page. From each of these strips, secondary Huygens wavelets spread out in all directions, as shown.

In Fig. 41–20b, a screen is placed at the right of the slit and P is one point on a line in the screen, the line being parallel to the long edges of the slit and perpendicular to the plane of the diagram. The light reaching a point on the line is calculated by applying the principle of superposition to all the wavelets arriving at the point, from all the elementary strips of the original wave surface. Because of the varying distances to the point, and the varying angles with the original direction of the light, the amplitudes and phases of the wavelets at the point will be different.

The problem is greatly simplified when the screen is sufficiently distant, or the slit sufficiently narrow, so that all rays from the slit to a point on the screen can be considered parallel, as in Fig. 41–20c. The former case, where the screen is relatively close to the slit (or the slit is relatively wide) is Fresnel diffraction, whereas the latter is Fraunhofer

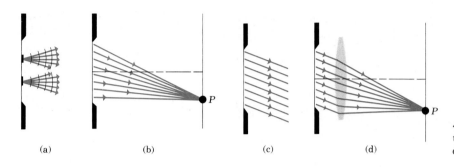

(a)   (b)   (c)   (d)

41-20 Diffraction by a single slit. Relation of Fresnel diffraction to Fraunhofer diffraction by a single slit.

diffraction. There is, of course, no difference in the *nature* of the diffraction process in the two cases, and Fresnel diffraction merges gradually into Fraunhofer diffraction as the screen is moved away from the slit, or as the slit width is decreased.

Fraunhofer diffraction occurs also if a lens is placed just beyond the slit, as in Fig. 41-20d, since the lens brings to a focus, in its second focal plane, all light traveling in a specified direction. That is, the lens forms in its focal plane a reduced *image* of the pattern that would appear on an infinitely distant screen in the absence of the lens.

When a lens is used in this manner to observe a diffraction pattern, the question arises whether the lens may introduce additional phase shifts that are different for different parts of the wave front. If it does, the interference phenomena that result may be quite different from those we have described. But it can be shown from very general considerations that *no* such additional phase shifts occur. For example, when a beam of parallel rays is brought to a focus by a converging lens or a concave mirror, the wavelets that are in phase across a plane perpendicular to the beam direction are still in phase when they arrive at the focal point after reflection or refraction. Alternatively, one may say that all rays require the same *time* to travel from a given cross-section plane perpendicular to the beam to the focal point. Thus the lenses in Figs. 41-20, 41-21, and 41-24 do not introduce additional phase shifts.

Some aspects of Fraunhofer diffraction from a single slit can be deduced easily. We first consider two narrow strips, one just below the top edge of the slit and one at its center. The difference in path length to point P in Fig. 41-21 is $(a/2) \sin \theta$, where a is the slit width. Suppose this

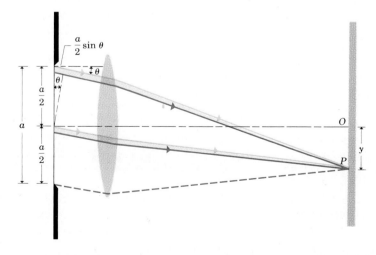

41-21 The wavefront is divided into a large number of narrow strips.

path difference happens to be equal to $\lambda/2$; then light from these two strips arrives at point P with a half-cycle phase difference, and cancellation occurs. Similarly, light from two strips just below these two will also arrive a half-cycle out of phase; and, in fact, light from *every* strip in the top half cancels out that from a corresponding strip in the bottom half, resulting in complete cancellation, and giving a dark fringe in the interference pattern. Thus, a dark fringe occurs whenever

$$\left(\frac{a}{2}\right)\sin\theta = \frac{\lambda}{2} \quad \text{or} \quad \sin\theta = \frac{\lambda}{a}. \tag{41-14}$$

We may also divide the screen into quarters, sixths, and so on, and use the above argument to show that a dark fringe occurs whenever $\sin\theta = 2\lambda/a$, $3\lambda/a$, and so on. Thus the condition for a *dark* fringe is

$$\sin\theta = \frac{n\lambda}{a}, \quad \text{where } n = 1, 2, 3, \ldots \tag{41-15}$$

For example, if the slit width is equal to ten wavelengths, dark fringes occur at $\sin\theta = \frac{1}{10}, \frac{2}{10}, \frac{3}{10}, \ldots$ Midway between the dark fringes are bright fringes. We also note that $\sin\theta = 0$ is a *bright* band, since then light from the entire slit arrives at P in phase. Thus the central bright fringe is twice as wide as the others, as Fig. 41-19 shows.

The complete intensity distribution for the single-slit pattern may be calculated by the same method used to obtain Eq. (41-9) for the two-slit pattern. We again imagine a plane wavefront at the slit subdivided into a large number of strips each of which sends out rays in *all* directions toward the lens, shown in Fig. 41-21. If we choose an arbitrary point P on a screen in the focal plane of the lens, only those rays making an angle θ with the axis of the lens will arrive at P. Point O on the screen is the special point at which all rays making the angle $\theta = 0$ arrive. Figure 41-22a is a phasor diagram, showing that when the slit is subdivided into 14 sections and each section emits a Huygens wavelet at the angle $\theta = 0$, all wavelets arrive in phase. The resultant amplitude at O is denoted by S.

With the same subdivision of the wavefront into 14 strips, the wavelets that make the angle θ and arrive at P have a slight phase difference between succeeding wavelets, and the corresponding phasor diagram is shown in Fig. 41-22b. The sum S is now the perimeter of a portion of a many-sided polygon and E_P, the amplitude of the resultant electric intensity at P, is the *chord*. The angle δ is the total phase difference between the wave from the bottom strip of Fig. 41-21 and the wave from the top strip.

In the limit, as the number of strips into which the slit is subdivided is increased, the phasor diagram becomes an *arc of a circle*, as shown in Fig. 41-22c, with the length of arc S equal to a constant. By constructing perpendiculars at A and B, the center of the circular arc C is found. The radius of the circle of which S is an arc is S/δ, and the resultant amplitude E_P (distance AB) is $2(S/\delta)\sin\delta/2$. We have then

$$E_P = S\frac{\sin\delta/2}{\delta/2},$$

where δ is the phase difference between the two rays at the extreme edges of the slit.

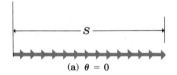

(a) $\theta = 0$

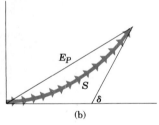

(b)

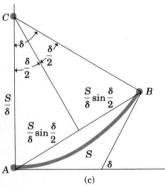

(c)

41-22 (a) Phasor diagram when all elementary electric intensities are in phase ($\theta = 0$, $\delta = 0$). (b) Phasor diagram when each elementary electric intensity differs in phase slightly from the preceding one. (c) Limit reached by the phasor diagram when the slit is subdivided infinitely.

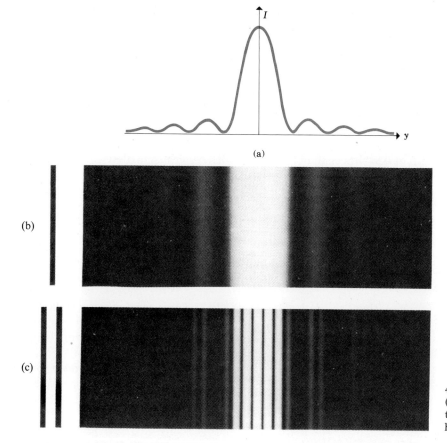

(a)

(b)

(c)

41-23 (a) Intensity distribution. (b) Photograph of the Fraunhofer diffraction pattern of a single slit. (c) Fraunhofer diffraction pattern of a double slit.

From Eq. (41–6), the phase difference is $2\pi/\lambda$ times the path difference. From Fig. 41–21, the path difference between the ray from the top of the slit and that from the bottom is $a \sin \theta$. Therefore

$$\delta = \frac{2\pi}{\lambda} a \sin \theta, \qquad (41\text{--}16)$$

and

$$E_P = S \frac{\sin[\pi a (\sin \theta)/\lambda]}{\pi a (\sin \theta)/\lambda}.$$

Since the intensity I is proportional to the *square* of the amplitude,

$$I = I_0 \left\{ \frac{\sin[\pi a (\sin \theta)/\lambda]}{\pi a (\sin \theta)/\lambda} \right\}^2, \qquad (41\text{--}17)$$

where I_0 is the intensity at O in Fig. 41–21 where $\theta = 0$.

Equation (41–17) is plotted in Fig. 41–23a, and a photograph of the diffraction pattern is shown directly underneath. Note that most of the light exists in the region close to the point where $\theta = 0$, or the geometrical focus. From Eq. (41–17), the smallest value of $\pi a (\sin \theta)/\lambda$ at which the intensity becomes zero is the value π. This corresponds to a value of θ equal to θ_1 and given by

$$\sin \theta_1 = \frac{\lambda}{a}.$$

Since λ is of the order of 5×10^{-5} cm, and a typical slit width is 10^{-2} cm, $\sin \theta_1$ is ordinarily so small that $\sin \theta_1 = \theta_1$, and

$$\theta_1 = \frac{\lambda}{a}.$$

When a is several centimeters, θ_1 is so small that one can consider practically all the light to be concentrated at the geometrical focus.

The photograph in Fig. 41-23c shows the Fraunhofer diffraction pattern of *two* slits each of the same width a as the one above, but separated by a distance $d = 4a$. Note that the interference fringes due to cooperation between the slits have intensities that follow the diffraction pattern of each separate slit. The "cosine-squared" interference fringes are modulated by the shape of the curve shown in part (a) of the figure, because of the finite width of the slits S_1 and S_2.

41-9 The diffraction grating

Suppose that, instead of a single slit, or two slits side by side as in Young's experiment, we have a very large number of parallel slits, all of the same width and spaced at regular intervals. Such an arrangement, known as a *diffraction grating,* was first constructed by Fraunhofer. The earliest gratings were of fine wires, 0.04 mm to 0.6 mm in diameter. Gratings are now made by using a diamond point to rule a large number of equidistant grooves on a glass or metal surface.

Let GG, in Fig. 41-24, represent the grating, the slits of which are perpendicular to the plane of the paper. While only five slits are shown in the diagram, an actual grating contains several thousand, with a grating spacing d of the order of 0.002 mm. Let a train of plane waves be incident normally on the grating from the left. The problem of finding the intensity of the light transmitted by the grating then combines the principles of interference and diffraction. That is, each slit gives rise to a diffracted beam whose nature, as we have seen, depends on the slit width. These diffracted beams then interfere with one another to produce the final pattern.

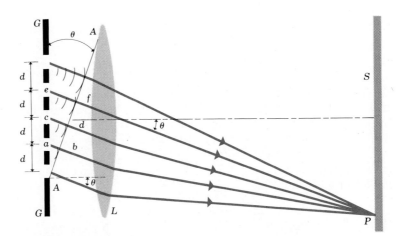

41-24 The plane diffraction grating.

Let us assume that the slits are so narrow that the diffracted beam from each spreads out over a sufficiently wide angle for it to interfere with all the other diffracted beams. Consider first the light proceeding from elements of infinitesimal width at the lower edges of each opening, and traveling in a direction making an angle θ with that of the incident beam, as in Fig. 41-24. A lens at the right of the grating forms in its focal plane a diffraction pattern similar to that which would appear on a screen at infinity.

Suppose the angle θ in Fig. 41-24 is taken so that the distance ab equals λ, the wavelength of the incident light. Then $cd = 2\lambda$, $ef = 3\lambda$, etc. The waves from all of these elements, since they are in phase at the plane of the grating, are also in phase along the plane AA and therefore reach the point P in phase. The same holds true for any set of elements in corresponding positions in the various slits.

If the angle θ is increased slightly, the disturbances from the grating elements no longer arrive at AA in phase with one another, and even an extremely small change in angle results in almost complete destructive interference among them, provided there is a large number of slits in the grating. Hence, the maximum at the angle θ is an extremely sharp one, differing from the rather broad maxima and minima that result from interference or diffraction effects with a small number of openings.

As the angle θ is increased still further, a position is eventually reached in which the distance ab in Fig. 41-24 becomes equal to 2λ. Then cd equals 4λ, cf equals 6λ, and so on. The disturbances at AA are again all in phase, the path difference between them now being 2λ, and another maximum results. Evidently still others will appear when $ab = 3\lambda, 4\lambda, \ldots$ Maxima will also be observed at corresponding angles on the opposite side of the grating normal, as well as along the normal itself, since in the latter position the phase difference between disturbances reaching AA is zero.

The angles of deviation for which the maxima occur may readily be found from Fig. 41-25. Consider the right angle Aba. Let d be the distance between successive grating elements, called the *grating spacing*. The necessary condition for a maximum is that $ab = m\lambda$, where $m = 0$, 1, 2, 3, etc. It follows that

$$\sin \theta = m\frac{\lambda}{d} \quad (m = 0, 1, 2, \ldots). \qquad (41\text{-}18)$$

is the necessary condition for a maximum. The angle θ is also the angle by which the rays corresponding to the maxima have been *deviated* from the direction of the incident light.

In practice, the parallel beam incident on the grating is usually produced by a lens having a narrow illuminated slit at its first focal point. Each of the maxima is then a sharp image of the slit, of the same color as that of the light illuminating the slit, assumed thus far to be monochromatic. If the slit is illuminated by light consisting of a mixture of wavelengths, the lens will form a number of images of the slit in different positions; each wavelength in the original light gives rise to a set of slit images deviated by the appropriate angles. If the slit is illuminated with white light, a continuous group of images is formed side by side or, in other words, the white light is dispersed into continuous spectra. In contrast with the single spectrum produced by a prism, a grating forms a

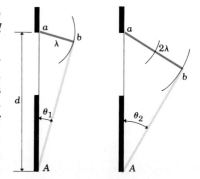

41-25 First-order maximum when $ab = \lambda$, second-order maximum when $ab = 2\lambda$.

number of spectra on either side of the normal. Those that correspond to $m = 1$ in Eq. (41–18) are called *first-order,* those that correspond to $m = 2$ are called *second-order,* and so on. Since for $m = 0$ the deviation is zero, all colors combine to produce a white image of the slit in the direction of the incident beam.

In order that an appreciable deviation of the light may be produced, it is necessary that the grating spacing be of the same order of magnitude as the wavelength of light. Gratings for use in or near the visible spectrum are ruled with from about 500 to 1500 lines per millimeter.

The diffraction grating is widely used in spectrometry, instead of a prism, as a means of dispersing a light beam into spectra. If the grating spacing is known, then from a measurement of the angle of deviation of any wavelength, the value of this wavelength may be computed. In the case of a prism this is not so; the angles of deviation are not related in any simple way to the wavelengths but depend on the characteristics of the material of which the prism is constructed. Since the index of refraction of optical glass varies more rapidly at the violet than at the red end of the spectrum, the spectrum formed by a prism is always spread out more at the violet end than it is at the red. Also, while a prism deviates red light the least and violet the most, the reverse is true of a grating, since in the latter case the deviation increases with increasing wavelength.

Example 1   The wavelengths of the visible spectrum are approximately 400 nm to 700 nm. Find the angular breadth of the first-order visible spectrum produced by a plane grating having 6,000 lines per centimeter, when light is incident normally on the grating.

Solution   The grating spacing d is

$$d = \frac{1}{6000 \text{ lines} \cdot \text{cm}^{-1}} = 1.67 \times 10^{-6} \text{ m.}$$

The angular deviation of the violet is

$$\sin \theta = \frac{400 \times 10^{-9} \text{ m}}{1.67 \times 10^{-6} \text{ m}} = 0.240,$$
$$\theta = 13.9°.$$

The angular deviation of the red is

$$\sin \theta = \frac{700 \times 10^{-9} \text{ m}}{1.67 \times 10^{-6} \text{ m}} = 0.420,$$
$$\theta = 24.8°.$$

Hence, the first-order visible spectrum includes an angle of

$$24.8° - 13.9° = 10.9°.$$

Example 2   Show that the violet of the third-order spectrum overlaps the red of the second-order spectrum.

Solution   The angular deviation of the third-order violet is

$$\sin \theta = \frac{(3)(400 \times 10^{-9} \text{ m})}{d}$$

and of the second-order red it is

$$\sin \theta = \frac{(2)(700 \times 10^{-9} \text{ m})}{d}.$$

Since the first angle is smaller than the second, whatever the grating spacing, the third order will *always* overlap the second.

*41-10 Diffraction of x-rays by a crystal

Although x-rays were discovered by Roentgen in 1895, it was not until 1913 that x-ray wavelengths were measured with any degree of precision. Experiments had indicated that these wavelengths might be of the order of 10^{-8} cm, which is about the same as the interatomic spacing in a solid. It occurred to Laue in 1913 that if the atoms in a crystal were arranged in a regular way, a crystal might serve as a three-dimensional diffraction grating for x-rays. The experiment was performed by Friederich and Knipping and it succeeded, thus verifying in a single stroke both the hypothesis that x-rays *are* waves (or, at any rate, wavelike in some of their properties) and that the atoms in a crystal *are* arranged in a regular manner. Since that time, the phenomenon of x-ray diffraction by a crystal has proved an invaluable tool of the physicist, both as a way to measure x-ray wavelengths and as a method of studying the structure of crystals.

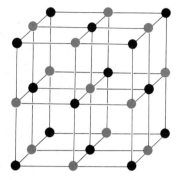

41-26 Model of arrangement of ions in a crystal of NaCl. Black circles, Na; color circles, Cl.

Figure 41-26 is a diagram of a simple type of crystal, that of sodium chloride (NaCl). The black circles represent the sodium, and the color circles the chlorine ions. Figure 41-27 is a diagram of a section through the crystal. Planes such as those parallel to *aa, bb, cc,* etc., can be constructed through the crystal in such a way that they pass through relatively large numbers of atoms.

Figure 41-28 is a photograph made by directing a narrow beam of x-rays at a thin section of a crystal of quartz and allowing the diffracted beams to strike a photographic plate. Each atom in the crystal scatters some of the incident radiation, and interference occurs between the waves scattered by the various atoms. Just as with the diffraction grating, nearly complete cancellation occurs for all but certain very specific directions for which constructive interference is possible; hence, the spots in Fig. 41-28. The interference pattern is often described in terms of *reflections* from various planes of atoms as in Fig. 41-27, but the basic phenomenon is one of *interference*.

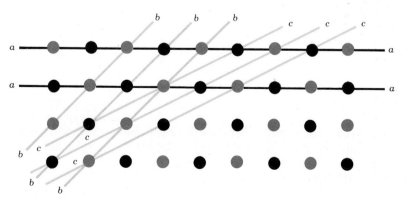

41-27 Crystal planes such as *aa, bb,* and *cc* serve as a three-dimensional diffraction grating for x-rays.

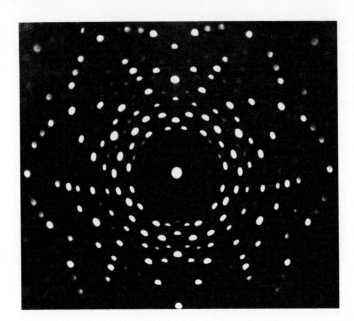

41-28 Laue diffraction pattern formed by directing a beam of x-rays at a thin section of quartz crystal. (Courtesy of Dr. B. E. Warren.)

*41-11 The resolving power of optical instruments

The expressions for the magnification of a telescope or a microscope involved (except for certain numerical factors) only the focal lengths of the lenses making up the optical system of the instrument. It appears, at first glance, as though any desired magnification might be attained by a proper choice of these focal lengths. Beyond a certain point, however, while the image formed by the instrument becomes larger (or subtends a larger angle) it *does not gain in detail*, even though all lens aberrations have been corrected. This limit to the useful magnification is set by the fact that light is a wave motion and the laws of geometrical optics do not hold strictly for a wave surface of limited extent. Physically, the image of a point source is not the intersection of *rays* from the source, but the diffraction pattern of those *waves* from the source that pass through the lens system.

It is an important experimental fact that the light from a point source, diffracted by a circular opening, is focused by a lens not as a geometrical point, but as a disk of finite radius surrounded by dark and bright rings. This diffraction pattern is, in fact, identical to the acoustic radiation pattern from a circular piston, which was discussed in Sec. 23-8. The larger the wave surface admitted (i.e., the larger the lenses or diaphragms in an optical system), the smaller the diffraction pattern of a point source and the closer together may two point sources be before their diffraction disks overlap and become indistinguishable. An optical system is said to be able to *resolve* two point sources if the corresponding diffraction patterns are sufficiently small or sufficiently separated to be distinguished. The numerical measure of the ability of the system to resolve two such points is called its *resolving power* or its *resolution*.

Figure 41-29a is a photograph of four point sources made with the camera lens "stopped down" to an extremely small aperture. The nature of the diffraction patterns is clearly evident, and it is obvious that further magnification of the picture would not aid in resolving the sources.

What is necessary is not to make the image *larger*, but to make the diffraction patterns *smaller*. Figures 41–29b and 41–29c show how the resolving power of the lens is increased by increasing its aperture. In (b) the diffraction patterns are sufficiently small for all four sources to be distinguished. In (c) the full aperture of the lens was utilized.

An arbitrary criterion proposed by Lord Rayleigh is that two point sources are just resolvable if the central maximum of the diffraction pattern of one source just coincides with the first *minimum* of the other. Let P_1 and P_2 in Fig. 41–30 be the central rays in the two parallel beams coming from two very distant point sources. Let P_1' and P_2' be the cen-

41-29 Diffraction patterns of four "point" sources, with a circular opening in front of the lens. In (a), the opening is so small that the patterns at the right are just resolved, by Rayleigh's criterion. Increasing the aperture decreases the size of the diffraction patterns, as in (b) and (c).

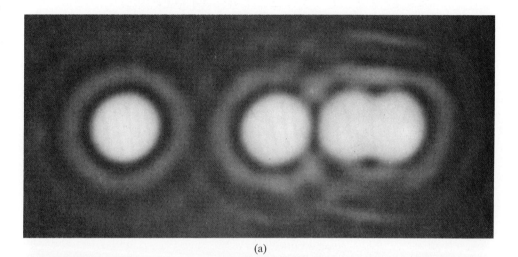

(a)

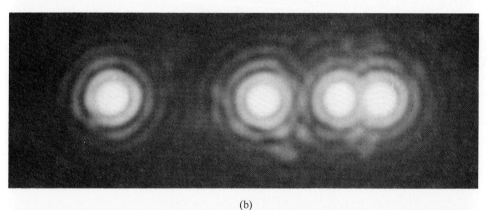

(b)

(c)

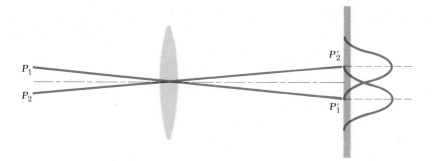

41-30 Two point sources are just resolved if the maximum of the diffraction pattern of one source just coincides with the first minimum of the other (Rayleigh's criterion).

ters of their diffraction patterns, formed by some optical instrument such as a microscope or telescope, which for simplicity is represented in the diagram as a single lens. If the images are just resolved, that is, if according to Rayleigh's criterion the first minimum of one pattern coincides with the center of the other, the separation of the centers of the patterns equals the radius of the central bright disk. From a knowledge of the focal lengths and separations of the lenses in any particular instrument, one can compute the corresponding distance between the two point objects. This distance, the minimum separation of two points that can just be resolved, is called the *limit of resolution* of the instrument. The smaller this distance, the greater is said to be the *resolving power*. The resolving power increases with the solid angle of the cone of rays intercepted by the instrument, and is inversely proportional to the wavelength of the light used. It is as if the light waves were the "tools" with which our optical system is provided, and the smaller the tools the finer the work the system can do.

*41-12 Holography

Holography is a technique for recording and reproducing an image of an object without the use of lenses or mirrors. Unlike the two-dimensional images recorded by an ordinary photograph or television system, a holographic image is truly three-dimensional. Such an image can be viewed from different directions to reveal different sides, and from various distances to reveal changing perspective.

The basic procedure for making a hologram is very simple in principle. A possible arrangement is shown in Fig. 41–31a. The object to be holographed is illuminated by monochromatic light, and a photographic film is located so that it is struck by scattered light from the object and also by direct light from the source. In practice the source must be a laser, for reasons to be discussed later. Interference between the direct and scattered light leads to the formation and recording of a complex interference pattern on the film.

To form the images, one simply projects laser light through the developed film, as shown in Fig. 41–31b. Two images are formed, a virtual image on the side of the film nearer the source, and a real image on the opposite side.

A complete analysis of holography is beyond our scope, but we can gain some insight into the process by examining how a single point is holographed and imaged. We consider the interference pattern formed

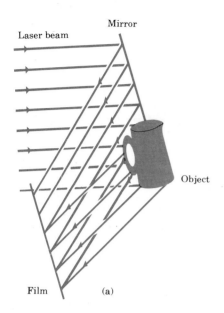

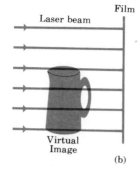

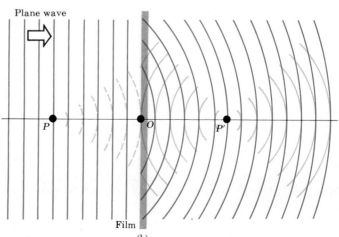

41-31 (a) The hologram is the record on film of the interference pattern formed with light directly from the source and light scattered from the object. (b) Images are formed when light is projected through the hologram.

on a screen by the superposition of an incident plane wave and a spherical wave, as shown in Fig. 41–32a. The spherical wave originates at a point source P a distance d_0 from the screen; P may in fact be a small object that scatters part of the incident plane wave. In any event, we assume that the two waves are monochromatic and coherent, and that the phase relation is such that constructive interference occurs at point O on the diagram. Then constructive interference will *also* occur at any point Q on the screen that is farther from P than O is, by an integer number of wavelengths. That is, if $d_n - d_0 = n\lambda$, where n is an integer, constructive interference occurs. The points where this condition is satisfied form circles centered at O, with radii r_n given by

$$d_n - d_0 = \sqrt{d_0^2 + r_n^2} - d_0 = n\lambda, \quad (n = 1, 2, 3, \ldots). \quad (41\text{-}19)$$

Solving this for r_n^2, we find

$$r_n^2 = \lambda\,(2nd_0 + n^2\lambda).$$

41-32 (a) Constructive interference of the plane and spherical waves occurs in the plane of the film at every point Q for which the distance d_n from P is greater than the distance d_0 from P to O by an integer number of wavelengths $n\lambda$. For the point shown, $n = 2$. (b) When a plane wave strikes the developed film, the diffracted wave consists of a wave converging to P' and then diverging again, and a diverging wave that appears to originate at P. These waves form the real and virtual images, respectively.

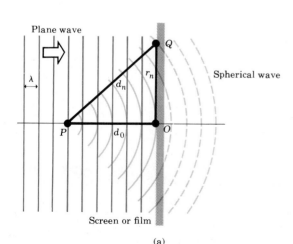

Ordinarily d_0 is very much larger than λ, so we neglect the second term in parentheses, obtaining

$$r_n = \sqrt{2n\lambda d_0}. \tag{41-20}$$

Since n may be any integer, the interference pattern consists of a series of concentric bright circular fringes, with the radii of the brightest regions given by Eq. (41-20). Between these bright fringes will be darker fringes. This pattern may be recorded on film by simply placing the film in the position of the screen.

Now the film is developed and a transparent positive print is made, so the bright-fringe areas have the greatest transparency on the film. It is then illuminated with monochromatic plane-wave light of the same wavelength as that used initially. In Fig. 41-32b, consider a point P' at a distance d_0 along the axis from the film; the centers of successive bright fringes differ in their distances from P' by an integer number of wavelengths, and therefore a strong *maximum* in the diffracted wave occurs at P'. That is, light converges to P' and then diverges from it on the opposite side, and P' is therefore a *real image* of point P.

This is not the entire diffracted wave, however; there is also a diverging spherical wave, which would represent a continuation of the wave originally emanating from P if the film had not been present. Thus the *total* diffracted wave is a superposition of a converging spherical wave forming a real image at P', and a diverging spherical wave shaped as though it had originated at P, forming a virtual image at P.

Because of the principle of linear superposition, what is true for the imaging of a single point is also true for the imaging of any number of points. The film records the superposed interference pattern from the various points, and when light is projected through the film the various image points are reproduced simultaneously. Thus the images of an extended object can be recorded and reproduced just as for a single point object.

Several practical problems must be overcome. First, the light used must be *coherent* over the distances large compared to the dimensions of the object and its distance from the film. Ordinary light sources *do not* satisfy this requirement, for reasons discussed in Sec. 41-1, and laser light is essential. Second, extreme mechanical stability is needed. If there is any relative motion of source, object, or film during exposure, even by as much as a wavelength, the interference pattern on the film is blurred enough to prevent satisfactory image formation. These obstacles are not insurmountable, however, and holography promises to become increasingly important in research, entertainment, and a wide variety of technological applications.

Questions

41-1 Could an experiment similar to Young's two-slit experiment be performed with sound? How might this be carried out? Does it matter that sound waves are longitudinal and electromagnetic waves transverse?

41-2 At points of constructive interference between waves of equal amplitude, the intensity is four times that of either individual wave. Does this violate energy conservation? If not, why not?

41-3 In using the superposition principle to calculate intensities in interference and diffraction patterns, could one add the intensities of the waves instead of their amplitudes? What's the difference?

41-4 A two-slit interference experiment is set up and the fringes displayed on a screen. Then the whole apparatus is immersed in the nearest swimming pool. How does the fringe pattern change?

41-5 Would the headlights of a distant car form a two-source interference pattern? If so, how might it be observed? If not, why not?

41-6 A student asserted that it is impossible to observe interference fringes in a two-source experiment if the distance between sources is less than half the wavelength of the wave. Do you agree? Explain.

41-7 An amateur scientist proposed to record a two-source interference pattern using only one source, placing it first in position S_1 in Fig. 41–1 and turning it on for a certain time, then placing it at S_2 and turning it on for an equal time. Does this work?

41-8 When a thin oil film spreads out on a puddle of water, the thinnest part of the film looks lightest in the resulting interference pattern. What does this tell you about the relative magnitudes of the refractive indexes of oil and water?

41-9 A glass windowpane with a thin film of water on it reflects less than when it is perfectly dry. Why?

41-10 In high-quality camera lenses, the resolution in the image is determined by diffraction effects. Is the resolution best when the lens is "wide open" or when it is "stopped down" to a smaller aperture? How does this behavior compare with the effect of aperture size on the depth of focus, that is, on the limit of resolution due to imprecise focusing?

41-11 If a two-slit interference experiment were done with white light, what would be seen?

41-12 Why is a diffraction grating better than a two-slit setup for measuring wavelengths of light?

41-13 Would the interference and diffraction effects described in this chapter still be seen if light were a longitudinal wave instead of transverse?

41-14 One sometimes sees rows of evenly spaced radio antenna towers. A student remarked that these act like diffraction gratings. What did he mean? Why would one *want* them to act like a diffraction grating?

41-15 Could x-ray diffraction effects with crystals be observed using visible light instead of x-rays? Why or why not?

41-16 Does a microscope have better resolution with red light or blue light?

41-17 How could an interference experiment, such as one using a Michelson interferometer or fringes caused by a thin air space between glass plates, be used to measure the refractive index of air?

Problems

41-1 Two slits are spaced 0.3 mm apart and are placed 50 cm from a screen. What is the distance between the second and third dark lines of the interference pattern when the slits are illuminated with light of 600-nm wavelength?

41-2 In Problem 41–1, suppose the entire apparatus is immersed in water. Then what is the distance between the second and third dark lines?

41-3 Young's experiment is performed with sodium light ($\lambda = 589$ nm). Fringes are measured carefully on a screen 100 cm away from the double slit, and the center of the twentieth fringe is found to be 11.78 mm from the center of the zeroth fringe. What is the separation of the two slits?

41-4 An FM radio station has a frequency of 100 MHz and uses two identical antennas mounted at the same elevation, 12 m apart. The resulting radiation pattern has maximum intensity along a horizontal line perpendicular to the line joining the antennas. At what other angles (measured from the line of maximum intensity) is the intensity maximum? At what angles is it zero?

41-5 An AM radio station has a frequency of 1000 kHz; it uses two identical antennas at the same elevation, 150 m apart.

a) In what directions is the intensity maximum, considering points in a horizontal plane?

b) Calling the maximum intensity in (a) I_0, determine in terms of I_0 the intensity in directions making angles of 30°, 45°, 60°, and 90° to the direction of maximum intensity.

41-6 Light from a mercury-arc lamp is passed through a filter that blocks everything except for one spectrum line in the green region of the spectrum. It then falls on two slits separated by 0.6 mm. In the resulting interference pattern on a screen 2.5 mm away, adjacent bright fringes are separated by 2.27 mm. What is the wavelength?

41-7 Light of wavelength 500 nm is incident perpendicularly from air on a film 10^{-4} cm thick and of refractive index 1.375. Part of the light is reflected from the first surface of the film, and part enters the film and is reflected back at the second face.

a) How many waves are contained along the path of this light in the film?

b) What is the phase difference between these waves as they leave the film?

41-8 A glass plate, 0.40 μm thick, is illuminated by a beam of white light normal to the plate. The index of refraction of the glass is 1.50. What wavelengths within the limits of the visible spectrum ($\lambda = 400$ nm to $\lambda = 700$ nm) will be intensified in the reflected beam?

41-9 Figure 41-33 shows an interferometer known as *Fresnel's biprism*. The magnitude of the prism angle A is extremely small

a) If S_0 is a very narrow source slit, show that the separation of the two virtual coherent sources S_1 and S_2 is given by $d = 2aA(n-1)$, where n is the index of refraction of the material of the prism.

b) Calculate the spacing of the fringes of green light of wavelength 500 nm on a screen 2 m from the biprism. Take $a = 20$ cm, $A = 0.005$ rad, and $n = 1.5$.

41-10 In Lloyd's mirror, the source slit S_0 and its virtual image S_1 lie in a plane 20 cm behind the left edge of the mirror. (See Fig. 41-8.) The mirror is 30 cm long and a ground-glass screen is placed at the right edge. Calculate the distance from this edge to the first light maximum, if the perpendicular distance from S_0 to the mirror is 2 mm and $\lambda = 7.2 \times 10^{-5}$ cm.

41-11 How far must the mirror M_2 of the Michelson interferometer be moved in order that 3000 fringes of krypton-86 light ($\lambda = 606$ nm) move across a line in the field of view?

41-12 Two rectangular pieces of plane glass are laid one upon the other on a table. A thin strip of paper is placed between them at one edge so that a very thin wedge of air is formed. The plates are illuminated by a beam of sodium light at normal incidence ($\lambda = 589$ nm). Interference fringes are formed, there being ten per centimeter length of wedge measured normal to the edges in contact. Find the angle of the wedge.

41-13 A sheet of glass 10 cm long is placed in contact with a second sheet, and is held at a small angle with it by a metal strip 0.1 mm thick placed under one end. The glass is illuminated from above with light of 546-nm wavelength. How many interference fringes are observed per cm in the reflected light?

41-14 The radius of curvature of the convex surface of a plano-convex lens is 120 cm. The lens is placed convex side down on a plane glass plate, and illuminated from above with red light of wavelength 650 nm. Find the diameter of the third bright ring in the interference pattern.

41-15 What is the thinnest film of 1.40 refractive index in which destructive interference of the violet component (400 nm) of an incident white beam in air can take place by reflection? What is, then, the residual color of the beam?

41-16 The surfaces of a prism of index 1.52 are to be made "nonreflecting" by coating them with a thin layer of transparent material of index 1.30. The thickness of the layer is such that, at a wavelength of 550 nm (in vacuum), light reflected from the first surface is $\frac{1}{2}$ wavelength out of phase with that reflected from the second surface. Find the thickness of the layer.

41-17

a) Is a thin film of quartz suitable as a nonreflecting coating for fabulite? (See Table 38-1.)

b) If so, how thick should the film be?

41-18 Parallel rays of green mercury light of wavelength 546 nm pass through a slit of width 0.437 mm covering a lens of focal length 40 cm. In the focal plane of the lens, what is the distance from the central maximum to the first minimum?

41-19 Light of wavelength 589 nm from a distant source is incident on a slit 1.0 mm wide, and the resulting diffraction pattern is observed on a screen 2.0 m away. What is the distance between the two dark fringes on either side of the central bright fringe?

41-20 In Problem 41-19, suppose the entire apparatus is immersed in water. Then what is the distance between the two dark fringes?

41-21 Plane monochromatic waves of wavelength 600 nm are incident normally on a plane transmission grating having 500 lines·mm^{-1}. Find the angles of deviation in the first, second, and third orders.

41-22 A plane transmission grating is ruled with 4000 lines·cm^{-1}. Compute the angular separation in degrees, in

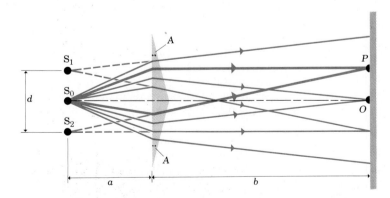

41-33 The Fresnel biprism.

the second-order spectrum, between the α and δ lines of atomic hydrogen, whose wavelengths are, respectively, 656 nm and 410 nm. Assume normal incidence.

41-23

a) What is the wavelength of light that is deviated in the first order through an angle of 20° by a transmission grating having 6000 lines·cm^{-1}?

b) What is the second-order deviation of this wavelength? Assume normal incidence.

41-24 A slit of width a was placed in front of a lens of focal length 80 cm. The slit was illuminated by parallel light of wavelength 600 nm and the diffraction pattern of Fig. 41–23b was formed on a screen in the second focal plane of the lens. If the photograph of Fig. 41–23b represents an enlargement to twice the actual size, what was the slit width?

41-25 The intensity of light in the Fraunhofer diffraction pattern of a single slit is

$$I = I_0(\sin \beta/\beta)^2,$$

where

$$\beta = (\pi a \sin \theta)/\lambda.$$

Show that the equation for the values of β at which I is a maximum is $\tan \beta = \beta$. How can you solve such an equation graphically?

41-26 What is the longest wavelength that can be observed in the fourth order for a transmission grating having 5000 lines · cm^{-1}? Assume normal incidence.

41-27 In Fig. 41–34, two point sources of light, a and b, at a distance of 50 m from lens L and 6 mm apart, produce images at c that are just resolved by Rayleigh's criterion. The focal length of the lens is 20 cm. What is the diameter of the diffraction circles at c?

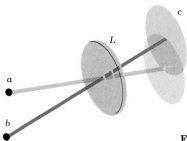

Figure 41-34

41-28 A telescope is used to observe two distant point sources 1 m apart. ($\lambda = 500$ nm.) The objective of the telescope is covered with a slit of width 1 mm. What is the maximum distance in meters at which the two sources may be distinguished?

41-29 A typical lens in a 35-mm camera has a focal length of 50 mm and a diameter of 25 mm ($f/2$). The resolution of such lenses is expressed as the number of lines per millimeter in the image that can be resolved. If the resolution of this lens is limited by diffraction effects, approximately what is its resolution in lines·mm^{-1}?

41-30 If a hologram is made using 600-nm light and then viewed using 500-nm light, how will the images look compared to those observed with 500-nm light?

41-31 A hologram is made using 600-nm light and is then viewed using continuous-spectrum white light from an incandescent bulb. What will be seen?

41-32 Ordinary photographic film reverses black and white, in the sense that the most brightly illuminated areas become blackest upon development (hence the term *negative*). Suppose a hologram negative is viewed directly, without making a positive transparency. How will the resulting images differ from those obtained with the positive hologram?

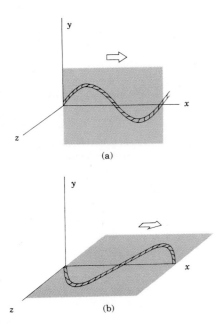

(a)

(b)

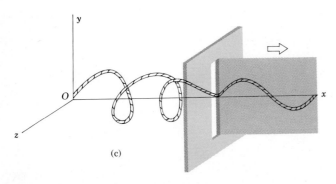

(c)

42 POLARIZATION

42–1 Polarization

Interference and diffraction effects occur with all kinds of waves, including sound waves and surface waves on a liquid as well as electromagnetic waves. These effects are, in general, independent of whether the waves are longitudinal or transverse. In this chapter we consider *polarization* phenomena, a group of wave phenomena that occur only with *transverse* waves. Our principal area of interest will be with electromagnetic waves, but we begin with a mechanical example.

To introduce basic ideas in a familiar context, we recall the discussion of mechanical waves on a string in Chapter 21. Displacements of points on the string are perpendicular to the length of the string, and to the direction of propagation of the wave, so the wave is transverse. For a string whose equilibrium position is along the x-axis, we considered principally displacements along the y-direction, in which case the string always lies entirely in the xy-plane; an example of such a wave is shown in Fig. 42–1a. But the displacement might instead be along the z-axis, so that the string lies always in the xz-plane, as in Fig. 42–1b. Furthermore, in view of the principle of linear superposition, any superposition of waves having displacements along the y- and z-directions is also possible; conversely, *any* transverse wave can be represented as a superposi-

42–1   (a) Transverse wave on a string, polarized in the y-direction. (b) Wave polarized in the z-direction. (c) Barrier with a frictionless vertical slot passes components polarized in the y-direction but blocks those polarized in the z-direction, acting as a polarizing filter.

tion of two component waves, one having displacements only along the y-direction, the other only along the z-direction.

The wave having only y-displacements in the above discussion is said to be *linearly polarized* in the y-direction, and the one with only z-displacements is linearly polarized in the z-direction. The qualifying words *linear* and *linearly* refer to the fact that each particle of the string moves back and forth along a straight line. These words are often omitted when it is clear from the context that linear polarization is meant.

Any transverse wave on a string may be regarded as a superposition of waves polarized in the y- and z-directions, respectively. It is also easy in principle to construct a mechanical *filter* that permits only waves with a certain polarization direction to pass. An example is shown in Fig. 42–1c; the string can slide vertically in the slot without friction, but no horizontal motion is possible. Thus this filter passes waves polarized in the y-direction but blocks those polarized in the z-direction.

This same language can be applied to light and other electromagnetic waves, which also exhibit polarization. An electromagnetic wave consists of fluctuating electric and magnetic fields, perpendicular to each other and to the direction of propagation, as described in detail in Chapter 37. By convention, the direction of polarization is taken to be that of the *electric*-field vector, not the magnetic field, partly because most mechanisms for detecting electromagnetic waves employ principally the electric-field forces on electrons in materials. The most common manifestations of electromagnetic radiation are due chiefly to the electric-field force, not the magnetic-field force.

Polarizing filters can be made for electromagnetic waves; the details of construction depend on the wavelength. For microwaves having a wavelength of a few centimeters, a grid of closely spaced, parallel conducting wires insulated from each other will pass waves whose E-fields are perpendicular to the wires but not those with E-fields parallel to the wires. For light, the most common polarizing filter is a material known by the trade name Polaroid, widely used for sunglasses and polarizing filters for camera lenses. This material works on a principle of preferential absorption, passing waves polarized parallel to a certain axis in the material (called the *polarizing axis*) with 80 percent or more transmission, but offering only one percent or less transmission to waves with polarization perpendicular to this axis. The action of such a polarizing filter is shown schematically in Fig. 42–2. Other polarizing filters are discussed in Sec. 42–5.

Waves emitted by a radio transmitter are usually linearly polarized; a vertical rod antenna of the type widely used for CB radios emits waves which, in a horizontal plane around the antenna, are polarized in the vertical direction (parallel to the antenna). Light from ordinary sources is *not* polarized, for a slightly subtle reason. The "antennas" that radiate light waves are the molecules of which light sources are composed. The electrically charged particles in the molecules acquire energy in some way, and radiate this energy as electromagnetic waves of short wavelength. The waves from any one molecule may be linearly polarized, like those from a radio antenna; but since any actual light source contains a tremendous number of molecules, oriented at random, the light emitted is a random mixture of waves linearly polarized in all possible transverse directions.

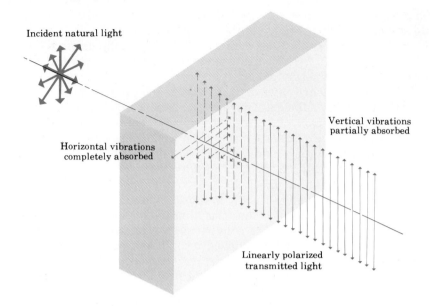

Incident natural light

Vertical vibrations
partially absorbed

Horizontal vibrations
completely absorbed

Linearly polarized
transmitted light

42-2 Linearly polarized light transmitted by a polarizing filter.

42-2 Malus' law and percent polarization

To clarify basic concepts, we introduce in this section the idea of an ideal polarizing filter or *polarizer* having the property that it passes 100 percent of incident light polarized in the direction of the filter's polarizing axis, but blocks completely all light polarized perpendicular to this axis. We recognize that such a polarizer is an unattainable idealization, like the frictionless plane and the zero-resistance wire, but it can be approximated closely by real devices and is helpful in clarifying the discussion.

For quantitative experiments, the intensity of transmitted light can be measured by means of a photocell connected to an amplifier and a current-measuring device. In Fig. 42–3, unpolarized light is incident on a polarizer whose axis is represented by the broken line. As the polarizer is rotated about an axis parallel to the incident ray, the intensity does not change. The polarizer transmits the components of the incident waves in which the *E*-vector is parallel to the transmission direction of the polarizer, and by symmetry the components are equal for all azimuths.

The transmitted intensity is found to be exactly one-half that of the incident light. This may be understood as follows: The incident light can always be resolved into components polarized parallel to the polarizer

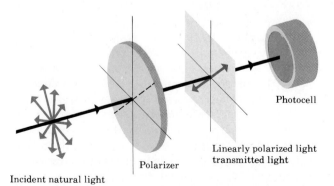

Photocell

Linearly polarized light
transmitted light

Polarizer

42-3 The intensity of the transmitted linearly polarized light is the same at all azimuths of the polarizer.

Incident natural light

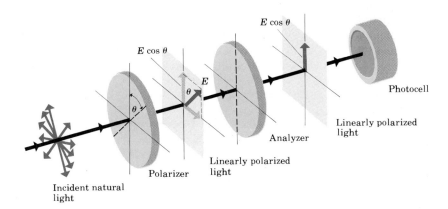

$E \cos \theta$

$E \cos \theta$

E

θ

θ

Photocell

Linearly polarized
light

Analyzer

Linearly polarized
light

Polarizer

Incident natural
light

42-4 The analyzer transmits only the
component parallel to its transmission
direction or polarizing axis.

axis and components polarized perpendicular to it. Because the incident
light is a random mixture of all states of polarization, these two components are, on the average, equal. Thus (in the ideal polarizer) exactly
half of the incident intensity, that corresponding to the component parallel to the polarizer axis, is transmitted.

Suppose now that a second polarizer is inserted between the first
polarizer and the photocell, as in Fig. 42–4. Let the transmission direction of the second polarizer, or *analyzer,* be vertical, and let that of the
first polarizer make an angle θ with the vertical. The linearly polarized
light transmitted by the polarizer may be resolved into two components
as shown, one parallel and the other perpendicular to the transmission
direction of the analyzer. Evidently only the parallel component, of
amplitude $E \cos \theta$, will be transmitted by the analyzer. The transmitted
intensity is maximum when $\theta = 0$, and is zero when $\theta = 90°$, or when
polarizer and analyzer are *crossed.* At intermediate angles, since the
quantity of energy is proportional to the *square* of the amplitude, we
have

$$I = I_{\max} \cos^2 \theta, \qquad (42–1)$$

where $I_{\max}$ is the maximum amount of light transmitted and I is the
amount transmitted at the angle θ. This relation, discovered experimentally by Etienne Louis Malus in 1809, is called *Malus' Law.*

The angle θ is the angle between the transmission directions of polarizer and analyzer. If either the analyzer or the polarizer is rotated, the
amplitude of the transmitted beam varies with the angle between them
according to Eq. (42–1).

Example   In Fig. 42–4, let the incident unpolarized light have intensity
I_0. Find the intensity transmitted by the first polarizer and by the second, if the angle θ is 30°.

Solution   As explained above, the intensity after the first filter is $(I_0/2)$.
According to Eq. (42–1), the second filter reduces the intensity by a
factor of $\cos^2 30° = \frac{3}{4}$. Thus the intensity transmitted by the second polarizer is $(I_0/2)(\frac{3}{4}) = 3I_0/8$.

If, in Fig. 42–3, the incident light is completely linearly polarized
rather than unpolarized, then of course the transmitted intensity varies
from zero to some maximum value as the polarizer axis is turned from

perpendicular to parallel to the polarization direction of the light. Or the incident light may be a mixture of linearly polarized and unpolarized, with intensities I_{pol} and I_{un}, respectively. The *percent polarization* is then defined as the ratio of the polarized intensity to the total intensity, times 100%:

$$\% \text{ polarization} = \frac{I_{pol}}{I_{pol} + I_{un}} \times 100\%. \qquad (42\text{--}2)$$

In this case, in Fig. 42–3 the transmitted intensity varies from a minimum (but not zero) when the polarizer axis is perpendicular to the polarized component of incident light, to a maximum when it is parallel. The percent polarization can be expressed in terms of these two intensities, which may be denoted as I_{min} and I_{max}. Recalling that half the unpolarized intensity is transmitted by the polarizer, we may write

$$I_{min} = \tfrac{1}{2}I_{un}, \qquad I_{max} = \tfrac{1}{2}I_{un} + I_{pol}. \qquad (42\text{--}3)$$

These equations can be solved for I_{pol} and I_{un} and the results substituted into Eq. (42–2). The details of this calculation are left as an exercise; the result is

$$\% \text{ polarization} = \frac{I_{max} - I_{min}}{I_{max} + I_{min}} \times 100\%. \qquad (42\text{--}4)$$

42-3 Polarization by reflection

There are various processes by which an initially unpolarized beam of light can yield a partially polarized beam. One of these is the familiar process of *reflection*. When natural light strikes a reflecting surface between two optical media, there is found to be a preferential reflection for those waves in which the electric-field vector is perpendicular to the plane of incidence (the plane containing the incident ray and the normal to the surface). An exception is that at *normal* incidence all directions of polarization are reflected equally. At one particular angle of incidence, known as the *polarizing angle* ϕ_p, no light whatever is reflected except that in which the electric vector is perpendicular to the plane of incidence. This case is illustrated in Fig. 42–5.

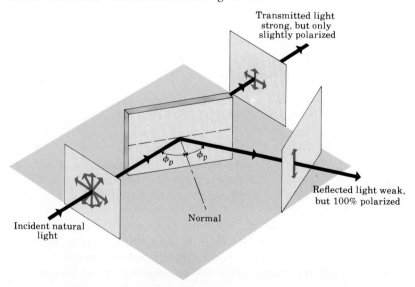

Transmitted light strong, but only slightly polarized

Reflected light weak, but 100% polarized

Normal

Incident natural light

42-5   When light is incident at the polarizing angle, the reflected light is linearly polarized.

When light is incident at the polarizing angle, *none* of the component parallel to the plane of incidence is reflected; this component is 100% transmitted in the *refracted* beam. Of the component perpendicular to the plane of incidence, about 15% is reflected if the reflecting surface is glass. The fraction reflected depends on the index of the reflecting material. Hence the *reflected* light is weak and *completely* linearly-polarized. The *refracted* light is a mixture of the parallel component, all of which is refracted, and the remaining 85% of the perpendicular component. It is therefore strong, but only *partially* polarized.

At angles of incidence other than the polarizing angle, some of the component parallel to the plane of incidence is reflected, so that, except at the polarizing angle, the reflected light is not completely linearly-polarized.

In 1812, Sir David Brewster noticed that when the angle of incidence is equal to ϕ_p, the reflected ray and the refracted ray are *perpendicular* to each other, as shown in Fig. 42–6. When this is the case, the angle of refraction ϕ' becomes the complement of ϕ_p, so that $\sin \phi' = \cos \phi_p$. Since

$$n \sin \phi_p = n' \sin \phi',$$

we find $n \sin \phi_p = n' \cos \phi_p$, and

$$\tan \phi_p = \frac{n'}{n}, \tag{42–5}$$

a relation known as *Brewster's law*.

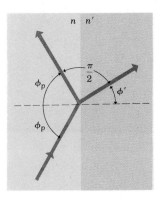

42-6 At the polarizing angle the reflected and transmitted rays are perpendicular to each other.

*42-4   Double refraction

The progress of a wave train through a homogeneous isotropic medium, such as glass, may be determined graphically by Huygens' construction. The secondary wavelets in such a medium are spherical surfaces. There exist, however, many transparent crystalline substances which, while homogeneous, are *anisotropic*. That is, the velocity of a light wave in them is not the same in all directions. Crystals having this property are said to be *doubly refracting,* or *birefringent. Two* sets of Huygens wavelets propagate from every wave surface in such a crystal, one set being spherical and the other ellipsoidal.

A consequence of this property of anisotropic crystals is that a ray of light striking such a crystal at normal incidence is broken up into two rays as it enters the crystal. The ray that corresponds to wave surfaces tangent to the spherical wavelets is undeviated and is called the *ordinary ray.* The ray corresponding to the wave surfaces tangent to the ellipsoids is deviated even though the incident ray is normal to the surface, and is called the *extraordinary ray.* If the crystal is rotated about the incident ray as an axis, the ordinary ray remains fixed but the extraordinary ray revolves around it, as shown in Fig. 42–7. Furthermore, for angles of incidence other than 0°, Snell's law (that is, $\sin \phi / \sin \phi' =$ constant) holds for the ordinary but *not* for the extraordinary ray, since the velocity of the latter is different in different directions.

The index of refraction for the extraordinary ray is therefore a function of direction. There is always one direction in the crystal for which there is no distinction between the *O*- and *E*-rays; this direction is called the *optic axis.* It is customary to state the index for the direction at right

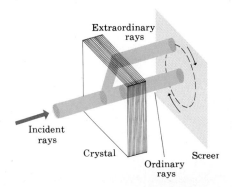

42-7 A narrow beam of natural light can be split into two beams by a doubly refracting crystal.

Table 42-1   Indexes of refraction of doubly refracting crystals (for light wavelength 589 nm)

Material	n_O	n_E
Calcite	1.658	1.486
Quartz	1.544	1.553
Tourmaline	1.64	1.62
Ice	1.306	1.307

angles to the optic axis, in which the velocity is a maximum or a minimum. Some values of n_O and n_E, the indices for the ordinary and extraordinary rays, are listed in Table 42-1.

42-5 Polarizers

Experiment shows that the ordinary and extraordinary waves in a doubly refracting crystal are linearly polarized in mutually perpendicular directions. Consequently, if some means can be found to separate one wave from the other, a doubly refracting crystal may be used to obtain plane-polarized light from natural light. There are several ways in which this separation may be accomplished:

1. One of the rays may be made to undergo internal reflection and be deflected to one side, allowing the other to proceed undeflected.

2. Both rays may be separated slightly so that, at sufficient distance from the separating prism, only one ray is intercepted.

3. One of the rays may be absorbed while the other is unaffected.

Many ingenious composite prisms of doubly refracting crystals have been developed as polarizers in the last 150 years. They are known by the names of the discoverers: Nicol, Rochon, Wollaston, Glan, Foucault, Ahrens, and many others. The best known is the Nicol prism, invented in 1828 by the Scottish physicist William Nicol. It is complicated and difficult to construct, uses a large amount of optically clear calcite, produces a lateral displacement of the transmitted beam, gives rise to a distorted image, and produces less than 100% linearly polarized light.

A prism originally developed by Glan and Thompson and later modified by Ammann and Massey (*Journal of the Optical Society of America,* November 1968) is constructed on a similar principle but avoids the defects of the Nicol prism. The modified Glan–Thompson prism consists of a glass prism with index of refraction 1.655 joined to a calcite prism, with a cement that has the same index of refraction, as shown in Fig. 42-8. The natural unpolarized light incident at the left is equivalent to two equal linearly polarized beams. The direction of polarization of one (represented by dots) is perpendicular to the page and that of the other (represented by short vertical lines) is parallel to the page. Both beams travel the same path with equal speed in the glass. In calcite, however, with the optic axis vertical as shown, perpendicular vibrations constitute

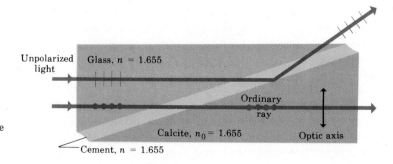

42-8   Glan–Thompson polarizing prism, as modified by Ammann and Massey. The thickness of the layer of cement is exaggerated.

the ordinary ray for which calcite has an index of refraction 1.6583. This value is so close to 1.655 that the perpendicular vibrations proceed without deflection from glass to cement to calcite and emerge into the air as 100% linearly polarized light.

The parallel vibrations, if they existed in the calcite crystal, would constitute the extraordinary ray with an index of refraction 1.4864. In traveling *in the cement* toward the calcite, this ray proceeds toward a medium in which its speed is greater, and therefore a critical angle exists. With the dimensions of the prism as shown, the angle of incidence of the ray with parallel polarization exceeds the critical angle, and this ray is therefore internally reflected and deflected from its original direction. This polarizing prism functions properly for all incident rays that make an angle of not more than 10° above or below the horizontal.

Certain doubly refracting crystals exhibit *dichroism;* that is, one of the polarized components is *absorbed* much more strongly than the other. Hence, if the crystal is cut of the proper thickness, one of the components is practically extinguished by absorption, while the other is transmitted in appreciable amount, as indicated in Fig. 42-2. Tourmaline is one example of such a dichroic crystal.

An early form of Polaroid, invented by Edwin H. Land in 1928, consists of a thin layer of tiny needlelike dichroic crystals of herapathite (iodoquinine sulfate), in parallel orientation, embedded in a plastic matrix and enclosed for protection between two transparent plates. A more recent modification, developed by Land in 1938 and known as an H-sheet, is a *molecular* polarizer. It consists of long polymeric molecules of polyvinyl alcohol (PVA) that have been given a preferred direction by stretching, and have been stained with an ink containing iodine that causes the sheet to exhibit dichroism. The PVA sheet is laminated to a support sheet of cellulose acetate butyrate.

Polaroid disks do not polarize all wavelengths equally. When two such disks are crossed, small amounts of red and of violet (the two ends of the visible spectrum) are transmitted. When white light passes through one Polaroid sheet, the transmitted light is slightly colored. The large area of such plates, however, and their moderate cost more than compensate for these small deficiencies. The existence of Polaroid sheets has stimulated the development and applications of polarized light to an extent that was out of the question when reflecting surfaces, Nicol prisms, and other costly devices had to be used.

Polaroid sheet is widely used in sunglasses where, from the standpoint of its polarizing properties, it plays the role of the analyzer in Fig. 42-4. We have seen that, when unpolarized light is reflected, there is a preferential reflection for light polarized perpendicular to the plane of incidence. When sunlight is reflected from a horizontal surface, the plane of incidence is vertical. Hence, in the reflected light there is a preponderance of light polarized in the horizontal direction, the proportion being greater the nearer the angle of incidence is to the polarizing angle. When such reflection occurs at smooth asphalt road surfaces, the surface of a lake, or a similar situation, it causes unwanted "glare," and vision is improved by eliminating it. The transmission direction of the Polaroid sheet in the sunglasses is vertical, so none of the horizontally polarized light is transmitted to the eyes.

Apart from this polarizing feature, these glasses serve the same purpose as any dark glasses, absorbing 50% of the incident light; in an unpolarized beam, half the light can be considered as polarized horizontally and half vertically, and only the vertically polarized light is transmitted. The sensitivity of the eye is independent of the state of polarization of the light.

*42-6 The scattering of light

The sky is blue. Sunsets are red. Skylight is partially polarized, as can readily be verified by looking at the sky directly overhead through a polarizing filter. It turns out that one phenomenon is responsible for all three of these effects.

In Fig. 42-9, sunlight (unpolarized) comes from the left along the z-axis and passes over an observer looking vertically upward along the y-axis. Molecules of the earth's atmosphere are located at point O. The electric field in the beam of sunlight sets the electric charges in the molecules in vibration. Since light is a transverse wave, the direction of the electric field in any component of the sunlight lies in the xy-plane and the motion of the charges takes place in this plane. There is no field, and hence no vibration, in the direction of the z-axis.

An arbitrary component of the incident light, vibrating at an angle θ with the x-axis, sets the electric charges in the molecules vibrating in the same direction, as indicated by the heavy line through point O. In the usual way, we can resolve this vibration into two components, one along the x-axis and the other along the y-axis. The result, then, is that each component in the incident light produces the equivalent of two molecular "antennas," oscillating with the frequency of the incident light, and lying along the x- and y-axis.

It has been explained in Sec. 37–7 that an antenna does not radiate in the direction of its own length. Hence, the antenna along the y-axis

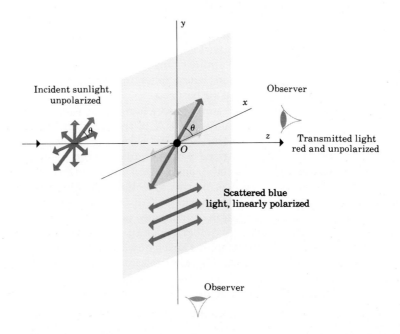

42-9 Scattered light is linearly polarized.

does not send any light to the observer directly below it. It does, of course, send out light in other directions. The only light reaching the observer comes from the component of vibration along the x-axis, and, as is the case with the waves from any antenna, this light is linearly polarized, with the electric field parallel to the antenna. The vectors on the y-axis below point O show the direction of polarization of the light reaching the observer.

The process described above is called *scattering*. The energy of the scattered light is removed from the original beam, which becomes weakened in the process. Analysis of the details of the scattering process shows that, for a given intensity of incident light, the intensity of the scattered light increases with increasing frequency. In other words, blue light is scattered more than red, with the result that the hue of the scattered light is blue.

Polarizers are commonly used in photography. Because skylight is partially polarized, the sky can be darkened in a photograph by appropriate orientation of the polarizer axis. The effect of atmosphere haze can be reduced in exactly the same way, and unwanted reflections can be controlled just as with polarizing sunglasses, discussed in Sec. 42-5.

Toward evening, when sunlight has to travel a large distance through the earth's atmosphere to reach a point over or nearly over an observer, a large proportion of the blue light in sunlight is removed from it by scattering. White light minus blue light is yellow or red in hue. Thus when sunlight, with the blue component removed, is incident on a cloud, the light reflected from the cloud to the observer has the yellow or red hue so commonly seen at sunset.

From the discussion above, it follows that if the earth had no atmosphere we would receive *no* skylight at the earth's surface, and the sky would appear as black in the daytime as it does at night. Thus, to an astronaut in a spaceship or on the moon, the sky appears black, not blue.

42-7 Circular and elliptical polarization

Up to this point we have discussed polarization phenomena in terms of *linearly polarized* light. This is the simplest kind of polarization, but not the only kind. Light (and all other electromagnetic radiation) may also have *circular* or *elliptical* polarization. To introduce these new concepts, we return again to mechanical waves on a stretched string.

In Fig. 42-1, suppose the two linearly polarized waves in parts (a) and (b) have equal amplitude and are *superimposed*; then each point in the string has simultaneously y- and z-displacements of equal magnitude, and a little thought shows that the resultant wave lies in a plane oriented at 45° to the xy-plane. The amplitude of the resultant wave is larger by a factor of $\sqrt{2}$ than that of either component wave, and the resultant wave is again linearly polarized. The motion of each point on the string is a superposition of two simple harmonic motions at right angles and in phase, which is a simple harmonic motion at 45° to the component motions.

But now suppose one component wave differs in phase by a quarter-cycle from the other. Then the resultant motion of each point corresponds to a superposition of two simple harmonic motions at right angles with a quarter-cycle phase difference. As a specific example, suppose that the y- and z-components of motion of a certain point are given by

$$y = A \sin \omega t,$$

$$z = A \sin \left(\omega t + \frac{\pi}{2} \right) = A \cos \omega t. \qquad (42\text{-}6)$$

Then z is maximum at times when y is zero, and conversely, and the resultant motion of the point is no longer back and forth along a line perpendicular to the direction of propagation of the wave. In fact, the point moves in a *circle*; to show this, we square the two equations and add, obtaining

$$y^2 + z^2 = A^2 \sin^2 \omega t + A^2 \cos^2 \omega t = A^2. \qquad (42\text{-}7)$$

Now $y^2 + z^2 = A^2$ is the equation of a circle of radius A in the yz-plane, and the circle is traversed with constant angular velocity equal to the angular frequency ω. Similarly, every other point in the string undergoes circular motion, but with phase differences depending on position relative to the end of the string.

The overall motion of the string then has the appearance of a rotating helix. This particular superposition of two linearly polarized waves is called *circular polarization*. By convention, the wave is said to be *right circularly polarized* when, as in the present instance, the sense of motion of a particle, to an observer looking *backwards* along the direction of propagation, is *clockwise,* and *left circularly polarized* when it appears counterclockwise to that observer. The latter would be the result if the phase difference between y- and z-components were opposite to that in our example.

More generally, the phase difference need not be $\pi/2$. Figure 42–10 shows the result of combining a horizontal and a vertical simple harmonic motion with equal amplitudes, but with the horizontal having various phase differences with respect to the vertical. We note the following results:

1. When the phase difference is 0, 2π, or any even multiple of π, the result is a linear vibration at 45° to both original vibrations.

2. When the phase difference is π, 3π, or any *odd* multiple of π, the result is also a linear vibration, but at right angles to that corresponding to even multiples of π.

3. When the phase difference is $\pi/2$, $3\pi/2$, or any odd multiple of $\pi/2$, the resulting vibration is *circular*.

4. At all other phase differences, the resulting vibration is *elliptical*.

For electromagnetic waves of radio frequencies, circular or elliptical polarization can be produced by using two antennas at right angles, fed from the same transmitter but with a phase-shifting circuit that intro-

42-10 Vibrations that result from the combination of a horizontal and a vertical simple harmonic motion of the same frequency and the same amplitude, for various values of the phase difference.

0	$\dfrac{\pi}{4}$	$\dfrac{\pi}{2}$	$\dfrac{3\pi}{4}$	π	$\dfrac{5\pi}{4}$	$\dfrac{3\pi}{2}$	$\dfrac{7\pi}{4}$	2π

duces the desired phase difference between the two waves. For light, the necessary phase shift can be introduced by use of a birefringent material. In Fig. 42–7, when the crystal faces are parallel to the optic axis of the crystal, the ordinary and extraordinary rays are not spatially separated, but they travel through the crystal with different *speeds*, corresponding to the two different indexes of refraction. Thus if the O-wave and E-wave are in phase as they enter the crystal, they are in general *not* in phase when they emerge.

Let us consider the optical apparatus shown in Fig. 42–11. After unpolarized light traverses the polarizer, it is linearly polarized, with the polarization direction parallel to the broken line on the polarizer. This linearly polarized light then enters a crystal plate, cut so that the light travels perpendicular to the optic axis and oriented so that the optic axis makes an angle of 45° with the direction of vibration of the linearly polarized light incident upon it. Since the E-vibration is, in this case, parallel to the optic axis, and the O-vibration is perpendicular to it, it follows that the amplitudes of the E- and O-beams are identical. The E- and O-beams travel through the crystal along the same path but with different speeds and, as they are about to emerge from the second crystal face, they combine to form one of the vibrations shown in Fig. 42–10, depending on the phase difference.

The phase difference between the E- and O-vibrations at the second face of the crystal depends on the following: (1) the frequency of the light, (2) the indexes of refraction of the crystal for E- and O-light, and (3) the thickness of the crystal.

If a given crystal has such a thickness as to give rise to a phase difference of $\pi/2$ for a given frequency, then, according to Fig. 42–10, a *circular* vibration results, and the light emerging from this crystal is said to be *circularly polarized light*. The crystal itself is called a *quarter-wave plate*. If the crystal plate shown in Fig. 42–11 is a quarter-wave plate, the intensity of light transmitted by the analyzer will remain constant as the analyzer is rotated. In other words, if an analyzer alone is used to analyze circularly polarized light, it will give the same result as when used to analyze unpolarized light.

A quarter-wave plate for, say, green light is not a quarter-wave plate for any other color. Other colors have different frequencies and different E- and O-indices. Hence the phase difference would not be $\pi/2$. (See Problem 42–19.)

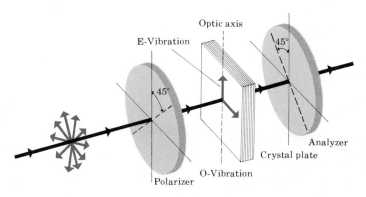

42-11 Crystal plate between crossed Polaroids. Since the optic axis makes an angle of 45° to the vibration direction of the linearly polarized light transmitted by the polarizer, the amplitudes of the O- and E-vibrations in the crystal are equal.

If the crystal has such a thickness as to give rise to a phase difference of π for a given frequency, then, according to Fig. 42–10, a linear vibration *perpendicular* to the incident vibration direction results. The light that emerges from this crystal is linearly polarized; and by rotating the analyzer, we may find a position of the analyzer at which this light will be completely stopped. A crystal plate of this sort is called a *half-wave plate* for the given frequency of light. For any other frequency it would not be a half-wave plate.

If the phase difference produced by a crystal plate is such as to produce an elliptical vibration, the emerging light is said to be *elliptically polarized*.

42-8 Optical stress analysis

When a polarizer and an analyzer are mounted in the "crossed" position, i.e., with their transmission directions at right angles to each other, no light is transmitted through the combination. But if a doubly refracting crystal is inserted between polarizer and analyzer, the light after passing through the crystal is, in general, elliptically polarized, and some light will be transmitted by the analyzer. Thus the field of view, dark in the absence of the crystal, becomes light when the crystal is inserted.

Some substances, such as glass and various plastics, while not normally doubly refracting, become so when subjected to mechanical stress. From a study of the specimen between crossed polarizers, much information regarding the stresses can be obtained. Improperly annealed glass, for example, may be internally stressed to an extent that might cause it later to develop cracks. It is evidently important that optical glass should be free from such a condition before it is subjected to expensive grinding and polishing. Hence, such glass is always examined between crossed polarizers before grinding operations are begun.

The double refraction produced by stress is the basis of the science of *photoelasticity*. The stresses in opaque engineering materials, such as girders, boiler plates, gear teeth, etc., can be analyzed by constructing a transparent model of the object, usually of a plastic, and examining it between a polarizer and an analyzer in the crossed position. Very complicated stress distributions, such as those around a hole or a gear tooth, which it would be practically impossible to analyze mathematically, may thus be studied by optical methods. Figure 42–12 is a photograph of a photoelastic model under stress.

Liquids are not normally doubly refracting, but some become so when an electric field is established with them. This phenomenon is known as the *Kerr effect*. The existence of the Kerr effect makes it possible to construct an electrically controlled "light valve." A cell with transparent walls contains the liquid between a pair of parallel plates. The cell is inserted between crossed Polaroid disks. Light is transmitted when an electric field is set up between the plates and is cut off when the field is removed.

42-12 Photoelastic stress analysis. (Courtesy of Dr. W. M. Murray, Massachusetts Institute of Technology.)

42-9 Optical activity

When a beam of linearly polarized light is sent through certain types of crystals and certain liquids, the direction of polarization of the emerging linearly polarized light is found to be different from the original direction. This phenomenon is called *rotation of the direction of polarization*, and substances that exhibit the effect are called *optically active*. Those that rotate the direction of polarization to the right, looking along the advancing beam, are called *dextrorotatory* or righthanded; those that rotate it to the left, *levorotatory* or lefthanded.

Optical activity may be due to an asymmetry of the molecules of a substance, or it may be a property of a crystal as a whole. For example, solutions of cane sugar are dextrorotatory, indicating that the optical activity is a property of the sugar molecule. The molecules of the sugars dextrose and levulose are mirror images, and their optical activities are opposite. The rotation of the plane of polarization by a sugar solution is used commercially as a method of determining the proportion of cane sugar in a given sample. Crystalline quartz is also optically active, some natural crystals being righthanded and others lefthanded. Here the optical activity is a consequence of the crystalline structure, since it disappears when the quartz is melted and allowed to resolidify into a glassy, noncrystalline state called fused quartz.

Questions

42-1 It has been proposed that automobile windshields and headlights should have polarizing filters, to reduce the glare of oncoming lights during night driving. Would this work? How should the polarizing axes be arranged? What advantages would this scheme have. What disadvantages?

42-2 A salesperson at a bargain counter claims that a certain pair of sunglasses has Polaroid filters; you suspect they are just tinted plastic. How could you find out for sure?

42-3 The fact that light consists of transverse rather than longitudinal waves was known even before the electromagnetic nature of light was firmly established. How was this determined?

42-4 When unpolarized light is incident on two crossed polarizers, no light is transmitted. A student asserted that if a third polarizer is inserted between the other two, some transmission may occur. Does this make sense? How can adding a third filter *increase* transmission?

42-5 How could you determine the direction of the polarizing axis of a single polarizer?

42-6 When unpolarized light is incident on a polarizer, only half the intensity is transmitted. What becomes of the energy of the untransmitted part? Is the answer different for the various polarizers described in Sec. 42-5?

42-7 In three-dimensional movies, two images are projected on the screen, and the viewers wear special glasses to sort them out. How does this work?

42-8 Suppose a two-slit interference experiment, as described in Chapter 41, is carried out with a polarizer in front of each slit. Would interference effects be observed when the two polarizing axes are parallel? When they are perpendicular? When they are at 45°?

42-9 The phenomenon of polarization is usually cited as evidence for the transverse nature of electromagnetic radiation. But how can we be sure it does not have *both* longitudinal and transverse components? How could one test for this possibility?

42-10 In Fig. 42-9, since the light scattered out of the incident beam is polarized, why is the transmitted beam not also partially polarized?

42-11 Light from the blue sky is strongly polarized because of the nature of the scattering process described in Sec. 42-6. But light from white clouds is usually *not* polarized. Why not?

42-12 When a sheet of plastic food wrap is placed between two crossed polarizers, no light is transmitted. When the sheet is stretched in one direction, some light passes through. What is happening?

42-13 Television transmission usually uses plane-polarized waves. It has been proposed to use circularly polarized waves to improve reception. Why? [*Hint:* See Problem 42-23.]

42-14 Given two polarizers and a quarter-wave plate, describe how one might determine completely the composition of a given light beam, with respect to its state of polarization, such as unpolarized, partially circularly polarized, completely elliptically polarized, and so on.

Problems

42–1 Unpolarized light of intensity I_0 is incident on a polarizing filter, and the emerging light strikes a second polarizing filter with its axis at 45° to that of the first. Determine

a) the intensity of the emerging beam, and

b) its state of polarization.

42–2 Three polarizing filters are stacked, with the polarizing axes of the second and third at 45° and 90°, respectively, with that of the first.

a) If unpolarized light of intensity I_0 is incident on the stack, find the intensity and state of polarization of light emerging from each filter.

b) If the second filter is removed, how does the situation change?

42–3 Three polarizing filters are stacked, with the polarizing axes of the second and third at angles θ and 90°, respectively, with that of the first. Unpolarized light of intensity I_0 is incident on the stack.

a) Derive an expression for the intensity of light transmitted through the stack, as a function of I_0 and θ.

b) Take the derivative of this expression with respect to θ to show that maximum transmission occurs when $\theta = 45°$.

42–4 Complete the derivation of Eq. (42–4) in the text.

42–5 It is desired to rotate the direction of polarization of linearly polarized light 90°, using two Polaroid filters. Explain how this can be done, and find the final intensity in terms of the incident intensity.

42–6 A polarizer and an analyzer are oriented so that the maximum amount of light is transmitted. To what fraction of its maximum value is the intensity of the transmitted light reduced when the analyzer is rotated through (a) 30°, (b) 45°, (c) 60°?

42–7 A beam of light is incident on a liquid of 1.40 refractive index. The reflected rays are completely polarized. What is the angle of refraction of the beam?

42–8 The critical angle of light in a certain substance is 45°. What is the polarizing angle?

42–9

a) At what angle above the horizontal must the sun be in order that sunlight reflected from the surface of a calm body of water shall be completely polarized?

b) What is the plane of the E-vector in the reflected light?

42–10 A parallel beam of "natural" light is incident at an angle of 58° on a plane glass surface. The reflected beam is completely linearly polarized.

a) What is the angle of refraction of the transmitted beam?

b) What is the refractive index of the glass?

42–11 The refractive index of a certain flint glass is 1.65. For what incident angle is light reflected from the surface of this glass completely polarized if the glass is immersed in (a) air? (b) water?

42–12 The Glan–Thompson prism shown in Fig. 42–8 has a horizontal length of 10 cm.

a) What is the critical angle for a ray in the cement approaching the calcite?

b) If a horizontal ray approaching the calcite is to make an angle of incidence 10° larger than the critical angle, what should be the vertical height of the prism?

42–13 A certain birefringent material has a refractive index of 1.71 for the ordinary ray and 1.74 for the extraordinary ray, for 600-nm light. What thickness of material is needed for a quarter-wave plate?

42–14 A parallel beam of linearly polarized light of wavelength 589 nm (in vacuum) is incident on a calcite crystal. Find the wavelengths of the ordinary and extraordinary waves in the crystal.

42–15 A beam of linearly polarized light strikes a calcite crystal, the direction of the electric vector making an angle of 60° with the optic axis.

a) What is the ratio of the amplitudes of the two refracted beams?

b) What is the ratio of their intensities?

42–16 Figure 42–13 represents a Wollaston prism made of two prisms of quartz cemented together. The optic axis of the righthand prism is perpendicular to the page, whereas that of the lefthand prism is parallel. The incident light is normal to the surface and gives rise to O- and E-beams that travel along the same path in the lefthand prism but with different speeds. Copy Fig. 42–13 and show on your diagram how the O- and E-beams are bent in going into the righthand prism and thence into the air.

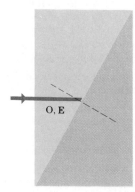

Figure 42–13

42–17 A beam of light, after passing through the Polaroid disk P_1 in Fig. 42–14, traverses a cell containing a scatter-

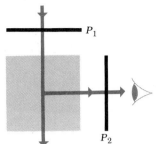

Figure 42-14

ing medium. The cell is observed at right angles through another Polaroid disk P_2. Originally, the disks are oriented until the brightness of the field as seen by the observer is a maximum.

a) Disk P_2 is now rotated through 90°. Is extinction produced?

b) Disk P_1 is now rotated through 90°. Is the field bright or dark?

c) Disk P_2 is then restored to its original position. Is the field bright or dark?

42-18 In Fig. 42-15, A and C are Polaroid sheets whose transmission directions are as indicated. B is a sheet of doubly refractive material whose optic axis is vertical. All three sheets are parallel. Unpolarized light enters from the left. Discuss the state of polarization of the light at points 2, 3, and 4.

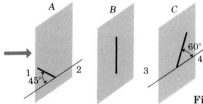

Figure 42-15

42-19 The phase difference δ between the E- and O-rays emerging from a birefringent crystal as in Fig. 42-11 is equal to 2π times the difference in the number of wavelengths in the path for the two rays. The wavelength for the O-ray is $\lambda_O = \lambda/n_O$, where λ is the wavelength in vacuum and n_O is the refractive index for the ordinary ray. Hence the number of wavelengths in a crystal of thickness t is tn_O/λ. A similar expression holds for the extraordinary ray.

a) Show that the phase difference is given by

$$\delta = \frac{2\pi}{\lambda} t(n_O - n_E).$$

b) Show that the minimum thickness of a quarter-wave plate is given by $t = \lambda/4(n_O - n_E)$.

c) What is this minimum thickness for a quarter-wave calcite plate and light of 400-nm wavelength?

42-20 What is the state of polarization of the light transmitted by a quarter-wave plate when the electric vector of the incident linearly polarized light makes an angle of 30° with the optic axis?

42-21 Assume the values of n_O and n_E for quartz to be independent of wavelength. A certain quartz crystal is a quarter-wave plate for light of wavelength 800 nm (in vacuum). What is the state of polarization of the transmitted light when linearly polarized light of wavelength 400 nm (in vacuum) is incident on the crystal, the direction of polarization making an angle of 45° with the optic axis?

42-22 Consider two vibrations, one along the y-axis,

$$y = a \sin (\omega t - \alpha),$$

and the other along the z-axis, of equal amplitude and frequency, but differing in phase,

$$z = a \sin (\omega t - \beta).$$

Let us write them as follows:

$$\frac{y}{a} = \sin \omega t \cos \alpha - \cos \omega t \sin \alpha, \qquad (1)$$

$$\frac{z}{a} = \sin \omega t \cos \beta - \cos \omega t \sin \beta. \qquad (2)$$

a) Multiply Eq. (1) by $\sin \beta$ and Eq. (2) by $\sin \alpha$ and then subtract the resulting equations.

b) Multiply Eq. (1) by $\cos \beta$ and Eq. (2) by $\cos \alpha$ and then subtract the resulting equations.

c) Square and add the results of (a) and (b).

d) Derive the equation $y^2 + z^2 - 2yz \cos \delta = a^2 \sin^2 \delta$, where $\delta = \alpha - \beta$.

e) Justify the diagrams in Fig. 42-10.

42-23 A beam of right circularly polarized light is reflected at normal incidence from a reflecting surface. Is the reflected beam right or left circularly polarized? Explain.

RELATIVISTIC MECHANICS

43

43-1 Invariance of physical laws

In preceding chapters we have stressed the importance of inertial frames of reference. Newton's laws of motion are valid only in inertial frames, but they are valid in *all* inertial frames. Any frame moving with constant velocity with respect to an inertial frame is itself an inertial frame, and all such frames are equivalent with respect to expressing the basic principles of mechanics. *The laws of mechanics are the same in every inertial frame of reference.*

Einstein proposed in 1905 that this principle should be extended to include *all* the basic laws of physics. This innocent-sounding proposition has far-reaching and startling consequences, a few of which have already been mentioned. If the principle of conservation of momentum is to be valid in all inertial systems, for example, the definition of momentum, for particles moving at speeds comparable to the speed of light, must be changed from mv to $mv/(1 - v^2/c^2)^{1/2}$, as discussed in Sec. 7–9. Even more fundamental are the modifications needed in the *kinematic* aspects of motion, as we shall see. Nevertheless, Einstein's *principle of relativity,* as it has come to be called, is now accepted as an essential requirement for a physical theory. The principle of relativity states that *the laws of physics are the same in every inertial frame of reference.*

A familiar example of this principle in electromagnetism is the induced electromotive force in a coil of wire resulting from motion of a nearby permanent magnet. In a frame of reference where the *coil* is stationary, the moving magnet causes a change of magnetic flux through the coil and hence an induced emf. In a frame where the *magnet* is stationary, the coil is moving through a magnetic field, and this motion causes magnetic-field forces on the mobile charges in the conductor, inducing an emf. According to the principle of relativity, both of these points of view have equal validity and both must predict the same result for the induced emf. As we have seen in Chapter 33, Faraday's law of electromagnetic induction does indeed satisfy this requirement.

Of equal significance is the prediction of the speed of light and other electromagnetic radiation, emerging from the development in Chapter

37. The principle of relativity requires that the speed of propagation of this radiation must be independent of the frame of reference and must be the same for all inertial frames.

Indeed, the speed of light plays a special role in the theory of relativity. Light travels in vacuum with speed c that is independent of the motion of the source. Its numerical value is known very precisely; to six significant figures it is

$$c = 2.99792 \times 10^8 \, \text{m} \cdot \text{s}^{-1}.$$

The approximate value $c = 3.00 \times 10^8 \, \text{m} \cdot \text{s}^{-1}$ is in error by less than one part in 1000 and is often used when greater precision is not required.

At one time it was thought that light traveled through a hypothetical medium called *the ether,* just as sound waves travel through air. During the late nineteenth and early twentieth centuries, intensive efforts were made to find experimental evidence for the existence of the ether. The Michelson–Morley experiment, described in Sec. 41–6, was an effort to detect motion of the earth relative to the ether. This and all similar experiments yielded consistently negative results, and it is now known that *there is no ether.* With this result in mind, let us suppose the speed of light is measured by two observers, one at rest with respect to the source, the other moving away from it. Both are in inertial frames of reference, and according to Einstein's principle of relativity, the laws of physics, and in particular the speed of light, must be the same in both frames.

For example, suppose a light source is located in a spaceship moving with respect to earth. An observer moving with the spaceship measures the speed of the light and obtains the value c. But after the light has left the source its motion cannot be influenced by the motion of the source, so an observer on earth measuring the speed of this same light must also obtain the value c, despite the fact that there is relative motion between the two observers. This conclusion may appear not to be consistent with common sense; but it is important to recognize that "common sense" is intuition based on everyday experience, and this does not usually include measurements of the speed of light. We must be prepared to accept results that seem to be in conflict with common sense when they involve realms far removed from everyday observation.

Thus the speed of light (in vacuum) is independent of the motion of the source and is the same in all frames of reference. To explore the consequences of this statement, we consider the newtonian relationship between two inertial frames, labeled S and S' in Fig. 43–1. Let the x-axes of the two frames lie along the same line, but let the origin O' of S' move relative to the origin O of S with constant velocity u along the common x-axis. If the two origins coincide at time $t = 0$, then their separation at a later time t is ut.

A point P may be described by the coordinates (x, y, z) in S or by coordinates (x', y', z') in S'. Reference to the figure shows that these are related by

$$x = x' + ut, \qquad y = y', \qquad z = z'. \tag{43-1}$$

These equations are called the *galilean coordinate transformation.*

If the point P moves in the x-direction, its velocity v relative to S is given by $v = \Delta x / \Delta t$, and its velocity v' relative to S' is $v' = \Delta x' / \Delta t$. Intu-

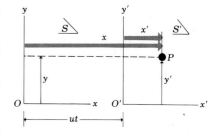

43-1 The position of point P can be described by the coordinates x and y in frame of reference S, or by x' and y' in S'. S' moves relative to S with constant velocity u along the common x–x' axis. The two origins O and O' coincide at time $t = t' = 0$.

itively it is clear that these are related by

$$v = v' + u. \tag{43-2}$$

This relation may also be obtained formally from Eqs. (43–1). Suppose the particle is at a point described by coordinate x_1 or x'_1 at time t_1, and at x_2 or x'_2 at time t_2. Then $\Delta t = t_2 - t_1$, and, from Eq. (43–1),

$$\Delta x = x_2 - x_1 = (x'_2 - x'_1) + u(t_2 - t_1) = \Delta x' + u\,\Delta t,$$
$$\frac{\Delta x}{\Delta t} = \frac{\Delta x'}{\Delta t} + u,$$

and, in the limit as $\Delta t \to 0$,

$$v = v' + u,$$

in agreement with Eq. (43–2).

A fundamental problem now appears. Applied to the speed of light, Eq. (43–2) says $c = c' + u$. Einstein's principle of relativity, supported by experimental observations, says $c = c'$. This is a genuine inconsistency, not an illusion, and it demands resolution. If we accept the principle of relativity, we are forced to conclude that Eqs. (43–1) and (43–2), intuitively appealing as they are, cannot be correct, but need to be modified to bring them into harmony with this principle.

The resolution involves modifications of our basic kinematic concepts. The first of these involves an assumption so fundamental that it might seem unnecessary, namely the assumption that the same *time scale* is used in frames S and S'. This may be stated formally by adding to Eqs. (43–1) a fourth equation $t = t'$. Obvious though this assumption may seem, it turns out not to be correct when the relative speed u of the two frames of reference is comparable to the speed of light. The difficulty lies in the concept of *simultaneity,* which we examine next.

43-2 Relative nature of simultaneity

Measuring times and time intervals involves the concept of *simultaneity*. When a person says he awoke at seven o'clock he means that two *events,* his awakening and the arrival of the hour hand of his clock at the number seven, occurred *simultaneously*. The fundamental problem in measuring time intervals is that, in general, two events that appear simultaneous in one frame of reference *do not* appear simultaneous in a second frame that is moving relative to the first, even if both are inertial frames.

The following thought experiment, devised by Einstein, illustrates this point. Consider a long train moving with uniform velocity, as shown in Fig. 43–2a. Two lightning bolts strike the train, one at each end. Each bolt leaves a mark on the train and one on the ground at the same instant. The points on the ground are labeled A and B in the figure, and the corresponding points on the train A' and B'. An observer on the ground is located at O, midway between A and B; another observer is at O', moving with the train and midway between A' and B'. Both observers use the light signals from the lightning to observe the events.

Suppose the two light signals reach the observer at O simultaneously; he concludes that the two events took place at A and B simultaneously. But the observer at O' is moving with the train, and the light pulse from B' reaches him before the light pulse from A' does; he con-

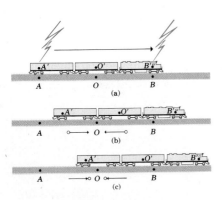

43-2 (a) To the stationary observer at point O, two lightning bolts appear to strike simultaneously. (b) The moving observer at point O' sees the light from the front of the train first and thinks that the bolt at the front struck first. (c) The two light pulses arrive at O simultaneously.

cludes that the event at the front of the train happened *earlier* than that at the rear. Thus the two events appear simultaneous to one observer, but not to the other. *Whether or not two events at different space points are simultaneous depends on the state of motion of the observer.* It follows that *the time interval between two events at different space points is in general different for two observers in relative motion.*

It might be argued that, in this example, the lightning bolts really *are* simultaneous, and that if the observer at O' could communicate with the distant points without time delay, he would realize this. But the finite speed of information transmission is not the problem. If O' is really midway between A' and B', then, in his frame of reference, the time for a signal to travel from A' to O' is the same as from B' to O'. Two signals arrive simultaneously at O' only if they were emitted simultaneously at A' and B'; in this example they *do not* arrive simultaneously at O', and so O' must conclude that the events at A' and B' were *not* simultaneous.

Furthermore, there is no basis for saying either that O is right and O' is wrong, or the reverse since, according to the principle of relativity, no inertial frame of reference is preferred over any other in the formulation of physical laws. Each observer is correct *in his own frame of reference,* but simultaneity is not an absolute concept. Whether or not two events are simultaneous depends on the frame of reference, and the time interval between two events depends on the frame of reference.

43-3 Relativity of time

To derive a quantitative relation between time intervals in different coordinate systems, we consider another thought experiment. As before, a frame of reference S' moves with velocity u relative to a frame S. An observer in S' directs a source of light at a mirror a distance d away, as shown in Fig. 43-3, and measures the time interval $\Delta t'$ for light to make the "round trip" to the mirror and back. The total distance is $2d$, so the time interval is

$$\Delta t' = \frac{2d}{c}. \tag{43-3}$$

As measured in frame S, the time for the round trip is a different interval Δt. During this time, the source moves relative to S a distance $u\,\Delta t$, and the total round-trip distance is not just $2d$ but is $2l$, where

$$l = \sqrt{d^2 + \left(\frac{u\,\Delta t}{2}\right)^2}.$$

The speed of light is the same for both observers, so the relation in S analogous to Eq. (43-3) is

$$\Delta t = \frac{2l}{c} = \frac{2}{c}\sqrt{d^2 + \left(\frac{u\,\Delta t}{2}\right)^2}. \tag{43-4}$$

To obtain a relation between Δt and $\Delta t'$ that does not contain d, we solve Eq. (43-3) for d and substitute the result into Eq. (43-4), obtaining

$$\Delta t = \frac{2}{c}\sqrt{\left(\frac{c\,\Delta t'}{2}\right)^2 + \left(\frac{u\,\Delta t}{2}\right)^2}.$$

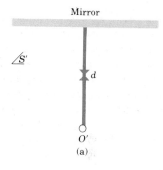

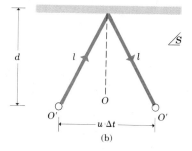

43-3 (a) Light pulse emitted from source at O' and reflected back along the same line, as observed in S'. (b) Path of the same light pulse, as observed in S. The positions of O' at the times of departure and return of the pulse are shown. The speed of the pulse is the same in S as in S', but the path is longer in S.

This may now be squared and solved for Δt; the result is

$$\Delta t = \frac{\Delta t'}{\sqrt{1 - u^2/c^2}}. \tag{43-5}$$

We may generalize this important result: If a time interval $\Delta t'$ separates two events occurring at the same space point in a frame of reference S' (in this case, the departure and arrival of the light signal at O'), then the time interval Δt between these two events as observed in S is *larger* than $\Delta t'$, and the two are related by Eq. (43-5). Thus when the rate of a clock at rest in S' is measured by an observer in S, the rate measured in S is *slower* than the rate observed in S'. This effect is called *time dilation*.

It is important to note that the observer in S measuring the time interval Δt cannot do so with a single clock. In Fig. 43-3 the points of departure and return of the light pulse are different space points in S, although they are the same point in S'. If S tries to use a single clock, the finite time of communication between two points will becloud the issue. To avoid this, S may use two clocks at the two relevant points. There is no difficulty in synchronizing two clocks in the same frame of reference; one procedure is to send a light pulse simultaneously to two clocks from a point midway between them, with the two operators setting their clocks to a predetermined time when the pulses arrive. In thought experiments it is often helpful to imagine a large number of synchronized clocks distributed conveniently in a single frame of reference. Only when a clock is moving relative to a given frame of reference do ambiguities of synchronization or simultaneity arise.

Example A spaceship flies past earth with a speed of $0.99c$ (about 2.97×10^8 m·s^{-1}). A high-intensity signal light (perhaps a pulsed laser) blinks on and off, each pulse lasting 2×10^{-6} s. At a certain instant the ship appears to an earthling observer to be directly overhead at an altitude of 1000 km, and to be traveling perpendicular to the line of sight. What is the duration of each light pulse, as measured by this observer, and how far does the ship travel relative to earth during each pulse?

Solution The observer does not see the pulse at the instant it is emitted, because the light signal requires a time equal to $(1000 \times 10^3 \text{ m})/(3 \times 10^8 \text{ m·s}^{-1})$, or $(1/300)$ s, to travel from the ship to earth. But if the distance from the spaceship to observer is essentially constant during the emission of a pulse, the time delays at beginning and end of the pulse are equal and the time *interval* is not affected.

Let S be the earth's frame of reference, S' that of the spaceship. Then, in the notation of Eq. (43-5), $\Delta t' = 2 \times 10^{-6}$ s. This interval refers to two events occurring at the same point relative to S', namely, the starting and stopping of the pulse. The corresponding interval in S is given by Eq. (43-5):

$$\Delta t = \frac{\Delta t'}{\sqrt{1 - u^2/c^2}} = \frac{2 \times 10^{-6} \text{ s}}{\sqrt{1 - (0.99)^2}} = 14.2 \times 10^{-6} \text{ s}.$$

Thus the time dilation in S is about a factor of seven. The distance D traveled in S during this interval is

$$D = u\,\Delta t = (0.99)(3 \times 10^8 \text{ m·s}^{-1})(14.2 \times 10^{-6} \text{ s})$$
$$= 4220 \text{ m} = 4.22 \text{ km}.$$

If the spaceship is traveling directly *toward* the observer, the time interval cannot be measured directly by a single observer because the time delay is not the same at the beginning and end of the pulse. One possible scheme, at least in principle, is to use *two* observers at rest in S, with synchronized clocks, one at the position of the ship when the pulse starts, the other at its position at the end of the pulse. These observers will again measure a time interval in S of 14.2×10^{-6} s.

Time-dilation effects are not observed in everyday life because the speeds of all practical modes of transportation are much less than the speed of light. For example, for a jet airplane flying at 600 mi·hr⁻¹ (about 270 m·s⁻¹),

$$\frac{v^2}{c^2} = \left(\frac{270 \text{ m·s}^{-1}}{3 \times 10^8 \text{ m·s}^{-1}}\right)^2 = 8.1 \times 10^{-13},$$

and the time-dilation factor in Eq. (43-5) is approximately $1 + (4 \times 10^{-13})$. Thus to observe time dilation in this situation requires a clock with a precision of the order of one part in 10^{13}. But, as noted in Sec. 1-2, atomic clocks capable of this precision have recently been developed, and in the past few years experiments with such clocks in jet airplanes have verified Eq. (43-5) directly.

From the derivation of Eq. (43-5) and the spaceship example, it can be seen that a time interval between two events occurring *at the same point* in a given frame of reference is a more fundamental quantity than an interval between events at different points. The term *proper time* is used to denote an interval between two events occurring at the same space point. Thus Eq. (43-5) may be used *only* when $\Delta t'$ is a proper time interval in S', in which case Δt is *not* a proper time interval in S. If, instead, Δt is proper in S, then Δt and $\Delta t'$ must be interchanged in Eq. (43-5).

When the relative velocity u of S and S' is very small, the factor $(1 - u^2/c^2)$ is very nearly equal to unity, and Eq. (43-5) approaches the newtonian relation $\Delta t = \Delta t'$ (i.e., the same time scale for all frames of reference). This assumption, therefore, retains its validity in the limit of small relative velocities.

43-4 Relativity of length

Just as the time interval between two events depends on the frame of reference, the *distance* between two points also depends on the frame of reference. To measure a distance one must, in principle, observe the positions of two points, such as the two ends of a ruler, simultaneously; but what is simultaneous in one frame is not in another.

To develop a relation between lengths in various coordinate systems, we consider another thought experiment. We attach a source of light pulses to one end of a ruler and a mirror to the other, as shown in Fig. 43-4. Let the ruler be at rest in S' and its length in this frame be l'. Then the time $\Delta t'$ required for a light pulse to make the round trip from source to mirror and back is given by

$$\Delta t' = \frac{2l'}{c}. \tag{43-6}$$

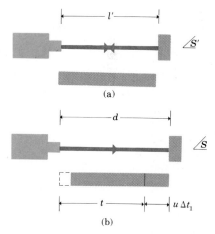

43-4 (a) A light pulse is emitted from a source at one end of a ruler, reflected from a mirror at the opposite end, and returned to the source position. (b) Motion of the light pulse as seen by an observer in S. The distance traveled from source to mirror is greater than the length l measured in S, by the amount $u \Delta t_1$, as shown.

This is a proper time interval, since departure and return occur at the same point in S'.

In S the ruler is displaced during this travel of the light pulse. Let the length of the ruler in S be l, and let the time of travel from source to mirror, as measured in S, be Δt_1. During this interval the mirror moves a distance $u\,\Delta t_1$, and the total length of path d from source to mirror is not l but

$$d = l + u\,\Delta t_1. \tag{43-7}$$

But since the pulse travels with speed c, it is also true that

$$d = c\,\Delta t_1. \tag{43-8}$$

Combining Eqs. (43-7) and (43-8) to eliminate d,

$$c\,\Delta t_1 = l + u\,\Delta t_1,$$

or

$$\Delta t_1 = \frac{l}{c - u}. \tag{43-9}$$

In the same way it can be shown that the time Δt_2 for the return trip from mirror to source is

$$\Delta t_2 = \frac{l}{c + u}. \tag{43-10}$$

The *total* time $\Delta t = \Delta t_1 + \Delta t_2$ for the round trip, as measured in S, is

$$\Delta t = \frac{l}{c - u} + \frac{l}{c + u} = \frac{2l}{c(1 - u^2/c^2)}. \tag{43-11}$$

It is also known that Δt and $\Delta t'$ are related by Eq. (43-5), since $\Delta t'$ is proper in S'. Thus Eq. (43-6) becomes

$$\Delta t\,\sqrt{1 - \frac{u^2}{c^2}} = \frac{2l'}{c}. \tag{43-12}$$

Finally, combining this with Eq. (43-11) to eliminate Δt, and simplifying we obtain

$$l = l'\,\sqrt{1 - \frac{u^2}{c^2}}. \tag{43-13}$$

Thus the length measured in S, in which the ruler is moving, is *shorter* than in S', where it is at rest. A length measured in the rest frame of the body is called a proper length; thus, l' above is a proper length in S', and the length measured in any other frame is less than l'. This effect is called *contraction of length*.

Example   In the spaceship example of Sec. 43-3, what distance does the spaceship travel during emission of a pulse, as measured in its rest frame?

Solution   The question is somewhat ambiguous since, of course, in its own frame of reference the ship is at rest. But suppose it leaves markers in space, such as small smoke bombs, at the instants when the pulse starts and stops, and measures the distance between these markers, with the aid of observers behind the ship but moving with it, each with a clock synchronized with that of the ship. The distance d between the markers is a proper length in the earth's frame S. In the spaceship's frame S', the distance d' is contracted by the factor given in Eq. (43–13):

$$d' = d \sqrt{1 - \frac{u^2}{c^2}} = (4220 \text{ m}) \sqrt{1 - (0.99)^2}$$

$$= 595 \text{ m}.$$

(Note that because d, not d', is a proper length, we must reverse the roles of l and l'.) An observer in the spaceship can calculate his speed relative to earth from this data:

$$u = \frac{d'}{\Delta t'} = \frac{595 \text{ m}}{2 \times 10^{-6} \text{ s}} = 2.97 \times 10^8 \text{ m} \cdot \text{s}^{-1},$$

which agrees with the initial data.

When u is very small compared to c, the contraction factor in Eq. (43–13) approaches unity, and in the limit of small speeds we recover the newtonian relation $l = l'$. This and the corresponding result for time dilation show that Eqs. (43–1) retain their validity in the limit of small speeds; only at speeds comparable to c are modifications needed.

Lengths measured perpendicular to the direction of motion are *not* contracted; this may be verified by constructing a thought experiment for measuring in S and S' the length of a ruler oriented perpendicular to the direction of relative motion. The details of this discussion are not essential for our further work and will not be given here.

*43–5 The Lorentz transformation

The galilean coordinate transformation given by Eqs. (43–1) is valid only in the limit when u is much smaller than c, but we are now in position to derive a more general transformation not subject to this limitation. The more general relations are called the *Lorentz transformation*. When u is small they reduce to the galilean transformation, but they may also be used when u is comparable to c.

The basic problem is this: When an event occurs at point (x, y, z) at time t, as observed in a frame of reference S, what are the coordinates (x', y', z') and time t' of the event as observed in a second frame S' moving relative to S with constant velocity u along the x-direction?

To derive the transformation equations, we return to Fig. 43–1. As before, we assume that the origins coincide at the initial time $t = t' = 0$. Then in S the distance from O to O' at time t is still ut. The coordinate x' is a proper length in S', so in S it appears contracted by the factor given in Eq. (43–13). Thus the distance x from O to P in S is given not simply by $x = ut + x'$ as in the galilean transformation, but by

$$x = ut + x' \sqrt{1 - \frac{u^2}{c^2}}. \tag{43–14}$$

Solving this equation for x', we obtain

$$x' = \frac{x - ut}{\sqrt{1 - u^2/c^2}}. \tag{43–15}$$

This is half of the Lorentz transformation; the other half is the equation giving t' in terms of x and t. To obtain this note that the principle of relativity requires that the *form* of the transformation from S to S' must be identical to that from S' to S, the only difference being a change in the sign of the relative velocity u. Thus, from Eq. (43–14), it must be true that

$$x' = -ut' + x\sqrt{1 - \frac{u^2}{c^2}}. \qquad (43\text{–}16)$$

We may now equate Eqs. (43–15) and (43–16) to eliminate x' from this expression, obtaining the desired relation for t' in terms of x and t. The algebraic details will be omitted; the result is

$$t' = \frac{t - ux/c^2}{\sqrt{1 - u^2/c^2}}. \qquad (43\text{–}17)$$

As remarked previously, lengths perpendicular to the direction of relative motion are not affected by the motion, so $y' = y$ and $z' = z$.

Collecting all the transformation equations, we have

$$
\begin{aligned}
x' &= \frac{x - ut}{\sqrt{1 - u^2/c^2}}, \\
y' &= y, \\
z' &= z, \\
t' &= \frac{t - ux/c^2}{\sqrt{1 - u^2/c^2}}.
\end{aligned}
\qquad (43\text{–}18)
$$

These are the *Lorentz tranformation equations*, the relativistic generalization of the galilean transformation, Eqs. (43–1). When u is much smaller than c, the two transformations become identical.

Next we consider the relativistic generalization of the velocity transformation relation, Eq. (43–2), which, as peviously noted, is valid only when u is very small. The relativistic expression can easily be obtained from the Lorentz transformation. Suppose that a body observed in S' is at point x_1' at time t_1' and point x_2' at time t_2'. Then its speed v' in S' is given by

$$v' = \frac{x_2' - x_1'}{t_2' - t_1'} = \frac{\Delta x'}{\Delta t'}. \qquad (43\text{–}19)$$

To obtain the speed in S we use Eqs. (43–18) to translate this expression into terms of the corresponding positions x_1 and x_2 and times t_1 and t_2 observed in S. We find

$$
\begin{aligned}
x_2' - x_1' &= \frac{x_2 - x_1 - u(t_2 - t_1)}{\sqrt{1 - u^2/c^2}} = \frac{\Delta x - u\,\Delta t}{\sqrt{1 - u^2/c^2}}, \\
t_2' - t_1' &= \frac{t_2 - t_1 - u(x_2 - x_1)/c^2}{\sqrt{1 - u^2/c^2}} = \frac{\Delta t - u\,\Delta x/c^2}{\sqrt{1 - u^2/c^2}}.
\end{aligned}
$$

Using these results in Eq. (43–19), we find

$$v' = \frac{\Delta x - u\,\Delta t}{\Delta t - u\,\Delta x/c^2} = \frac{(\Delta x/\Delta t) - u}{1 - (u/c^2)(\Delta x/\Delta t)}.$$

Now $\Delta x/\Delta t$ is just the velocity v measured in S, so we finally obtain

$$v' = \frac{v - u}{1 - uv/c^2}. \qquad (43\text{-}20)$$

We note that when u and v are much smaller than c, the denominator becomes equal to unity, and we obtain the nonrelativistic result $v' = v - u$. The opposite extreme is the case $v = c$; then we find

$$v' = \frac{c - u}{1 - uc/c^2} = c.$$

That is, anything moving with speed c relative to S also has speed c relative to S', despite the relative motion of the two frames. This result demonstrates the consistency of Eq. (43-20) with our initial assumption that the speed of light is the same in all frames of reference.

Equation (43-20) may also be rearranged to give v in terms of v'. The algebraic details are left as a problem; the result is

$$v = \frac{v' + u}{1 + uv'/c^2}. \qquad (43\text{-}21)$$

Example   A spaceship moving away from earth with a speed $0.9c$ fires a missile in the same direction as its motion, with a speed $0.9c$ relative to the spaceship. What is the missile's speed relative to earth?

Solution   Let the earth's frame of reference be S, the spaceship's S'. Then $v' = 0.9c$ and $u = 0.9c$. The nonrelativistic velocity addition formula would give a velocity relative to earth of $1.8\ c$. The correct relativistic result, obtained from Eq. (43-21), is

$$v = \frac{0.9c + 0.9c}{1 + (0.9c)(0.9c)/c^2} = 0.994c.$$

When u is less than c, a body moving with a speed less than c in one frame of reference also has a speed less than c in *every other* frame of reference. This is one reason for thinking that no material body may travel with a speed greater than that of light, relative to any frame of reference. The relativistic generalizations of energy and momentum, to be considered next, give further support to this hypothesis.

43-6 Momentum

We have discussed the fact that Newton's laws of motion are *invariant* under the galilean coordinate transformation, but that to satisfy the principle of relativity this transformation must be replaced by the more general Lorentz transformation. This requires corresponding generalizations in the laws of motion and the definitions of momentum and energy.

The principle of conservation of momentum states that *when two bodies collide, the total momentum is constant,* provided there is no interaction except that of the two bodies with each other. However, when one considers a collision in one coordinate system S, in which momentum is conserved, and then uses the Lorentz transformation to obtain the velocities in a second system S', it is found that if the newtonian definition of momentum ($p = mv$) is used, momentum is *not* conserved in the second system. Thus if the Lorentz transformation is correct, and if we believe in the principle of relativity (i.e., momentum conservation must hold in *all* systems), the only choice remaining is to modify the *definition* of momentum.

Deriving the correct relativistic generalization is beyond our scope, and we simply quote the result, which has already been mentioned in Sec. 7–9:

$$p = \frac{mv}{\sqrt{1 - v^2/c^2}}. \tag{43-22}$$

We note that, as usual when the particle's speed v is much less than c, this reduces to the newtonian expression $p = mv$, but that in general the momentum is greater in magnitude than mv.

In newtonian mechanics the second law of motion can be stated in the form

$$F = \frac{dp}{dt}. \tag{43-23}$$

That is, force equals time rate of change of momentum. Experiment shows that this result is still valid in relativistic mechanics, provided we use the relativistic momentum given by Eq. (43–22). That is, the relativistically correct generalization of Newton's second law is found to be

$$F = \frac{d}{dt} \frac{mv}{\sqrt{1 - v^2/c^2}}. \tag{43-24}$$

This relation has the consequence that constant force no longer causes constant acceleration. For example, when force and velocity are both along a single line, causing straight-line motion, it is not difficult to show that Eq. (43–24) becomes

$$F = \frac{m}{(1 - v^2/c^2)^{3/2}} \frac{dv}{dt},$$

or

$$\frac{dv}{dt} = a = \frac{F}{m} (1 - v^2/c^2)^{3/2}. \tag{43-25}$$

The derivation of Eq. (43–25) is left as a problem; the equation shows that as a particle's speed increases, the acceleration for a given force continuously *decreases*. As the speed approaches c, the acceleration approaches zero, no matter how great the force. Thus it is impossible to accelerate a particle from a state of rest to a speed equal to or greater than c, and the speed of light is sometimes referred to as "the ultimate speed."

43–7 Work and energy

The work–energy relation developed in Chapter 6 made use of Newton's laws of motion. Since these must be generalized to bring them into accord with the principle of relativity, it is not surprising that the work–energy relation also requires generalization.

Because a constant force on a body no longer causes a constant acceleration (except at very small velocities), even the simplest dynamics problems require the use of the calculus. We may, however, follow the

same pattern as that of Sec. 6–3 in deriving the relativistic generalization of the work–energy principle. We begin with the definition of work:
$W = \int F \, dx$. According to Eq. (43–23), $F = dp/dt$. By repeated application of the chain rule for derivatives, we obtain

$$F = \frac{dp}{dt} = \frac{dp}{dv}\frac{dv}{dx}\frac{dx}{dt} = \frac{dp}{dv}\frac{dv}{dx}v. \qquad (43\text{–}26)$$

Thus if the particle has speed v_1 at point x_1 and v_2 at x_2, the work done
by F during the motion from x_1 to x_2 may be expressed as

$$W = \int_{x_1}^{x_2} F \, dx = \int_{x_1}^{x_2} \frac{dp}{dv}\frac{dv}{dx} v \, dx.$$

This may now be converted into an integral on v, with $dv = (dv/dx) \, dx$:

$$W = \int_{v_1}^{v_2} \frac{dp}{dv} v \, dv. \qquad (43\text{–}27)$$

An expression for dp/dv may be obtained by taking the derivative of
Eq. (43–22). The reader may verify that the result is

$$\frac{dp}{dv} = \frac{m}{(1 - v^2/c^2)^{3/2}}. \qquad (43\text{–}28)$$

Substituting this result in Eq. (43–27) and integrating, we find

$$W = \int_{v_1}^{v_2} \frac{mv \, dv}{(1 - v^2/c^2)^{3/2}}$$

$$= \frac{mc^2}{\sqrt{1 - v_2^2/c^2}} - \frac{mc^2}{\sqrt{1 - v_1^2/c^2}}, \qquad (43\text{–}29)$$

where v_1 and v_2 are the initial and final velocities of the particle.
    This result suggests defining kinetic energy as

$$\frac{mc^2}{\sqrt{1 - v^2/c^2}}. \qquad (43\text{–}30)$$

But this expression is not zero when $v = 0$; instead it becomes equal to
mc^2. Thus the correct relativistic generalization of kinetic energy K is:

$$K = \frac{mc^2}{\sqrt{1 - v^2/c^2}} - mc^2. \qquad (43\text{–}31)$$

This expression, if correct, must reduce to the newtonian expression
$K = \frac{1}{2}mv^2$ when v is much smaller than c. It is not obvious that this is
the case; to demonstrate that it is so, we can expand the radical using
the binomial theorem:

$$\left(1 - \frac{v^2}{c^2}\right)^{-1/2} = 1 + \frac{1}{2}\frac{v^2}{c^2} + \frac{3}{8}\frac{v^4}{c^4} + \frac{5}{16}\frac{v^6}{c^6} + \cdots.$$

Combining this with Eq. (43–31), we find

$$K = mc^2\left(1 + \frac{1}{2}\frac{v^2}{c^2} + \frac{3}{8}\frac{v^4}{c^4} + \cdots\right) - mc^2$$

$$= \frac{1}{2}mv^2 + \frac{3}{8}m\frac{v^4}{c^2} + \cdots. \qquad (43\text{–}32)$$

In each expression the dots stand for omitted terms. When v is much smaller than c, all the terms in the series except the first are negligibly small, and we obtain the classical $\frac{1}{2}mv^2$.

But what is the significance of the term mc^2 that had to be subtracted in Eq. (43–31)? Although Eq. (43–30) does not give the *kinetic energy* of the particle, perhaps it represents some kind of *total* energy, including both the kinetic energy and an additional energy mc^2, which the particle possesses even when it is not moving. Calling this total energy E and using Eq. (43–31), we find

$$E = K + mc^2 = \frac{mc^2}{\sqrt{1 - v^2/c^2}}. \tag{43–33}$$

The hypothetical energy mc^2 associated with mass rather than motion may be called the *rest energy* of the particle. This speculation does not prove that the concept of rest energy is meaningful, but it points the way toward further investigation.

There is in fact direct experimental evidence of the existence of rest energy. The simplest example is the decay of the π^0 meson, an unstable particle that "decays"; in the decay process, the particle disappears and electromagnetic radiation appears. When the particle is at rest (and therefore with no kinetic energy) before its decay, the total energy of the radiation produced is found to be exactly equal to mc^2. There are many other examples of fundamental particle transformations in which the total mass of the system changes, and in every case there is a corresponding energy change consistent with the assumption of a rest energy mc^2 associated with a mass m.

Although the principles of conservation of mass and of energy originally developed quite independently, the theory of relativity shows that they are but two special cases of a single broader conservation principle, the *principle of conservation of mass and energy*. There are physical phenomena where neither mass nor energy is separately conserved, but where the changes in these quantities are governed by the more general relation that a change m in the mass of the system must be accompanied by an opposite change mc^2 in its energy.

The term *mass* as used here always means the rest mass of a particle, the inertial mass m measured through Eq. (43–22). For a given particle, m is a constant, independent of the state of motion of the particle. It is sometimes useful to introduce the concept of a variable, velocity-dependent relativistic mass. This concept is not needed in the present analysis, and it is not used in this discussion.

The possibility of conversion of mass into energy is the fundamental principle involved in the generation of power through nuclear reactions, a subject to be discussed in later chapters. When a uranium nucleus undergoes fission in a nuclear reactor, the total mass of the resulting fragments is less than that of the parent nucleus, and the total kinetic energy of the fragments is equal to this mass deficit times c^2. This kinetic energy can be used to produce steam to operate turbines for electric power generators or in a variety of other ways.

The total energy E (kinetic plus rest) of a particle is related simply to its momentum, as shown by combining Eqs. (43–22) and (43–33) to eliminate the particle's velocity. This is most easily accomplished by rewriting these equations in the following forms:

$$\left(\frac{E}{mc^2}\right)^2 = \frac{1}{1 - v^2/c^2}; \qquad \left(\frac{p}{mc}\right)^2 = \frac{v^2/c^2}{1 - v^2/c^2}.$$

Subtracting the second of these from the first and rearranging, we find

$$E^2 = (mc^2)^2 + (pc)^2. \qquad (43\text{--}34)$$

Again we see that for a particle at rest ($p = 0$), $E = mc^2$. Equation (43–34) also suggests that a particle may have energy and momentum even when it has no rest mass. In such a case, $m = 0$ and

$$E = pc. \qquad (43\text{--}35)$$

Massless particles, including photons, the quanta of electromagnetic radiation, and others, were mentioned in Sec. 7–9. The existence of such particles is well established, and they will be discussed in greater detail in later chapters. These particles always travel with the speed of light; they are emitted and absorbed during changes of state of atomic systems, accompanied by corresponding changes in the energy and momentum of these systems.

43-8 Relativity and newtonian mechanics

The sweeping changes required by the principle of relativity go to the very roots of newtonian mechanics, including the concepts of length and time, the equations of motion, and the conservation principles. Thus it may appear that foundations on which Newton's mechanics are built have been destroyed. While this is true in one sense, it is essential to keep in mind that the newtonian formulation still retains its validity whenever speeds are small compared with the speed of light. In such cases time dilation, length contraction, and the modifications of the laws of motion are so small that they are unobservable. In fact, every one of the principles of newtonian mechanics survives as a special case of the relativistic formulation.

Relativity does not *contradict* the older mechanics but *generalizes* it. After all, Newton's laws rest on a very solid base of experimental evidence, and it would be very strange indeed to advance a new theory inconsistent with this evidence. So it always is with the development of physical theory. Whenever a new theory is in partial conflict with an older, established theory, it nevertheless must yield the same predictions as the old in areas where the old theory is supported by experimental evidence. Every new physical theory must pass this test, called the *correspondence principle*, which has come to be regarded as a fundamental procedural rule in all physical theory. There are many problems for which newtonian mechanics is clearly inadequate, including all situations where particle speeds approach that of light or there is direct conversion of mass to energy. But there is still a large area, including nearly all the behavior of macroscopic bodies in mechanical systems, in which newtonian mechanics is still perfectly adequate.

At this point it is legitimate to ask whether the relativistic mechanics just discussed is the final word on this subject or whether *further* generalizations are possible or necessary. For example, inertial frames of reference have occupied a privileged position in all our discussion thus far. Should the principle of relativity be extended to non-inertial frames as well?

Here is an example to illustrate some implications of this question. A man decides to go over Niagara Falls while enclosed in a large wooden box. During his free fall over the falls he can, in principle, perform experiments inside the box. An object released inside the box does not fall to the floor because both the box and the object are in free fall with a downward acceleration of $9.8 \text{ m} \cdot \text{s}^{-2}$. But an alternative interpretation, from this man's point of view, is that the force of gravity has suddenly been turned off. Provided he remains in the box and it remains in free fall, he cannot tell whether he is indeed in free fall or whether the force of gravity has vanished. A similar problem appears in a space station in orbit around the earth. Objects in the spaceship appear weightless, but without going outside the ship there is no way to determine whether gravity has disappeared or the spaceship is in an accelerated (i.e., non-inertial) frame of reference.

These considerations form the basis of Einstein's *general theory of relativity*. If one cannot distinguish experimentally between a gravitational field and an accelerated reference system, then there can be no real distinction between the two. Pursuing this concept, we may try to represent *any* gravitational field in terms of special characteristics of the coordinate system. This turns out to require even more sweeping revisions of our space–time concepts than did the special theory of relativity, and we find that, in general, the geometric properties of the space are non-euclidean.

The basic ideas of the general theory of relativity are now well established, but some of the details remain speculative in nature. Its chief application is in cosmological investigations of the structure of the universe, the formation and evolution of stars, and related matters. It is not believed to have any relevance for atomic or nuclear phenomena or macroscopic mechanical problems of less than astronomical dimensions.

Questions

43-1 What do you think would be different in everyday life if the speed of light were $10 \text{ m} \cdot \text{s}^{-1}$ instead of its actual value?

43-2 The average life span in the United States is about 70 years. Does this mean that it is impossible for an average person to travel a distance greater than 70 light-years away from the earth? (A light-year is the distance light travels in a year.)

43-3 What are the fundamental distinctions between an inertial frame of reference and a noninertial frame?

43-4 A physicist claimed that it is impossible to define what is meant by a rigid body in a relativistically correct way. Why?

43-5 Two events occur at the same space point in a particular frame of reference and appear simultaneous in that frame. Is it possible that they may not appear simultaneous in another frame?

43-6 If simultaneity is not an absolute concept, does this also destroy the concept of *causality*? If event A is to *cause* event B, A must occur first. Is it possible that in some frames A may appear to cause B, and in others B appears to cause A?

43-7 A social scientist who has done distinguished work in fields far removed from physics has written a book purporting to refute the special theory of relativity. He begins with a premise that might be paraphrased as follows: "Either two events occur at the same time, or they don't; that's just common sense." How would you respond to this in the light of our discussion of the relative nature of simultaneity?

43-8 When an object travels across an observer's field of view at a relativistic speed, it appears not only foreshortened but also slightly rotated, with the side toward the observer shifted in the direction of motion relative to the side away from him. How does this come about?

43-9 According to the famous "twin paradox," if one twin stays on earth while the other takes off in a spaceship at relativistic speed and then returns, one will be older than the other. Which is older? Does this really happen? Can you think of a practical experiment, perhaps using two very precise atomic clocks, that would test this conclusion?

43-10 When a monochromatic light source moves toward an observer, its wavelength appears to be shorter than the value measured when the source is at rest. Does this contradict the hypothesis that the speed of light is the same for all observers? What about the apparent frequency of light from a moving source?

43-11 A student asserted that a massive particle must always have a speed less than that of light, while a massless particle must always travel at exactly the speed of light. Is he correct? If so, how do massless particles such as photons and neutrinos acquire this speed? Can't they start from rest and accelerate?

43-12 The theory of relativity sets an upper limit on the speed a particle can have. Are there also limits on its energy and momentum?

43-13 In principle, does a hot gas have more mass than the same gas when it is cold? Explain. In practice would this be a measurable effect?

43-14 Why do you think the development of newtonian mechanics preceded the more refined relativistic mechanics by so many years?

Problems

43-1 The π^+ meson, an unstable particle, lives, on the average, about 2.6×10^{-8} s (measured in its own frame of reference) before decaying.

a) If such a particle is moving with respect to the laboratory with a speed of $0.8\,c$, what lifetime is measured in the laboratory?

b) What distance, measured in the laboratory, does the particle move before decaying?

43-2 The μ^+ meson (or positive muon) is an unstable particle with a lifetime of about 2.3×10^{-6} s (measured in the rest frame of the muon).

a) If a muon travels with a speed of $0.99\,c$ relative to a laboratory, what is the lifetime as measured in the laboratory?

b) What distance, measured in the laboratory, does the particle travel during its lifetime?

43-3 For the two trains discussed in Sec. 43-2, suppose the two lightning bolts appear simultaneous to an observer on the train. Show that they *do not* appear simultaneous to an observer on the ground. Which appears to come first?

43-4 Solve Eqs. (43-18) to obtain x and t in terms of x' and t', and show that the resulting transformation has the same form as the original one except for a change of sign for u.

43-5 A light pulse is emitted at the origin of a frame of reference S' at time $t' = 0$. Its distance x' from the origin after a time t' is given by $x'^2 = c^2 t'^2$. Use the Lorentz transformation to transform this equation to an equation in x and t, and show that the result is $x^2 = c^2 t^2$; that is, the motion appears exactly the same in the frame of reference S of x and t, as it does in S'.

43-6 Two events observed in a frame of reference S have positions and times given by (x_1, t_1) and (x_2, t_2), respectively. Show that in a frame S' moving just fast enough so the two events occur at the same point in S', the time interval $\Delta t'$ between the two events is given by

$$\Delta t' = \sqrt{(\Delta t)^2 - \left(\frac{\Delta x}{c}\right)^2},$$

where $\Delta x = x_2 - x_1$, and $\Delta t = t_2 - t_1$. Hence show that, if $\Delta x \geq c\,\Delta t$, there is *no* frame S' in which the two events occur at the same point. The interval $\Delta t'$ is sometimes called the *proper time interval* for the events; is this term appropriate?

43-7 For the two events in Problem 43-6, show that if $\Delta x > c\,\Delta t$ there is a frame of reference S' in which the two events occur *simultaneously*. Find the distance between the two events in S'. This distance is sometimes called a *proper length*; is this term appropriate?

43-8 Two events are observed in a frame of reference S to occur at the same space point, the second occurring 2 s after the first. In a second frame S' moving relative to S, the second event is observed to occur 3 s after the first. What is the distance between the positions of the two events as measured in S'?

43-9 Two events are observed in a frame of reference S to occur simultaneously, at points separated by a distance of 1 m. In a second frame S' moving relative to S along the line joining the two points in S, the two events appear to be separated by 2 m. What is the time interval between the events, as measured in S'?

43–10 A particle is said to be in the *extreme relativistic range* when its kinetic energy is much larger than its rest energy.

a) What is the speed of a particle (expressed as a fraction of c) such that the total energy is ten times the rest energy?

b) For such a particle, what percent error in the energy-momentum relation of Eq. (43–34) results if the term $(mc^2)^2$ is neglected?

43–11 A photon of energy E is emitted by an atom of mass m, which recoils in the opposite direction. Assuming that the atom can be treated nonrelativistically, compute the recoil velocity of the atom. Hence, show that the recoil velocity is much smaller than c whenever E is much smaller than the rest energy mc^2 of the atom.

43–12 Two particles emerge from a high-energy accelerator in opposite directions, each with a speed $0.6\,c$. What is the relative velocity of the particles?

43–13 At what speed is the momentum of a particle twice as great as the result obtained from the nonrelativistic expression mv?

43–14

a) At what speed does the momentum of a particle differ from the value obtained using the nonrelativistic expression mv by 1 percent?

b) Is the correct relativistic value greater or less than that obtained from the nonrelativistic expression?

43–15 The mass of an electron is 9.11×10^{-31} kg. Comparing the classical definition of momentum with its relativistic generalization, by how much is the classical expression in error if (a) $v = 0.01\,c$; (b) $v = 0.5\,c$; (c) $v = 0.9\,c$?

43–16 A radioactive isotope of cobalt, ^{60}Co, emits an electromagnetic photon (γ ray) of wavelength 0.932×10^{-12} m. The cobalt nucleus contains 27 protons and 33 neutrons, each with a mass of about 1.66×10^{-27} kg.

a) If the nucleus is at rest before emission, what is the speed afterward?

b) Is it necessary to use the relativistic generalization of momentum?

43–17 In Problem 43–16, suppose the cobalt atom is in a metallic crystal containing 0.01 mol of cobalt (about 6.02×10^{21} atoms) and that the entire crystal recoils as a unit, rather than just the single nucleus. Find the recoil velocity. (This recoil of the entire crystal rather than a single nucleus is called the *Mössbauer effect*, in honor of its discoverer, who first observed it in 1958.)

43–18 What is the speed of a particle whose kinetic energy is equal to its rest energy?

43–19 At what speed is the kinetic energy of a particle equal to $10\,mc^2$?

43–20 How much work must be done to accelerate a particle from rest to a speed $0.1\,c$? From a speed $0.9\,c$ to a speed $0.99\,c$?

43–21 In *positron annihilation*, an electron and a positron (a positively charged electron) collide and disappear, producing electromagnetic radiation. If each particle has a mass of 9.1×10^{-31} kg and they are at rest just before the annihilation, find the total energy of the radiation.

43–22 The total consumption of electrical energy per year in the United States is of the order of 10^{19} joules. If matter could be converted completely into energy, how many kilograms of matter would have to be converted to produce this much energy?

43–23 Compute the kinetic energy of an electron (mass 9.11×10^{-31} kg) using both the nonrelativistic and relativistic expressions, and compare the two results, for speeds of (a) 1.0×10^8 m·s^{-1}; (b) 2.0×10^8 m·s^{-1}.

43–24 In a hypothetical nuclear-fusion reactor, two deuterium nuclei combine or "fuse" to form one helium nucleus. The mass of a deuterium nucleus, expressed in atomic mass units (u), is 2.0147 u; that of a helium nucleus is 4.0039 u. (1 u = 1.66×10^{-27} kg.)

a) How much energy is released when 1 kg of deuterium undergoes fusion?

b) The annual consumption of electrical energy in the United States is of the order of 10^{19} J. How much deuterium must react to produce this much energy?

43–25 A nuclear bomb containing 20 kg of plutonium explodes. The rest mass of the products of the explosion is less than the original rest mass by one part in 10^4.

a) How much energy is released in the explosion?

b) If the explosion takes place in 1 μs, what is the average power developed by the bomb?

c) How much water could the released energy lift to a height of 1 km?

43–26 Construct a right triangle in which one of the angles is α, where $\sin \alpha = v/c$. (v is the speed of a particle, c the speed of light.) If the base of the triangle (the side adjacent to α) is the rest energy mc^2, show that

a) the hypotenuse is the total energy and

b) the side opposite α is c times the relativistic momentum.

c) Describe a simple graphical procedure for finding the kinetic energy K.

43–27 An electron in a certain x-ray tube is accelerated from rest through a potential difference of 180,000 V in going from the cathode to the anode. When it arrives at the anode, what is its

a) kinetic energy in eV,

b) its total energy,

c) its velocity?

d) What is the velocity of the electron, calculated classically?

43-28 The Cambridge electron accelerator accelerates electrons through a potential difference of 6.5×10^9 V, so that their kinetic energy is 6.5×10^9 eV.

a) What is the ratio of the speed v of an electron having this energy to the speed of light, c?

b) What would the speed be if computed from the principles of classical mechanics?

43-29 A particle moves along a straight line under the action of a force lying along the same line. Show that the acceleration $a = dv/dt$ of the particle is given by Eq. (43–25).

43-30 A particle moves along the x-axis with speed v. A force in the $+y$-direction, having magnitude F, is applied. Show that the magnitude of the acceleration initially is

$$a = \frac{F}{m}(1 - v^2/c^2)^{1/2}.$$

This result and that of Problem 43–29 were sometimes interpreted in the early days of relativity as meaning that a particle has a "longitudinal mass" given by $m/(1 - v^2/c^2)^{3/2}$ and a "transverse mass" given by $m/(1 - v^2/c^2)^{1/2}$. These terms are no longer in common use.

43-31 Calculate, relativistically, the amount of work in MeV that must be done

a) to bring an electron from rest to a velocity of $0.4c$, and

b) to increase its velocity from $0.4c$ to $0.8c$.

c) What is the ratio of the kinetic energy of the electron at the velocity of $0.8c$ to that of $0.4c$ when computed (1) from relativistic values and (2) from classical values?

44

PHOTONS, ELECTRONS, AND ATOMS

44–1 Emission and absorption of light

The past several chapters have been concerned with understanding various phenomena associated with the propagation of light, on the basis of an electromagnetic wave theory. The work of Hertz established the existence of electromagnetic waves and the fact that light is an electromagnetic wave. The phenomena of interference, diffraction, and polarization are easily understood on the basis of a wave model; and, when interference effects can be neglected, the further simplification of ray optics permits analysis of the behavior of lenses and mirrors. These phenomena are collectively referred to as *classical optics*, and insofar as the phenomena of classical optics are concerned, the electromagnetic wave theory is *complete*.

There are, however, many phenomena that are *not* so readily understood on this basis. An example is the emission of light from matter. The electromagnetic waves of Hertz, with frequencies of the order of 10^8 Hz, were produced by oscillations in a resonant L–C circuit similar to those studied in Chapter 34. Frequencies of visible light are much larger, of the order of 10^{15} Hz, far higher than the highest frequencies attainable with conventional electronic equipment.

In the mid-nineteenth century it was speculated that visible light might be produced by motion of electric charge within individual atoms rather than in macroscopic circuits. In fact, in 1862 Faraday placed a light source in a strong magnetic field in an attempt to determine whether the emitted radiation was changed by the field. He was not able to detect any change, but when his experiments were repeated thirty years later by Zeeman with greatly improved equipment, changes *were* observed.

Particularly puzzling was the existence of *line spectra*. Light emitted from atoms heated in a flame, or excited electrically in a glow tube such as the familiar neon sign or mercury-vapor light, does not contain a continuous spread of wavelengths, but only certain well-defined wavelengths. In a spectrometer using a narrow slit in conjunction with a prism or a diffraction grating, the spectrum pattern appears as a series of bright lines, and hence has come to be known as a *line spectrum*. It was learned early in the nineteenth century that each element emits a *characteristic spectrum*, suggesting that there is a direct relation between the characteristics and internal structure of an atom and its spectrum. Attempts to understand this relation on the basis of newtonian mechanics and classical electricity and magnetism were not successful, however.

There were other mysteries. The *photoelectric effect*, discovered by Hertz in 1887 during his investigations of electromagnetic wave propagation, is the liberation of electrons from the surface of a conductor when light strikes the surface. This phenomenon can be understood qualitatively on the basis that when light is absorbed by the surface it transfers energy to electrons near the surface and that some of the electrons acquire enough energy to surmount the potential-energy barrier at the surface and escape from the material into space. More detailed investigation revealed some puzzling features that could *not* be understood on the basis of classical optics.

Still another area of unsolved problems centered around the production and scattering of *x-rays*, electromagnetic radiation with wavelengths shorter than those of visible light by a factor of the order of 10^4 and with correspondingly greater frequencies. These rays were produced in high-voltage glow discharge tubes, but the details of this process eluded understanding. Even worse, when these rays collided with matter, the scattered ray sometimes had a longer wavelength than the original ray. This is like directing a beam of blue light at a mirror and having it reflect red!

All these phenomena, and several others, pointed forcefully to the conclusion that classical optics, successful though it was in explaining ray optics, interference, and polarization, nevertheless had its limitations. Understanding the phenomena cited above would require at least some generalization of the classical theory. In fact, it has required something much more radical than that. All these phenomena are concerned with the *quantum* theory of radiation, which includes the assumption that despite the *wave* nature of electromagnetic radiation, it nevertheless has some properties akin to those of *particles*. In particular, the *energy* conveyed by an electromagnetic wave is always carried in units whose magnitude is proportional to the frequency of the wave. These units of energy are called *photons* or *quanta*.

Thus, electromagnetic radiation emerges as an entity with a dual nature, having both wave and particle aspects. The remainder of this chapter will be devoted to the applications of this duality to some of the phenomena mentioned above, and to a study of this seemingly (but not actually) inconsistent nature of electromagnetic radiation.

44-2 Thermionic emission

As a prelude to the discussion of the photoelectric effect, we consider briefly a related phenomenon, *thermionic emission*, discovered by Thomas Edison in 1883 during his experiments on electric lightbulbs. A glassblower had sealed into the bulb of an ordinary filament lamp an extra metal electrode or plate, shown in Fig. 44-1. The glass bulb was then evacuated and the filament heated as usual. When the plate was

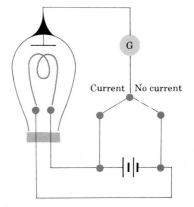

44-1 Edison's original thermionic-emission experiment.

connected through a galvanometer to the *positive* terminal of the 110-V dc source, a galvanometer deflection indicated the existence of a current, despite the fact that there was no conducting path from the plate to the other terminal of the battery. When the plate was connected to the *negative* terminal there was no current. Edison was very much interested in this phenomenon at the time but could not explain it.

The effect is caused by the escape of electrons from the hot filament. Ordinarily electrons in a conductor are prevented from escaping from the surface by a potential-energy barrier. When an electron starts to move away from the surface it induces a corresponding positive charge in the material, which tends to pull it back into the surface. To escape, the electron must somehow acquire enough energy to surmount this energy barrier; the minimum energy needed to escape is called the *work function*, and varies from one material to another. Typical work functions are of the order of a few electronvolts. The electronvolt (eV) is defined in Sec. 26–7, and it would do no harm for the reader to review that section in preparation for this chapter.

At ordinary temperatures almost none of the electrons can acquire enough energy to escape, but when the filament is very hot the electron energies are greatly increased by thermal motion, and at sufficiently high temperatures considerable numbers are able to escape. Once out of the material, the electrons are attracted to the plate if it is positively charged, but repelled if it is negatively charged. The liberation of electrons from a hot surface is called *thermionic emission*.

The electrons that have escaped from the hot conductor form a cloud of negative charge near it, called a *space charge*. If a second conductor (the plate) is maintained by a battery at a higher potential than the first, the electrons in the cloud are attracted to it, and so long as the potential difference between the conductors is maintained, there will be a steady drift of electrons from the emitter or *cathode* to the other body, which is called the plate or *anode*.

In the common thermionic tube the cathode and anode (and often other electrodes as well) are enclosed within an evacuated glass or metal container, and leads to the various electrodes are brought out through the base or walls of the tube. The simplest thermionic tube, in which the only electrodes are the cathode and anode, is called a *diode*.

The diode is shown schematically in Fig. 44–2. The cathode and plate are represented by K and P. The cathode is often in the form of a hollow cylinder, which is heated by a fine resistance wire H within it. Electrons emitted from the outer surface of the cathode are attracted to the plate, which is a larger cylinder surrounding the cathode and coaxial with it. The electron current to the plate is read on the milliammeter MA. The potential difference between plate and cathode can be controlled by the slide wire and read on voltmeter V.

If the potential difference between cathode and anode is small (a few volts), only a few of the emitted electrons reach the plate, the majority penetrating a short distance into the cloud of space charge and then returning to the cathode. As the plate potential is increased, more and more electrons are drawn to it, and with sufficiently high potentials (of the order of 100 V), *all* of the emitted electrons arrive at the plate. Further increase of plate potential does not increase the plate current, which is then said to become *saturated*.

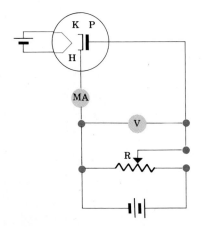

44-2 Circuit for measuring plate current and plate voltage in a vacuum diode.

A graph of plate current, I_p, versus plate potential, V_p, is shown in Fig. 44–3a. Note that I_p is not zero even when V_p is zero. This is because the electrons leave the cathode with an initial speed and the more rapidly moving ones may penetrate the cloud of space charge and reach the plate even with no accelerating field. In fact, a *retarding* field is necessary to prevent their reaching the plate, an effect which may be used to measure their speeds of emission.

The saturation current I_s, in Fig. 44–3a, is equal to the current from the cathode, and for a given tube its magnitude increases markedly with cathode temperatue. Figure 44–3b shows three plate current curves at three different temperatures, where $T_3 > T_2 > T_1$.

The work function ϕ of a surface may be considerably reduced by the presence of impurities. A small amount of thorium, for example, reduces the work function of pure tungsten by about 50%. Since the smaller the work function the larger the current density at a given temperature (or the lower the temperature at which a given emission can be attained), most vacuum tubes now use cathodes having composite surfaces.

Because thermionic emission occurs only at the heated cathode, not at the anode, electron flow occurs in only one direction, from cathode to anode. Most circuit applications of the diode make use of this one-directional characteristic; the simplest example is a *rectifier* for converting alternating current to direct current.

Lee de Forest discovered in 1907 that the electron flow can be modified by inserting a third electrode called a *grid* between cathode and anode. The grid is usually an open structure, either a screen or a coil of wire. When the grid is negative with respect to the cathode the resulting field enhances the space charge near the cathode and decreases the electron flow through the grid to the plate. Because of the grid's proximity to the cathode, the current is much more sensitive to changes in grid voltage than to changes in anode voltage, and thus the device can function as an *amplifier*. In the form just described it is called a *triode*. This basic device and various elaborations of it are very widely used in electronic equipment, although in the past two decades they have been supplanted in many applications by transistors, which will be discussed in Chapter 46.

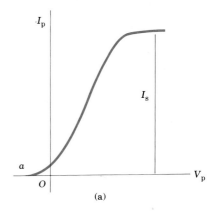

(a)

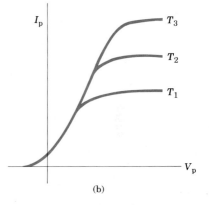

(b)

44-3 (a) Plate current-plate voltage characteristic of a diode. (b) Plate current curves at three different cathode temperatures, $T_3 > T_2 > T_1$.

44-3 The photoelectric effect

In the thermionic emission of electrons from metals, the energy needed by an electron to escape from the metal surface is furnished by the energy of thermal agitation. Electrons may also acquire enough energy to escape from a metal, even at low temperatures, if the metal is illuminated by light of sufficiently short wavelength. This phenomenon is called the *photoelectric effect*. It was first observed by Heinrich Hertz in 1887, who noticed that a spark would jump more readily between two spheres when their surfaces were illuminated by the light from another spark. The effect was investigated in detail in the following years by Hallwachs and Lenard.

A modern phototube is shown schematically in Fig. 44-4. A beam of light, indicated by the arrows, falls on a photosensitive surface K called the *cathode*. The battery or other source of potential difference creates an electric field in the direction from A (called the *collector* or *anode*)

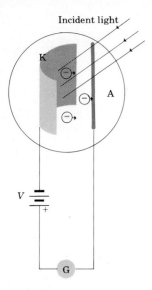

Incident light

K

A

V

G

44-4 Schematic diagram of a photocell circuit.

toward K, and electrons emitted from K are pushed by this field to the anode A. Anode and cathode are enclosed in an evacuated container; the photoelectric current is measured by the galvanometer G.

It is found that with a given material as emitter, no photoelectrons at all are emitted unless the wavelength of the light is *shorter* than some critical value. The corresponding *minimum* frequency is called the *threshold frequency* of the particular surface. The threshold frequency for most metals is in the ultraviolet (critical wavelength 200 to 300 nm), but for potassium and cesium oxide it lies in the visible spectrum (400 to 700 nm).

Just as in the case of thermionic emission, the photoelectrons form a cloud of space charge around the emitter. Some of the electrons are emitted with an initial speed; this is shown by the fact that, even with *no* emf in the external circuit, a few electrons penetrate the cloud of space charge and reach the collector, causing a small current in the external circuit. Indeed, even when the polarity of the potential difference V is reversed and the associated electric-field force on the electrons is back toward the cathode, some electrons still reach the anode. Only when the reversed potential V is made large enough so that the potential energy eV is greater than the maximum kinetic energy $\frac{1}{2}mv_{max}^2$ with which the electrons leave the cathode, does the electron flow stop completely. This critical reversed potential is called the *stopping potential,* denoted by V_0, and it provides a direct measurement of the maximum kinetic energy with which electrons leave the cathode, through the relation

$$\tfrac{1}{2}mv_{max}^2 = eV_0. \tag{44-1}$$

Surprisingly, it turns out that the maximum kinetic energy of a photoelectron does not depend on the intensity of the incident light, but does depend on its wavelength. When the light intensity increases, the photoelectric current also increases, but only because the *number* of emitted electrons increases, not the energy of an individual electron.

The correct explanation of the photoelectric effect was given by Einstein in 1905. Extending a proposal made two years earlier by Planck, Einstein postulated that a beam of light consisted of small bundles of energy that are now called *light quanta* or *photons.* The energy E of a photon is proportional to its frequency f, or is equal to its frequency multiplied by a constant. That is,

$$E = hf, \tag{44-2}$$

where h is a universal constant, called *Planck's constant,* whose value is 6.626×10^{-34} J·s. When a photon collides with an electron at or just within the surface of a metal, it may transfer its energy to the electron. This transfer is an "all-or-none" process, the electron getting all the photon's energy or none at all. The photon then simply drops out of existence. The energy acquired by the electron may enable it to escape from the surface of the metal if it is moving in the right direction.

In leaving the surface of the metal, the electron loses an amount of energy ϕ (the work function of the surface). Some electrons may lose more than this if they start at some distance below the metal surface, but the *maximum* energy with which an electron can emerge is the energy gained from a photon minus the work function. Hence the maximum kinetic energy of the photoelectrons ejected by light of frequency f is

$$\tfrac{1}{2}mv_{\text{max}}^2 = hf - \phi. \tag{44-3}$$

Combining this with Eq. (44–1) leads to the relation

$$eV_0 = hf - \phi. \tag{44-4}$$

Thus by measuring the stopping potential V_0 required for each of several values of frequency f for a given cathode material, one can determine both the work function ϕ for the material and the value of the quantity h/e. Thus this experiment provides a direct measurement of the value of Planck's constant in addition to a direct confirmation of Einstein's interpretation of photoelectric emission.

Example   For a certain cathode material used in a photoelectric-effect experiment, a stopping potential of 3.0 V was required for light of wavelength 300 nm, 2.0 V for 400 nm, and 1.0 V for 600 nm. Determine the work function for this material, and the value of Planck's constant.

Solution   According to Eq. (44–4), a graph of V_0 as a function of f should be a straight line. We rewrite this equation as

$$V_0 = \frac{h}{e}f - \frac{\phi}{e}.$$

In this form we see that the *slope* of the line is h/e and the *intercept* on the vertical axis (corresponding to $f = 0$) is at $-\phi/e$. The frequencies, obtained from $f = c/\lambda$ and $c = 3.00 \times 10^8$ m·s^{-1}, are 0.5, 0.75, and 1.0 $\times$ 10^{15} s^{-1}, respectively. The graph is shown in Fig. 44–5. From it we find

$$-\frac{\phi}{e} = -1.0 \text{ V}, \qquad \phi = 1.0 \text{ eV},$$

and

$$\frac{h}{e} = \frac{1.0 \text{ V}}{0.25 \times 10^{15} \text{ s}^{-1}} = 4 \times 10^{-15} \text{ J·C}^{-1}\text{·s},$$
$$h = (4.0 \times 10^{-15} \text{ J·C}^{-1}\text{·s})(1.6 \times 10^{-19} \text{ C})$$
$$= 6.4 \times 10^{-34} \text{ J·s}.$$

The particle-like nature of electromagnetic radiation, used in the above analysis of the photoelectric effect, has been established beyond any reasonable doubt. A photon of electromagnetic radiation of frequency f and corresponding wavelength $\lambda = c/f$ has energy E given by

$$E = hf = \frac{hc}{\lambda}. \tag{44-5}$$

Furthermore, according to relativity theory, every particle having energy must also have momentum, even if it has no rest mass. Photons have zero rest mass; according to Eq. (43–35), the momentum p of a photon of energy E has magnitude p given by

$$E = pc. \tag{44-6}$$

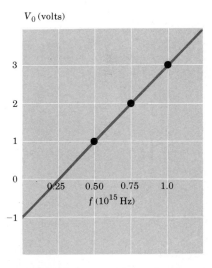

44-5 Stopping potential as a function of frequency. For a different cathode material having a different work function, the line would be displaced up or down but would have the same slope.

Thus the wavelength of a photon and its momentum are related simply by

$$p = \frac{h}{\lambda}. \tag{44-7}$$

These relations will be used frequently in the remainder of this chapter.

44-4 Line spectra

The quantum hypothesis, used in the preceding section for the analysis of the photoelectric effect, also plays an important role in the understanding of atomic spectra. We have seen how a prism or grating spectrograph functions to disperse a beam of light into a *spectrum*. If the light source is an incandescent solid or liquid, the spectrum is *continuous*; that is, light of all wavelengths is present. If, however, the source is a gas through which an electrical discharge is passing, or a flame into which a volatile salt has been introduced, the spectrum is of an entirely different character. Instead of a continuous band of color, only a few colors appear, in the form of isolated parallel lines. (Each "line" is an image of the spectrograph slit, deviated through an angle dependent on the frequency of the light forming the image.) A spectrum of this sort is termed a *line spectrum*. The wavelengths of the lines are characteristic of the element emitting the light. That is, hydrogen always gives a set of lines in the same position, sodium another set, iron still another, and so on. The line structure of the spectrum extends into both the ultraviolet and infrared regions, where photographic or other means are required for its detection.

It might be expected that the frequencies of the light emitted by a particular element would be arranged in some regular way. For instance, a radiating atom might be analogous to a vibrating string, emitting a fundamental frequency and its harmonics. At first sight there does not seem to be any semblance of order or regularity in the lines of a typical spectrum; for many years unsuccessful attempts were made to correlate the observed frequencies with those of a fundamental and its overtones. Finally, in 1885, Johann Jakob Balmer (1825–1898) found a simple formula which gave the frequencies of a group of lines emitted by atomic hydrogen. Since the spectrum of this element is relatively simple, and fairly typical of a number of others, we shall consider it in more detail.

Under the proper conditions of excitation, atomic hydrogen may be made to emit the sequence of lines illustrated in Fig. 44–6. This sequence is called a *series*. There is evidently a certain order in this spectrum, the

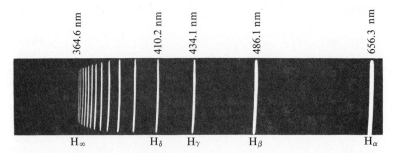

44-6 The Balmer series of atomic hydrogen. (Reproduced by permission from *Atomic Spectra and Atomic Structures* by Gerhard Herzberg. Copyright 1937 by Prentice-Hall, Inc.)

lines becoming crowded more and more closely together as the limit of the series is approached. The line of longest wavelength or lowest frequency, in the red, is known as H_α, the next, in the blue-green, as H_β, the third as H_γ, and so on. Balmer found that the wavelengths of these lines were given accurately by the simple formula

$$\frac{1}{\lambda} = R\left(\frac{1}{2^2} - \frac{1}{n^2}\right), \qquad (44\text{--}8)$$

where λ is the wavelength, R is a constant called the Rydberg constant, and n may have the integral values 3, 4, 5, etc. If λ is in meters,

$$R = 1.097 \times 10^7 \text{ m}^{-1}.$$

Letting $n = 3$ in Eq. (44–5), one obtains the wavelength of the H_α-line:

$$\frac{1}{\lambda} = 1.097 \times 10^7 \text{ m}^{-1}\left(\frac{1}{4} - \frac{1}{9}\right)$$
$$= 1.524 \times 10^6 \text{ m}^{-1},$$

whence

$$\lambda = 656.3 \text{ nm}.$$

For $n = 4$, one obtains the wavelength of the H_β-line, etc. For $n = \infty$, one obtains the limit of the series, at $\lambda = 364.6$ nm. This is the *shortest* wavelength in the series.

Other series spectra for hydrogen have since been discovered. These are known, after their discoverers, as the Lyman, Paschen, Brackett, and Pfund series. The formulas for these are

Lyman series:

$$\frac{1}{\lambda} = R\left(\frac{1}{1^2} - \frac{1}{n^2}\right), \qquad n = 2, 3, \ldots,$$

Paschen series:

$$\frac{1}{\lambda} \; R\left(\frac{1}{3^2} - \frac{1}{n^2}\right), \qquad n = 4, 5, \ldots,$$

Brackett series:

$$\frac{1}{\lambda} = R\left(\frac{1}{4^2} - \frac{1}{n^2}\right), \qquad n = 5, 6, \ldots,$$

Pfund series:

$$\frac{1}{\lambda} = R\left(\frac{1}{5^2} - \frac{1}{n^2}\right), \qquad n = 6, 7, \ldots.$$

The Lyman series is in the ultraviolet, and the Paschen, Brackett, and Pfund series are in the infrared. The Balmer series evidently fits into the scheme between the Lyman and the Paschen series.

The Balmer formula, Eq. (44–8), may also be written in terms of the frequency of the light, by use of the relation

$$c = f\lambda \qquad \text{or} \qquad \frac{1}{\lambda} = \frac{f}{c}.$$

Thus, Eq. (44–8) becomes

$$f = Rc \left(\frac{1}{2^2} - \frac{1}{n^2} \right) \tag{44-9}$$

or

$$f = \frac{Rc}{2^2} - \frac{Rc}{n^2}. \tag{44-10}$$

Each of the fractions on the right side of Eq. (44–10) is called a *term,* and the frequency of every line in the series is given by the difference between two terms.

There are only a few elements (hydrogen, singly ionized helium, doubly ionized lithium) whose spectra can be represented by a simple formula of the Balmer type. Nevertheless, it is possible to separate the more complicated spectra of other elements into series, and to express the frequency of each line in the series as the difference of two *terms*. The first term is constant for any one series, while the various values of the second term can be labeled by values of an integer index n analogous to the n appearing in Eq. (44–10). In a few simple cases the numerical values of the terms can be *calculated* from theoretical considerations, as we shall see in Chapter 45. For complex atoms, however, the term values must be determined experimentally by analysis of spectra, often an extremely complex problem.

44-5 Energy levels

Every element has a characteristic line spectrum, which must somehow result from the characteristics and structure of the *atoms* of that element. Part of the key to understanding the relation of atomic structure to atomic spectra was supplied in 1913 by the Danish physicist Niels Bohr, who applied to spectra the same concept of light quanta or *photons* that Einstein had used earlier in analysis of the photoelectric effect.

Bohr's hypothesis was as follows: Each atom, as a result of its internal structure (and presumably internal motion), can have a variable amount of *internal energy.* But the energy of an atom cannot change by any arbitrary amount; rather, each atom has a series of discrete *energy levels,* such that an atom can have an amount of internal energy corresponding to any one of these levels, but it cannot have an energy *intermediate* between two levels. All atoms of a given element have the same set of energy levels, but atoms of different elements have different sets. While an atom is in one of these states corresponding to a definite energy, it does not radiate, but an atom can make a transition from one energy level to a lower level by emitting a photon, whose energy is equal to the energy difference between the initial and final states. If E_i is the initial energy of the atom, before such a transition, and E_f its final energy, after the transition, then, since the photon's energy is hf, we have

$$hf = E_i - E_f. \tag{44-11}$$

For example, a photon of orange light of wavelength 600 nm has a frequency f given by

$$f = \frac{c}{\lambda} = \frac{3.00 \times 10^8 \, \text{m} \cdot \text{s}^{-1}}{600 \times 10^{-9} \, \text{m}} = 5.00 \times 10^{14} \, \text{Hz}$$

$$= 5.00 \times 10^{14} \, \text{s}^{-1}.$$

The corresponding photon energy is

$$E = hf = (6.63 \times 10^{-34} \text{ J} \cdot \text{s})(500 \times 10^{14} \text{ s}^{-1})$$
$$= 3.31 \times 10^{-19} \text{ J} = 2.07 \text{ eV}.$$

Thus, this photon must be emitted in a transition between two states of the atom differing in energy by 2.07 eV.

The Bohr hypothesis, if correct, would shed new light on the analysis of spectra on the basis of *terms,* as described in Sec. 44-4. For example, Eq. (44-10) gives the frequencies of the Balmer series in the hydrogen spectrum. Multiplied by Planck's constant h, this becomes

$$hf = \frac{Rch}{2^2} - \frac{Rch}{n^2}. \tag{44-12}$$

If we now compare Eqs. (44-11) and (44-12), identifying $-Rch/n^2$ with the initial energy of the atom E_i and $-Rch/2^2$ with its final energy, E_f, before and after a transition in which a photon of energy $hf = E_i - E_f$ is emitted, then Eq. (44-12) takes on the same form as Eq. (44-11). More generally, if we assume that the possible energy levels for the hydrogen atom are given by

$$E_n = -\frac{Rch}{n^2}, \qquad n = 1, 2, 3, \ldots, \tag{44-13}$$

then *all* the series spectra of hydrogen can be understood on the basis of transitions from one energy level to another. For the Lyman series the final state is always $n = 1$, for the Paschen series it is $n = 3$, and so on. Similarly, complex spectra of other elements, represented by *terms,* are understood on the basis that each term corresponds to an energy level; and a frequency, represented as a difference of two terms, corresponds to a transition between the two corresponding energy levels.

In 1914 a series of experiments by Franck and Hertz provided more direct experimental evidence for energy levels in atoms. In studying the motion of electrons through mercury vapor, under the action of an electric field, they found that a spectrum line at 254 nm was emitted by the vapor when the electron kinetic energy was greater than 4.9 eV but not when it was less. This strongly suggests the existence of an energy level 4.9 eV above the ground state; a mercury atom is excited to this level by collision with an electron, and subsequently decays to the ground state by emitting a photon. The energy of the photon, according to Eq. (44-2), should be

$$E = hf = \frac{hc}{\lambda} = \frac{(6.63 \times 10^{-34} \text{ J} \cdot \text{s})(3.00 \times 10^8 \text{ m} \cdot \text{s}^{-1})}{254 \times 10^{-9} \text{ m}}$$
$$= 7.82 \times 10^{-19} \text{ J} = \frac{7.82 \times 10^{-19} \text{ J}}{1.60 \times 10^{-19} \text{ J} \cdot \text{eV}^{-1}}$$
$$= 4.9 \text{ eV},$$

in excellent agreement with the measured electron energy.

Although the Bohr hypothesis permits partial understanding of line spectra on the basis of energy levels of atoms, it is not yet complete because it provides no basis for *predicting* what the energy levels for any particular kind of atom should be. We shall return to this problem in

Chapter 45, where the new mechanical principles required for the understanding of the structure and energy levels of atoms will be developed. These principles constitute the subject of *quantum mechanics.*

44-6 Atomic spectra

As we have seen, the key to the understanding of atomic spectra is the concept of atomic *energy levels.* Every spectrum line corresponds to a specific transition between two energy levels of an atom, and the corresponding frequency is given in each case by Eq. (44–11).

Thus the fundamental problem of the spectroscopist is to determine the energy levels of an atom from the measured values of the wavelengths of the spectral lines emitted when the atom proceeds from one energy level to another. In the case of complicated spectra emitted by the heavier atoms, this is a task requiring tremendous ingenuity. Nevertheless, almost all atomic spectra have been analyzed, and the resulting energy levels have been tabulated or plotted with the aid of diagrams similar to the one shown for sodium in Fig. 44–7.

As we shall see, every atom has a lowest energy level, representing the *minimum* energy the atom can have. This lowest energy level is called the *ground state,* and all higher levels are called *excited states.* As we have seen, a spectral line is emitted when an atom proceeds from an excited state to a lower state. The only means discussed so far for raising the atom from the normal state to an excited state has been with the aid of an electric discharge. Let us consider now another method, involving absorption of radiant energy.

From Fig. 44–7 it may be seen that a sodium atom emits the characteristic yellow light of wavelengths 589.0 and 589.6 nm (the D_1- and D_2-lines) when it undergoes the transitions from the two levels marked *resonance levels* to the ground state. Suppose a sodium atom in the ground state were to *absorb* a quantum of radiant energy of wavelength 589.0 or 589.6 nm. It would then undergo a transition in the opposite direction and be raised to one of the resonance levels. After a short time, the average value of which is called the *lifetime* of the excited state, the atom returns to the ground state and emits this quantum. For the resonance levels of the sodium atom, the lifetime is about 1.6×10^{-8} s.

This emission process is called *resonance radiation* and may be easily demonstrated as follows. A strong beam of the yellow light from a sodium arc is concentrated on a glass bulb that has been highly evacuated and into which a small amount of pure metallic sodium has been distilled. If the bulb is warmed to increase the sodium vapor pressure, resonance radiation will take place throughout the whole bulb, which glows with the yellow light characteristic of sodium.

A sodium atom in the ground state may absorb radiant energy of wavelengths other than the yellow resonance lines. All wavelengths corresponding to spectral lines *emitted* when the sodium atom returns to its normal state may also be absorbed. Thus, from Fig. 44–7, wavelengths 330.2 nm, 285.3 nm, etc., may be *absorbed* by a normal sodium atom. If, therefore, the continuous-spectrum light from a carbon arc is sent through an absorption tube containing sodium vapor, and then examined with a spectroscope, there will be a series of dark lines corresponding to the wavelengths absorbed, as shown in Fig. 44–8. This is known as an *absorption spectrum.*

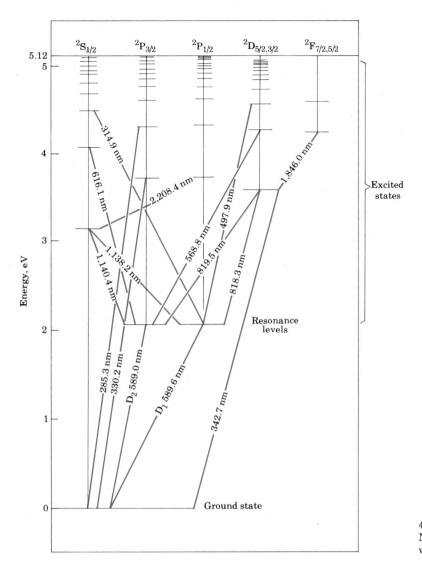

44-7 Energy levels of the sodium atom. Numbers on the lines between levels are wavelengths.

The sun's spectrum is an absorption spectrum. The main body of the sun emits a continuous spectrum, whereas the cooler vapors in the sun's atmosphere emit line spectra corresponding to all the elements present. When the intense light from the main body of the sun passes through the cooler vapors, the lines of these elements are absorbed. The light *emitted* by the cooler vapors is so small compared with the unabsorbed continuous spectrum that the continuous spectrum appears to be crossed by many faint *dark* lines. These were first observed by Fraunhofer and are therefore called *Fraunhofer lines*. They may be observed with any student spectroscope pointed toward any part of the sky.

44-8 Absorption spectrum of sodium.

44-7 The laser

If the energy difference between the normal and the first excited state of an atom is E, the atom is capable of absorbing a photon whose frequency f is given by the Planck equation $E = hf$. The *absorption* of a photon by a normal atom A is depicted schematically in Fig. 44-9a. After absorbing the photon, the atom becomes an excited atom A^*. A short time later, *spontaneous emission* takes place and the excited atom becomes normal again by emitting a photon of the same frequency as that which was originally absorbed but in a random direction and with a random phase, as shown in Fig. 44-9b. There is also a third process, first proposed by Einstein, called *stimulated emission,* shown schematically in Fig. 44-9c. *Stimulated emission takes place when a photon encounters an excited atom and forces it to emit another photon of the same frequency, in the same direction, and in the same phase.* The two photons go off together as *coherent* radiation.

Consider an absorption cell containing a large number of atoms of the type depicted in Fig. 44-9. In the absence of an external beam of radiation, most of the atoms are in the ground state; there are only a few excited atoms present in the cell. The ratio of the number n_E of excited atoms to the number n_0 of normal atoms is extremely small.

Now suppose a beam of radiation is sent through the cell with frequency f corresponding to the energy difference E. The ratio of the numbers n_E and n_0, that is, the ratio of the *populations* of the energy levels, is increased. Since the population of the normal state was so much larger than that of the excited state, an enormously intense beam of light would be required to increase the population of the excited state to a value comparable to or greater than that of the normal state. Therefore, the rate at which energy is extracted from the beam by absorption of normal atoms far outweighs the rate at which energy is added to the beam by stimulated emission of excited atoms.

If a condition can be created in which n_E is substantially increased compared to the normal equilibrium value, creating a condition known as *population inversion,* the rate of energy radiation by stimulated emission may *exceed* the rate of absorption. The system then acts as a *source* of radiation with photon energy E. Furthermore, since the photons are

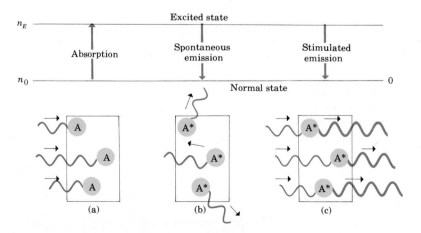

44-9 Three interaction processes between an atom and radiation.

the result of stimulated emission, they all have the same frequency, phase, polarization, and direction. The resulting radiation is therefore very much more *coherent* than light from ordinary sources, in which the emissions of individual atoms are *not* coordinated.

The necessary population inversion can be achieved in a variety of ways. As an example we consider the helium–neon laser, a simple, inexpensive laser available in many undergraduate laboratories. A mixture of helium and neon, each typically at a pressure of the order of 10^2 Pa (or 10^{-3} atm), is sealed in a glass enclosure provided with two electrodes. When a sufficiently high voltage is applied, a glow discharge occurs. Collisions between ionized atoms and electrons carrying the discharge current excite atoms to various energy states.

Figure 44–10 shows an energy-level diagram for the system. The notation used to label the various energy levels, such as $1s2s$ or $5s$, will be discussed in Sec. 46–1, and need not concern us here. Helium atoms excited to the $1s2s$ state cannot return to the ground state by emitting a 20.61-eV photon, as might be expected, because both the states have zero total angular momentum, while a photon must carry away at least one unit $(h/2\pi)$ of angular momentum. Such a state, in which radiative decay is impossible, is called a *metastable state.*

The helium atoms *can,* however, lose energy by energy-exchange collisions with neon atoms initially in the ground state. A $1s2s$ helium atom, with its internal energy of 20.61 eV and a little additional kinetic

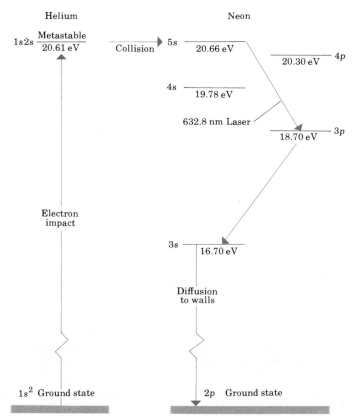

44–10 Energy-level diagram for helium–neon laser.

energy, can collide with a neon atom in the ground state, exciting it to the 5s excited state at 20.66 eV and leaving the helium atom in the $1s^2$ ground state. Thus, we have the necessary mechanism for a population inversion in neon, with the population in the 5s state substantially enhanced. Stimulated emission from this state then results in the emission of highly coherent light at 632.8 nm, as shown on the diagram. In practice the beam is sent back and forth through the gas many times by a pair of parallel mirrors, so as to stimulate emission from as many excited atoms as possible. One of the mirrors is partially transparent, so a portion of the beam emerges as an external beam.

The net effect of all the processes taking place in a laser tube is a beam of radiation that is (1) very intense, (2) almost perfectly parallel, (3) almost monochromatic, and (4) spatially *coherent* at all points within a given cross section. To understand this fourth characteristic, we recall the simple double-slit interference experiment. A mercury arc placed directly behind the double slit would not give rise to interference fringes because the light issuing from the two slits would come from different points of the arc and would not retain a constant phase relationship. In the use of the usual laboratory arc-lamp sources, it is necessary to use the light from a very small portion of the source to illuminate the double slit. The slightly diverging beam from a laser, however, may be allowed to fall directly on a double slit (or other interferometer) because the light rays from any two points of a cross section are in phase, and are said to exhibit "spatial coherence."

In recent years lasers have found a wide variety of practical applications. The high intensity of a laser beam makes it a convenient drill. A very small hole can be drilled in a diamond for use as a die in drawing very small-diameter wire. The ability of a laser beam to travel long distances without appreciable spreading makes it a very useful tool for surveyors, especially in situations where great precision is required over long distances, as in the case of a long tunnel drilled from both ends.

Lasers are finding increasing application in medical science. A laser can produce a very *narrow* beam with extremely *high intensity,* high enough to vaporize anything in its path. This property is used in the treatment of a detached retina; a short burst of radiation damages a small area of the retina, and the resulting scar tissue "welds" the retina back to the choroid from which it has become detached. Laser beams are also used in surgery; blood vessels cut by the beam tend to seal themselves off, making it easier to control bleeding. The use of laser radiation in the treatment of skin cancer is an active area of research.

44-8 X-ray production and scattering

X-rays are produced when rapidly moving electrons that have been accelerated through potential differences of the order of 10^3 to 10^6 V are allowed to strike a metal target. They were first observed by Wilhelm K. Röntgen (1845–1923) in 1895, and were originally called *Röntgen rays.*

X-rays are of the same nature as light or any other electromagnetic wave and, like light waves, they are governed by quantum relations in their interaction with matter. Hence, one may speak of x-ray photons or quanta, the energy of such a photon being given by the relation $E = hf$. Wavelengths of x-rays range from approximately 0.001 to 1 nm (10^{-12} to

10^{-9} m). X-ray wavelengths can be measured quite precisely by crystal diffraction techniques, described in Sec. 41–10.

A common x-ray tube is the Coolidge type, invented by W. D. Coolidge of the General Electric laboratories in 1913. A Coolidge tube is shown in Fig. 44–11. A thermionic cathode and an anode are enclosed in a glass tube that has been pumped down to an extremely low pressure. Electrons emitted from the cathode can then travel directly to the anode with only a small probability of a collision on the way, and they reach the anode with a speed corresponding to the full potential difference across the tube. X-radiation is emitted from the anode surface as a consequence of its bombardment by the electron stream.

Two distinct processes are involved in x-ray emission. Some of the electrons are stopped by the target and their kinetic energy is converted directly to x-radiation. Others transfer their energy in whole or in part to the atoms of the target, which retain it temporarily as "energy of excitation" but very shortly emit it as x-radiation. The latter is characteristic of the material of the target, while the former is not.

The atomic energy levels associated with x-ray excitation are rather different in character from those associated with visible spectra. To understand them we need some understanding of the arrangement of electrons in complex atoms, a topic to be discussed in greater detail in Chapter 46. For the present we state simply that in a many-electron atom the electrons are always arranged in concentric *shells* at increasing distances from the nucleus. These shells are labeled K, L, M, N, etc., the K shell being closest to the nucleus, the L shell next, and so on. For any given atom in the ground state there is a definite number of electrons in each shell.

For reasons to be discussed later, each shell has a maximum number of electrons it can accommodate, and we may speak of *filled shells* and *partially filled shells*. The K shell can contain at most two electrons. The next, the L shell, can contain eight. The third, the M shell, has a capacity for 18 electrons, while the N shell may hold 32. The sodium atom, for example, which contains 11 electrons, has two in the K shell, eight in the L shell, and a single electron in the M shell. Molybdenum, with 42 electrons, has two in the K shell, eight in the L shell, 18 in the M shell, 13 in the N shell, and one in the O shell.

The *outer* electrons of an atom are the ones responsible for the optical spectra of the elements. Relatively small amounts of energy suffice to remove these to excited states, and on their return to their normal states, wavelengths in or near the visible region are emitted. The inner electrons, being closer to the nucleus, are more tightly bound, and much more energy is required to displace them from their normal levels. As a result, we would expect a photon of much larger energy, and hence much higher frequency, to be emitted when the atom returns to its normal state after the displacement of an inner electron. This is, in fact, the case, and it is the displacement of the inner electrons that gives rise to the emission of x-rays.

On colliding with the atoms of the anode, some of the electrons accelerated in an x-ray tube, provided they have acquired sufficient energy, will dislodge one of the inner electrons of a target atom, say one of the K electrons. This leaves a vacant space in the K shell, which is immediately filled by an electron from either the L, M, or N shell. The readjustment of the electrons is accompanied by a decrease in the energy of the atom, and an x-ray photon is emitted with energy just equal to this decrease. Since the energy change is perfectly definite for atoms of a given element, the emitted x-rays should have definite frequencies. In other words, the x-ray spectrum should be a *line spectrum*. We can predict further that there should be just three lines in the series, corresponding to the three possibilities that the vacant space may have been filled by an L, M, or N electron.

This is precisely what is observed. Figure 44–12 illustrates the so-called K series of the elements tungsten, molybdenum, and copper. Each series consists of three lines, known as the K_α-, K_β-, and K_γ-lines. The K_α-line is produced by the transition of an L electron to the vacated space in the K shell, the K_β-line by an M electron, and the K_γ-line by an N electron.

In addition to the K series, there are other series known as the L, M, and N series, produced by the ejection of electrons from the L, M, and N shells rather than the K shell. As would be expected, the electrons in these outer shells, being farther away from the nucleus, are not held as firmly as those in the K shell. Consequently, the other series may be excited by more slowly moving electrons, and the photons emitted are of lower energy and longer wavelength.

In addition to the x-ray *line* spectrum there is a background of *continuous* x-radiation from the target of an x-ray tube. This is due to the sudden deceleration of those "cathode rays" (bombarding electrons) that do not happen to eject an atomic electron. The remarkable feature of the continuous spectrum is that while it extends indefinitely toward the *long* wavelength end, it is cut off very sharply at the *short* wavelength end. The quantum theory furnishes a simple explanation of the short-wave limit of the continuous x-ray spectrum.

A bombarding electron may be brought to rest in a single process if the electron happens to collide head on with an atom of the target; or

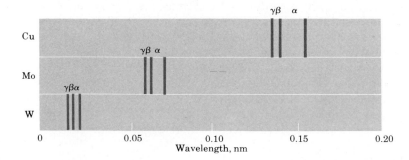

44-12 Wavelengths of the K_α, K_β, and K_γ lines of copper, molybdenum, and tungsten.

it may make a number of collisions before coming to rest, giving up part of its energy each time. If we assume that the energy lost at each collision is radiated as an x-ray photon, these photons may have any energy up to a certain maximum, namely, that of an electron that gives up all of its energy in a single collision. Hence there is a short-wave limit to the spectrum. The frequency of this limit is found by setting the energy of the electron equal to the energy of the x-ray photon:

$$hf = \tfrac{1}{2}mv^2. \tag{44-14}$$

This is precisely the same equation as that for the photoelectric effect except for the work-function term, which is negligible here since the energies of the x-ray photons are so large. In fact, the emission of x-rays may be described as an *inverse photoelectric effect.* In photoelectric emission the energy of a photon is transformed into kinetic energy of an electron; here, the kinetic energy of an electron is transformed into that of a photon.

Example Compute the potential difference through which an electron must be accelerated in order that the short-wave limit of the continuous x-ray spectrum shall be exactly 0.1 nm.

Solution The frequency corresponding to 0.1 nm (10^{-10} m) is given by

$$f = \frac{c}{\lambda} = \frac{3 \times 10^8 \text{ m} \cdot \text{s}^{-1}}{10^{-10} \text{ m}} = 3 \times 10^{18} \text{ s}^{-1} = 3 \times 10^{18} \text{ Hz.}$$

The energy of the photon is

$$hf = (6.63 \times 10^{-34} \text{ J} \cdot \text{s})(3 \times 10^{18} \text{ s}^{-1}) = 19.9 \times 10^{-16} \text{ J.}$$

This must equal the kinetic energy of the electron, $\tfrac{1}{2}mv^2$, which is also equal to the product of the electronic charge and the accelerating voltage, V:

$$\tfrac{1}{2}mv^2 = eV = 19.9 \times 10^{-16} \text{ J.}$$

Since

$$e = 1.60 \times 10^{-19} \text{ C,}$$

$$V = \frac{19.9 \times 10^{-16} \text{ J}}{1.60 \times 10^{-19} \text{ C}} = 12{,}400 \text{ V.}$$

A phenomenon called *Compton scattering,* first observed in 1924 by A. H. Compton, provides additional direct confirmation of the quantum nature of electromagnetic radiation. When x-rays impinge on matter, some of the radiation is *scattered,* just as visible light falling on a rough surface undergoes diffuse reflection. Observation shows that some of the scattered radiation has smaller frequency and longer wavelength than the incident radiation, and that the change in wavelength depends on the angle through which the radiation is scattered. Specifically, if the scattered radiation emerges at an angle ϕ with respect to the incident direction, and if λ and λ' are the wavelengths of the incident and scattered radiation, respectively, it is found that

$$\lambda' - \lambda = \frac{h}{mc}(1 - \cos \phi) \tag{44-15}$$

where m is the electron mass.

Compton scattering cannot be understood on the basis of classical electromagnetic theory. On the basis of classical principles, the scattering mechanism is induced motion of electrons in the material, caused by the incident radiation. This motion must have the same frequency as that of the incident wave, and so the scattered wave radiated by the oscillating charges should have the same frequency. There is no way the frequency can be *shifted* by this mechanism.

The quantum theory, by contrast, provides a beautifully simple explanation. We imagine the scattering process as a collision of two *particles,* the incident photon and an electron initially at rest, as in Fig. 44-13. The photon gives up some of its energy and momentum to the electron, which recoils as a result of this impact; and the final photon has less energy, smaller frequency, and longer wavelength than the initial one.

Equation (44-15) can be derived from the principles of conservation of energy and of momentum. We sketch the derivation below; the details of the calculation are left as a problem. Because relativistic energies may be involved, we use relativistic energy and momentum relations for the electron; its initial energy is mc^2, and its final energy E is given by $E^2 = (mc^2)^2 + (Pc)^2$. The energies of the incident and scattered photons are respectively pc and $p'c$; Thus energy conservation yields the relation

$$pc + mc^2 = p'c + E,$$

or

$$(pc - p'c - mc^2)^2 = E^2 = (mc^2)^2 + (Pc)^2. \qquad (44\text{-}16)$$

We may eliminate the electron momentum $\boldsymbol{P}$ from this equation by using momentum conservation:

$$\boldsymbol{p} = \boldsymbol{p'} + \boldsymbol{P},$$

or

$$\boldsymbol{p} - \boldsymbol{p'} = \boldsymbol{P}. \qquad (44\text{-}17)$$

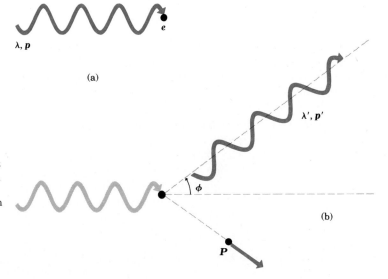

(a)

(b)

44-13 Schematic diagram of Compton scattering showing (a) electron initially at rest with incident photon of wavelength λ and momentum $\boldsymbol{p}$; (b) scattered photon with longer wavelength λ' and momentum $\boldsymbol{p'}$ and recoiling electron with momentum $\boldsymbol{P}$. The direction of the scattered photon makes an angle ϕ with that of the incident photon, and the angle between $\boldsymbol{p}$ and $\boldsymbol{p'}$ is also ϕ.

We take the scalar product of this quantity with itself, noting that $\boldsymbol{p} \cdot \boldsymbol{p}' = pp' \cos \phi$; we obtain

$$P^2 = p^2 + p'^2 - 2pp' \cos \phi. \qquad (44\text{--}18)$$

This expression for P^2 may now be substituted into Eq. (44–16) and the left side multiplied out. A common factor c^2 is divided out; several terms cancel, and when the resulting equation is divided through by (pp'), the result is

$$\frac{mc}{p'} - \frac{mc}{p} = 1 - \cos \phi. \qquad (44\text{--}19)$$

Finally, we substitute $p = h/\lambda$ and $p' = h/\lambda'$ and rearrange again to obtain Eq. (44–15).

Questions

44-1 In analyzing the photoelectric effect, how can we be sure that each electron absorbs only *one* proton?

44-2 In what ways do photons resemble other particles such as electrons? In what ways do they differ? Do they have mass? Electric charge? Can they be accelerated? What mechanical properties do they have?

44-3 Considering a two-slit interference experiment, if the photons are not synchronized with each other (i.e., are not coherent) and if half go through each slit, how can they possibly interfere with each other? Is there any way out of this paradox?

44-4 Can you devise an experiment to measure the work function of a material?

44-5 How might the energy levels of an atom be measured directly, i.e., without recourse to analysis of spectra?

44-6 Would you expect quantum effects to be generally more important at the low-frequency end of the electromagnetic spectrum (radio waves) or at the high-frequency end (x-rays and gamma rays)? Why?

44-7 Most black-and-white photographic film (with the exception of some special-purpose films) is less sensitive at the far red end of the visible spectrum than at the blue end, and has almost no sensitivity to infrared. How can these properties be understood on the basis of photons?

44-8 Human skin is relatively insensitive to visible light, but ultraviolet radiation can be quite destructive. Does this have anything to do with photon energies?

44-9 Does the concept of photon energy shed any light (no pun intended) on the question of why x-rays are so much more penetrating than visible light?

44-10 The phosphorescent materials that coat the inside of a fluorescent lamp tube convert ultraviolet radiation (from the mercury-vapor discharge inside the tube) to visible light. Could one also make a phosphor that converts visible light to ultraviolet?

44-11 As a body is heated to very high temperature and becomes self-luminous, the apparent color of the emitted radiation shifts from red to yellow and finally to blue as the temperature increases. Why the color shift?

44-12 Elements in the gaseous state emit line spectra with well-defined wavelengths; but hot solid bodies usually emit a continuous spectrum, that is, a continuous smear of wavelengths. Can you account for this difference?

44-13 Could Compton scattering occur with protons as well as electrons? Suppose, for example, one directed a beam of x-rays at a liquid-hydrogen target. What similarities and differences in behavior would be expected?

Problems

$$e = 1.602 \times 10^{-19} \text{ C}$$
$$m = 9.109 \times 10^{-31} \text{ kg}$$
$$h = 6.626 \times 10^{-34} \text{ J} \cdot \text{s}$$
$$N_0 = 6.022 \times 10^{23} \text{ atoms} \cdot \text{mol}^{-1}$$
Energy equivalent of $1 \text{ u} = 931.5 \text{ MeV}$

$$\frac{e}{m} = 1.758 \times 10^{11} \text{ C} \cdot \text{kg}^{-1}$$
$$k = 1.381 \times 10^{-23} \text{ J} \cdot \text{K}^{-1}$$
$$1 \text{ eV} = 1.602 \times 10^{-19} \text{ J}$$
$$1 \text{ u} = 1.661 \times 10^{-27} \text{ kg}$$
$$\varepsilon_0 = 8.854 \times 10^{-12} \text{ C}^2 \cdot \text{N}^{-1} \cdot \text{m}^{-2}$$

44-1 In the photoelectric effect, what is the relation between the threshold frequency f_0 and the work function ϕ?

44-2 The photoelectric threshold wavelength of tungsten is 2.73×10^{-5} cm. Calculate the maximum kinetic energy of the electrons ejected from a tungsten surface by ultraviolet radiation of wavelength 1.80×10^{-5} cm. (Express the answer in electron volts.)

44-3 A photoelectric surface has a work function of 4.00 eV. What is the maximum velocity of the photoelectrons emitted by light of frequency 3×10^{15} Hz?

44-4 When ultraviolet light of wavelength 2.54×10^{-5} cm from a mercury arc falls upon a clean copper surface, the retarding potential necessary to stop the emission of photoelectrons is 0.59 V. What is the photoelectric threshold wavelength for copper?

44-5 When a certain photoelectric surface is illuminated with light of different wavelengths, the stopping potentials in the table below are observed:

Wavelength, nm	Stopping potential, V
366	1.48
405	1.15
436	0.93
492	0.62
546	0.36
579	0.24

Plot the stopping potential as ordinate against the frequency of the light as abscissa. Determine

a) the threshold frequency,

b) the threshold wavelength,

c) the photoelectric work function of the material, and

d) the value of Planck's constant h (the value of e being known).

44-6 The photoelectric work function of potassium is 2.0 eV. If light having a wavelength of 360 nm falls on potassium, find

a) the stopping potential,

b) the kinetic energy in electron volts of the most energetic electrons ejected, and

c) the velocities of these electrons.

44-7 What will be the change in the stopping potential for photoelectrons emitted from a surface if the wavelength of the incident light is reduced from 400 nm to 360 nm?

44-8 The photoelectric work functions for particular samples of certain metals are as follows: cesium, 2.00 eV; copper, 4.00 eV; potassium, 2.25 eV; and zinc, 3.60 eV.

a) What is the threshold wavelength for each metal?

b) Which of these metals could not emit photoelectrons when irradiated with visible light?

44-9 The light-sensitive compound on most photographic films is silver bromide, AgBr. A film is "exposed" when the light energy absorbed dissociates this molecule into its atoms. (The actual process is more complex, but the quantitative result does not differ greatly.) The energy of or heat of dissociation of AgBr is 23.9 kcal·mol^{-1}. Find

a) the energy in electron volts,

b) the wavelength, and

c) the frequency of the photon that is just able to dissociate a molecule of silver bromide.

d) What is the energy in electron volts of a quantum of radiation having a frequency of 100 MHz?

e) Explain the fact that light from a firefly can expose a photographic film, whereas the radiation from a TV station transmitting 50,000 W at 100 MHz cannot.

f) Will photographic films stored in a light-tight container be ruined (exposed) by the radio waves constantly passing through them? Explain.

44-10

a) Show that the energy E (in electron volts) of a photon of wavelength λ (in nanometers) is given by E (eV) $= (1240/\lambda)$(nm).

b) What is the energy in electron volts of a photon having a wavelength of 91.2 nm?

44-11 If 5% of the energy supplied to an incandescent lightbulb is radiated as visible light, how many visible quanta are emitted per second by a 100-W bulb? Assume the wavelength of all the visible light to be 560 nm.

44-12 The directions of emission of photons from a source of radiation are random. According to the wave theory, the intensity of radiation from a point source varies inversely as the square of the distance from the source. Show that the number of photons from a point source passing out through a unit area is also given by an inverse-square law.

44-13 Calculate (a) the frequency, and (b) the wavelength of the H_β-line of the Balmer series for hydrogen. This line is emitted in the transition from $n = 4$ to $l = 2$.

44-14

a) What is the least amount of energy in electron volts that must be given to a hydrogen atom so that it can emit the H_β-line (see Problem 44–13) in the Balmer series?

b) How many different possibilities of spectral line emission are there for this atom when the electron goes from $n = 4$ to the ground state?

44-15 In Fig. 44–10, compute the energy difference for the $5s$–$3p$ transition; express your result in electron volts and in joules. Compute the wavelength of a photon having this energy, and compare your result with the observed wavelength of the laser light.

44-16 In the helium–neon laser, what wavelength corresponds to the $3p$–$3s$ transition in neon? Why is this not observed in the beam with the same intensity as the 632.8-nm laser line?

44-17

a) What is the minimum potential difference between the filament and the target of an x-ray tube if the tube is to produce x-rays of wavelength 0.05 nm?

b) What is the shortest wavelength produced in an x-ray tube operated at 2×10^6 V?

44-18 An electron in a certain x-ray tube is accelerated from rest through a potential difference of 180,000 V in going from the cathode to the anode. When it arrives at the anode, what is

a) its kinetic energy in electron volts,

b) its relativistic velocity?

c) What is the velocity of the electron, calculated classically?

44-19 An x-ray tube is operating at 150,000 V and 10 mA.

a) If only 1% of the electric power supplied is converted into x-rays, at what rate is the target being heated in calories per second?

b) If the target has a mass of 300 g and a specific heat of 0.035 cal·g^{-1}·°C^{-1}, at what average rate would its temperature rise if there were no thermal losses?

c) What must be the physical properties of a practical target material? What would be some suitable target elements?

44-20 If electrons in a metal had the same energy distribution as molecules in a gas at the same temperature (which is not actually the case), at what temperature would the average electron kinetic energy equal 1 eV, typical of work functions of metals?

44-21 If hydrogen were monatomic, at what temperature would the average translational kinetic energy be equal to the energy required to raise a hydrogen atom from the ground state to the $n = 2$ excited state?

44-22 Complete the derivation of the Compton-scattering formula, Eq. (44-15), following the outline given in Eqs. (44-16) through (44-19).

44-23 Calculate the maximum increase in x-ray wavelength that can occur during Compton scattering.

44-24 X-rays with initial wavelength 0.5×10^{-10} m undergo Compton scattering. For what scattering angle is the wavelength of the scattered x-rays greater than that of the incident x-rays by one percent?

44-25 X-rays are produced in a tube operating at 50 kV. After emerging from the tube, some x-rays strike a target and are Compton-scattered through an angle of 20°.

a) What is the original x-ray wavelength?

b) What is the wavelength of the scattered x-rays?

c) What is the energy of the scattered x-rays (in electron-volts)?

44-26 What is the energy (in electronvolts) of the smallest-energy x-ray photon for which Compton scattering could result in doubling the original wavelength?

QUANTUM MECHANICS

45

45-1 The Bohr atom

We have seen in the preceding chapter that some aspects of atomic spectra can be understood on the basis of the quantum or photon concept of electromagnetic radiation, together with the concept of discrete energy levels of atoms. Of course, any theory of atomic spectra that aspires to completeness should also offer some means of *predicting,* on theoretical grounds, the values of these energy levels for any given atom. At the same time that Bohr advanced his hypothesis about the relation of spectrum-line frequencies to energy levels, he also proposed a mechanical model of the simplest atom, hydrogen; he was able to calculate the energy levels of hydrogen and obtain agreement with values determined from spectra.

Bohr's was not by any means the first attempt to understand the internal structure of atoms. Starting in 1906, Rutherford and his co-workers had performed experiments on the scattering of alpha particles (helium nuclei emitted from radioactive elements) by thin metallic foils. These experiments, which will be discussed in Chapter 47, showed that each atom contains a massive nucleus whose size (of the order of 10^{-15} m) is very much *smaller* than the overall size of the atom (of the order of 10^{-10} m). The nucleus is surrounded by a swarm of electrons.

To account for the fact that the electrons remain at relatively large distances from the positively charged nucleus despite the electrostatic attraction the nucleus exerts on the electrons, Rutherford postulated that the electrons *revolve* about the nucleus in orbits, more or less as the planets in the solar system revolve around the sun, but with the electrical attraction providing the necessary centripetal force.

This assumption, however, has an unfortunate consequence. A body moving in a circle is continuously accelerated toward the center of the circle and, according to classical electromagnetic theory, an accelerated electron radiates energy. The total energy of the electrons would therefore gradually decrease, their orbits would become smaller and smaller, and eventually they would spiral into the nucleus and come to rest. Furthermore, according to classical theory, the *frequency* of the electromagnetic waves emitted by a revolving electron is equal to the frequency of revolution. As the electrons radiated energy, their angular velocities would change continuously and they would emit a *continuous* spectrum (a mixture of all frequencies), in contradiction to the *line* spectrum actually observed.

Faced with the dilemma that electromagnetic theory predicted an unstable atom emitting radiant energy of all frequencies, while observation showed stable atoms emitting only a few frequencies, Bohr concluded that, in spite of the success of electromagnetic theory in explaining large-scale phenomena, it could not be applied to processes on an atomic scale. He therefore postulated that an electron in an atom can revolve in certain stable orbits, each having a definite associated energy, *without* emitting radiation, contrary to the predictions of classical electromagnetic theory. According to Bohr, an atom radiates only when it makes a transition from one of these special orbits to another, emitting (or absorbing) a photon of appropriate energy, given by Eq. (44–8), at the same time.

To determine the radii of the "permitted" orbits, Bohr introduced what must be regarded in hindsight as a brilliant intuitive guess. He noted that the *units* of Planck's constant h, usually written as J·s, are the same as the units of angular momentum, usually written as kg·m^2·s^{-1}, and he postulated that only those orbits are permitted for which the angular momentum is an integer multiple of $h/(2\pi)$. We recall from Chapter 9 that the angular momentum of a particle of mass m, moving with tangential speed v in a circle of radius r, is mvr. Hence the above condition may be stated as

$$mvr = n\frac{h}{2\pi}, \qquad (45\text{–}1)$$

where $n = 1, 2, 3$, etc.

We now incorporate this condition into the analysis of the hydrogen atom. This atom consists of a single electron of charge $-e$, revolving about a single proton of charge $+e$. The proton, being nearly 2000 times as massive as the electron, will be assumed stationary in this discussion. The electrostatic force of attraction between the charges,

$$F = \frac{1}{4\pi\varepsilon_0}\frac{e^2}{r^2},$$

provides the centripetal force and, from Newton's second law,

$$\frac{1}{4\pi\varepsilon_0}\frac{e^2}{r^2} = \frac{mv^2}{r}. \qquad (45\text{–}2)$$

When Eqs. (45–1) and (45–2) are solved simultaneously for r and v, we obtain

$$r = \varepsilon_0\frac{n^2h^2}{\pi me^2}, \qquad (45\text{–}3)$$

$$v = \frac{1}{\varepsilon_0}\frac{e^2}{2nh}. \qquad (45\text{–}4)$$

Let

$$\varepsilon_0\frac{h^2}{\pi me^2} = r_0. \qquad (45\text{–}5)$$

Then Eq. (45–3) becomes

$$r = n^2 r_0,$$

and the permitted, nonradiating orbits have radii r_0, $4r_0$, $9r_0$, etc. The appropriate value of n is called the *quantum number* of the orbit.

The numerical values of the quantities on the left side of Eq. (45–5) are

$$\varepsilon_0 = 8.854 \times 10^{-12} \, \text{C}^2 \cdot \text{N}^{-1} \cdot \text{m}^{-2},$$
$$h = 6.626 \times 10^{-34} \, \text{J} \cdot \text{s},$$
$$m = 9.109 \times 10^{-31} \, \text{kg},$$
$$e = 1.602 \times 10^{-19} \, \text{C}.$$

Hence r_0, the radius of the first Bohr orbit, is

$$r_0 = \frac{(8.854 \times 10^{-12} \, \text{C}^2 \cdot \text{N}^{-1} \cdot \text{m}^{-2})(6.626 \times 10^{-34} \, \text{J} \cdot \text{s})^2}{(3.14)(9.109 \times 10^{-31} \, \text{kg})(1.602 \times 10^{-19} \, \text{C})^2}$$
$$= 0.53 \times 10^{-10} \, \text{m} = 0.53 \times 10^{-8} \, \text{cm}.$$

This is in good agreement with atomic diameters as estimated by other methods, namely, about 10^{-8} cm.

The kinetic energy of the electron in any orbit is

$$K = \tfrac{1}{2}mv^2 = \frac{1}{\varepsilon_0{}^2} \frac{me^4}{8n^2h^2},$$

and the potential energy is

$$U = -\frac{1}{4\pi\varepsilon_0} \frac{e^2}{r} = -\frac{1}{\varepsilon_0{}^2} \frac{me^4}{4n^2h^2}.$$

The total energy, E, is therefore

$$E = K + U = -\frac{1}{\varepsilon_0{}^2} \frac{me^4}{8n^2h^2}. \tag{45–6}$$

The total energy has a negative sign because the reference level of potential energy is taken with the electron at an infinite distance from the nucleus. Since we are interested only in energy *differences,* this is not of importance.

The energy of the atom is least when its electron is revolving in the orbit for which $n = 1$, for then E has its largest negative value. For $n = 2, 3, \ldots$, the absolute value of E is smaller; hence the energy is progressively larger in the outer orbits. The *normal* state of the atom, called the *ground state,* is that of lowest energy, with the electron revolving in the orbit of smallest radius, r_0. As a result of collisions with rapidly moving electrons in an electrical discharge, or for other causes, the atom may temporarily acquire sufficient energy to raise the electron to some outer orbit. The atom is then said to be in an *excited* state. This state is an unstable one, and the electron soon falls back to a state of lower energy, emitting a photon in the process.

Let n be the quantum number of some excited state, and l the quantum number of the lower state to which the electron returns after the emission process. Then E_i, the initial energy, is

$$E_i = -\frac{1}{\varepsilon_0{}^2} \frac{me^4}{8n^2h^2},$$

and E_f, the final energy, is

$$E_f = -\frac{1}{\varepsilon_0^2} \frac{me^4}{8l^2h^2}.$$

The decrease in energy, $E_i - E_f$, which we place equal to the energy hf of the emitted photon, is

$$E_i - E_f = hf = -\frac{1}{\varepsilon_0^2} \frac{me^4}{8n^2h^2} + \frac{1}{\varepsilon_0^2} \frac{me^4}{8l^2h^2},$$

or

$$f = \frac{1}{\varepsilon_0^2} \frac{me^4}{8h^3} \left(\frac{1}{l^2} - \frac{1}{n^2} \right). \qquad (45\text{--}7)$$

This equation is of precisely the same form as the Balmer formula (Eq. 44–6) for the frequencies in the hydrogen spectrum if we place

$$\frac{1}{\varepsilon_0^2} \frac{me^4}{8h^3} = Rc, \qquad (45\text{--}8)$$

and let $l = 1$ for the Lyman series, $l = 2$ for the Balmer series, etc. The Lyman series is therefore the group of lines emitted by electrons returning from some excited state to the ground state. The Balmer series is the group emitted by electrons returning from some higher state, but stopping in the *second orbit* instead of falling at once to that of lowest energy. That is, an electron returning from the third orbit ($n = 3$) to the second orbit ($l = 2$) emits the H_α-line. One returning from the fourth orbit ($n = 4$) to the second ($l = 2$) emits the H_β-line, etc. These transitions are shown in Fig. 45–1.

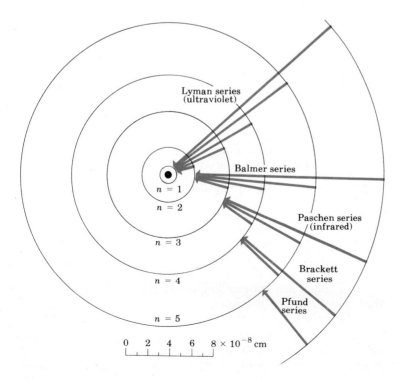

45-1 "Permitted" orbits of an electron in the Bohr model of the hydrogen atom. The transitions responsible for some of the lines of the various series are indicated by arrows.

Every quantity in Eq. (45–8) may be determined quite independently of the Bohr theory, and apart from this theory we have no reason to expect these quantities to be related in this particular way. The quantities m and e, for instance, are found from experiments on free electrons, h may be found from the photoelectric effect, and R by measurements of wavelengths, while c is the speed of light. However, if we substitute in Eq. (45–8) the values of these quantities, obtained by such diverse means, we find that it *does* hold exactly, within the limits of experimental error, providing direct confirmation of Bohr's theory.

The interaction of atoms with a magnetic field can be analyzed with the Bohr model. It was suggested first by Faraday that shifts in spectrum wavelengths might occur when atoms are placed in a magnetic field. Faraday's spectroscopic techniques were not refined enough to observe such shifts, but Zeeman, using improved instruments, *was* able to detect shifts, an effect now called the *Zeeman effect*.

The interaction is most easily treated using the concept of magnetic moment, introduced in Sec. 31–3. Let us consider an electron in the first Bohr orbit ($n = 1$). The orbiting charge is equivalent to a current loop of radius r and area πr^2. The average charge per unit time passing a point of the orbit is the average current I, and this is given by e/τ, where τ is the time for one revolution: $\tau = 2\pi r/v$. The magnetic moment, which we denote by μ to avoid confusion with the electron mass m, is given by

$$\mu = IA = \tfrac{1}{2}evr.$$

But according to Bohr's theory, the angular momentum mvr is equal to $h/2\pi$, or

$$vr = \frac{h}{2\pi m}.$$

Therefore

$$\mu = \frac{h}{4\pi} \cdot \frac{e}{m}$$

$$= \frac{(6.626 \times 10^{-34}\,\text{J·s})(1.758 \times 10^{11}\,\text{C·kg}^{-1})}{12.57}$$

$$= 9.27 \times 10^{-24}\,\text{A·m}^2.$$

Example   Find the interaction potential energy when the hydrogen atom described above is placed in a magnetic field of 2 T.

Solution   According to Eq. (31–12), the interaction energy U when $\alpha = 0$ is

$$U = -\mu B = -(9.27 \times 10^{-24}\,\text{A·m}^2)(2\,\text{T})$$

$$= -1.85 \times 10^{-23}\,\text{J}$$

$$= -1.16 \times 10^{-4}\,\text{eV}.$$

When $\boldsymbol{\mu}$ and $\boldsymbol{B}$ are antiparallel, the energy is $+1.16 \times 10^{-4}$ eV. We note that these energies are much *smaller* than the energy of the electron.

Although the Bohr model was successful in predicting the energy levels of the hydrogen atom, it raised as many questions as it answered. It combined elements of classical physics with new postulates inconsistent with classical ideas. It provided no insight into what happens during a transition from one orbit to another, and the stability of certain orbits was bought at the expense of discarding the only picture available at the time of the mechanism of radiation of energy. There was no clear justification (except that it led to the right answer) for restricting the angular momentum to multiples of $h/2\pi$. Furthermore, attempts to extend the model to atoms with two or more electrons were not successful. We shall see in the next section that an even more radical departure from classical concepts was required before the understanding of atomic structure could progress further.

45-2 Wave nature of particles

The next advance in atom building came in 1923, about 10 years after the Bohr theory. This was a suggestion by de Broglie that, since light is dualistic in nature, behaving in some aspects like waves and in others like particles, the same might be true of matter. That is, electrons and protons, which until that time had been thought to be purely corpuscular, might in some circumstances behave like *waves*. Specifically, de Broglie postulated that a free electron of mass m, moving with speed v, should have a wavelength λ given by

$$\lambda = \frac{h}{mv}, \tag{45-9}$$

where h is the same Planck's constant that appears in the frequency–energy relation for photons.

This wave hypothesis, unorthodox though it seemed at the time, almost immediately received very direct experimental confirmation. We have described in Sec. 41–10 how the layers of atoms in a crystal serve as a diffraction grating for x-rays. An x-ray beam is strongly reflected when it strikes a crystal at such an angle that the waves scattered from the atomic layers combine to reinforce one another. The essential point here is that the existence of these strong reflections is evidence of the *wave* nature of x-rays.

In 1927, Davisson and Germer, working in the Bell Telephone Laboratories, were studying the nature of the surface of a crystal of nickel by directing a beam of *electrons* at the surface and observing the electrons reflected at various angles. It might be expected that even the smoothest surface attainable would still look rough to an electron, and that the electron beam would therefore be diffusely reflected. But Davisson and Germer found that the electrons were reflected in almost the same way that x-rays would be reflected from the same crystal. The wavelengths of the electrons in the beam were computed from their known speed, with the help of Eq. (45–9); and the angles at which strong reflection took place were found to be the same as those at which x-rays of the same wavelength would be reflected. This result gave strong support to de Broglie's hypothesis.

This wave hypothesis clearly requires sweeping revisions of our fundamental concepts regarding the description of matter. What we are accustomed to calling a *particle* actually looks like a particle only if we do not look too closely. In general, a particle has to be regarded as a spread-out entity that is not entirely localized in space; and at least in some cases this spreading out appears as a periodic pattern suggesting wavelike properties. The wave and particle aspects are not inconsistent, but the particle model is an *approximation* of a more general wave picture. We are reminded of the ray picture of geometrical optics, a special case of the more general wave picture of physical optics; indeed, there is a very close analogy between optics and the description of particles.

Within a few years, after 1923, the wave hypothesis of de Broglie was developed by Heisenberg, Schrödinger, and many others, into a complete theory called *wave mechanics* or *quantum mechanics*. In the following sections we shall attempt to outline the main lines of thought in a non-mathematical way, describe some of the experimental evidence of the wave nature of material particles, and show how the quantum numbers that were introduced in such an artificial way by Bohr now enter naturally into the theory of atomic structure.

One of the essential features of quantum mechanics is that material particles are no longer regarded as geometrical points, localized in space, but are intrinsically spread-out entities. The spatial distribution of a *free* electron may have a recurring pattern characteristic of a wave that propagates through space. Electrons *in atoms* are visualized as diffuse clouds surrounding the nucleus. The idea that the electrons in an atom move in definite orbits such as those in Fig. 45–1 has been abandoned. The orbits themselves, however, were never an essential part of Bohr's theory, since the quantities that determine the frequencies of the emitted photons are the *energies* corresponding to the orbits. The new theory still assigns definite energy states to an atom. In the hydrogen atom the energies are the same as those given by Bohr's theory; in more complicated atoms where the Bohr theory did not work, the quantum mechanical picture is in excellent agreement with observation.

We shall illustrate how quantization arises in atomic structure by an analogy with the classical mechanical problem of a vibrating string fixed at its ends. When the string vibrates, the ends must be nodes, but nodes may occur at other points also, and the general requirement is that the length of the string shall equal some *integer* number of half-wavelengths. The point of interest is that the solution of the problem of the vibrating string leads to the appearance of *integral numbers*.

In a similar way, the principles of quantum mechanics lead to a wave equation (Schrödinger's equation) that must be satisfied by an electron in an atom, subject also to certain boundary conditions. Let us think of an electron as a wave extending in a circle around the nucleus. In order that the wave may "come out even," the circumference of this circle must include some *integer number* of wavelengths. The wavelength of a particle of mass m, moving with speed v, is given, according to wave mechanics, by Eq. (45–9), $\lambda = h/mv$. Then if r is the radius and $2\pi r$ the circumference of the circle occupied by the wave, we must have $2\pi r = n\lambda$, where $n = 1, 2, 3$, etc. Since $\lambda = h/mv$, this equation becomes

$$2\pi r = n\frac{h}{mv}, \qquad mvr = n\frac{h}{2\pi}. \qquad (45\text{--}10)$$

But mvr is the angular momentum of the electron, so we see that the wave-mechanical picture leads naturally to Bohr's postulate that the angular momentum equals some integral multiple of $h/2\pi$.

To be sure, the idea of wrapping a wave around in a circular orbit is a rather vague notion. But the agreement of Eq. (45–10) with Bohr's hypothesis is much too remarkable to be a coincidence; and it strongly suggests that the wave properties of electrons do indeed have something to do with atomic structure.

45-3 The electron microscope

The shorter the wavelength, the smaller the limit of resolution of a microscope. The wavelengths of electron waves can easily be made very much *shorter* than the wavelengths of visible light. Hence, the limit of resolution of a microscope may be extended to a value several hundred times smaller than that obtainable with an optical instrument, by using electrons, rather than light waves, to form an image of the object being examined.

A beam of electrons can be focused by either a magnetic or an electric field of the proper configuration, and both types are used in electron microscopes. Figure 45–2 illustrates an electrostatic lens. Two hollow cylinders are maintained at different potentials. A few of the equipotentials are indicated, and the trajectories of a beam of electrons traveling from left to right are shown by the colored lines. The optical analog of this electrostatic lens is shown in Fig. 45–3. It will be evident, without going into further details, that by the proper design of such lenses the elements of an optical microscope, such as its condensing lens, objective, and eyepiece, can all be duplicated electrically.

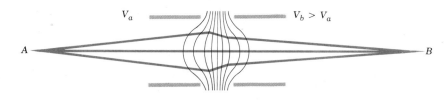

45-2 An electrostatic electron lens. The cylinders are at different potentials V_a and V_b. A beam of electrons diverging from point A is focused at point B.

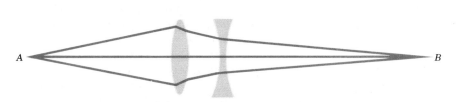

45-3 Optical analog of the electron lens in Fig. 45–2.

The source of electrons in an electron microscope is a heated filament. Electrons emitted by the filament are accelerated by an electron gun and strike the object to be examined. This must necessarily be a thin section so that some of the electrons can pass through it. The thicker portions of the section absorb more of the electron stream than do the thinner portions, just as would a slide in a slide projector. The entire apparatus must, of course, be evacuated.

The final image may be formed on a photographic plate, or on a fluorescent screen that can be examined visually or photographed with a still further gain in magnification. Commercial electron microscopes give satisfactory definition at an overall magnification (electronic, followed by photographic) as great as 50,000 times. Figure 45–4 is an electron micrograph of aluminum oxide, magnified 53,500 times.

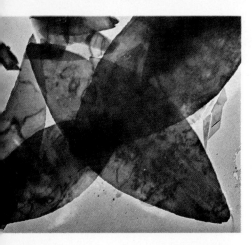

45-4   Electron micrograph of aluminum oxide, magnified 53,500 times. (Courtesy of Radio Corporation of America.)

It should be pointed out that the ability of the electron microscope to form an image does *not* depend on the wave properties of the electrons; their trajectories can be computed by treating them as charged particles, deflected by the electric or magnetic fields through which they move. It is only when considerations of *resolving power* arise that the electron wavelengths come into the picture. The situation is analogous to that in the optical microscope. The paths of light rays through an optical microscope can be computed by the principles of geometrical optics, but the resolving power of the microscope is determined by the wavelength of the light used.

Example   What is the speed of an electron having a wavelength of 10^{-10} m (the order of magnitude of atomic spacing in crystals)? Through what potential difference must an electron be accelerated to acquire this speed?

Solution   From the de Broglie relation, Eq. (45-9), we have

$$v = \frac{h}{m\lambda} = \frac{6.626 \times 10^{-34}\,\text{J}\cdot\text{s}}{(9.109 \times 10^{-31}\,\text{kg})(10^{-10}\,\text{m})} = 7.27 \times 10^6\,\text{m}\cdot\text{s}^{-1}.$$

To find the required potential difference V, we equate the gain in kinetic energy, $\frac{1}{2}mv^2$, to the loss in potential energy, eV:

$$V = \frac{mv^2}{2e} = \frac{(9.109 \times 10^{-31}\,\text{kg})(7.27 \times 10^6\,\text{m}\cdot\text{s}^{-1})^2}{(2)(1.602 \times 10^{-19}\,\text{C})} = 150\,\text{V}.$$

*45-4 Probability and uncertainty

As we have seen, the entities that we are accustomed to calling *particles* can, in some experiments, behave like *waves*. Figure 45-5 shows a *wave packet* or *wave pulse* that has both wave and particle properties. The regular spacing λ_{av} between successive maxima is characteristic of a wave, but there is also a particle-like localization in space. To be sure, the wave pulse is not localized at a single point, but in any experiment that detects only dimensions much larger than Δx, the wave pulse appears to be a localized particle. Although it would be simplistic to say that Fig. 45-5 is a picture of an electron, the figure does suggest that wave and particle properties are not necessarily incompatible.

It must be emphasized that it would *not* be correct to regard the wave of Fig. 45-5 as having only a single wavelength. A sinusoidal wave with a definite wavelength has no beginning and no end; to make a wave *pulse* it is necessary to superpose many sinusoidal waves having various wavelengths. Thus Fig. 45-5 is a wave having such a distribution of wavelengths, of which λ_{av} is an average value. This distribution has additional important implications, which will be explored at the end of this section.

The discovery of the dual wave–particle nature of matter has forced a drastic revision of the language used to describe the behavior of a particle. In classical newtonian mechanics we think of a particle as an idealized geometrical point, which, at any instant of time, has a perfectly definite location in space and is moving with a perfectly definite velocity. As we shall see, such a specific description is, in general, not possible; on a sufficiently small scale, there are fundamental limitations on the precision with which the position and velocity of a particle can be described;

Δx

λ_{av}

Direction of motion

45-5   A wave pulse or packet. There is an average wavelength λ_{av}, the distance between adjacent peaks, but the wave is localized at any instant in a region with length of the order of Δx. The broken lines are called the envelope of the pulse; all the peaks lie on the envelope curves, which approach zero at both ends of the pulse.

and some aspects of a particle's behavior can be stated only in terms of *probabilities.*

To illustrate the nature of the problem, let us consider again the single-slit diffraction experiment described in Sec. 41–8. As Fig. 41–23 suggests, most of the intensity in the diffraction pattern is concentrated in the central maximum; the angular size of this maximum is determined by the positions of the first intensity minimum on either side of the central maximum. Using Eq. (41–15), with $n = 1$, we find that the angle θ in Fig. 45–6 between the central peak and the minimum on either side is given by $\sin \theta = \lambda/a$, where a is the slit width. If λ is much smaller than a, then θ is very small, $\sin \theta$ is very nearly equal to θ, and this may be simplified further to

$$\theta = \frac{\lambda}{a}. \qquad (45\text{--}11)$$

Now we perform the same experiment again, but using a beam of *electrons* instead of a beam of monochromatic light. The apparatus must be evacuated to avoid collisions of electrons with air molecules; and there are other experimental details that need not concern us. The electron beam can be produced with a setup similar in principle to the electron gun in a cathode-ray oscilloscope, described in Sec. 26–8, which produces a narrow beam of electrons all having the same direction and speed, and therefore also the same wavelength. Such an experiment is shown schematically in Fig. 45–6.

The result of this experiment, as again recorded on photographic film or by means of more sophisticated detectors, is a diffraction pattern identical to that shown in Fig. 41–23, providing additional direct evidence of the wave nature of electrons. Most of the electrons strike the film in the vicinity of the central maximum, but a few strike farther from the center, near the edges of that maximum and also in the subsidiary maxima on both sides. Thus the wave behavior in this experiment presents no surprises.

Interpreted in terms of *particles,* however, this experiment poses very serious problems. First, although the electrons all have the same initial state of motion, they do not all follow the same path; and in fact the trajectory of an individual electron cannot be predicted from knowledge of its initial state. The best we can do is to say that most of the electrons go to a certain region, fewer to other regions, and so on; alter-

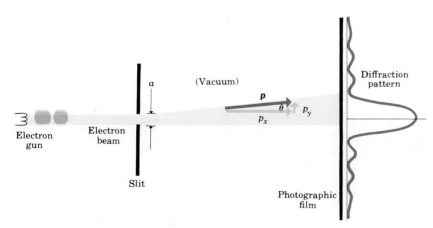

45-6 An electron diffraction experiment. The graph at the right shows the degree of blackening of the film, which in any region is proportional to the number of electrons striking that region. The components of momentum of an electron striking the outer fringe of the central maximum are shown.

natively, we can describe the *probability* for an individual electron to strike each of various areas on the film. This fundamental indeterminacy has no counterpart in newtonian mechanics, where the motion of a particle or a system is always completely predictable if the initial position and motion are known with sufficient precision.

Second, there are fundamental *uncertainties* in both position and momentum of an individual particle, and these two uncertainties are related inseparably. To illustrate this point, we note that, in Fig. 45–6, an electron striking the film at the outer edge of the central maximum, at angle θ, must have a component of momentum p_y in the y-direction, as well as a component p_x in the x-direction, despite the fact that initially the beam was directed along the x-axis. From the geometry of the situation, the two components are related by $p_y/p_x = \tan \theta$; and if θ is small we may approximate $\tan \theta = \theta$, obtaining

$$p_y = p_x \theta. \tag{45–12}$$

Neglecting any electrons striking the film outside the central maximum (that is, at angles greater than λ/a), we see that the y-component of momentum may be as large as

$$p_y = p_x\left(\frac{\lambda}{a}\right). \tag{45–13}$$

Hence the *uncertainty* Δp_y in the y-component of momentum is of the order of $p_x \lambda/a$. Thus the narrower the slit width a, the broader is the diffraction pattern and the greater the uncertainty in the y-component of momentum.

Now the electron wavelength λ is related to the momentum $p_x = mv_x$ by the de Broglie relation, Eq. (45–9), which may be rewritten as $\lambda = h/p_x$. Using this result in Eq. (45–13) and simplifying, we find

$$\Delta p_y = p_x\left(\frac{h}{ap_x}\right) = \frac{h}{a},$$

or

$$\Delta p_y \, a = h. \tag{45–14}$$

To interpret this result, we note that the slit width a represents the uncertainty in *position* of an electron as it passes through the slit; we do not know through which particular part of the slit each particle passes. Thus the y-components of *both* position and momentum have uncertainties, and the two uncertainties are related by Eq. (45–14). We can reduce the *momentum* uncertainty only by increasing the slit width, which increases the *position* uncertainty; and conversely, when we decrease the position uncertainty by narrowing the slit, the diffraction pattern broadens and the corresponding momentum uncertainty increases.

All of this may be bitter medicine for a reader steeped in the tradition of nearly three centuries of newtonian mechanics; but the weight of experimental evidence leaves us no alternatives. To those who protest that the lack of a definite position and momentum is contrary to common sense, we reply that what we call common sense is based on familiarity gained through experience, and that our usual experience includes very little contact with the microscopic behavior of particles. Thus we must sometimes be prepared to accept conclusions that seem not to make sense when we are dealing with areas far removed from everyday experience.

Equation (45–14) is one form of the *Heisenberg uncertainty principle*; it states that, in general, neither the momentum nor the position of a particle can be known with arbitrarily great precision, but that the two play complementary roles as described above. It might be protested that greater precision could be attained by using more sophisticated particle detectors in various areas of the slit or by other means, but this turns out not to be possible. To detect a particle the detector must *interact* with it, and this interaction unavoidably changes the state of motion of the particle. A more detailed analysis of such hypothetical experiments shows that the uncertainty is a fundamental and intrinsic one that cannot be circumvented even in principle by any experimental technique, no matter how sophisticated.

Additional insight into the uncertainty principle is provided by the wave packet shown in Fig. 45–5. Such a packet can be constructed by superposing several sinusoidal waves (each of which has no beginning or end but extends indefinitely in both directions) with various wavelengths, choosing the wavelengths and the amplitude of each component wave so that constructive interference occurs only in a small region of width Δx, as shown in the figure, and the interference is destructive everywhere else. It turns out that a rather broad wave packet can be obtained by superposing waves with a relatively small range of wavelengths; but to make a narrow packet requires a wider range of wavelengths. But a range of wavelengths also means a corresponding range of values of momentum, because of the de Broglie relation. So again we see that a small uncertainty in position must be accompanied by a large uncertainty in momentum, and conversely.

There is also an uncertainty principle involving *energy*. It turns out that the energy of a system as well as its position and momentum always has uncertainty. The uncertainty ΔE is found to depend on the time interval Δt during which the system remains in the given state. The relation is

$$\Delta E \, \Delta t = \frac{h}{2\pi}. \tag{45–15}$$

Thus a system that remains in a certain state for a very long time can have a very well-defined energy, but if it remains in that state for only a short time, the uncertainty in energy must be correspondingly greater.

Example A sodium atom in one of the "resonance levels" shown in Fig. 44–7 remains in that state for an average time of 1.6×10^{-8} s before making a transition to the ground state by emitting a photon of wavelength 589 nm and energy 2.109 eV. What is the uncertainty in energy of the resonance level?

Solution From Eq. (45–15),

$$\Delta E = \frac{h}{2\pi \, \Delta t} = \frac{(6.626 \times 10^{-34} \text{ J·s})}{(2\pi)(1.6 \times 10^{-8} \text{ s})}$$
$$= 6.59 \times 10^{-27} \text{ J} = 4.11 \times 10^{-8} \text{ eV}.$$

The atom remains an indefinitely long time in the ground state, so there is *no* uncertainty there; the uncertainty of the resonance level energy and of the photon energy amounts to about two parts in 10^8. This irreducible uncertainty is called the *natural line width* of this particular spectrum line. Ordinarily the natural line width is much smaller than line broadening from other causes such as collisions among atoms.

*45-5 Wave functions

As we have seen, the dual wave-particle nature of electrons and other fundamental particles requires generalization of the kinematic language used to describe the position and motion of a particle. The classical notion of a particle as a point having at each instant a definite position in space (described by three coordinates) and a definite velocity (described by three components) must be replaced by a more general language.

In making the needed generalizations we are guided by the language of classical wave motion. In Chapter 21, in studying transverse waves on a string, we described the motion of the string by specifying the position of each point on the string at each instant of time. This was done by means of a *wave function,* introduced in Sec. 21–3. If y represents the displacement from equilibrium at time t of a point on the string whose equilibrium position is a distance x from the origin, then the function $y = f(x, t)$ or, more briefly, $y(x, t)$ represents the displacement of point x at time t. If we know the wave function for a given motion, we know everything there is to know about the motion; from the function the shape of the string at any time, the slope at each point, the velocity and acceleration of each point, and any other needed information can be obtained.

Similarly, in Chapter 23 we discussed a sound wave propagating in the x-direction. Letting p represent the variation in air pressure from its equilibrium value at any point, we write $p(x, t)$ as the pressure variation at any point x at any time t; again this is a *wave function.* If the wave is three-dimensional, we can describe p at a space point with coordinates (x, y, z) at any time t by means of a wave function $p(x, y, z, t)$, which contains all the space coordinates and time. The same pattern reappears once more in the description of *electromagnetic* waves in Sec. 37–5, where we use two wave functions to describe the electric and magnetic fields at any point in space, at any time.

Thus it is natural to use a wave function as the central element of our generalized language for describing particles. The symbol usually used for this wave function is Ψ, and it is, in general, a function of all the space coordinates and time. Just as the wave function $y(x, t)$ for mechanical waves on a string provides a complete description of the motion, the wave function $\Psi(x, y, z, t)$ for a particle contains all the information that can be known about the particle.

Two questions immediately arise. First, what is the *meaning* of the wave function Ψ for a particle? Second, how is Ψ determined for any given physical situation? Our answers to both these questions must be qualitative and incomplete. The wave function describes the distribution of the particle in space. It is related to the *probability* of finding the particle in each of various regions; the particle is most likely to be found in regions where Ψ is large, and so on. If the particle has charge, the wave function can be used to find the *charge density* at any point in space. In addition, from Ψ one can calculate the *average* position of the particle, its average velocity, and dynamic quantities such as momentum, energy, and angular momentum. The required techniques are far beyond the scope of this discussion, but they are well established and no longer subject to any reasonable doubt.

The answer to the second question is that the wave function must be one of a set of solutions of a certain differential equation called the *Schrödinger equation,* developed by Erwin Schrödinger in 1925. One can set up a Schrödinger equation for any given physical situation, such as the electron in a hydrogen atom; the functions that are solutions of this equation represent various possible physical states of the system. Furthermore, it turns out for some systems that acceptable solutions exist only when some physical quantity such as the energy of the system has certain special values. Thus the solutions of the Schrödinger equation are also associated with *energy levels.* This discovery is of the utmost importance; before the discovery of the Schrödinger equation there was no way to predict energy levels from any fundamental theory, except for the very limited success of the Bohr model for hydrogen.

Soon after its discovery the Schrödinger equation was applied to the problem of the hydrogen atom. The predicted energy levels turned out to be identical to those from the Bohr model, Eq. (45–6), and thus to agree with experimental values from spectrum analysis. The energy levels are labeled with the quantum number n. In addition, the solutions have *quantized* values of angular momentum; that is, only certain discrete values of the magnitude of angular momentum and its components are possible. We recall that angular momentum quantization was put into the Bohr model as an *ad hoc* assumption with no fundamental justification; with the Schrödinger equation it comes out automatically!

Specifically, it is found that the magnitude L of the angular momentum of an electron in the hydrogen atom in a state with energy E_n and principal quantum number n must be given by

$$L = \sqrt{l(l + 1)}\left(\frac{h}{2\pi}\right), \tag{45–16}$$

where l is zero or a positive integer no larger than $n - 1$. The *component* of L in a given direction, say the z-component L_z, can have only the set of values

$$L_z = mh/2\pi, \tag{45–17}$$

where m can be zero or a positive or negative integer up to but no larger than l.

The quantity $h/2\pi$ appears so often in quantum mechanics that it is given a special symbol, $\hbar$. That is,

$$\hbar = \frac{h}{2\pi} = 1.054 \times 10^{-34} \text{ J·s.}$$

In terms of $\hbar$, the two preceding equations become

$$L = \sqrt{l(l + 1)}\hbar \qquad (l = 0, 1, 2, \ldots, n - 1), \tag{45–18}$$

and

$$L_z = m\hbar \qquad (m = 0, \pm 1, \pm 2, \ldots, \pm l). \tag{45–19}$$

We note that the component L_z can never be quite as large as L. For example, when $l = 4$ and $m = 4$, we find

$$L = \sqrt{4(4 + 1)}\hbar = 4.47\hbar,$$
$$L_z = 4\hbar.$$

This inequality arises from the uncertainty principle, which makes it impossible to know the *direction* of the angular momentum vector with complete certainty. Thus the component of L in a given direction can never be quite as large as the magnitude L, except when $l = 0$ and both L and L_z are zero. Unlike the Bohr model, the Schrödinger equation gives values for the magnitude L of angular momentum which are *not* integer multiples of $\hbar$.

Another interesting feature of Eqs. (45-18) and (45-19) is that there are states for which the angular momentum is *zero*. This is a result that has no classical analog; in the Bohr model the electron always moved in an orbit and thus had nonzero angular momentum; but in the new mechanics we find states having zero angular momentum.

The possible wave functions for the hydrogen atom can be labeled according to the values of the three integers n, l, and m, called, respectively, the *principal* quantum number, the *angular momentum* quantum number, and the *magnetic* quantum number. For each energy level E_n there are several distinct states with the same energy but different values of l and m, the only exception being the ground state $n = 1$, for which only $l = 0$, $m = 0$ is possible. Examination of the spatial extent of the wave functions shows that, in each case, the wave function is confined primarily to a region of space around the nucleus having a radius of the same order of magnitude as the corresponding Bohr radius. For increasing values of n the electron is, on the average, farther away from the nucleus, and hence has less negative potential energy and becomes less tightly bound.

The new mechanics described above is called *quantum mechanics,* and it is clearly much more complex, both conceptually and mathematically, than newtonian mechanics. To a reader who yearns for the comparative simplicity of newtonian mechanics, we can only reply that quantum mechanics enables us to understand physical phenomena and to analyze physical problems for which classical mechanics is completely powerless. Added complexity is the price we pay for this greatly expanded understanding.

45-6 Electron spin

One additional concept, that of *electron spin,* is needed for the analysis of more complex atoms, and even for certain details of the spectrum of hydrogen. To illustrate this concept, we consider the motion of the earth around the sun. The earth travels in a nearly circular orbit and at the same time *rotates* on its axis. There is angular momentum associated with each motion, called the *orbital* and *spin* angular momentum, respectively, and the total angular momentum of the system is the sum of the two. If we were to model the earth as a single point, no spin angular momentum would be possible; but a more refined model including an earth of finite size includes the possibility of spin angular momentum.

This discussion can be translated to the language of the Bohr model. Suppose the electron is not a point charge moving in an orbit but a small spinning sphere in orbit. Then again there is additional angular momentum associated with the spin motion. Because the sphere carries an electric charge, the spinning motion leads to current loops and to a magnetic moment, as discussed in Sec. 31-3. If this magnetic moment really exists, then when the atom is placed in a magnetic field there is an interaction with the field. Associated with the interaction is a potential energy that causes shifts in the energy levels of the atom and thus in the wavelengths of the spectrum lines.

Such shifts *are* indeed observed in precise spectroscopic analysis; this and other experimental evidence have shown conclusively that the electron *does* have angular momentum and magnetic moment that are not related to the orbital motion but are intrinsic to the particle itself. Like orbital angular momentum, spin angular momentum (usually denoted by S) is found to be *quantized*. Denoting the z-component of S by S_z, we find that the only possible values are

$$S_z = \pm\tfrac{1}{2}\hbar. \tag{45-20}$$

This relation is reminiscent of Eq. (45–19) for the z-component of orbital angular momentum, but the component is one-half of $\hbar$ instead of an *integer* multiple.

In quantum mechanics, where the Bohr orbits are superseded by wave functions, it is not so easy to picture electron spin. If the wave functions are visualized as clouds surrounding the nucleus, then one can imagine many tiny arrows distributed through the cloud, all pointing in the same direction, either all $+z$ or all $-z$. Of course, this picture should not be taken too seriously; there is no hope of actually seeing an atom's structure because it is thousands of times smaller than wavelengths of light and because interactions with light photons would seriously disturb the very structure one is trying to observe.

In any event, the concept of electron spin is well established by a variety of experimental evidence. To label completely the state of the electron in a hydrogen atom, we now need a fourth quantum number s, which specifies the electron spin orientation. If s can take the values $+1$ or -1, then the z-component of spin angular momentum is given by

$$S_z = \tfrac{1}{2}s\hbar. \tag{45-21}$$

The corresponding component of magnetic moment, which we may denote by μ_z, turns out to be related to S_z, in a simple way:

$$\mu_z = \frac{e}{m_e}S_z, \tag{45-22}$$

where e and m_e are the charge and mass of the electron, respectively. The ratio of μ to S, in this case (e/m_e), is called the *gyromagnetic ratio*.

In this connection it should be mentioned that there is also magnetic moment associated with the *orbital* motion of the electron; in this case the gyromagnetic ratio is found to be just *half* as great as for electron spin, $e/2m_e$. Thus the z-component of magnetic moment associated with the orbital angular momentum component L_z is given by

$$\mu_z = \frac{e}{2m_e}L_z = m\frac{e\hbar}{2m_e}. \tag{45-23}$$

(Care must be taken to distinguish between the electron mass m_e and the magnetic quantum number m.) This magnetic moment leads to additional interactions with a magnetic field and to further contributions to shifts and splitting of spectrum lines.

Questions

45-1 In analyzing the absorption spectrum of hydrogen at room temperature, one finds absorption lines corresponding to wavelengths in the Lyman series but not to those in the Balmer series. Why not?

45-2 A singly ionized helium atom has one of its two electrons removed, and the energy levels of the remaining electron are closely related to those of the hydrogen atom. The nuclear charge for helium is $+2e$ instead of just $+e$; exactly how are the energy levels related to those of hydrogen? How is the size of the ion in the ground state related to that of the hydrogen atom?

45-3 Consider the line spectrum emitted from a gas discharge tube such as a neon sign or a sodium-vapor or mercury-vapor lamp. It is found that when the pressure of the vapor is increased the spectrum lines spread out, i.e., are less sharp and less monochromatic. Why?

45-4 Suppose a two-slit interference experiment is carried out using an electron beam. Would the same interference pattern result if one slit at a time is uncovered instead of both at once? If not, why not? Doesn't each electron go through one slit or the other? Or does every electron go through both slits? Does the latter possibility make sense?

45-5 Is the wave nature of electrons significant in the function of a television picture tube? For example, do diffraction effects limit the sharpness of the picture?

45-6 A proton and an electron have the same speed. Which has longer wavelength?

45-7 A proton and an electron have the same kinetic energy. Which has longer wavelength?

45-8 Does the uncertainty principle have anything to do with marksmanship? That is, is the accuracy with which a bullet can be aimed at a target limited by the uncertainty principle?

45-9 Is the Bohr model of the hydrogen atom consistent with the uncertainty principle?

45-10 If the energy of a system can have uncertainty, as stated by Eq. (45-15), does this mean that the principle of conservation of energy is no longer valid?

45-11 If quantum mechanics replaces the language of newtonian mechanics, why do we not have to use wave functions to describe the motion of macroscopic objects such as baseballs and cars?

45-12 Why is the analysis of the helium atom much more complex than that of the hydrogen atom, either in a Bohr-type model or using the Schrödinger equation?

45-13 Do gravitational forces play a significant role in atomic structure?

Problems

$$e = 1.602 \times 10^{-19} \text{ C}$$
$$m = 9.109 \times 10^{-31} \text{ kg}$$
$$h = 6.626 \times 10^{-34} \text{ J·s}$$
$$N_0 = 6.022 \times 10^{23} \text{ atoms·mol}^{-1}$$

Energy equivalent of 1 u = 931.5 MeV

$$\frac{e}{m} = 1.758 \times 10^{11} \text{ C·kg}^{-1}$$
$$k = 1.381 \times 10^{-23} \text{ J·K}^{-1}$$
$$1 \text{ eV} = 1.602 \times 10^{-19} \text{ J}$$
$$1 \text{ u} = 1.661 \times 10^{-27} \text{ kg}$$
$$\varepsilon_0 = 8.854 \times 10^{-12} \text{ C}^2 \text{·N}^{-1} \text{·m}^{-2}$$

45-1 For a hydrogen atom in the ground state, determine, in electron volts:

a) the kinetic energy;

b) the potential energy;

c) the total energy;

d) the energy required to remove the electron completely.

45-2 Referring to Problem 45-1, what are the energies if the atom is a singly ionized helium atom (i.e., a helium atom with one electron removed)?

45-3 A singly ionized helium ion (a helium atom with one electron removed) behaves very much like a hydrogen atom, except that the nuclear charge is twice as great.

a) How do the energy levels differ in magnitude from those of the hydrogen atom?

b) Which spectral series for He$^+$ have lines in the visible spectrum?

45-4 Refer to the example in Sec. 45-1. Suppose a hydrogen atom makes a transition from the $n = 3$ state to the $n = 2$ state (the Balmer Hα line at 656.3 nm) while in a magnetic field of magnitude 2 T. If the magnetic moment of the atom is parallel to the field in both the initial and final states,

a) by how much is each enery level shifted from the zero-field value;

b) by how much is the wavelength of the spectrum line shifted?

45-5 The negative μ-meson (or muon) has a charge equal to that of an electron but a mass about 207 times as great. Consider a hydrogenlike atom consisting of a proton and a muon.

a) What is the ground-state energy?

b) What is the radius of the $n = 1$ Bohr orbit?

c) What is the wavelength of the radiation emitted in the transition from the $n = 2$ state to the $n = 1$ state?

45-6 According to Bohr, the Rydberg constant R is equal to $me^4/8\varepsilon_0^2 h^3 c$. Calculate R in m^{-1} and compare with the experimental value.

45-7

a) Show that the frequency of revolution of an electron in its circular orbit in the Bohr model of the hydrogen atom is $f = me^4\varepsilon_0^2 n^3 h^3$.

b) Show that when n is very large, the frequency of revolution equals the radiated frequency calculated from Eq. (45-7) for a transition from

$$n = n' + 1 \qquad \text{to} \qquad l = n'.$$

(This problem illustrates Bohr's *correspondence principle,* which is often used as a check on quantum calculations. When n is small, quantum physics gives results that are very different from those of classical physics. When n is large, the differences are not significant and the two methods then "correspond.")

45-8 A 10-kg satellite circles the earth once every 2 hr in an orbit having a radius of 8000 km.

a) Assuming that Bohr's angular-momentum postulate applies to satellites just as it does to an electron in the hydrogen atom, find the quantum number of the orbit of the satellite.

b) Show from Bohr's first postulate and Newton's law of gravitation that the radius of an earth-satellite orbit is directly proportional to the square of the quantum number, $r = kn^2$, where k is the constant of proportionality.

c) Using the result from part (b), find the distance between the orbit of the satellite in this problem and its next "allowed" orbit.

d) Comment on the possibility of observing the separation of the two adjacent orbits.

e) Do quantized and classical orbits correspond for this satellite? Which is the "correct" method for calculating the orbits?

45-9 A hydrogen atom initially in the ground state absorbs a photon, which excites it to the $n = 4$ state. Determine the wavelength and frequency of the photon.

45-10 In Problem 45-3,

a) how much energy would be required to remove the remaining electron completely from a helium ion?

b) What wavelength would a photon of this energy have?

c) In what region of the electromagnetic spectrum would it lie?

45-11 Calculate the energy of the longest-wavelength line in the Balmer series for hydrogen, and of the shortest-wavelength line.

45-12 Approximately what range of photon energies (in electron volts) corresponds to the visible spectrum? Approximately what range of wavelengths would electrons in this energy range have?

45-13 For crystal diffraction experiments, wavelengths of the order of 0.1 nm are often appropriate. Find the energy, in electron volts, for a particle with this wavelength if the particle is

a) a photon;

b) an electron.

45-14

a) An electron moves with a velocity of $3 \times 10^8 \text{ cm} \cdot \text{s}^{-1}$. What is its de Broglie wavelength?

b) A proton moves with the same velocity. Determine its de Broglie wavelength.

45-15 The average kinetic energy of a thermal neutron is $(3/2)kT$. What is the de Broglie wavelength associated with the neutrons in thermal equilibrium with matter at 300 K? (The mass of a neutron is approximately 1 u.)

45-16

a) What is the de Broglie wavelength of an electron that has been accelerated through a potential difference of 200 V?

b) Would this electron exhibit particlelike or wavelike characteristics on meeting an obstacle or opening 1 mm in diameter?

45-17 What is the de Broglie wavelength of an electron accelerated through 20,000 V, a typical voltage for color-television picture tubes?

45-18 Suppose the uncertainty of position of an electron is equal to the radius of the $n = 1$ Bohr orbit, about 0.5×10^{-10} m. Find the uncertainty in its momentum, and compare this with the magnitude of the momentum of the electron in the $n = 1$ Bohr orbit.

45-19 Suppose that the uncertainty in position of a particle is equal to its de Broglie wavelength; show that in this case the uncertainty in its momentum is equal to its momentum.

45-20 In a certain television picture tube the accelerating voltage is 20,000 V and the electron beam passes through an aperture 0.5 mm in diameter to a screen 0.3 m away.

a) What is the uncertainty in position of the point where the electrons strike the screen?

b) Does this uncertainty affect the clarity of the picture significantly?

45–21 A certain atom has an energy level 2.0 eV above the ground state. When excited to this state it remains on the average 2.0×10^{-6} s before emitting a photon and returning to the ground state.

a) What is the energy of the photon? Its wavelength?

b) What is the uncertainty in energy of the photon? Of its wavelength?

45–22 The π^0 meson is an unstable particle produced in high-energy particle collisions. Its mass is about 264 times that of the electron, and it exists for an average lifetime of 0.8×10^{-16} s before decaying into two gamma-ray photons. Assuming that the mass and energy of the particle are related by the Einstein relation $E = mc^2$, find the uncertainty in the mass of the particle, and express it as a fraction of the mass.

45–23 Suppose an unstable particle produced in a high-energy collision has a mass three times that of the proton and an uncertainty in mass that is 1% of the particle's mass. Assuming mass and energy are related by $E = mc^2$, find the lifetime of the particle.

45–24 A hydrogen atom is placed in a magnetic field of magnitude 1.2 T. Find the interaction energy with the field due to the electron spin.

45–25 Make a chart showing all the possible sets of quantum numbers for the states of the electron in the hydrogen atom when $n = 3$.

45–26 Show that the total number of hydrogen-atom states for a given value of the principal quantum number n is $2n^2$.

ATOMS, MOLECULES, AND SOLIDS 46

46-1 The exclusion principle

The hydrogen atom, discussed in Chapter 45, is the simplest of all atoms, containing one electron and one proton. Analysis of atoms with more than one electron increases in complexity very rapidly; each electron interacts not only with the positively charged nucleus but also with all the other electrons. In principle the motion of the electrons is governed by the Schrödinger equation, mentioned in Sec. 45–5, but the mathematical problem of finding appropriate solutions of this equation is so complex that it has not been accomplished exactly even for the helium atom, with two electrons.

Various approximation schemes can be used; the simplest (and most drastic) is to ignore completely the interactions between electrons, and regard each electron as influenced only by the electric field of the nucleus, considered to be a point charge. A less drastic and more useful approximation is to think of all the electrons together as making up a charge cloud that is, on the average, *spherically symmetric,* and to think of each individual electron as moving in the total electric field due to the nucleus and this averaged-out electron cloud. This is called the *central-field* approximation; it provides a useful starting point for the understanding of atomic structure.

An additional principle is also needed, the *exclusion principle.* To understand the need for this principle, we consider the lowest-energy state or *ground state* of a many-electron atom. The central-field model suggests that each electron has a lowest energy state (roughly corresponding to the $n = 1$ state for the hydrogen atom). It might be expected that, in the ground state of a complex atom, all the electrons should be in this lowest state. If this is the case, then when we examine the behavior of atoms with increasing numbers of electrons, we should find gradual changes in physical and chemical properties of elements as the number of electrons in the atoms increases.

A variety of evidence shows conclusively that this is *not* what happens at all. For example, the elements of fluorine, neon, and sodium have, respectively, 9, 10, and 11 electrons per atom. Fluorine is a halogen and tends strongly to form compounds in which each atom acquires an extra electron. Sodium, an alkali metal, forms compounds in which it loses an electron, and neon is an inert gas, forming no compounds at all. This and many other observations show that, in the ground state of a complex atom, the electrons *cannot* all be in the lowest energy states.

The key to this puzzle, discovered by the Swiss physicist Wolfgang Pauli in 1925, is called the *Pauli exclusion principle.* Briefly, it states that *no two electrons can occupy the same quantum-mechanical state.* Since different states correspond to different spatial distributions, including different distances from the nucleus, this means that in a complex atom there is not room for all the electrons in states near the nucleus; some are forced into states farther away, having higher energies.

To apply the Pauli principle to atomic structure, we first review some results quoted in Sec. 45–5. The quantum-mechanical state of the electron in the hydrogen atom is identified by the four quantum numbers n, l, m, and s, which determine the energy, angular momentum, and components of orbital and spin angular momentum in a particular direction. It turns out that this scheme can also be used when the electron moves not in the electric field of a point charge, as in the hydrogen atom, but in the electric field of *any spherically symmetric charge distribution,* as in the central-field approximation. One important difference is that the energy of a state is no longer given by Eq. (45–6); in general the energy corresponding to a given set of quantum numbers depends on *both* n and l, usually increasing with increasing l for a given value of n.

We can now make a list of the possible sets of quantum numbers and thus of the possible states of electrons in an atom. Such à list is given in Table 46–1, which also indicates two alternative notations. It is customary to designate the value of l by a letter, according to this scheme:

$$l = 0; \quad s \text{ state}$$
$$l = 1; \quad p \text{ state}$$
$$l = 2; \quad d \text{ state}$$
$$l = 3; \quad f \text{ state}$$
$$l = 4; \quad g \text{ state}$$

The origins of these letters are rooted in the early days of spectroscopy, and need not concern us. A state for which $n = 2$ and $l = 0$ is called a $2s$ state, and so on, as shown in Table 46–1. This table also shows the relation between values of n and the x-ray levels ($K, L, M, \ldots$) described in Sec. 44–8. The $n = 1$ levels are designated as K, $n = 2$ as L, and so on. Because the average electron distance from the nucleus increases with n, each value of n corresponds roughly to a region of space around the nucleus in the form of a spherical shell. Hence, one speaks of the *L shell* as that region occupied by the electrons in the $n = 2$ states, and so on. States with the same n but different l are said to form *subshells,* such as the $3p$ subshell.

We are now ready for a more precise statement of the exclusion principle: *In any atom, only one electron can occupy any given quantum state.* That is, no two electrons in an atom can have all four quantum numbers the same. Since each quantum state corresponds to a certain distribution of the electron "cloud" in space, the principle says in effect: "Not more than two electrons (with opposite values of the spin quantum

Table 46-1

n	l	m	Spectroscopic notation	Maximum number of electrons		Shell
1	0	0	$1s$	2		K
2	0	0	$2s$	2		
2	1	-1			8	L
2	1	0	$2p$	6		
2	1	1				
3	0	0	$3s$	2		
3	1	-1				
3	1	0	$3p$	6		
3	1	1				
3	2	-2			18	M
3	2	-1				
3	2	0	$3d$	10		
3	2	1				
3	2	2				
4	0	0	$4s$	2		
4	1	-1				
4	1	0	$4p$	6	32	N
4	1	1				
	etc.					

number s) can occupy the same region of space." This statement must be
interpreted rather broadly, since the clouds and associated wave functions describing electron distributions do not have definite, sharp boundaries; but the exclusion principle limits the degree of overlap of electron
wave functions that is permitted. The maximum numbers of electrons in
each shell and subshell are shown in Table 46–1.

The exclusion principle plays an essential role in the understanding
of the structure of complex atoms. In the next section we shall see how
the periodic table of the elements can be understood on the basis of this
principle.

46-2 Atomic structure

The number of electrons in an atom in its normal state is called the
atomic number, denoted by Z. The nucleus contains Z protons and some
number of neutrons. The proton and electron charges have the same
magnitude but opposite sign, so in the normal atom the net electric
charge is zero. Because the electrons are attracted to the nucleus, we
expect the quantum states corresponding to regions near the nucleus to
have lowest energy. We may imagine starting with a bare nucleus with Z
protons, and adding electrons one by one until the normal complement
of Z electrons for a neutral atom is reached. We expect the lowest-energy
states, ordinarily those with the smallest values of n and l, to fill first,
and we use successively higher states until all electrons are accommodated.

The chemical properties of an atom are determined principally by
interactions involving the outermost electrons, so it is of particular in-

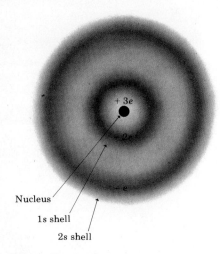

46-1 Schematic representation of charge distribution in lithium atom. The nucleus has a charge $3e$; the two $1s$ electrons are closer to the nucleus than the $2s$ electron, which moves in a field approximately equal to that of a point charge $3e - 2e$, or simply e.

terest to find out how these are arranged. For example, when an atom has one electron considerably farther from the nucleus (on the average) than the others, this electron will be rather loosely bound; the atom will tend to lose this electron and form what chemists call an *electrovalent* or *ionic* bond, with valence $+1$. This behavior is characteristic of the alkali metals lithium, sodium, potassium, and so on.

We now proceed to describe ground-state electron configurations for the first few atoms (in order of increasing Z). For hydrogen the ground state is $1s$; the single electron is in the state $n = 1$, $l = 0$, $m = 0$, and $s = \pm 1$. In the helium atom ($Z = 2$) *both* electrons are in $1s$ states, with opposite spins; this state is denoted as $1s^2$. For helium, the K shell is completely filled and all others are empty.

Lithium ($Z = 3$) has three electrons; in the ground state two are in the $1s$ state, and one in a $2s$ state. We denote this state as $1s^2 2s$. On the average, the $2s$ electron is considerably farther from the nucleus than the $1s$ electrons, as shown schematically in Fig. 46–1. Thus, according to Gauss's law, the *net* charge influencing the $2s$ electron is $+e$, rather than $+3e$ as it would be without the $1s$ electrons present. Thus the $2s$ electron is loosely bound, as the chemical behavior of lithium suggests. An alkali metal, it forms ionic compounds with a valence of $+1$, in which each atom loses an electron.

Next is beryllium ($Z = 4$); its ground-state configuration is $1s^2 2s^2$, with two electrons in the L shell. Beryllium is the first of the *alkaline-earth* elements, forming ionic compounds with a valence of $+2$.

Table 46–2 shows the ground-state electron configurations of the first twenty elements. The L shell can hold a total of eight electrons, as the reader should verify from the rules in Sec. 46–1; at $Z = 10$ the K and L shells are filled and there are no electrons in the M shell. This is

Table 46-2   Ground-state electron configurations

Element	Symbol	Atomic number (Z)	Electron configuration
Hydrogen	H	1	$1s$
Helium	He	2	$1s^2$
Lithium	Li	3	$1s^2 2s$
Beryllium	Be	4	$1s^2 2s^2$
Boron	B	5	$1s^2 2s^2 2p$
Carbon	C	6	$1s^2 2s^2 2p^2$
Nitrogen	N	7	$1s^2 2s^2 2p^3$
Oxygen	O	8	$1s^2 2s^2 2p^4$
Fluorine	F	9	$1s^2 2s^2 2p^5$
Neon	Ne	10	$1s^2 2s^2 2p^6$
Sodium	Na	11	$1s^2 2s^2 2p^6 3s$
Magnesium	Mg	12	$1s^2 2s^2 2p^6 3s^2$
Aluminum	Al	13	$1s^2 2s^2 2p^6 3s^2 3p$
Silicon	Si	14	$1s^2 2s^2 2p^6 3s^2 3p^2$
Phosphorus	P	15	$1s^2 2s^2 2p^6 3s^2 3p^3$
Sulfur	S	16	$1s^2 2s^2 2p^6 3s^2 3p^4$
Chlorine	Cl	17	$1s^2 2s^2 2p^6 3s^2 3p^5$
Argon	Ar	18	$1s^2 2s^2 2p^6 3s^2 3p^6$
Potassium	K	19	$1s^2 2s^2 2p^6 3s^2 3p^6 4s$
Calcium	Ca	20	$1s^2 2s^2 2p^6 3s^2 3p^6 4s^2$

expected to be a particularly stable configuration, with little tendency to gain or lose electrons, and in fact this element is neon, an inert gas with no known compounds. The next element after neon is sodium ($Z = 11$), with filled K and L shells and one electron in the M shell. Thus its "filled-shell-plus-one-electron" structure resembles that of lithium; both are alkali metals. The element *before* neon is fluorine, with $Z = 9$. It has a vacancy in the L shell and might be expected to have an affinity for an electron, forming ionic compounds with a valence of -1. This behavior is characteristic of the *halogens* (fluorine, chlorine, bromine, iodine, astatine), all of which have "filled-shell-minus-one" configurations.

By similar analysis, one can understand all the regularities in chemical behavior exhibited by the periodic table of the elements, on the basis of electron configurations. With the M and N shells there is a slight complication because the $3d$ and $4s$ subshells ($n = 3$, $l = 2$, and $n = 4$, $l = 0$, respectively) overlap in energy. Thus argon ($Z = 18$) has all the $1s$, $2s$, $2p$, $3s$, and $3p$ states filled, but in potassium ($Z = 19$) the additional electron goes into a $4s$ rather than $3d$ level. The next several elements have one or two electrons in the $4s$ states and increasing numbers in the $3d$ states. These elements are all metals with rather similar properties, and form the first *transition series,* starting with scandium ($Z = 21$) and ending with zinc ($Z = 30$), for which the $3d$ levels are filled.

Thus, the similarity of elements in each *group* of the periodic table reflects corresponding similarity in electron configuration. All the inert gases (helium, neon, argon, krypton, xenon, and radon) have filled-shell configurations. All the alkali metals (lithium, sodium, potassium, rubidium, cesium, and francium) have "filled-shell-plus-one" configurations. All the alkaline-earth metals (beryllium, magnesium, calcium, strontium, barium, and radium) have "filled-shell-plus-two" configurations, and all the halogens (fluorine, chlorine, bromine, iodine, and astatine) have "filled-shell-minus-one" structures. And so it goes.

This whole theory can, of course, be refined to account for differences between elements in a group and to account for various aspects of chemical behavior. Indeed, some physicists like to claim that all of chemistry is contained in the Schrödinger equation! This statement may be a bit extreme, but we can see that even this qualitative discussion of atomic structure takes us a considerable distance toward understanding the atomic basis of many chemical phenomena. Next we shall examine in somewhat more detail the nature of the chemical bond.

46-3 Diatomic molecules

As indicated in Sec. 46-2, the study of electron configurations in atoms provides valuable insight into the nature of the chemical bond, that is, the interaction that holds atoms together to form stable structures such as molecules and crystalline solids. There are a number of types of chemical bonds; the simplest to understand is the *ionic* bond, also called the *electrovalent* or *heteropolar* bond. The most familiar example is sodium chloride (NaCl), in which the sodium atom gives its $3s$ electron to the chlorine atom, filling the vacancy in the $3p$ subshell of chlorine. Energy is required to make this transfer if the atoms are far apart, but the result is two ions, one positively charged and one negatively, which attract each other. As they come together, their potential energy decreases, so that the final bound state of Na^+Cl^- has lower total energy than the state in which the two atoms are separated and neutral.

Removing the $3s$ electron from the sodium requires 5.1 eV of energy; this is called the *ionization energy* or *ionization potential*. Chlorine actually has an electron *affinity* of 3.8 eV. That is, the neutral chlorine atom can attract an extra electron which, once it takes its place in the $3p$ level, requires 3.8 eV for its removal. Thus, creating the separate Na^+ and Cl^- ions requires a net expenditure of only $5.1 - 3.8 = 1.3$ eV. The potential energy associated with the mutual attraction of the ions is about -5 eV, more than enough to repay the initial energy investment of creating the ions. This potential energy is determined by the closeness with which the ions can approach each other, and this in turn is detemined by the exclusion principle, which forbids extensive overlap of the electron clouds of the two atoms.

Ionic bonds can involve more than one electron per atom. The alkaline-earth elements form ionic compounds in which each atom loses *two* electrons; an example is $Mg^{++}Cl_2^-$. Loss of more than two electrons is relatively rare, and instead a *different kind of bond* comes into operation.

The *covalent* or *homopolar* bond is characterized by a more nearly symmetric participation of the two atoms, as contrasted with the complete asymmetry involved in the electron-transfer process of the ionic bond. The simplest example of the covalent bond is the hydrogen molecule, a structure containing two protons and two electrons. As a preliminary to understanding this bond we consider first the interaction of two electric dipoles, as in Fig. 46–2. In (a) the dipoles are far apart; in (b) the like charges are farther part than the unlike charges, and there is a net *attractive* force. In (c) and (d) the interaction is repulsive.

In a molecule the charges are, of course, not at rest, but Fig. 46–2b at least makes it seem plausible that if the electrons are in the region between the protons, the attractive force the electrons exert on each proton may more than counteract the repulsive interactions of the protons on each other and the electrons on each other. That is, in the covalent bond in the hydrogen molecules, the attractive interaction is supplied by a pair of electrons, one contributed by each atom, whose charge clouds are concentrated primarily in the region between the two atoms. Thus this bond may also be thought of as a shared-electron or electron-pair bond.

According to the exclusion principle, two electrons can occupy the same region of space only when they have opposite spin orientations. When the spins are parallel, the state that would be most favorable from energy considerations is forbidden by the exclusion principle, and the lowest-energy state permitted is one in which the electron clouds are concentrated *outside* the central region between atoms. The nuclei then repel each other and the interaction is repulsive rather than attractive. Thus, opposite spins are an essential requirement for an electron-pair bond, and no more than two electrons can participate in such a bond.

This is not to say, however, that an atom cannot have several electron-pair bonds. On the contrary, an atom having several electrons in its outermost shell can form covalent bonds with several other atoms. The bonding of carbon and hydrogen atoms, of central importance in organic

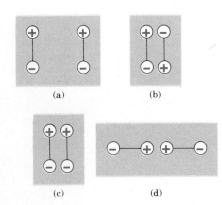

(a)

(b)

(c)

(d)

46-2 When two dipoles (a) are brought together, the interaction may be (b) attractive, or (c and d) repulsive.

chemistry, is an example. In the *methane* molecule (CH_4) the carbon atom is at the center of a regular tetrahedron, with a hydrogen atom at each corner. The carbon atom has four electrons in its L shell, and one of these electrons forms a covalent bond with each of the four hydrogen atoms, as shown in Fig. 46–3. Similar patterns occur in more complex organic molecules.

Both the ionic and covalent bonds are important in the structure of solids. In many respects a crystalline solid is really a very large molecule held together by chemical bonds. A third type of bonding, the *metallic* bond, is also important in the structure of some solids; we return to this subject in Sec. 46–5.

46-4 Molecular spectra

The energy levels of atoms are associated with the kinetic and potential energies of electrons with respect to the nucleus. Energy levels of molecules have additional features resulting from relative motion of the nuclei of the atoms, and there are characteristic spectra associated with these levels.

For the sake of simplicity we consider only diatomic molecules. Viewed as a rigid dumbbell, a diatomic molecule can *rotate* about an axis through its center of mass. According to classical mechanics, the kinetic energy of a rotating body may be expressed as $\frac{1}{2}I\omega^2$, as given by Eq. (9–13). This energy may also be expressed in terms of the magnitude L of angular momentum, which is given by $L = I\omega$. Combining this with the preceding expression, we find that the energy E is given by

$$E = \frac{L^2}{2I}. \tag{46-1}$$

Now in a quantum-mechanical discussion of this motion, it is reasonable to assume that the angular momentum is quantized in the same way as for electrons in an atom, as given by Eq. (45–18):

$$L^2 = l(l + 1)\hbar^2 \qquad (l = 0, 1, 2, 3, \ldots). \tag{46-2}$$

Combining Eqs. (46–1) and (46–2), we obtain the result

$$E = l(l + 1)\left(\frac{\hbar^2}{2I}\right) \qquad (l = 0, 1, 2, 3, \ldots). \tag{46-3}$$

A more detailed analysis, using the Schrödinger equation, confirms that the angular momentum is indeed given by Eq. (46–2) and the energy levels by Eq. (46–3).

Considering the magnitudes involved, we note that, for an oxygen molecule, $I = 5 \times 10^{-46}$ kg·m². Thus, the constant $\hbar^2/2I$ in Eq. (46–3) is approximately equal to

$$\frac{\hbar^2}{2I} = \frac{(1.05 \times 10^{-34}\text{ J·s})^2}{2(5 \times 10^{-46}\text{ kg·m}^2)} = 1.11 \times 10^{-23}\text{ J} = 0.69 \times 10^{-4}\text{ eV}.$$

We note that this energy is much *smaller* than typical atomic energy levels associated with optical spectra. The energies of photons emitted and absorbed in transitions among rotational levels are correspondingly small, and they fall in the far infrared region of the spectrum.

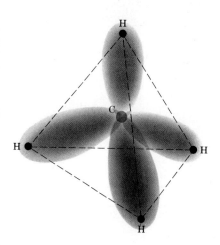

46-3 Methane (CH_4) molecule, showing four covalent bonds. The electron cloud between the central carbon atom and each of the four hydrogen nuclei represents the two electrons of a covalent bond. The hydrogen nuclei are at the corners of a regular tetrahedron.

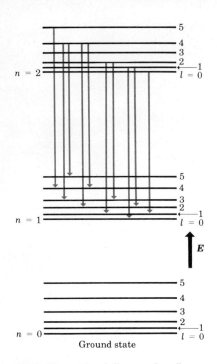

46–4 Energy-level diagram for vibrational and rotational energy levels of a diatomic molecule. For each vibrational level (n) there is a series of more closely spaced rotational levels (l). Several transitions corresponding to a single band in a band spectrum are shown.

The rigid-dumbbell model of a diatomic molecule suggests that the distance between the two nuclei is fixed; in fact, a more realistic model would represent the connection as a *spring* rather than a rigid rod. The atoms can undergo *vibrational* motion relative to the center of mass, and there is additional kinetic and potential energy associated with this motion. Application of the Schrödinger equation shows that the corresponding energy levels are given by

$$E = \left(n + \frac{1}{2}\right)hf \qquad (n = 0, 1, 2, 3, \ldots), \qquad (46\text{–}4)$$

where f is the frequency of vibration. For typical diatomic molecules this turns out to be of the order of 10^{13} Hz; thus the constant hf in Eq. (46–4) is of the order of

$$hf = (6.6 \times 10^{-34}\,\text{J·s})(10^{13}\,\text{s}^{-1}) = 6.6 \times 10^{-21}\,\text{J}$$
$$= 0.041\,\text{eV}.$$

Thus the vibrational energies, while still much smaller than those of the optical spectra, are typically considerably *larger* than the rotational energies.

An energy-level diagram for a diatomic molecule has the general appearance of Fig. 46–4. For each value of n, there are many values of l, leading to series of closely spaced levels. Transition between different pairs of n values gives different series of spectrum lines, and the resulting spectrum has the appearance of a series of *bands;* each band corresponds to a particular vibrational transition, and each individual line in a band, to a particular rotational transition. A typical band spectrum is shown in Fig. 46–5.

The same considerations can be applied to more complex molecules. A molecule with three or more atoms has several different kinds or *modes* of vibratory motion, each with its own set of energy levels related to the frequency by Eq. (46–4). The resulting energy-level scheme and associated spectra can be quite complex, but the general considerations discussed above still apply. In nearly all cases the associated radiation lies in the infrared region of the electromagnetic spectrum. Analysis of molecular spectra has proved to be an extremely valuable analytical tool, providing a great deal of information about the strength and rigidity of molecular bonds and the structure of complex molecules.

46–5 Typical band spectrum. (Courtesy of R. C. Herman.)

46–5 Structure of solids

At ordinary temperatures and pressures most materials are in the solid state, a condensed state of matter characterized by interatomic interactions of sufficient strength to give the material a definite volume and shape, which change relatively little with applied stress. The separation of adjacent atoms in a solid is of the same order of magnitude as the diameter of the electron cloud around each atom.

A solid may be *amorphous* or *crystalline*. Crystalline solids have been discussed briefly in Sec. 20–7, and it would do no harm for the reader to review that discussion. A crystalline solid is characterized by *long-range order,* a recurring pattern in the arrangement of atoms; this pattern is called the *crystal structure* or the *lattice structure*. Amorphous solids have no long-range order but only short-range order, a state of affairs in which each atom has other atoms arranged around it in a more or less regular pattern, but without the recurring pattern characteristic of crystals. Amorphous solids have more in common with liquids than with crystalline solids; liquids also have short-range order but not long-range order.

The forces responsible for the regular arrangement of atoms in a crystal are, in some cases, the same as those involved in molecular binding. Corresponding to the classes of chemical bonds, there are ionic and covalent crystals. The alkali halides, of which ordinary salt (NaCl) is the most common variety, are the most familiar ionic crystals. The positive sodium ions and the negative chlorine ions occupy alternate positions in a cubic crystal lattice, as in Fig. 46–6. The forces are the familiar Coulomb's-law forces between charged particles; these forces are not directional, and the particular arrangement in which the material crystallizes is determined by the relative size of the two ions.

The simplest example of a *covalent* crystal is the *diamond structure,* a structure found in the diamond form of carbon and also in silicon, germanium, and tin, all elements in Group IV of the periodic table, with four electrons in the outermost shell. Each atom in this structure is situated at the center of a regular tetrahedron, with four nearest-neighbor atoms at the corners, and it forms a covalent bond with each of these atoms. These bonds are strongly directional because of the asymmetrical electron distribution, leading to the tetrahedral structure.

A third crystal type, which is less directly related to the chemical bond than ionic or covalent crystals, is the *metallic crystal*. In this structure the outermost electrons are not localized at individual lattice sites but are detached from their parent atoms and free to move through the crystal. The corresponding charge clouds extend over many atoms. Thus a metallic crystal can be visualized roughly as an array of positive ions from which one or more electrons have been removed, immersed in a sea of electrons whose attraction for the positive ions holds the crystal together. This sea has many of the properties of a gas, and indeed one speaks of the *electron-gas model* of metallic solids.

In a metallic crystal the situation is as though the atoms would like to form shared-electron bonds but there are not enough valence electrons. Instead, electrons are shared among *many* atoms. This bonding is not strongly directional in nature, and the shape of the crystal lattice is determined primarily by considerations of *close packing,* i.e., the maximum number of atoms that can fit into a given volume. The two most common metallic crystal lattices, the face-centered cubic and the hexagonal close-packed, are shown in Fig. 46–7. In each of these, each atom has 12 nearest neighbors.

The discussion in this section has centered around *perfect crystals,* crystals in which the crystal lattice extends uninterrupted through the

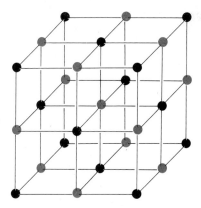

46-6 Symbolic representation of a sodium chloride crystal, with exaggerated distances between ions.

46-7 Close-packed crystal lattice structures. (a) Face-centered cubic; (b) hexagonal close-packed.

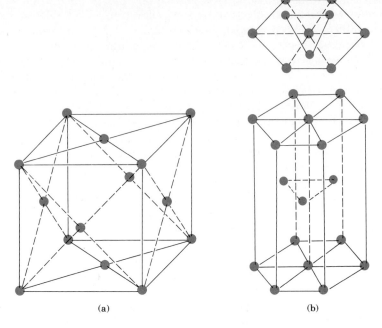

(a) (b)

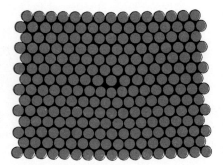

46-8 A dislocation. The concept is seen most easily by viewing the figure from various directions at a grazing angle with the page.

entire material. Real crystals exhibit a variety of departures from this idealized structure. Materials are often *polycrystalline,* composed of many small single crystals bonded together at *grain boundaries.* Within a single crystal there may be *interstitial* atoms in places where they do not belong, and *vacancies,* lattice sites that should be occupied by an atom but are not. An imperfection of particular interest in semiconductors, to be discussed in Sec. 46–7, is the *impurity atom,* a foreign atom (e.g., arsenic in a silicon crystal) occupying a regular lattice site.

A more complex kind of imperfection is the *dislocation,* illustrated schematically in Fig. 46–8, in which one plane of atoms slips relative to another. The mechanical properties of metallic crystals are influenced strongly by the presence of dislocations; the ductility and malleability of some metals depends on the presence of dislocations that move through the lattice during plastic deformations.

46-6 Properties of solids

Many *macroscopic* properties of solids, including mechanical, thermal, electrical, magnetic, and optical properties, can be understood by considering their relation to the *microscopic* structure of the material, and various aspects of the relation of structure to properties are part of the vigorous program of research in the physics of solids being carried out throughout the world. Although these topics cannot be discussed in detail in a book such as this one, a few examples will indicate the kinds of insights to be gained through study of the microscopic structure of solids.

We have already discussed, in Sec. 20–8, the subject of specific heat capacities of crystals, using the same principle of equipartition of energy as in the kinetic theory of gases. A simple analysis permits understanding of the empirical rule of Dulong and Petit on the basis of a microscopic model. This analysis has its limitations, to be sure; it does not include the energy of electron motion, which in metals makes a small additional contribution to specific heat, nor does it predict the temperature dependence of specific heats resulting from the quantization of the lattice-vibration energy discussed in Sec. 20-8. But these additional refinements *can* be included in the model to permit more detailed comparison of observed macroscopic properties with theoretical predictions.

Electrical conductivity is understood on the basis of the mobility or lack of mobility of electrons in the material. In a metallic crystal the valence electrons are not bound to individual lattice sites but are free to move through the crystal. Thus we expect metals to be good conductors, and they usually are. In a covalent crystal the valence electrons are involved in the bonds responsible for the crystal structure and are therefore *not* free to move. Thus there are no mobile charges available for conduction, and such materials are expected to be insulators. Similarly, an ionic crystal such as NaCl has no charges that are free to move, and solid NaCl is an insulator. However, when salt is melted, the ions are no longer locked to their individual lattice sites but are free to move, and *molten* NaCl is a good conductor.

There are, of course, no perfect conductors or insulators, although the resistivity of good insulators is greater than that of good conductors by an enormous factor, the order of at least 10^{15}. This great difference is one of the factors that make extremely precise electrical measurements possible. In addition, the resistivity of all materials depends on temperature; in general, the small conductivity of insulators *increases* with temperature, but that of good conductors usually *decreases* at increased temperatures.

Two competing effects are responsible for this difference. In metals the *number* of electrons available for conduction is nearly independent of temperature, and the resistivity is determined by the frequency of collisions of electrons with the lattice. Roughly speaking, lattice vibrations increase with increased temperature, and the ion cores present a larger target area for collisions with electrons. In insulators, what little conduction does take place is due to electrons that have gained enough energy from thermal motion of the lattice to break away from their "home" atoms and wander through the lattice. The number of electrons able to acquire the needed energy is very strongly temperature-dependent; a twofold increase for a 10° temperature rise is typical. There is also increased scattering at higher temperatures, as with metals, but the increased number of carriers is a far larger effect; thus insulators invariably become better conductors at higher temperatures.

A similar analysis can be made for *thermal* conductivity, which involves transport of microscopic mechanical energy rather than electric charge. The wave motion associated with lattice vibrations is one mechanism for energy transfer, and in metals the mobile electrons also carry kinetic energy from one region to another. This effect turns out to be much larger than that of the lattice vibrations, with the result that metals are usually much better thermal conductors than are electrical insulators, which have at most very few free electrons available to transport energy.

Optical properties are also related directly to microscopic structure. Good electrical conductors cannot be transparent to electromagnetic waves, for the electric fields of the waves induce currents in the material, and these dissipate the wave energy into heat as the electrons collide with the atoms in the lattice. All transparent materials are very good insulators. Metals are, however, good *reflectors* of radiation, and again the reason is the presence of free electrons at the surface of the material, which can move in response to the incident wave and generate a reflected wave. Reflection from the polished surface of an insulator is a somewhat more subtle phenomenon, dependent on polarization of the material, but again it is possible to relate macroscopic properties to microscopic structure.

46-7 Semiconductors

As the name implies, the electrical resistivity of a semiconductor is intermediate between that of good conductors and of good insulators. This is, however, only one aspect of the behavior of this important class of materials, so vital to present-day electronics. This discussion will be confined, for simplicity, to the elements germanium and silicon, the simplest semiconductors, but these examples serve to illustrate the principal concepts.

Silicon and germanium, having four electrons in the outermost subshell, crystallize in the diamond structure described in Sec. 46-5; each atom lies at the center of a regular tetrahedron, with four nearest neighbors at the corners and a covalent bond with each. Thus all the electrons are involved in the bonding and the materials should be insulators. However, an unusually small amount of energy is needed to break one of the bonds and set an electron free to roam around the lattice, 1.1 eV for silicon and only 0.7 eV for germanium. Thus even at room temperature a substantial number of electrons are dissociated from their parent atoms, and this number increases rapidly with temperature.

Furthermore, when an electron is removed from a covalent bond, it leaves a positively charged vacancy where there would ordinarily be an electron. An electron in a neighboring atom can drop into this vacancy, leaving the neighbor with the vacancy. In this way the vacancy, usually called a *hole,* can travel through the lattice and serve as an additional current-carrier. In a pure semiconductor, holes and electrons are always present in equal numbers; the resulting conductivity is called *intrinsic conductivity* to distinguish it from conductivity due to impurities, to be discussed later.

An analogy is often used to clarify the mechanism of conduction in a semiconductor. A crystal with no bonds broken is like a completely filled floor of a parking garage. No cars (electrons) can move because there is nowhere for them to go. But if one car is removed to the empty floor above, it can move around freely, and the vacancy it leaves also permits cars to move on the nearly filled floor; this motion is most easily described in terms of *motion of the vacant space* from which the car has been removed. The analogy can be drawn even more closely by considering the quantum states available to electrons in the solid, using the concept of energy bands, groups of closely spaced energy levels separated by "forbidden bands."

Now suppose we mix into melted germanium a small amount of arsenic, a Group V element having five electrons in its outermost subshell. When one of these electrons is removed, the remaining electron structure is essentially that of germanium; the only difference is that it is scaled down in size by the insignificant factor $\frac{32}{33}$ because the arsenic nucleus contains a charge $+33e$ rather than $+32e$. Thus an arsenic atom can take the place of a germanium atom in the lattice. Four of its five valence electrons form the necessary four covalent bonds with the nearest neighbors; the fifth is very loosely bound (energy only about 0.01 eV) and even at ordinary temperature can very easily escape and wander about the lattice. The corresponding positive charge is associated with the nuclear charge and is *not* free to move, in contrast to the situation with electrons and holes in pure germanium.

Because at ordinary temperatures only a very small fraction of the valence electrons are able to escape their sites and participate in the conduction process, a concentration of arsenic atoms as small as one part in 10^{10} can increase the conductivity so drastically that conduction due to impurities becomes by far the dominant mechanism. In such a case the conductivity is due almost entirely to *negative* charge motion; the material is called an *n-type* semiconductor or is said to have *n-type* impurities.

Adding atoms of an element in Group III, with only three electrons in its outermost subshell, has an analogous effect. An example is gallium; placed in the lattice, the gallium atom would like to form four covalent bonds, but it has only three outer electrons. It can, however, steal an electron from a neighboring germanium atom to complete the bonding. This leaves the neighbor with a *hole* or missing electron, and this hole can then move through the lattice just as with intrinsic conductivity. In this case the corresponding negative charge is associated with the deficiency of positive charge of the gallium nucleus ($+31e$ instead of $+32e$), so it is not free to move. This state of affairs is characteristic of *p-type* semiconductors, materials with *p-type* impurities. The two types of impurities, n and p, are also called *donors* and *acceptors*, respectively.

The assertion that in n and p semiconductors the current *is* actually carried by electrons and holes, respectively, can be verified using the Hall effect, discussed in Sec. 31-2. The direction of the Hall emf is opposite in the two cases, and measurements of the Hall effect in various semiconductor materials confirm the above analysis of the conduction mechanisms.

46-8 Semiconductor devices

The tremendous importance of semiconductor devices in practical electronic equipment results directly from the fact that the conductivity can be controlled within wide limits and changed from one region of a device to another, by varying impurity concentrations. The simplest example is the *p-n junction,* a crystal of germanium or silicon with *p*-type impurities in one region and *n*-type in the other, with the two separated by a boundary region called the *junction*. The details of how such a crystal is produced need not concern us; one way, not necessarily the most economical, is to grow a crystal by pulling a seed crystal very slowly away from the surface of a melted semiconductor. By varying the concentration of impurities in the melt while the crystal is grown, one can make a crystal with two or more regions of varying conductivity.

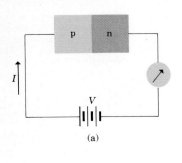

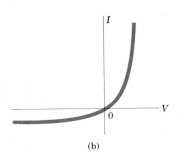

46-9 (a) A semiconductor *p-n* junction in a circuit; (b) graph showing the asymmetric voltage–current relationship given by Eq. (46-5).

When a *p-n* junction is connected in an external circuit as shown in Fig. 46-9a and the potential V across the device is varied, the behavior of the current is as shown in Fig. 46-9b. The device conducts much more readily in the direction $p \to n$ than in the reverse, in striking contrast to the behavior of materials that obey Ohm's law. In the language of electronics, such a one-way device is called a *diode*; a different kind of diode was discussed in Sec. 44-2 in connection with thermionic emission.

The behavior of a *p-n* junction diode can be understood at least roughly on the basis of the conductivity mechanisms in the two regions. When the *p* region is at higher potential than the *n*, holes in the *p* region flow into the *n* region and electrons in the *n* region into the *p* region, so both contribute substantially to current. When the polarity is reversed, the resulting electric fields tend to push electrons from *p* to *n* and holes from *n* to *p*. But there are very few electrons in the *p* region, only those associated with intrinsic conductivity and some that diffuse over from the *n* region. A similar condition prevails in the *n* region, and the current is much smaller than with the opposite polarity.

A more detailed analysis of this process, taking into account the effects of drift under the applied field and the diffusion that would take place even in the absence of a field, shows that the voltage–current relationship is given by

$$I = I_0(e^{eV/kT} - 1), \qquad (46\text{-}5)$$

where I_0 is a constant characteristic of the device, e is the electron charge, k is Boltzmann's constant, and T is absolute temperature.

The *transistor* includes two *p-n* junctions in a "sandwich" configuration which may be either *p-n-p* or *n-p-n*. One of the former type is shown in Fig. 46-10. The three regions are usually called the emitter, base, and collector, as shown. In the absence of current in the left loop of the circuit, there is only a very small current through the resistor R because the voltage across the base–collector junction is in the "reverse" direction, i.e., the direction of small current flow, as in a simple *p-n* junction. But when a voltage is applied between emitter and base, as shown, the holes traveling from emitter to base can travel *through* the base to the second junction, where they come under the influence of the collector-to-base potential difference and thus flow on around through the resistor.

Thus the current in the collector circuit is *controlled* by the current in the emitter circuit. Furthermore, since V_c may be considerably larger than V_e, the power dissipated in R may be much larger than that supplied to the emitter circuit by the battery V_e. Thus the device functions as a power amplifier. If the potential drop across R is greater than V_e, it may also be a voltage amplifier.

In this configuration the *base* is the common element between the "input" and "output" sides of the circuit. Another widely used arrangement is the *common-emitter* circuit, shown in Fig. 46-11. In this circuit the current in the collector side of the circuit is much larger than in the base side, and the result is current amplification.

Transistors can perform many of the functions for which vacuum tubes have been used ever since the invention of the triode by de Forest in 1907; and where applicable they offer many advantages over vacuum tubes. These include mechanical ruggedness (since no vacuum container

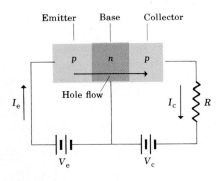

46-10 Schematic diagram of a *p-n-p* transistor and circuit. When $V_e = 0$ the current in the collector circuit is very small. When a potential V_e is applied between emitter and base, holes travel from emitter to base, as shown; when V_c is sufficiently large, most of them continue into the collector. The collector current I_c is controlled by the emitter current I_e.

or fragile electrodes are involved), small size, long life (because they operate at relatively low temperatures and deteriorate very little with age), and efficiency (because no power is wasted heating a cathode for thermionic emission). Since their invention in 1948, semiconductor devices have completely revolutionized the electronics industry, including applications to communications, computer systems, control systems, and many other areas.

A further refinement in semiconductor technology is the *integrated circuit*. By successive depositing of layers of material and etching patterns to define current paths, one can combine the functions of several transistors, capacitors, and resistors on a single square of semiconductor material that may be only a few millimeters on a side. A further elaboration of this idea leads to the *large-scale integrated circuit*. On a base consisting of a silicon chip, various layers are built up, including evaporated metal layers for conducting paths and silicon-dioxide layers for insulators and for dielectric layers in capacitors. Etching appropriate patterns into each layer by use of photosensitive etch-resistant materials and optically reduced patterns makes it possible to build up a circuit containing the functional equivalent of thousands of transistors, diodes, resistors, and capacitors on a single chip a few millimeters on a side. These so-called MOS (metal-oxide-semiconductor) chips are the heart of pocket calculators and microprocessers, as well as of many larger-scale computers.

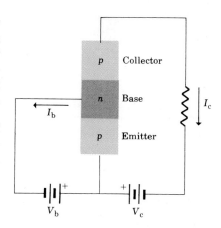

46-11 A common-emitter circuit. When $V_b = 0$, I_c is very small and most of the voltage V_c appears across the base–collector junction. As V_b increases, the base–collector potential decreases and more holes can diffuse into the collector; thus I_c increases. Ordinarily I_c is much larger than I_b.

Questions

46-1 In the ground state of the helium atom, the electrons must have opposite spins. Why?

46-2 The central-field approximation is more accurate for alkali metals than for transition metals (Group IV of the periodic table). Why?

46-3 The outermost electron in the potassium atom is in a 4s state. What does this tell you about the relative positions of the 3d and 4s states for this atom?

46-4 A student asserted that any filled shell (i.e., all the levels for a given n occupied by electrons) must have zero total angular momentum and hence must be spherically symmetric. Do you believe this? What about a filled *subshell* (all values of m for given values of n and l)?

46-5 What factors determine whether a material is a conductor of electricity or an insulator?

46-6 The nucleus of a gold atom contains 79 protons. How would you expect the energy required to remove a 1s electron completely from a gold atom to compare with the energy required to remove the electron from a hydrogen atom? In what region of the electromagnetic spectrum would a photon of the appropriate energy lie?

46-7 Elements can be identified by their visible spectra. Could analogous techniques be used to identify compounds from their molecular spectra? In what region of the electromagnetic spectrum would the appropriate radiation lie?

46-8 The ionization energies of the alkali metals (i.e., the energy required to remove an outer electron) are in the range from 4 to 6 eV, while those of the inert gases are in the range from 15 to 25 eV. Why the difference?

46-9 The energy required to remove the 3s electron from a sodium atom in its ground state is about 5 eV. Would you expect the energy required to remove an additional electron to be about the same, or more or less? Why?

46-10 Electrical conductivities of most metals decrease gradually with increasing temperature, but the intrinsic conductivity of semiconductors always *increases* rapidly with increasing temperature. Why the difference?

46-11 What are some advantages of transistors compared to vacuum tubes, for electronic devices such as amplifiers? What are some disadvantages? Are there any situations in which vacuum tubes *cannot* be replaced by solid-state devices?

Problems

46-1 Make a list of the four quantum numbers for each of the six electrons in the ground state of the carbon atom, and for each of the 14 electrons in the silicon atom.

46-2 Cobalt ($Z = 27$) has two electrons in $4s$ states. How many does it have in $3d$ states?

46-3 For germanium ($Z = 32$) make a list of the number of electrons in each state ($1s$, $2s$, $2p$, etc.)

46-4 Calculate the magnitude of the potential energy ($e^2/4\pi\varepsilon_0 r$) of interaction of a charge $+e$ with a charge $-e$ (where e is the electron charge magnitude) at a distance of 2×10^{-10} m, roughly the magnitude of the lattice spacing in the NaCl crystal.

46-5 For the sodium chloride molecule discussed at the beginning of Sec. 46-3, what is the maximum separation of the ions for stability if they may be regarded as point charges? (The potential energy of interaction must be at least -1.3 eV.)

46-6 For magnesium, the first ionization potential is 7.6 eV; the second (additional energy required to remove a second electron) is almost exactly twice this, 15 eV, and the third ionization potential is much larger, about 80 eV. How can these numbers be understood?

46-7 The ionization potentials of the alkali metals are:

Li	5.4 eV	Rb	4.2 eV
Na	5.1	Cs	3.9
K	4.3		

Which of these would you expect to be most active chemically?

46-8 The rotation spectrum of HCl contains the following wavelengths:

$$60.4 \ \mu m$$
$$69.0$$
$$80.4$$
$$96.4$$
$$120.4$$

Find the moment of inertia of the HCl molecule.

46-9 Show that the frequencies in a pure rotation spectrum (disregarding vibrational levels) are all integer multiples of the quantity $h/2\pi I$.

46-10 If the distance between atoms in a diatomic oxygen molecule is 2×10^{-10} m, calculate the moment of inertia about an axis through the center of mass perpendicular to the line joining the atoms.

46-11 The dissociation energy of the hydrogen molecule (i.e., the energy required to separate the two atoms) is 4.72 eV. At what temperature is the average kinetic energy of a molecule equal to this energy?

46-12 The vibrational frequency of the hydrogen molecule is 1.29×10^{14} Hz.

a) What is the spacing of adjacent vibrational energy levels, in electron volts?

b) What is the wavelength of radiation emitted in the transition from the $n = 2$ to $n = 1$ vibrational state of the hydrogen molecule?

c) From what initial values of n do transitions to the ground state of vibrational motion yield radiation in the visible spectrum?

46-13 The distance between atoms in a hydrogen molecule is 0.074 nm.

a) Calculate the moment of inertia of a hydrogen molecule about an axis perpendicular to the line joining the nuclei, at its center.

b) Find the energies of the $l = 0$, $l = 1$, and $l = 2$ rotational states.

c) Find the wavelength and frequency of the photon emitted in the transition from $l = 2$ to $l = 0$.

46-14 The spacing of adjacent atoms in a sodium-chloride crystal is 0.282 nm. Calculate the density of sodium chloride.

46-15 In the hexagonal close-packed crystal structure, regarding the atoms as rigid spheres, show that an atom can accommodate twelve nearest neighbors (all touching it) if six are placed around it in a plane hexagonal array, with three additional atoms above and three below the plane.

46-16 Suppose a piece of very pure germanium is to be used as a light detector by observing the increase in conductivity resulting from generation of electron–hole pairs by absorption of photons. If each pair requires 0.7 eV of energy, what is the maximum wavelength that can be detected? In what portion of the spectrum does it lie?

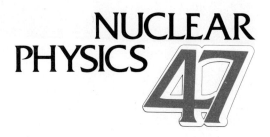

NUCLEAR PHYSICS 47

47-1 The nuclear atom

In preceding chapters we have frequently made use of the fact that every atom contains a massive, positively charged nucleus, much smaller than the overall dimensions of the atom but nevertheless containing most of the total mass of the atom. It is instructive to review the earliest experimental evidence for the existence of the nucleus, the *Rutherford scattering experiments*. The experiments were carried out in 1910–1911 by Sir Ernest Rutherford and two of his students, Hans Geiger and Ernest Marsden, at Cambridge, England.

The electron had been "discovered" in 1897 by Sir J. J. Thomson, and by 1910 its mass and charge were quite accurately known. It had also been well established that, with the sole exception of hydrogen, all atoms contain more than one electron. Thomson had proposed a model of the atom consisting of a relatively large sphere of positive charge (about 2 or 3×10^{-8} cm in diameter) within which were embedded, like plums in a pudding, the electrons.

What Rutherford and his coworkers did was to project other particles at the atoms under investigation; and, from observations of the way in which the projected particles were deflected or *scattered,* they drew conclusions about the distribution of charge within the "target" atoms.

At this time the high-energy particle accelerators now in common use for nuclear physics research had not yet been developed, and Rutherford had to use as projectiles the particles produced in natural radioactivity, to be discussed later in this chapter. Some radioactive disintegrations result in the emission of *alpha particles;* these particles are now known to be identical with the nuclei of helium atoms, each consisting of two protons and two neutrons bound together, but without the two electrons normally present in a neutral helium atom. Alpha particles are ejected from unstable nuclei with speeds of the order of 10^7 m·s^{-1}, and they can travel several centimeters through air, or on the order of 0.1 mm through solid matter, before they are brought to rest by collisions.

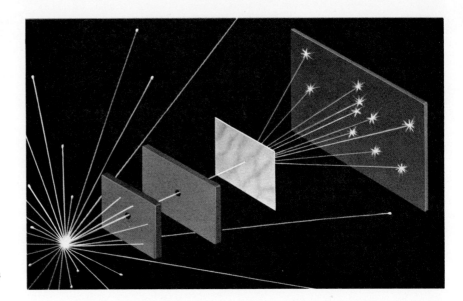

47-1 The scattering of alpha particles by a thin metal foil.

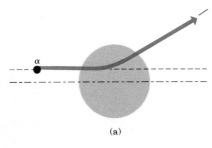

(a)

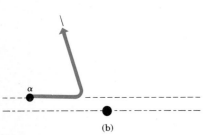

(b)

47-2 (a) Alpha particle scattered through a small angle by the Thomson atom. (b) Alpha particle scattered through a large angle by the Rutherford nuclear atom.

The experimental setup is shown schematically in Fig. 47–1. A radioactive source at the left emits alpha particles. Thick lead screens stop all particles except those in a narrow *beam* defined by small holes. The beam then passes through a thin metal foil (gold, silver, and copper were used) and strikes a plate coated with zinc sulfide. A momentary flash or scintillation can be observed visually on the screen whenever it is struck by an alpha particle, and the number of particles that have been deflected through any angle from their original direction can therefore be determined.

According to the Thomson model, the atoms of a solid are packed together like marbles in a box. The experimental fact that an alpha particle can pass right through a sheet of metal foil forces one to conclude, if this model is correct, that the alpha particle is capable of actually penetrating the spheres of positive charge. Granted that this is possible, we can compute the deflection it would undergo. The Thomson atom is electrically neutral, so outside the atom no force would be exerted on the alpha particle. Within the atom, the electrical force would be due in part to the electrons and in part to the sphere of positive charge. However, the mass of an alpha particle is about 7400 times that of an electron, and from momentum considerations it follows that the alpha particle can suffer only a negligible scattering as a consequence of forces between it and the much less massive electrons. It is only interactions with the *positive* charge, which makes up most of the atomic mass, that can deviate the alpha particle.

When electric charge is distributed uniformly inside a spherical volume, the electric field at points inside the sphere is proportional to the distance from the center of the sphere; this can be proved using Gauss's law, as in Problem 25–25. Thus the positively charged alpha particle inside the Thomson atom would be *repelled* from the center of the sphere with a force proportional to its distance from the center, and its trajectory can be computed for any initial direction of approach such as that in Fig. 47–2a. On the basis of such calculations, Rutherford predicted the number of alpha particles that should be scattered at any angle with respect to the original direction.

The experimental results did not agree with the calculations based on the Thomson atom. In particular, many more particles were scattered through large angles than were predicted. To account for the observed large-angle scattering, Rutherford concluded that the positive charge, instead of being spread through a sphere of atomic dimensions (2 or 3×10^{-8} cm) was concentrated in a much *smaller* volume, which he called a *nucleus*. When an alpha particle approaches the nucleus, the entire nuclear charge exerts a repelling effect on it down to extremely small separations, with the consequence that much larger deviations can be produced. Figure 47–2b shows the trajectory of an alpha particle deflected by a Rutherford nuclear atom, for the same original path as that in part (a) of the figure.

Rutherford again computed the expected number of particles scattered through any angle, assuming an inverse-square law of force between the alpha particle and the nucleus of the scattering atom. Within the limits of experimental accuracy, the computed and observed results were in agreement down to distances of approach of about 10^{-12} cm. These experiments thus indicate that the size of the nucleus is no larger than the order of 10^{-12} cm.

47-2 Properties of nuclei

As indicated in Sec. 47–1, the most obvious feature of the atomic nucleus is its size, of the order of 20,000 to 200,000 times smaller than the atom itself. Since Rutherford's initial experiments, many additional scattering experiments have been performed, using high-energy protons, electrons, and neutrons as well as alpha particles. Although the "surface" of a nucleus is not a sharp boundary, these experiments determine an approximate radius for each nucleus. The radius is found to depend on the mass, which in turn depends on the total number A of neutrons and protons, usually called the *mass number*. The radii of most nuclei are represented fairly well by the empirical equation

$$r = r_0 A^{1/3} \qquad (47–1)$$

where r_0 is an empirical constant equal to 1.2×10^{-15} m, the same for all nuclei.

Since the volume of a sphere is proportional to r^3, Eq. (47–1) shows that the *volume* of a nucleus is proportional to A (i.e., to the total mass) and therefore that the mass per unit volume (proportional to A/r^3) is the same for all nuclei. That is, *all nuclei have approximately the same density*. This fact is of crucial importance in understanding nuclear structure.

Two additional important properties of nuclei are angular momentum and magnetic moment. The particles in the nucleus are in motion, just as the electrons in an atom are in motion; there is angular momentum associated with this motion, and, because circulating charge constitutes a current, there is also magnetic moment. Experimental evidence for the existence of nuclear angular momentum (often called *nuclear spin*) and magnetic moment came originally from spectroscopy. Some spectrum lines are found to be split into series of very closely spaced lines, called *hyperfine structure,* which can be understood on the basis of interactions between the electrons in an atom and the magnetic field produced by a nuclear magnetic moment. Detailed analysis of this phenomenon indicates that the nuclear angular momentum is *quantized,* just as it is for electrons and for molecular rotation; the component of angular momentum in a specified axis direction is a multiple of $h/2\pi$, but some nuclei seem to require *integer* multiples (as with orbital angular momentum of electrons) and some *half-integer* multiples, as with electron spin. As we shall see later, the nuclear spin can be understood on the basis of the particles making up the nucleus.

Although it was once believed that nuclei were made of protons and electrons, the discovery of the neutron (discussed in Sec. 47–10) and many other experiments have established that the basic building blocks of the nucleus are the proton and the neutron. The total number of protons, equal in a neutral atom to the number of electrons, is the *atomic number Z.* The total number of *nucleons* (protons and neutrons) is called the mass number A. The number of neutrons, denoted by N, is called the *neutron number.* For any nucleus these are related by

$$A = Z + N \qquad (47\text{-}2)$$

Table 47–1 lists values of A, Z, and N for several nuclei. As the table shows, some nuclei have the same Z but different N. Since the electron structure, which determines the chemical properties, is determined by the charge of the nucleus, these are nuclei of the same element, but they have different masses and can be distinguished in precise experiments. The experimental investigation of *isotopes* through mass spectroscopy is discussed in Sec. 30–6, and the reader would do well to review this section. Nuclei of a given element having different mass numbers are called *isotopes* of the element, and a single nuclear species (single values of both Z and N) is called a *nuclide.*

Table 47–1 also shows the notation usually used to denote individual nuclides; the symbol of the element is used, with a pre-subscript equal to

Table 47-1   Nuclear particles

Nucleus	Mass number (total number of nuclear particles), A	Atomic number (number of protons), Z	Neutron number, $N = A - Z$
$_1\text{H}^1$	1	1	0
$_1\text{D}^2$	2	1	1
$_2\text{He}^4$	4	2	2
$_3\text{Li}^6$	6	3	3
$_3\text{Li}^7$	7	3	4
$_4\text{Be}^9$	9	4	5
$_5\text{B}^{10}$	10	5	5
$_5\text{B}^{11}$	11	5	6
$_6\text{C}^{12}$	12	6	6
$_6\text{C}^{13}$	13	6	7
$_7\text{N}^{14}$	14	7	7
$_8\text{O}^{16}$	16	8	8
$_{11}\text{Na}^{23}$	23	11	12
$_{29}\text{Cu}^{65}$	65	29	36
$_{80}\text{Hg}^{200}$	200	80	120
$_{92}\text{U}^{235}$	235	92	143
$_{92}\text{U}^{238}$	238	92	146

the atomic number Z and a post-superscript equal to the mass number A. This is of course redundant, since the element is determined by the atomic number, but the notation is a useful aid to memory.

The total mass of a nucleus is always *less* than the total mass of its constituent parts because of the mass equivalent ($E = mc^2$) of the negative potential energy associated with the attractive forces that hold the nucleus together. This mass difference is sometimes called the *mass defect*. In fact, the best way to determine the total potential energy, or *binding energy,* is to compare the mass of a nucleus with the masses of its constituents. The proton and neutron masses are

$$m_p = 1.673 \times 10^{-27} \text{ kg},$$

$$m_n = 1.675 \times 10^{-27} \text{ kg}.$$

Since these are nearly equal, it is not surprising that many nuclear masses are approximately integer multiples of the proton or neutron mass.

This observation suggests defining a new mass unit equal to the proton or neutron mass. Instead, it has been found more convenient to define the new unit, called the *atomic mass unit* (u), as $\frac{1}{12}$ the mass of the neutral carbon atom having mass number $A = 12$. It is found that

$$1 \text{ u} = 1.660566 \times 10^{-27} \text{ kg}.$$

In atomic units, the masses of the proton, neutron, and electron are found to be:

$$m_p = 1.007276 \text{ u},$$

$$m_n = 1.008665 \text{ u},$$

$$m_e = 0.000549 \text{ u}.$$

The masses of some common atoms, including their electrons, are shown in Table 47–2. The masses of the bare nuclei are obtained by subtracting Z times the electron mass. The energy equivalent of 1 u is found from the relation $E = mc^2$:

$$E = (1.660566 \times 10^{-27} \text{ kg})(2.998 \times 10^8 \text{ m} \cdot \text{s}^{-1})^2$$

$$= 1.492 \times 10^{-10} \text{ J} = 931.5 \text{ MeV}.$$

Example   Find the mass defect, the total binding energy, and the binding energy per nucleon for carbon (12).

Solution   The mass of the neutral carbon atom, including the nucleus and the six electrons, is, according to Table 47–2, 12.00000 u. The mass of the bare nucleus is obtained by subtracting the mass of the six electrons:

$$m = 12.00000 \text{ u} - (6)(0.000549 \text{ u}) = 11.996706 \text{ u}.$$

The total mass of the six protons and six neutrons in the nucleus is

$$(6)(1.007276 \text{ u}) + (6)(1.008665 \text{ u}) = 12.095646 \text{ u}.$$

The mass defect is therefore

$$12.095646 \text{ u} - 11.996706 \text{ u} = 0.09894 \text{ u}.$$

The energy equivalent of this mass is

$$(0.09894 \text{ u})(931.5 \text{ MeV} \cdot \text{u}^{-1}) = 92.16 \text{ MeV}.$$

Table 47-2   Atomic data*

Element	Atomic number, Z	Neutron number N	Atomic mass, u	Mass number, A
Hydrogen H^1	1	0	1.00783	1
Deuterium H^2	1	1	2.01410	2
Helium He^3	2	1	3.01603	3
Helium He^4	2	2	4.00260	4
Lithium Li^6	3	3	6.01513	6
Lithium Li^7	3	4	7.01601	7
Beryllium Be^8	4	4	8.00508	8
Beryllium Be^9	4	5	9.01219	9
Boron B^{10}	5	5	10.01294	10
Boron B^{11}	5	6	11.00931	11
Carbon C^{12}	6	6	12.00000	12
Carbon C^{13}	6	7	13.00335	13
Nitrogen N^{14}	7	7	14.00307	14
Nitrogen N^{15}	7	8	15.00011	15
Oxygen O^{16}	8	8	15.99491	16
Oxygen O^{17}	8	9	16.99913	17
Oxygen O^{18}	8	10	17.99916	18

* American Institute of Physics Handbook, 1963

Thus the total binding energy for the twelve nucleons is 92.16 MeV. To pull the carbon nucleus completely apart into twelve separate nucleons would require a minimum of 92.16 MeV. The binding energy *per nucleon* is $\frac{1}{12}$ of this, or 7.68 MeV per nucleon. Nearly all stable nuclei, from the lightest to the most massive, have binding energies in the range of 6 to 8 MeV per nucleon.

Because the nucleus is a composite structure, it can have internal motion, with corresponding energy levels. Thus each nucleus has a set of allowed energy levels, corresponding to a *ground state* (state of lowest energy) and several *excited states*. Because of the strength of nuclear interactions, excitation energies of nuclei are typically of the order of 1 MeV, compared with a few eV for atomic energy levels. In ordinary physical and chemical transformations the nucleus always remains in its ground state. When a nucleus is placed in an excited state, either by bombardment by high-energy particles or by a radioactive transformation, it can decay to the ground state by emission of one or more photons, called in this case *gamma rays* or *gamma photons*.

The forces that hold protons and neutrons together in the nucleus despite the electrical repulsion of the protons are not of any familiar sort but are unique to the nucleus. Some aspects of the nuclear force are still incompletely understood, but several qualitative features can be described. First, it does not depend on charge, since neutrons as well as protons must be bound. Second, it must be of short range; otherwise the nucleus would pull in additional protons and neutrons. But within the range it must be stronger than electrical forces; otherwise the nucleus could never be stable. Third, the nearly constant density of nuclear matter and the nearly constant binding energy per nucleon indicate that a given nucleon cannot interact simultaneously with all the other nucleons but only with those few in its immediate vicinity. This is again in contrast to the behavior of electrical forces, in which *every* proton in the nucleus repels every other one. This limitation on the maximum number of nucleons with which a nucleon can interact is called *saturation;* it is analogous in some respects to covalent bonding in solids. Finally, the nuclear forces appear to favor binding of pairs of particles (e.g., two protons with opposite spin, or two neutrons with opposite spin) and of *pairs of pairs,* a pair of protons and a pair of neutrons, with total spin zero. Thus the alpha particle is an exceptionally stable nuclear structure.

These qualitative features of the nuclear force are helpful in understanding the various kinds of nuclear instability, to be discussed in the following sections.

47-3 Natural radioactivity

In studying the fluorescence and phosphorescence of compounds irradiated with visible light, Becquerel, in 1896, performed a crucial experiment that led to a deeper understanding of the properties of the nucleus of an atom. After illuminating some pieces of uranium–potassium sulfate with visible light, Becquerel wrapped them in black paper and separated the package from a photographic plate by a piece of silver. After several hours' exposure, the photographic plate was developed and showed a blackening due to something that must have been emitted from the compound and was able to penetrate both the black paper and the silver.

Rutherford showed later that the emanations given off by uranium sulfate were capable of ionizing the air in the space between two oppositely charged metallic plates (an ionization chamber). The current registered by a galvanometer in series with the circuit was taken to be a measure of the "activity" of the compound.

A systematic study of the activity of various elements and compounds led Mme. Curie to the conclusion that this activity was an atomic phenomenon; and by the methods of chemical analysis, she and her husband, Pierre Curie, found that "ionizing ability" or "activity" was associated not only with uranium but with two other elements that they discovered, radium and polonium. The activity of radium was found to be more than a million times that of uranium. Since the pioneer days of the Curies, many more radioactive substances have been discovered.

The activity of radioactive material may be easily shown to be the result of three different kinds of emanations. An early experiment is shown in Fig. 47-3. A small piece of radioactive material is placed at the bottom of a long groove in a lead block. At some distance above the lead block a photographic plate is placed, and the whole apparatus is highly evacuated. A strong magnetic field is applied at right angles to the plane of the diagram. When the plate is developed, three distinct spots are found, one in the direct line of the groove in the lead block, one deflected to one side, and one to the other side. From a knowledge of the direction of the magnetic field, it is concluded that one of the emanations is positively charged (alpha particles), one is negatively charged (beta particles), and one is neutral (gamma rays).

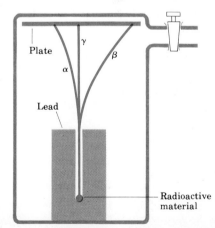

47-3 The three emanations from a radioactive material and their paths in a magnetic field perpendicular to the plane of the diagram.

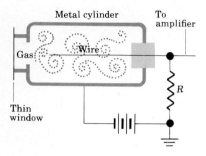

47-4 Schematic diagram of a Geiger counter.

Futher investigation showed that not all three emanations are emitted simultaneously by all radioactive substances. Some elements emit alpha particles, others emit beta particles, while gamma rays sometimes accompany one and sometimes the other. Furthermore, no simple macroscopic physical or chemical process, such as raising or lowering the temperature, chemical combination with other nonradioactive substances, etc., could change or affect in any way the activity of a given sample. As a result, it was suspected from the beginning that radioactivity is a *nuclear* process and that the emission of a charged particle from the nucleus of an atom results in leaving behind a different atom, occupying a different place in the periodic table. In other words, radioactivity involves the *transmutation of elements*.

The first measurement of the charge of the alpha particle used a device called a *Geiger counter,* still an important instrument of modern physics.

As shown in Fig. 47–4, a Geiger counter consists of a metal cylinder and a wire along its axis. The cylinder contains a gas such as air or argon, at a pressure of from 50 to 100 mm of mercury. A difference of potential slightly less than that necessary to produce a discharge is maintained between the wire and the cylinder wall. Alpha particles (or, for that matter, any particles to be studied) can enter through a thin glass or mica window. The particle entering the counter produces ionization of the gas molecules. These ions are accelerated by the electric field and produce more ions by collisions, causing the ionization current to build up rapidly. The current, however, decays rapidly since the circuit has a small time constant. There is therefore a momentary surge of current or a momentary potential surge across R, which may be amplified and made to advance an electronic counter, or to activate a counting-rate meter.

Placing a known mass of radium at a known distance from the window of a Geiger counter, Rutherford and Geiger counted the number of alpha particles emitted in a known time interval. They found that 3.57×10^{10} alpha particles were emitted per second per gram of radium. They then allowed the alpha particles from the same source to fall upon a plate and measured its rate of increase of charge. Dividing the rate of increase of charge by the number emitted per second, Rutherford and Geiger determined the charge on an alpha particle to be 3.19×10^{-19} C, or practically twice the charge on an electron, but opposite in sign.

The next problem was to determine the mass of an alpha particle. This was accomplished by measuring the ratio of charge to mass by the electric and magnetic deflection method described in Chapter 30. The ratio was found by Rutherford and Robinson to be 4.82×10^7 C·kg^{-1}. Combining this result with the charge of an alpha particle, the mass was found to be 6.62×10^{-27} kg, almost exactly four times the mass of a hydrogen atom.

Since a helium atom has a mass four times that of a hydrogen atom and, stripped of its two electrons (as a bare nucleus), has a charge equal in magnitude and opposite in sign to two electrons, it seemed certain that alpha particles were helium nuclei. To make the identification certain, however, Rutherford and Royds collected the alpha particles in a glass discharge tube over a period of about six days and then established an electric discharge in the tube. Examining the spectrum of the emitted light, they identified the characteristic helium spectrum and established without doubt that alpha particles *are* helium nuclei.

The speed of an alpha particle emitted from a given radioactive source such as radium may be measured by observing the radius of the circle traversed by the particle in a magnetic field perpendicular to the motion. Such experiments show that alpha particles are emitted with very high speeds. For example, the alpha particles emitted by radium, $_{88}Ra^{226}$, have speeds of about 1.5×10^7 m·s^{-1}. The corresponding kinetic energy is

$$K = \tfrac{1}{2}(6.62 \times 10^{-27}\text{ kg})(1.5 \times 10^7\text{ m·s}^{-1})^2$$
$$= 7.4 \times 10^{-13}\text{ J} = 4.7 \times 10^6\text{ eV}$$
$$= 4.7\text{ MeV.}$$

We note that the speed, although large, is only five percent of the speed of light, so the nonrelativistic kinetic-energy expression may be used. We also note that the energy is larger than typical energies of atomic electrons by a factor of the order of a million. Because of these large energies, alpha particles are capable of traveling several centimeters in air, or a few tenths or hundredths of a millimeter through solids, before they are brought to rest by collisions.

Beta particles are *negatively* charged and are therefore deflected in an electric or magnetic field. Deflection experiments similar to those described in Chapter 30 prove conclusively that beta particles have the same charge and mass as electrons. They are emitted with tremendous speeds, some reaching a value of 0.9995 that of light. Thus relativistic relations must be used in analyzing their motion. Unlike alpha particles, which are emitted from a given nucleus with one speed or a few definite speeds, beta particles are emitted with a *continuous* range of speeds, from zero up to a maximum that depends on the nature of the emitting nucleus. If the principles of conservation of energy and of momentum are to hold in nuclear processes, it is necessary to assume that the emission of a beta particle is accompanied by the emission of another particle with no charge. This particle, called a *neutrino,* has zero rest mass and zero charge and therefore, even in traversing the densest matter, produces very little measurable effect. In spite of this, Reines and Cowan, in 1953 and again in 1956, succeeded in detecting its existence in a series of extraordinary experiments. Subsequent investigation has shown that in fact there are at least three varieties of neutrinos, the one associated with beta decay and others associated with the decay of unstable particles.

Since gamma rays are not deflected by a magnetic field, they cannot consist of charged particles. They are, however, diffracted at the surface of a crystal in a manner similar to that of x-rays but with extremely small angles of diffraction. Experiments of this sort lead to the conclusion that gamma rays are actually electromagnetic waves of extremely short wavelength, about $\frac{1}{100}$ that of x-rays.

The gamma-ray spectrum of any one element is a line spectrum, suggesting that a gamma-ray photon is emitted when a nucleus proceeds from a state of higher to a state of lower energy. This view is substantiated in the case of radium by the following facts. When alpha particles are emitted from radium they are found to consist of two groups, those with a kinetic energy of 4.879 MeV and those with an energy of 4.695 MeV. When a radium atom emits an alpha particle of the smaller energy, the resulting nucleus (which corresponds to the element *radon*) has a *greater* amount of energy than if the higher-speed alpha particle had been emitted. This represents an excited state of the radon nucleus. If now the radon nucleus undergoes a transition from this excited state to the lower energy state, a gamma-ray photon of energy

$$(48.79 - 46.95) \times 10^5 = 1.84 \times 10^5 \text{ eV}$$

should be emitted. The *measured* energy of the gamma-ray photon emitted by radium is $1.89 = 10^5$ eV, in excellent agreement.

Thus, by correlating alpha-particles energies and gamma-ray energies, it is possible in some cases to construct *nuclear energy-level diagrams* similar to atomic energy-level diagrams.

When a radioactive nucleus decays by alpha or beta emission, the resulting nucleus may also be unstable, and there may be a chain of successive decays until a stable configuration is reached. The most abundant radioactive nucleus found in nature is that of uranium $_{92}\text{U}^{238}$, which undergoes a series of 14 decays, including eight alpha emissions and six beta emissions, terminating at the stable isotope of lead, $_{82}\text{Pb}^{206}$. Decay series are discussed further in Sec. 47–5.

In alpha decay the neutron number N and the charge number Z each decrease by two, and the mass number A decreases by four, corresponding to the values $N = 2$, $Z = 2$, $A = 4$ for the alpha particle. The situation in beta decay is less obvious; one might well question how a nucleus composed of protons and neutrons can emit an *electron*. The answer is that, in beta decay, a neutron in the nucleus is transformed into a proton, an electron, and a neutrino. Such transformations of fundamental particles are discussed further in Sec. 47–10. The effect is to *increase* the charge number Z by one, decrease the neutron number N by one, and leave the mass number A unchanged. Finally, gamma emission leaves all three numbers unchanged. Both alpha and beta emission are often accompanied by gamma emission.

47–4 Nuclear stability

Of about 1500 different nuclides now known, only about one-fifth are stable. The others are radioactive, with lifetimes ranging from a small fraction of a second to many years. The stable nuclei are indicated by dots on the graph in Fig. 47–5, where the neutron number N is plotted against the charge number Z. Such a graph is called a *Segre chart*, after its inventor.

Since the mass number A is the sum $N + Z$, a curve of constant A is a straight line perpendicular to the line $N = Z$. In general, lines of constant A pass through only one or two stable nuclei (see $A = 20$, $A = 40$, $A = 60$, etc.), but there are four cases when such lines pass through three stable isotopes, namely at $A = 96$, 124, 130, and 136. Only four stable nuclei have both odd Z and odd N:

$$_1\text{H}^2, \quad _3\text{Li}^6, \quad _5\text{B}^{10}, \quad _7\text{N}^{14};$$

these are called *odd-odd nuclei*. Also, there is *no* stable nucleus with $A = 5$ or $A = 8$.

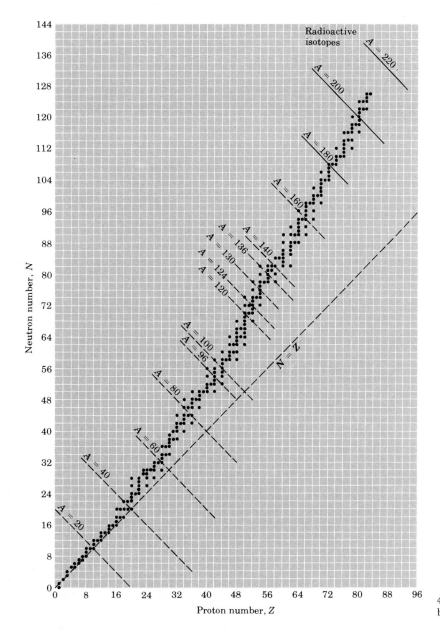

47-5 Segre chart, showing neutron number and proton number for stable nuclei.

The points representing stable isotopes define a rather narrow stability region. For low mass numbers $N/Z = 1$. This ratio increases and becomes about 1.6 at large mass numbers. Points to the right of the stability region represent nuclei that have an excess of protons or a deficiency of neutrons; to the left of the stability region are the points representing nuclei with an excess of neutrons or a deficiency of protons. The graph also shows that there is a maximum value of A; no nucleus with A greater than 209 is stable.

The stability of nuclei can be understood qualitatively on the basis of the nature of the nuclear force and the competition between the attractive nuclear force and the repulsive electrical force. As pointed out in Sec. 47-2, the nuclear force favors pairs of nucleons, and pairs of pairs. In the absence of electrical interactions, the most stable nuclei would be those having equal numbers of neutrons and protons, $N = Z$. The electrical repulsion shifts the balance to favor greater numbers of neutrons, but a nucleus with *too many* neutrons is unstable because not enough of them are paired with protons. A nucleus with too many *protons* has too much repulsive electrical interaction, compared with the attractive nuclear interaction, to be stable.

Furthermore, as the number of nucleons increases, the total energy of electrical interaction increases faster than that of the nuclear interaction. To understand this, we recall the discussion of electrostatic energy in Sec. 27–4. The energy of a capacitor with a charge Q is proportional to Q^2. It can be shown that to place a total charge Q on the surface of a sphere of radius a requires a total energy $Q^2/8\pi\epsilon_0 a$.

Thus the (positive) electric potential energy in the nucleus increases approximately as Z^2, while the (negative) nuclear potential energy increases approximately as A, with corrections for pairing effects. Thus the competition of electric and nuclear forces accounts for the fact that the neutron–proton ratio in stable nuclei increases with Z, and also for the fact that there is a maximum A (and a maximum Z) for stability. At large A the electric energy *per nucleon* grows faster than the nuclear energy per nucleon, until the point is reached where stability is impossible. Unstable nuclei respond to these conditions in various ways; the next several sections discuss various types of decay of unstable nuclei.

47-5 Radioactive transformations

The number of radioactive nuclei in any sample of radioactive material decreases continuously as some of the nuclei disintegrate. The *rate* at which the number decreases, however, varies widely for different nuclei. Let N represent the number of radioactive nuclei in a sample at time t, and dN the number that undergo transformations in a short time interval dt. Since every transformation results in a *decrease* in the number N, the corresponding change in N is $-dN$ and the rate of change of N is $-dN/dt$. The larger the number of nuclei in the sample, the larger the number that will undergo transformations, so that the rate of change of N is proportional to N, or is equal to a constant λ multiplied by N. Therefore

$$\frac{dN}{dt} = -\lambda N. \tag{47-3}$$

If we denote by N_0 the number of nuclei at time $t = 0$, then the solution of this differential equation is an exponential function:

$$N = N_0 e^{-\lambda t}. \tag{47-4}$$

A graph of this function is shown in Fig. 47–6.

The *half-life* $T_{1/2}$ of a radioactive sample is defined as the time at which the number of radioactive nuclei has decreased to one-half the number at $t = 0$. At this time,

$$e^{-\lambda T_{1/2}} = \tfrac{1}{2}.$$

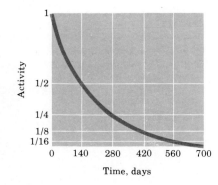

47-6 Decay curve for the radioactive element polonium. Polonium has a half-life of 140 days.

Taking natural logarithms of both sides and solving for $T_{1/2}$, we find

$$\lambda T_{1/2} = \ln 2,$$

$$T_{1/2} = \frac{\ln 2}{\lambda} = \frac{0.693}{\lambda}. \qquad (47\text{-}5)$$

Half of the original nuclei in a radioactive sample decay in a time interval $T_{1/2}$, half of those remaining at this time decay in a second interval $T_{1/2}$, and so on; thus the fraction remaining after successive intervals of $T_{1/2}$ is $\frac{1}{2}, \frac{1}{4}, \frac{1}{8}$, and so on. We also note that when $t = 1/\lambda$, N has decreased to $1/e$ of its initial value.

The *activity* of a sample is defined to be the number of disintegrations per unit time. A commonly used unit is the *curie,* defined to be 3.70×10^{10} decays per second. This is approximately equal to the activity of one gram of radium. Since the number of disintegrations is proportional to the number of radioactive nuclei in the sample, the activity decreases exponentially with time in the same way as the number N. Figure 47–6 is a graph of the activity of polonium, $_{84}\text{Po}^{210}$, which has a half-life of 140 days.

In studying radioactivity the following questions are relevant:

1. What is the parent nucleus?
2. What particle is emitted from this nucleus?
3. What is the half-life of the parent nucleus?
4. What is the resulting nucleus (often called the *daughter nucleus*)?
5. Is the daughter nucleus radioactive and if so, what are the answers to questions 2, 3, and 4 for this nucleus?

Very extensive investigations have been carried on in the last 75 years, and these questions have been answered for many nuclides. The results are most conveniently presented on a Segre chart such as that shown in Fig. 47–7. The neutron number N is plotted along the vertical axis and the atomic number (or charge number) Z on the horizontal axis. Unit increase of Z with unit decrease of N indicates beta emission; decrease of two in both N and Z indicates alpha emission. The half-lives are given either in years (y), days (d), hours (h), minutes (m), or seconds (s).

Figure 47–7 shows the uranium decay series, which begins with the common uranium isotope U^{238}. Each arrow represents a decay in which an alpha or beta particle is emitted. The decays can also be represented in equation form; the first two decays in the series are written as

$$\text{U}^{238} \longrightarrow \text{Th}^{234} + \alpha,$$
$$\text{Th}^{234} \longrightarrow \text{Pa}^{234} + \beta,$$

or, even more briefly, as

$$\text{U}^{238} \xrightarrow{\ \alpha\ } \text{Th}^{234},$$

$$\text{Th}^{235} \xrightarrow{\ \beta\ } \text{Pa}^{234}.$$

In the second decay a gamma photon is emitted following the beta; the reason for this is that the beta decay leaves the daughter nucleus not in

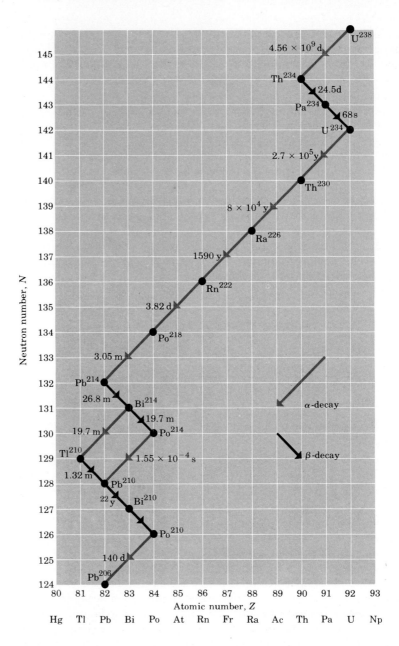

47–7 Segre chart showing the uranium U^{238} decay series, terminating with the stable nuclide $_{82}Pb^{206}$.

its lowest energy state but in an excited state, from which it decays to the ground state by emitting a photon.

An interesting feature of the U^{238} decay series is the branching that occurs at Bi^{214}; this nuclide decays to Pb^{210} by emission of an alpha and a beta, which can occur in either order. We also note that the series includes unstable isotopes of several elements that also have stable isotopes, including Tl, Pb, and Bi. The unstable isotopes of these elements that occur in the U^{238} series all have too many neutrons to be stable, as discussed in Sec. 47–4.

Three other decay series are known; two of these occur in nature, one starting with the uncommon isotope U^{235}, the other with thorium

(Th^{232}). The last series starts with neptunium (Np^{237}), an element not found in nature but produced in nuclear reactors. In each case the series continues until a stable nucleus is reached; for these series the final members are Pb^{206}, Pb^{207}, Pb^{208}, and Bi^{209}, respectively.

An interesting application of radioactivity is the dating of archeological and geological specimens by measurement of the concentration of radioactive isotopes. The most familiar example is carbon dating; C^{14}, an unstable isotope, is produced by nuclear reactions in the atmosphere caused by cosmic-ray bombardment, and there is a small portion of C^{14} in the CO_2 in the atmosphere. Plants that obtain their carbon from this source contain the same proportion of C^{14} as the atmosphere, but when a plant dies it stops taking in carbon, and the C^{14} it has already taken in decays, with a half-life of 5,568 years. Thus by measuring the proportion of C^{14} in the remains, one can determine how long ago the organism died. Similar techniques are used with other isotopes for dating of geologic specimens. A difficulty with carbon dating is that the C^{14} concentration in the atmosphere changes with time, over long intervals.

47-6 Nuclear reactions

The nuclear disintegrations described up to this point have consisted exclusively of a natural, uncontrolled emission of either an alpha or beta particle. Nothing was done to initiate the emission, and nothing could be done to stop it. It occurred to Rutherford in 1919 that it ought to be possible to penetrate a nucleus with a massive high-speed particle such as an alpha particle and thereby either produce a nucleus with greater atomic number and mass number or induce an artificial nuclear disintegration. Rutherford was successful in bombarding nitrogen with alpha particles and obtaining, as a result, an oxygen nucleus and a proton, according to the reaction

$$_2He^4 + {_7}N^{14} \rightarrow {_8}O^{17} + {_1}H^1. \qquad (47\text{-}6)$$

Note that the sum of the initial atomic numbers is equal to the sum of the final atomic numbers, a condition imposed by conservation of charge. The sum of the initial mass numbers is also equal to the sum of the final mass numbers, but the initial rest mass is *not* equal to the final rest mass.

The difference between the rest masses corresponds to the *nuclear reaction energy,* according to the mass–energy relation $E = mc^2$. If the sum of the final rest masses *exceeds* the sum of the initial rest masses, energy is *absorbed* in the reaction. Conversely, if the final sum is less than the initial sum, energy is released in the form of kinetic energy of the final particles. (1 u = 931 MeV.)

For example, in the nuclear reaction represented by Eq. (47-6), the rest masses of the various particles, in u, are found from Table 47-2 to be

$$
\begin{array}{ll}
_2He^4 = 4.00260\text{ u} & _8O^{17} = 16.99913\text{ u} \\
\underline{_7N^{14} = 14.00307\text{ u}} & \underline{_1H^1 = 1.00783\text{ u}} \\
\phantom{_7N^{14} = }18.00567\text{ u} & 18.00696\text{ u}
\end{array}
$$

(These values include nine electron masses in each case.) The total rest mass of the final products exceeds that of the initial particles by

0.00129 u, which is equivalent to 1.20 MeV. This amount of energy is *absorbed* in the reaction. If the initial particles do not have this much kinetic energy, the reaction cannot take place.

On the other hand, in the proton bombardment of lithium and consequent formation of two alpha particles,

$$_1H^1 + _3Li^7 \rightarrow _2He^4 + _2He^4, \tag{47-7}$$

the sum of the final masses is *smaller* than the sum of the initial values, as shown by the following data:

$_1H^1 = 1.00783$ u	$_2He^4 = 4.00260$ u
$_3Li^7 = 7.01601$ u	$_2He^4 = 4.00260$ u
8.02384 u	8.00520 u

(Four electron masses on each side are included.) Since the decrease in mass is 0.01864 u, 17.4 MeV of energy is liberated and appears as kinetic energy of the two separating alpha particles. This computation may be verified by observing the distance the alpha particles travel in air at atmospheric pressure before being brought to rest by collisions with molecules. The distance is found to be 8.31 cm. A series of independent experiments is then performed in which the range of alpha particles of known energy is measured. These experiments show that, in order to travel 8.31 cm, an alpha particle must have an initial kinetic energy of 8.64 MeV. The energy of the two alphas together is therefore $2 \times 8.64 = 17.28$ MeV, in excellent agreement with the value 17.4 MeV obtained from the mass decrease.

Alpha particles and protons are not the only particles used to initiate artificial nuclear disintegration. The nucleus of a deuterium atom, known as the *deuteron* and represented by the symbol $_1H^2$, may be accelerated to high energy in one of a variety of particle accelerators. In order that positively charged particles such as the alpha particle, the proton, and the deuteron can be used to penetrate the nuclei of other atoms, they must travel with very high speeds to avoid being repelled or deflected by the positive charge of the nucleus they are approaching.

Absorption of neutrons by nuclei is an important class of nuclear reactions. Heavy nuclei bombarded by neutrons in a nuclear reactor may undergo a series of neutron absorptions alternating with beta decays, in which the mass number A increases by as much as 25. The *transuranic* elements, having Z larger than 92, are produced in this way; they do not occur in nature. Thirteen transuranic elements, having Z up to 105 and A up to about 260, have been identified.

The analytical technique of *neutron activation analysis* uses similar reactions. When stable nuclei are bombarded by neutrons, some absorb neutrons and then undergo beta decay. The energies of the beta and gamma emissions depend on the parent nucleus and hence provide a means of identifying it. The presence of quantities of elements far too small for conventional chemical analysis can be detected in this way.

47-7 Particle accelerators

Many important experiments in nuclear and high-energy physics during the last 60 years or so have made use of beams of charged particles, such as protons or electrons, which have been accelerated to high speeds. Any device that uses electric and magnetic fields to guide and accelerate a beam of charged particles to high speed is called a *particle accelerator.* In a sense, the cathode-ray tubes of Thomson and his contemporaries were the first accelerators. In more recent times, accelerators have grown enormously in size, complexity, and energy range.

The *cyclotron,* developed in 1931 by Lawrence and Livingston at the University of California, is important historically because it was the first accelerator to use a magnetic field to guide particles in a nearly circular path so that they could be accelerated repeatedly by an electric field in a cyclic process. Lawrence was awarded the Nobel prize for physics in 1939 for his invention of the cyclotron.

The heart of the cyclotron is a pair of hollow metal chambers, labeled D_1 and D_2 in Fig. 47–8, resembling the halves of a hollow cylinder cut parallel to its axis. These are called *dees,* referring to their shape. A source of ions, often simply protons (ionized hydrogen atoms), is located near the midpoint of the gap between the dees, and the dees are connected to the terminals of a circuit generating an alternating voltage. The potential between the dees thus alternates rapidly, some millions of times per second; so the electric field in the gap between the dees is directed first toward one and then toward the other. But, because of the electrical shielding effect of the dees, the space *within* each is a region of nearly zero electric field.

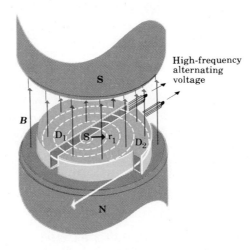

47–8   Schematic diagram of a cyclotron.

The two dees are enclosed within, but insulated from, a somewhat larger cylindrical metal container from which the air is exhausted, and the whole apparatus is placed between the poles of a powerful electromagnet, which provides a uniform magnetic field perpendicular to the ends of the cylindrical container.

Consider an ion of charge $+q$ and mass m, emitted from the ion source S at an instant when D_1 in Fig. 47–8 is positive. The ion is accelerated by the electric field in the gap between the dees and then enters the region within D_2, where $E = 0$, with a speed v_1. Since its motion is at right angles to the magnetic field, the ion travels in a circular arc of radius

$$r_1 = \frac{mv_1}{Bq}.$$

If, now, in the time required for the ion to complete a half-circle, the *electric* field has reversed so that its direction is toward D_1, the ion is again accelerated as it crosses the gap between the dees and enters D_1 with a greater speed v_2. It therefore moves in a half circle of larger radius within D_1 to emerge again into the gap.

The *angular* velocity ω of the ion is

$$\omega = \frac{v}{r} = B\left(\frac{q}{m}\right). \tag{47-8}$$

Hence the angular velocity is *independent of the speed of the ion and of the radius of the circle* in which it travels, depending only on the magnetic field and the charge-to-mass ratio (q/m) of the ion. If, therefore, the electric field reverses at regular intervals, each equal to the time required for the ion to make a half-revolution, the field in the gap will always be in the proper direction to accelerate an ion each time the gap is crossed. It is this feature of the motion, that the time of rotation is independent of the radius, that makes the cyclotron feasible, since the regularly timed reversals are accomplished automatically by the oscillator circuit to which the dees are connected.

The path of an ion is a sort of spiral, composed of semicircular arcs of progressively larger radius connected by short segments along which the radius is increasing. If R represents the outside radius of the dees, and v_{max} the speed of the ion when traveling in a path of this radius, then

$$v_{max} = BR\left(\frac{q}{m}\right),$$

and the corresponding kinetic energy of the ion is

$$K = \frac{1}{2}mv_{max}{}^2 = \frac{1}{2}m\left(\frac{q}{m}\right)^2 B^2 R^2. \tag{47-9}$$

If the ions are protons,

$$\frac{q}{m} = \frac{1.60 \times 10^{-19}\ \text{C}}{1.67 \times 10^{-27}\ \text{kg}} = 9.58 \times 10^7\ \text{C}\cdot\text{kg}^{-1}.$$

Taking values typical of early-day cyclotrons, we assume $B = 1.5$ T and $R = 0.5$ m. Then the maximum kinetic energy is, from Eq. (47-9),

$$K = \tfrac{1}{2}(1.67 \times 10^{-27}\ \text{kg})(9.58 \times 10^7\ \text{C}\cdot\text{kg}^{-1})^2(1.5\ \text{T})^2(0.5\ \text{m})^2$$
$$= 4.31 \times 10^{-12}\ \text{J} = 2.69 \times 10^7\ \text{eV} = 26.7\ \text{MeV}.$$

This energy, considerably larger than the average binding energy per nucleon, is sufficient to cause a variety of interesting nuclear reactions.

The energy attainable with the cyclotron is limited by relativistic effects. For Eq. (47-8) to be relativistically correct, m should be replaced by $m/(1 - v^2/c^2)^{1/2}$. As the particles speed up, the angular velocity *decreases;* if the decrease is appreciable, the particle motion is no longer in the correct phase relative to the alternating dee voltage. In a variation called the *synchrocyclotron,* the particles are accelerated in bursts, and for each burst the frequency of the alternating voltage is decreased at just the right rate to maintain the correct phase relation with the particles' motion. A practical limitation of the cyclotron is the expense of building very large electromagnets. The largest synchrocyclotron ever built has a vacuum chamber about 8 m in diameter and accelerates protons to energies of about 600 MeV.

To attain higher energies, another type of machine, called the *synchrotron,* is more practical. In a synchrotron, the vacuum chamber in which the particles move is in the form of a thin doughnut, called the *accelerating ring.* The particles are forced to move within this chamber by a series of magnets placed around it. As the particles speed up, the magnetic field is increased so that the particles retrace the same trajectory over and over. The synchrotron located at the Fermi National Accelerator Laboratory (Fermilab) in Batavia, Illinois, can accelerate protons to an energy of 500 GeV (500×10^9 eV), and modifications are underway which will permit a maximum energy of 1000 GeV. It uses an accelerating ring 2 km in diameter, costs about \$400 million to build, and has an operating budget of over \$50 million per year. In each machine cycle, of a few seconds' duration, it accelerates approximately 10^{13} protons.

As higher and higher energies are sought in the attempt to investigate new phenomena in particle interactions, a new problem emerges. In an experiment where a beam of high-energy particles collides with a stationary target (such as protons onto a liquid-hydrogen target), not all the kinetic energy of the incident particles is available to form new particle states. Because momentum must be conserved, the particles emerging from the collision must have some motion and thus some kinetic energy. The energy E_a available for creating new particles or particle configurations is the *difference* between initial and final kinetic energies.

In the extreme-relativistic range, where the kinetic energies of the particles are large compared to their rest energies, this is a severe limitation. When bombarding and target particles have equal mass, as with protons onto a hydrogen target, it can be shown from relativistic mechanics that the available energy E_a is related to the total energy E of the bombarding particle and to its mass m by

$$E_a = \sqrt{2mc^2E}. \tag{47-10}$$

For example, for the proton, $mc^2 = 931$ MeV $= 0.931$ GeV. If $E = 500$ GeV, as for the Fermilab accelerator, then

$$E_a = \sqrt{2(0.931 \text{ GeV})(500 \text{ GeV})} = 30.5 \text{ GeV},$$

and if $E = 1000$ GeV, $E_a = 43.1$ GeV. Thus, increasing the beam energy by 500 GeV increases the available energy by only 12.6 GeV.

This limitation may be circumvented in part by *colliding-beam* experiments, in which there is no stationary target but in which beams of particles circulate in opposite directions in arrangements called *storage rings.* In regions where the rings intersect, the beams are focused sharply onto one another, and collisions can occur. Because the total momentum in such a two-particle collision is zero, the available energy E_a is the total kinetic energy of the two particles.

An example is the storage ring facility at the Stanford Linear Accelerator Center (SLAC), where electron and positron beams collide with total available energy of up to 36 GeV. Some other storage-ring facilities are located at DESY in Hamburg, West Germany (E_a up to 38 GeV in e^+e^- collisions), and at the Cornell Electron Storage Ring Facility (CESR), where the maximum energy *per beam* is 8 GeV. At Brookhaven

National Laboratory there is under construction a machine called ISA-BELLE that is designed to have 400 GeV per beam, in a proton–proton storage-ring facility with high intensity in each beam.

47-8 Nuclear fission

Up to this point, all nuclear reactions considered have involved the ejection of relatively light particles, such as alpha particles, beta particles, protons, or neutrons. That this is not always the case was discovered by Hahn and Strassman in Germany in 1939. These scientists bombarded uranium ($Z = 92$) with neutrons, and after a careful chemical analysis discovered barium ($Z = 56$) and krypton ($Z = 36$) among the products. Cloud-chamber photographs showed the two heavy particles traveling in opposite directions with tremendous speed. The uranium nucleus is said to undergo *fission*. Measurement showed that an enormous amount of energy, 200 MeV, is released when uranium splits up in this way. The rest mass of a uranium atom exceeds the sum of the rest masses of the fission products. Thus it follows, from the Einstein mass–energy relation, that the extra energy released during fission is transformed into kinetic energy of the fission fragments. Uranium fission may be accomplished by either fast or slow neutrons. Of the two most abundant isotopes of uranium, $_{92}U^{238}$ and $_{92}U^{235}$, both may be split by a fast neutron, whereas only $_{92}U^{235}$ is split by a slow neutron.

When uranium undergoes fission, barium and krypton are not the only products. Over 100 different isotopes of more than 20 different elements have been detected among fission products. All of these atoms are, however, in the middle of the periodic table, with atomic numbers ranging from 34 to 58. Because the neutron–proton ratio needed for stability in this range is much *smaller* than that of the original uranium nucleus, the *fission fragments,* as the residual nuclei are called, always have too many neutrons for stability. A few free neutrons are liberated during fission, and the fission fragments undergo a series of beta decays (each of which increases Z by one and decreases N by one) until a stable nucleus is reached. During decay of the fission fragments, an average of 15 MeV of additional energy is liberated.

Discovery of the facts that 200 MeV of energy is released when uranium undergoes fission, and that other neutrons are liberated from the uranium nucleus during fission, suggested the possibility of a *chain reaction,* that is, a self-sustaining series of events which, once started, will continue until much of the uranium in a given sample is used up (provided the sample stays together). In the case of a uranium chain reaction, a neutron causes one uranium atom to undergo fission, during which a large amount of energy and several neutrons are emitted. These neutrons then cause fission in neighboring uranium nuclei, which also give out energy and more neutrons. The chain reaction may be made to proceed slowly and in a controlled manner, and the device for accomplishing this is called a *nuclear reactor.* If the chain reaction is fast and uncontrolled, the device is a bomb.

In a nuclear reactor, the fissionable element is contained in *fuel elements,* whose configuration must be designed so that sufficient neutrons are slowed down in the surrounding material (often water) and cause further fissions, rather than escaping from the fuel region. On the average each fission produces about 2.5 free neutrons, so 40% of the neutrons are needed to sustain a chain reaction. The *rate* of the reaction is controlled by inserting or withdrawing *control rods* made of elements (often cadmium) whose nuclei *absorb* neutrons without undergoing any additional chain reaction.

The most familiar application of nuclear reactors is for the generation of electric power. To illustrate some of the numbers involved, we consider a hypothetical nuclear power plant with a generating capacity of 1000 MW; this figure is typical of large plants currently being built. As noted above, the fission energy appears as kinetic energy of the fission fragments, and its immediate result is to heat the fuel elements and the surrounding water. This heat generates steam to drive turbines, which in turn drive the electrical generators. The turbines, being heat engines, are subject to the efficiency limitations imposed by the second law of thermodynamics, as discussed in Chapter 19. In modern nuclear plants the overall efficiency is about one-third, so 3000 MW of thermal power from the fission reaction are needed for 1000 MW of electrical power.

It is easy to calculate how much uranium must undergo fission per unit time to provide 3000 MW of thermal power. Each second we need 3000 MJ or 3000×10^6 J. Each fission provides 200 MeV, which is

$$200 \text{ MeV} = (200 \text{ MeV})(1.6 \times 10^{-13} \text{ J} \cdot \text{MeV}^{-1}) = 3.2 \times 10^{-11} \text{ J}.$$

Thus the number of fissions needed per second is

$$\frac{3000 \times 10^6 \text{ J}}{3.2 \times 10^{-11} \text{ J}} = 0.94 \times 10^{20}.$$

Each uranium atom has a mass of about $(235)(1.67 \times 10^{-27} \text{ kg}) = 3.9 \times 10^{-25}$ kg, so the mass of uranium needed per second is

$$(0.94 \times 10^{20})(3.9 \times 10^{-25} \text{ kg}) = 3.7 \times 10^{-5} \text{ kg} = 37 \text{ mg}.$$

In one day (86,400 seconds) the consumption of uranium is

$$(3.7 \times 10^{-5} \text{ kg} \cdot \text{s}^{-1})(86{,}400 \text{ s} \cdot \text{d}^{-1}) = 3.2 \text{ kg} \cdot \text{d}^{-1}.$$

For comparison, we note that the 1000-MW coal-fired power plant discussed in Sec. 19–11 burns 10,600 tons of coal per day!

Nuclear fission reactors have several other practical uses; among these are the production of artificial radioactive isotopes for medical and other research; production of high-intensity neutron beams for research in nuclear structure, and production of fissionable transuranic elements such as plutonium from the common isotope U^{238}. The last is the function of so-called *breeder reactors.*

As noted above, about 15 MeV of the energy liberated as a result of fission of a U^{235} nucleus comes from the subsequent beta decay of the fission fragments rather than the kinetic energy of the fragments themselves. This fact poses a serious problem with respect to control and safety of reactors. Even after the chain fission reaction has been completely stopped by insertion of control rods into the core, heat continues to be evolved by the beta decays, which cannot be stopped. For the 1000-MW reactor described above, this heat power amounts to about 200 MW, which, in the event of total loss of cooling water, is more than enough to cause a catastrophic "melt-down" of the reactor core and possible penetration of the containment vessel. The difficulty in achieving a "cold shut-down" following the accident at Three Mile Island in March 1979 was due to the continued evolution of heat due to beta decays.

Fission appears to set an upper limit on the production of transuranic nuclei, discussed in Sec. 47–6. When a nucleus with $Z = 105$ is bombarded with neutrons, fission occurs essentially instantaneously; no $Z = 106$ nucleus is formed even for a short time. There are theoretical reasons to expect that nuclei in the vicinity of $Z = 114$, $N = 184$, might be stable with respect to spontaneous fission. These numbers correspond to *filled shells* in the nuclear energy-level structure, analogous to the filled shells of electrons in the inert gases, as discussed in Sec. 46–2. Such nuclei, called *superheavy nuclei,* would still be unstable with respect to alpha emission, but they might live long enough to be identified. Attempts to produce superheavy nuclei in the laboratory have not been successful; whether or not they exist in nature is still an open question.

47-9 Nuclear fusion

There are two types of nuclear reactions in which large amounts of energy may be liberated. In both types, the rest mass of the products is less than the original rest mass. The fission of uranium, already described, is an example of one type. The other involves the combination of two light nuclei to form a nucleus that is more complex but whose rest mass is less than the sum of the rest masses of the original nuclei. Examples of such energy-liberating reactions are as follows:

$$_1H^1 + {}_1H^1 \rightarrow {}_1H^2 + {}_1e^0,$$

$$_1H^2 + {}_1H^1 \rightarrow {}_2He^3 + \gamma\text{-radiation},$$

$$_2He^3 + {}_2He^3 \rightarrow {}_2He^4 + {}_1H^1 + {}_1H^1.$$

In the first, two protons combine to form a deuteron and a positron (a positively charged electron, to be discussed in Sec. 47–10). In the second, a proton and a deuteron unite to form the light isotope of helium. For the third reaction to occur, the first two reactions must occur twice, in which case two nuclei of light helium unite to form ordinary helium. These reactions, known as the *proton–proton chain,* are believed to take place in the interior of the sun and also in many other stars that are known to be composed mainly of hydrogen.

The positrons produced during the first step of the proton–proton chain collide with electrons; annihilation takes place, and their energy is converted into gamma radiation. The net effect of the chain, therefore, is the combination of four hydrogen nuclei into a helium nucleus and gamma radiation. The net amount of energy released may be calculated from the mass balance as follows:

Mass of four hydrogen atoms (including electrons)	= 4.03132 u
Mass of one helium plus two additional electrons	= 4.00370 u
Difference in mass	= 0.02762 u
	= 25.7 MeV

In the case of the sun, 1 g of its mass contains about 2×10^{23} protons. Hence, if all of these protons were fused into helium, the energy released would be about 57,000 kWh. If the sun were to continue to radiate at its present rate, it would take about 30 billion years to exhaust its supply of protons.

For fusion to occur, the two nuclei must come together to within the range of the nuclear force, typically of the order of 2×10^{-15} m. To do this they must overcome the electrical repulsion of their positive charges; for two protons at this distance the corresponding potential energy is of the order of 1.1×10^{-13} J or 0.7 MeV, which thus represents the initial *kinetic* energy the fusing nuclei must have.

Such energies are available at extremely high temperatures. According to Sec. 20–4, the average translational kinetic energy of a gas molecule at temperature T is $3kT/2$, where k is Boltzmann's constant. For this to be equal to 1.1×10^{-13} J, the temperature must be of the order of 5×10^9 K. Of course, not all the nuclei have to have this energy, but this calculation shows that the temperature must be of the order of millions of kelvins if any appreciable fraction of the nuclei are to have enough kinetic energy to surmount the electrical repulsion and achieve fusion.

Such temperatures occur in stars as a result of gravitational contraction and its associated liberation of gravitational potential energy. When the temperature gets high enough, the reactions occur, more energy is liberated, and the pressure of the resulting radiation prevents further contraction. Only after most of the hydrogen has been converted into helium will further contraction and an accompanying increase of temperature result. Conditions are then suitable for the formation of heavier elements.

Temperatures and pressures similar to those in the interior of stars may be achieved on earth at the moment of explosion of a uranium or plutonium fission bomb. If the fission bomb is surrounded by proper proportions of the hydrogen isotopes, these may be caused to combine into helium and liberate still more energy. This combination of uranium and hydrogen is called a "hydrogen bomb."

Intensive efforts are underway in many laboratories to achieve controlled fusion reactions, which potentially represent an enormous new energy resource. In one kind of experiment, a plasma is heated to extremely high temperature by an electrical discharge, while being contained by appropriately shaped magnetic fields. In another, pellets of the material to be fused are heated by a high-intensity laser beam. Reactions being studied include the following:

$$\mathrm{_1H^2 + \ _1H^2 \rightarrow \ _1H^3 + \ _1H^1 + 4 \ MeV,} \tag{1}$$

$$\mathrm{_1H^3 + \ _1H^2 \rightarrow \ _2He^4 + \ _0n^1 + 17.6 \ MeV,} \tag{2}$$

$$\mathrm{_1H^2 + \ _1H^2 \rightarrow \ _2He^3 + \ _0n^1 + 3.3 \ MeV,} \tag{3}$$

$$\mathrm{_2He^3 + \ _1H^2 \rightarrow \ _2He^4 + \ _1H^1 + 18.3 \ MeV.} \tag{4}$$

In the first, two deuterons combine to form tritium and a proton. In the second, the tritium nucleus combines with another deuteron to form helium and a neutron. The result of both of these reactions together is the conversion of three deuterons into a helium-4 nucleus, a proton, and a neutron, with the liberation of 21.6 MeV of energy. Reactions (3) and (4) together achieve the same conversion. In a plasma containing deuterium, the two pairs of reactions occur with roughly equal probability. As yet no one has succeeded in producing these reactions under controlled conditions in such a way as to yield a net surplus of usable energy, but the practical problems do not appear to be insurmountable.

47-10 Fundamental particles

The physics of fundamental particles has been a recognized field of research only in the past 40 years. The electron and the proton were known by the turn of the century, but the existence of the neutron was not established definitely until 1930; its discovery is an interesting story and a useful illustration of nuclear reactions.

In 1930, Bothe and Becker in Germany observed that when beryllium, boron, or lithium was bombarded by fast alpha particles, the bombarded material emitted something, either particles or electromagnetic waves, of much greater penetrating power than the original alpha particles. Further experiments in 1932 by Curie and Joliot in Paris confirmed these results, but all attempts to explain them in terms of gamma rays were unsuccessful. Chadwick in England repeated the experiments and found that they could be satisfactorily interpreted on the assumption that *uncharged* particles of mass approximately equal to that of the proton were emitted from the nuclei of the bombarded material. He called the particles *neutrons*. The emission of a neutron from a beryllium nucleus takes place according to the reaction

$$_2\text{He}^4 + {_4}\text{Be}^9 \rightarrow {_6}\text{C}^{12} + {_0}\text{n}^1, \tag{47-10}$$

where $_0\text{n}^1$ is the symbol for a neutron.

Since neutrons have no charge, they produce no ionization in their passage through gases. They are not deflected by the electric field around a nucleus and can be stopped only by colliding with a nucleus in a direct hit, in which case they may either undergo an elastic impact or penetrate the nucleus. It was shown in Chapter 7 that, if an elastic body strikes a motionless elastic body of the same mass, the first is stopped and the second moves off with the same speed as the first. Since the proton and neutron masses are almost the same, fast neutrons are slowed down most effectively by collisions with the hydrogen atoms in hydrogenous materials such as water or paraffin. The usual laboratory method of obtaining slow neutrons is to surround the fast neutron source with water or blocks of paraffin.

Once the neutrons are moving slowly, they may be detected by means of the alpha particles they eject from the nucleus of a boron atom, according to the reaction

$$_0\text{n}^1 + {_5}\text{B}^{10} \rightarrow {_3}\text{Li}^7 + {_2}\text{He}^4. \tag{47-11}$$

The ejected alpha particle then produces ionization that may be detected in a Geiger counter or an ionization chamber.

The discovery of the neutron gave the first real clue to the structure of the nucleus. Before 1930 it had been thought that the total mass of a nucleus was due to protons only. We now know that a nucleus consists of both protons and neutrons (except hydrogen, whose nucleus consists only of one proton) and that (1) the mass number A equals the total number of nuclear particles and (2) the atomic number Z equals the number of protons.

The study of *cosmic rays* has been a very fertile field of particle physics. It has been known since the early years of this century that air and other gases are slightly ionized, and therefore slightly conductive, at all times, even in the absence of obvious causes of ionization such as x-rays, ultraviolet light, or radioactivity. For example, if a charged electroscope is left standing, it will eventually lose its charge no matter how well it is insulated. Ionization of air inside a vessel is decreased slightly if the vessel is lowered into a lake, but increases considerably if the vessel is transported in a balloon high into the stratosphere. Hess suggested that the ionization is due to some kind of penetrating waves or particles from outer space, and called them *cosmic rays.*

It is fairly certain that cosmic rays consist largely of high-speed protons with energies of the order of billions of electron volts. A collision between such a proton and the nucleus of a nitrogen or oxygen atom in the upper atmosphere gives rise to so many interesting secondary phenomena that the study of cosmic rays has become one of the richest sources of knowledge of the behavior and properties of fundamental particles.

An instrument called the *bubble chamber* has been used extensively not only to study cosmic rays but also to render visible the paths of particles involved in nuclear and fundamental-particle reactions; this instrument was described in Sec. 17-9. Bubble chambers are sometimes made several feet in diameter; particle tracks are photographed with two or three cameras to obtain stereoscopic pictures. Normally the chamber is placed in a magnetic field so that a charged particle will be deflected; by measuring the radius of curvature of the path we may determine the particle's momentum. The *cloud chamber,* also described in Sec. 17-9, was an important historical predecessor of the bubble chamber.

The positive electron or *positron* was first observed during the course of an investigation of cosmic rays by Dr. Carl D. Anderson in 1932, in a historic cloud-chamber photograph reproduced in Fig. 47-9. The photograph was made with the cloud chamber in a magnetic field perpendicular to the plane of the paper. A lead plate crosses the chamber and evidently the particle has passed through it. Since the curvature of the track is greater above the plate than below it, the velocity is less above than below; the inference is that the particle was moving upward, since it could not have *gained* energy going through the lead.

The *density* of droplets along the path is the same as would be expected if the particle were an electron. But the direction of the magnetic field and the direction of motion are consistent only with a particle of *positive* charge. Hence Anderson concluded that the track had been made by a positive electron or *positron.* Since the time of this discovery, the positron's existence has been definitely established. Its mass equals that of a negative electron and its charge is equal in magnitude but of opposite sign to that of the electron.

Positrons have only a transitory existence and do not form a part of ordinary matter. They are produced in high-energy collisions of charged particles or gamma photons with matter in a process called *pair production,* in which an ordinary electron and a positron are produced simultaneously. Electric charge is conserved in this process, but enough energy must be available to account for the energy equivalent of the rest masses

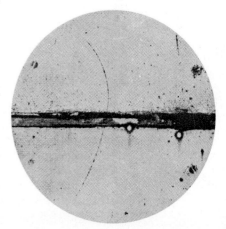

47-9 Track of a positive electron traversing a lead plate 6 mm thick. (Photograph by C. D. Anderson.)

of the two particles, about 0.5 MeV each. The inverse process, e^+e^- *anni-hilation,* occurs when a positron collides with an ordinary electron. Both particles disappear and two or three gamma photons appear, with total energy $2mc^2$, where m is the electron rest mass. Decay into a *single* gamma photon is impossible because such a process cannot possibly conserve both energy and momentum.

Positrons also occur in the decay of some unstable nuclei. We recall that nuclei having too many neutrons for stability often emit a beta particle (electron), decreasing N by one and increasing Z by one. Similarly, a nucleus having *too few neutrons* for stability may respond by emitting a positron, increasing N by one and decreasing Z by one. Such nuclides do not occur in nature, but they can be produced artificially by neutron bombardment of stable nuclides in nuclear reactors. An example is the unstable nuclide $_{11}Na^{22}$, which has one less neutron than the stable $_{11}Na^{23}$. It emits a positron, leaving the stable nuclide $_{10}Ne^{22}$.

In 1935, the Japanese physicist Hideki Yukawa inferred, from theoretical considerations, the existence of a particle of mass intermediate between that of the electron and the proton. A particle of intermediate mass, but *not* identical with that predicted by Yukawa, was discovered one year later by Anderson and Neddermeyer as a component of cosmic radiation. This particle is now known as a μ *meson* (or *muon*). The μ^- has charge equal to that of the electron, and its antiparticle the μ^+ has a positive charge of equal magnitude. The two particles have equal mass, about 207 times the electron mass. The muons are unstable; each decays into an electron of the same sign, plus two neutrinos, with a half-life of about 2.3×10^{-6} s.

Yukawa first proposed the mesons as a basis for nuclear forces; he suggested that nucleons could interact by emitting and absorbing unstable particles, just as two basketball players interact by tossing the ball back and forth, or by snatching it away from each other. It was established soon after the discovery of the muons that they could not be Yukawa's particles because their interactions with nuclei were far too weak. But in 1947 another family of mesons was discovered; called π *mesons* or *pions,* these can be positive, negative, and neutral. The charged pions have masses of about 273 times the electron mass and decay into muons with the same sign, plus a neutrino, with a half-life of about 2.6×10^{-8} s. The neutral pion has a smaller mass, about 264 electron masses, and decays, with a very short half-life of about 0.8×10^{-16} s, into two gamma-ray photons.

In the years since 1947, *high-energy physics* has emerged as a distinct branch of physics. These years have witnessed the attainment of higher and higher energies in particle accelerators, the discovery of a whole array of new particles, and intensive efforts to understand the properties of these new particles and their interactions.

*47-11 High-energy physics

It was recognized, even in the early years of high-energy physics, that the fundamental particles are not *permanent* entities but can be created and destroyed in interactions with other particles. The earliest such interaction to be observed was that of creation and destruction of electron–positron pairs. Such pairs are *created* in collisions of high-energy cosmic-ray particles with stationary targets; when an electron and a positron collide, both *disappear* and two or three gamma-ray photons are created to carry away the energy. This transitory nature of the fundamental particles may seem disturbing, but in one sense it is a welcome development. We have seen that photons and electrons (and indeed all particles) share the dual wave-particle nature discussed in Sec. 45-2, and photons are known to be created and destroyed (or emitted and absorbed) in atomic transitions. Thus it seems natural that other particles can also be created and destroyed.

As an example, it was speculated as early as 1932 that there might be an *antiproton*, bearing the same relation to the ordinary proton as the positron does to the electron, that is, a particle with the same mass as the proton but negatively charged. Finally in 1955 proton–antiproton pairs were created by impact on a stationary target of a beam of protons with kinetic energy 6 GeV (6×10^9 eV) from the Bevatron at the University of California at Berkeley.

In the years after 1960, as higher-energy accelerators and more sophisticated detectors were developed, a veritable blizzard of new unstable particles occurred. Initially they were classified according to *mass*. The electrons, muons, and their associated neutrinos were called *leptons;* particles with masses between those of muons and nucleons were called *mesons,* and the nucleons and more massive particles were called *baryons.* Particles were further classified according to electric charge, spin, and two additional quantum numbers, *isospin* (the number determining the number of different charges for a particular type of particle) and *strangeness* (a number needed to account for the production and decay modes of certain particles). A partial list of some of the particles known around 1965 is shown in Table 47–3.

During this period it became clear that particles could also be classified in terms of the types of *interactions* in which they participate, and in terms of the conservation laws associated with these interactions. There appear to be four classes of interactions; in order of decreasing strength, they are

1. Strong interactions;
2. Electromagnetic interactions;
3. Weak interactions;
4. Gravitational interactions.

Particles which experience strong interactions are called *hadrons;* these include all the mesons and baryons in Table 47–3. The strong interactions are responsible for the emission and absorption of pions and heavy mesons by nucleons, and hence for the nuclear force. They are also responsible for the *creation* of pions, heavy mesons, and hyperons in high-energy collisions. Electrons, muons, and neutrinos, collectively called *leptons,* have *no* strong interactions.

The *electromagnetic* interactions are those associated directly with electric charge; as noted previously, the electromagnetic interaction between two protons is weaker at distances of the order of nuclear dimensions than the strong interaction, but it has longer range. Neutral particles have no electromagnetic interactions, with the exception of effects due to the magnetic moments of neutral baryons; these magnetic mo-

Table 47-3   Some particles known in 1965

	Particle	Mass (MeV)	Charge	Spin	Isospin	Strange- ness	Lifetime (s)	Typical decay modes	Quark content
leptons	e^-	0.511	-1	$\frac{1}{2}$	—	0	stable	—	—
	ν_e	0	0	$\frac{1}{2}$	—	0	stable	—	—
	μ^-	105.7	-1	$\frac{1}{2}$	—	0	2.2×10^{-6}	$e^- \bar{\nu}_e \nu_\mu$	—
	ν_μ	0	0	$\frac{1}{2}$	—	0	stable	—	—
	τ^-	1784	-1	$\frac{1}{2}$	—	0	$< 2.3 \times 10^{-12}$	$\mu^- \bar{\nu}_\mu \nu_\tau$	—
	ν_τ	0	0	$\frac{1}{2}$	—	0	stable	—	—
mesons	π^0	135.0	0	0	1	0	0.83×10^{-16}	$\gamma\gamma$	$(u\bar{u} - d\bar{d})/\sqrt{2}$
	π^+	139.6	$+1$	0	1	0	2.6×10^{-8}	$\mu^+ \nu_\mu$	$u\bar{d}$
	π^-	139.6	-1	0	1	0	2.6×10^{-8}	$\mu^- \bar{\nu}_\mu$	$\bar{u}d$
	K^+	493.7	$+1$	0	$\frac{1}{2}$	$+1$	1.24×10^{-8}	$\mu^+ \nu_\mu$	$u\bar{s}$
	K^-	493.67	-1	0	$\frac{1}{2}$	-1	1.24×10^{-8}	$\mu^- \bar{\nu}_\mu$	$\bar{u}s$
	η^0	548.8	0	0	0	0	$\sim 10^{-18}$	$\gamma\gamma$	$(u\bar{u} + d\bar{d} - 2s\bar{s})/\sqrt{6}$
baryons	p	938.3	$+1$	$\frac{1}{2}$	$\frac{1}{2}$	0	stable	—	uud
	n	939.6	0	$\frac{1}{2}$	$\frac{1}{2}$	0	917	$pe^- \bar{\nu}_e$	udd
	Λ	1115	0	$\frac{1}{2}$	0	-1	2.63×10^{-10}	$p\pi^-$ or $n\pi^0$	uds
	Σ^+	1189	$+1$	$\frac{1}{2}$	1	-1	0.80×10^{-10}	$p\pi^0$ or $n\pi^+$	uus
	Δ^{++}	1232	$+2$	$\frac{3}{2}$	$\frac{3}{2}$	0	$\sim 10^{-23}$	$p\pi^+$	uuu
	Ξ^-	1321	-1	$\frac{1}{2}$	$\frac{1}{2}$	-2	1.64×10^{-10}	$\Lambda\pi^-$	dss
	Ω^-	1672	-1	$\frac{3}{2}$	0	-3	0.82×10^{-10}	ΛK^-	sss
	Λ_c^+	2273	1	$\frac{1}{2}$	0	0	$\sim 7 \times 10^{-13}$	$\Lambda\pi\pi$	udc

ments are believed to be associated with the emission and absorption of charged pions and heavy mesons.

The *weak* interaction is responsible for beta decay such as the conversion of a neutron into a proton, an electron, and a neutrino. It is also responsible for the *decay* of many unstable particles (pions into muons, muons into electrons, Λ particles into protons, and so on). The *gravitational* interaction, although of central importance for the large-scale structure of celestial bodies, is not believed to be of significance in the analysis of fundamental-particle interactions. For example, the gravitational attraction of two electrons is smaller than their electrical repulsion by a factor of about 2.4×10^{-43}.

Several conservation laws are believed to be obeyed by *all* of the above interactions. These include the laws growing out of classical physics: energy, momentum, angular momentum, and electric charge. In addition, several new quantities having no classical analog have been introduced to help characterize the properties of particles. These include *baryon number* (the number of baryons minus the number of antibaryons), *isospin* (used also to describe the charge-independence of nuclear forces), *parity* (the comparative behavior of two systems that are mirror images of each other), and *strangeness* (a quantum number used to classify particle production and decay reactions). Baryon number is conserved in *all* interactions; isospin is conserved in strong but not in electromagnetic or weak interactions. Parity and strangeness are conserved in strong and electromagnetic but not in weak interactions. Thus, the new conservation laws are not absolute but, instead, serve as a means for *classifying* interactions.

The large number of supposedly fundamental particles discovered in the last twenty years (well over a hundred) suggests strongly that these particles *do not* represent the most fundamental level of the structure of matter, but that there is at least one additional level of structure. There is now fairly general agreement among physicists concerning the nature of this level; the theory is based on a proposal made initially in 1964 by Gell-Mann and his collaborators. We cannot discuss this theory in detail, but the following is a very brief sketch of some of its features.

Leptons are indeed fundamental particles. In addition to the electrons and muons and their associated neutrinos, there is a third kind of lepton called the τ, having spin 1/2 and mass of about 1780 MeV. Its associated neutrino is the τ-neutrino, ν_τ. These six leptons are definitely established; there may be others, yet undiscovered.

Hadrons are *not* fundamental particles but are composite structures whose constituents are truly fundamental particles called *quarks*. In the original quark picture there were three quarks labeled *u*, *d*, and *s* ("up," "down," and "strange"), with three corresponding antiquarks $\bar{u}$, $\bar{d}$, and $\bar{s}$. Some properties of these quarks, and of three others to be discussed later, are shown in Table 47-4. We note that the charge and baryon number of each quark are fractions, unlike any particle observed in nature.

Each meson is a quark–antiquark pair; for example, the π^+ is $u\bar{d}$, and the K^+ is $u\bar{s}$. Each baryon is a combination of three quarks; the proton is *uud*, the neutron is *udd*, the Σ^+ is *uus*, and so on. The quark structure of each of the hadrons listed in Table 47-3 is shown in the last column of that table. Leptons, on the other hand, are *not* made of quarks, and no quark structure is shown for them.

Quarks are expected to be very massive, although their masses are not known at present. They are expected to have a very strong affinity for each other, so individual quarks should be seen at most rarely. As noted at the end of Sec. 26-6, there has been at least one report of observation of a fractionally charged particle. On the other hand, many theo-

Table 47-4   **Properties of quarks**

Quark	Charge	Baryon number	Strangeness	Isospin	Charm
u	$\frac{2}{3}$	$\frac{1}{3}$	0	$\frac{1}{2}$	0
d	$-\frac{1}{3}$	$\frac{1}{3}$	0	$\frac{1}{2}$	0
s	$-\frac{1}{3}$	$\frac{1}{3}$	-1	0	0
c	$\frac{2}{3}$	$\frac{1}{3}$	0	0	1
b	$-\frac{1}{3}$	$\frac{1}{3}$	0	0	0
t	$\frac{2}{3}$	$\frac{1}{3}$	0	0	0

For each antiquark, the values of charge, baryon number, strangeness, and charm have the same magnitude but opposite sign, compared to the corresponding quark. States involving the *t* quark have not yet been observed.

rists believe that phenomena associated with creation of quark–antiquark pairs make it impossible *ever* to observe an individual quark. Thus the direct observation of a single quark is still an open question.

In the quark model the strong interaction between quarks is mediated by particles called *gluons,* which play a role in the strong interaction analogous to that of *photons* for the electromagnetic interaction. The weak and gravitational interactions also have corresponding mediating particles, called the *weak boson* and the *graviton,* respectively.

Although three quarks sufficed to describe the particles known in 1970, more recent discoveries require the existence of additional quarks. It is now believed that there are six quarks, each existing in three varieties or *colors.* The three additional quarks are called the *charmed* quark c and the *top* t and *bottom* b (or *truth* and *beauty*) quarks. The existence of c was discovered in 1974 in experiments at Brookhaven and at SLAC, where the Ψ particle (3095 MeV) and some of its excited states were observed. These are believed to be states of the system $c\bar{c}$. Subsequently, charmed mesons such as D^+ (a $c\bar{d}$ state) have been observed through their decay products. More recently, evidence has been found at Fermilab for the existence of $b\bar{b}$ states. States involving t have not yet been observed.

The fact that all known particles can be represented as either leptons or combinations of these few quarks is indeed reason to believe that they are truly fundamental particles. Experimental and theoretical work is proceeding in many different directions. In a theoretical scheme called the Weinberg–Salam model, the weak and electromagnetic interactions are represented as two aspects of a single fundamental interaction. Some theorists are searching for a "grand unification scheme" that would unify *all* of the four basic interactions. Although such schemes are still speculative in nature, the entire area is a very active field of present-day theoretical and experimental research.

47–12 Radiation and the life sciences

The interaction of radiation with living organisms is a topic that grows daily in interest and usefulness. In this context we construe *radiation* to include radiation emitted as a result of nuclear instability (alpha, beta, gamma, and neutrons) as well as electromagnetic radiation such as microwaves and x-rays. Space permits only brief mention of a few examples; the two general classes of phenomena to be considered here are: (1) the use of radioactive isotopes as an analytical tool, and (2) the beneficial and harmful effects of radiation in changing living tissue.

Radioactive isotopes of an element have the same electron configuration as the stable isotopes and therefore exhibit the same chemical behavior. The location and concentration of radioactive isotopes can, however, be detected easily, even at a distance, by measuring the radiation emitted. For example, an unstable isotope of iodine, I^{131}, can be used to study thyroid function. It is known that nearly all the iodine in food that is not eliminated eventually reaches the thyroid; by feeding the patient minute, measured quantities of I^{131} and subsequently measuring the radiation from the thyroid, one can measure the activity of this organ.

More subtle applications occur in complex chemical reactions. By use of radioactive tracers one can "tag" specific parts of molecules and "follow" the radioactive atoms through complex reactions.

A different analytical technique, neutron activation analysis, makes use of the fact that when stable nuclei are bombarded with neutrons, some nuclei absorb neutrons, becoming radioactive beta-emitters. Each element has characteristic energies for the gamma radiation that usually follows beta emission, and measurement of these energies permits determination of the original elements present, even if the quantities are extremely minute.

The interactions of radiation with living tissue are extremely complex. It has been known for many years that excessive exposure to radiation, including sunlight, x-rays, and all the nuclear radiations, can cause destruction of tissues. In mild cases this destruction is manifested as a burn, as with common sunburn; greater exposure can cause very severe illness or death by a variety of mechanisms, one of which is the destruction of the components in bone marrow that produce red blood cells.

On the other hand, radiation is present everywhere from sunlight, cosmic rays, and natural radioactivity, and some exposure to radiation is unavoidable. Exactly what constitutes a *safe* level of radiation exposure is open to considerable question, but the available evidence has been interpreted to show that exposure to the extent of 10 to 100 times that from natural sources is very rarely harmful.

There has been a great deal of hysteria concerning alleged radiation hazards from nuclear power plants. It is certainly true that the radiation level from these plants is *not* zero. But to make a meaningful evaluation of hazards one must compare these levels with the alternatives, such as coal-powered plants. The health hazards of coal smoke are serious and well documented, and even the radioactivity in the smoke from a coal-fired power plant is believed to be greater than that from a properly operating nuclear plant of similar power capacity. It is clearly impossible to eliminate *all* hazards to health, and the next best alternative is an intelligent approach to the problem of *minimizing* hazards.

Radiation also has many *beneficial* effects; the great usefulness of x-rays in medical diagnosis is well known, and there is no doubt that, when x-rays are used properly, the benefits from this diagnostic usefulness almost always outweigh the small radiation hazard. Higher-energy radiation is used for intentional selective destruction of tissue, such as cancerous tumors. The hazards are considerable, but if the disease would be fatal without treatment, considerable hazard may be tolerable.

For sources of radiation for the treatment of cancers and related diseases, artifically produced isotopes are often used. One of the most commonly used is an isotope of cobalt, Co^{60}. This is prepared by bombarding the stable isotope Co^{59} with neutrons in a nuclear reactor. Neutron absorption leads to the unstable Co^{60}; with $Z = 27$ and $N = 33$, this is an "odd–odd" nucleus, which decays to Ni^{60} by beta and gamma emission, with a half-life of about 5 years. Such artificial sources have several advantages over naturally radioactive sources. Having shorter half-lives, they are more *intense* sources. They do not emit alpha particles, which are usually not wanted, and the electrons emitted are easily stopped by thin metal sheets without appreciably attenuating the intensity of the desired gamma radiation.

Questions

47-1 How can one be sure that nuclei are not made of protons and electrons, rather than protons and neutrons?

47-2 In calculations of nuclear binding energies such as those in the examples of Sec. 47-2 and 47-6, should the binding energies of the *electrons* in the atoms be included?

47-3 If different isotopes of the same element have the same chemical behavior, how can they be separated?

47-4 In beta decay a neutron becomes a proton, an electron, and a neutrino. This decay also occurs with free neutrons, with a half-life of about 15 minutes. Could a free *proton* undergo a similar decay?

47-5 Since lead is a stable element, why doesn't the U^{238} decay series shown in Fig. 47-7 stop at lead, Pb^{214}?

47-6 In the U^{238} decay chain shown in Fig. 47-7, some nuclides in the chain are found much more abundantly in nature than others, despite the fact that every U^{238} nucleus goes through every step in the chain before finally becoming Pb^{206}. Why are the abundances of the intermediate nuclides not all the same?

47-7 Radium has a half-life of about 1600 years. If the universe was formed five billion or more years ago, why is there any radium left now?

47-8 Why is the decay of an unstable nucleus unaffected by the *chemical* situation of the atom, such as the nature of the molecule in which it is bound, and so on?

47-9 Fusion reactions, which liberate energy, occur only with light nuclei, and fission reactions only with heavy nuclei. A student asserted that this shows that the binding energy per nucleon increases with A at small A but decreases at large A and hence must have a maximum somewhere in between. Do you agree?

47-10 Nuclear power plants use nuclear fission reactions to generate steam to run steam-turbine generators. How does the nuclear reaction produce heat?

47-11 There are cases where a nucleus having too few neutrons for stability can capture one of the electrons in the K shell of the atom. What is the effect of this process on N, A, and Z? Is this the same effect as that of β^+ emission? Might there be situations where K-capture is energetically possible while β^+ emission is not? Explain.

47-12 Is it possible that some parts of the universe contain antimatter whose atoms have nuclei made of antiprotons and antineutrons, surrounded by positrons? How could we detect this condition without actually going there? What problems might arise if we actually *did* go there?

47-13 When x-rays are used to diagnose stomach disorders such as ulcers, the patient first drinks a thick mixture of (insoluble) barium sulfate and water. What does this do? What is the significance of the choice of barium for this purpose?

47-14 Why are there so many health hazards associated with fission fragments that are produced during fission of heavy nuclei?

Problems

47-1 Find the potential energy of an alpha particle 10^{-14} m away from a gold nucleus. Express your results in joules and in electronvolts.

47-2 A 4.7-MeV alpha particle from a radium Ra^{226} decay makes a head-on collision with a gold nucleus. What is the distance of closest approach of the alpha particle to the center of the nucleus?

47-3 The radius of the uranium U^{238} nucleus is about 7.4×10^{-15} m. What energy is required to push an alpha particle to the "surface" of this nucleus?

47-4 Compute the approximate density of nuclear matter, and compare your result with typical densities of ordinary matter.

47-5 How much energy would be required to add a proton to a nucleus with $Z = 91$ and $A = 234$? Express your results in joules, and in MeV.

47-6 Using the data in Table 47-2, calculate the binding energy of deuterium.

47-7 Calculate the mass defect, the binding energy, and the binding energy per nucleon, for the common isotope of oxygen, O^{16}.

47-8 Calculate the energy in MeV of an electron whose wavelength is of the order of nuclear dimensions, say 2×10^{-15} m. This calculation is one argument used to show that there cannot be any electrons in the nucleus.

47-9 Tritium is an unstable isotope of hydrogen, $_1H^3$; its mass, including one electron, is 3.01647 u.

a) Show that it must be unstable with respect to beta decay because $_2He^3$ plus an emitted electron has less total mass.

b) Determine the total kinetic energy of the decay products, taking care to account for the electron masses correctly.

47-10 A carbon specimen found in a cave believed to have been inhabited by cavemen contained $\frac{1}{8}$ as much C^{14} as an equal amount of carbon in living matter. Find the approximate age of the specimen.

47-11 Radium (Ra^{226}) undergoes alpha emission, leading to radon (Rn^{222}). The masses, including all electrons in each atom, are 226.0254 u and 222.0163 u, respectively. Find the maximum kinetic energy that the emitted alpha particle can have.

47-12 The common isotope of uranium, U^{238}, has a half-life of 4.50×10^9 years, decaying by alpha emission.

a) What is the decay constant?

b) What mass of uranium would be required for an activity of one curie?

c) How many alpha particles are emitted per second by 1 g of uranium?

47-13 The unstable isotope K^{40} is used for dating of rock samples. Its half-life is 2.4×10^8 years.

a) How many decays occur per second in a sample containing 2×10^{-6} g of K^{40}?

b) What is the activity of the sample, in curies?

47-14 An unstable isotope of cobalt, Co^{60}, has one more neutron in its nucleus than the stable Co^{59}, and is a beta emitter with a half-life of 5.3 years. This isotope is widely used in medicine. A certain radiation source in a hospital contains 0.01 g of Co^{60}.

a) What is the decay constant for this isotope?

b) How many atoms are in the source?

c) How many decays occur per second?

d) What is the activity of the source, in curies? How does this compare with the activity of an equal mass of radium?

47-15 A free neutron decays into a proton, an electron, and a neutrino, with a half-life of about 15 minutes. Calculate the total kinetic energy of the decay products.

47-16 The magnetic field in a cyclotron that is accelerating protons is 1.5 T.

a) How many times per second should the potential across the dees reverse?

b) The maximum radius of the cyclotron is 0.35 m. What is the maximum velocity of the proton?

c) Through what potential difference would the proton have to be accelerated to give it the maximum cyclotron velocity?

47-17 Deuterons in a cyclotron describe a circle of radius 32.0 cm just before emerging from the dees. The frequency of the applied alternating voltage is 10 MHz. Find (a) the magnetic field, and (b) the energy and speed of the deuterons upon emergence.

47-18 In the fission of U^{238}, 200 MeV of energy is released. Express this energy in joules per mole, and compare with typical heats of combustion.

47-19 Consider the fusion reaction.

$$H^2 + H^2 \rightarrow He^4 + \text{energy}.$$

Compute the energy liberated in this reaction, in MeV and in joules. Compute the energy *per mole* of deuterium, remembering that the gas is diatomic, and compare with the heat of combustion of hydrogen, about 2.9×10^5 J·mol^{-1}.

47-20 Why do elements with mass numbers of 210 and above decay by emission of alpha particles rather than single protons or neutrons.

47-21 If two gamma-ray photons are produced in positron annihilation, find the energy, frequency, and wavelength of each photon.

47-22 A neutral pion at rest decays into two gamma-ray photons. Find the energy, frequency, and wavelength of each photon.

APPENDIX A THE INTERNATIONAL SYSTEM OF UNITS

The Système International d'Unités, abbreviated SI, is the system developed by the General Conference on Weights and Measures and adopted by nearly all the industrial nations of the world. It is based on the mksa (meter-kilogram-second-ampere) system. The following material is reproduced from Publication SP–7012 of the Scientific and Technical Information Office of the National Aeronautics and Space Administration.

Quantity	Name of unit	Symbol	
SI Base Units			
length	meter	m	
mass	kilogram	kg	
time	second	s	
electric current	ampere	A	
thermodynamic temperature	kelvin	K	
luminous intensity	candela	cd	
amount of substance	mole	mol	
SI Derived Units			
area	square meter	m^2	
volume	cubic meter	m^3	
frequency	hertz	Hz	s^{-1}
mass density (density)	kilogram per cubic meter	$kg \cdot m^{-3}$	
speed, velocity	meter per second	$m \cdot s^{-1}$	
angular velocity	radian per second	$rad \cdot s^{-1}$	
acceleration	meter per second squared	$m \cdot s^{-2}$	
angular acceleration	radian per second squared	$rad \cdot s^{-2}$	
force	newton	N	$kg \cdot m \cdot s^{-2}$
pressure (mechanical stress)	pascal	Pa	$N \cdot m^{-2}$
kinematic viscosity	square meter per second	$m^2 \cdot s^{-1}$	
dynamic viscosity	newton-second per square meter	$N \cdot s \cdot m^{-2}$	
work, energy, quantity of heat	joule	J	$N \cdot m$
power	watt	W	$J \cdot s^{-1}$
quantity of electricity	coulomb	C	$A \cdot s$
potential difference, electromotive force	volt	V	$W \cdot A^{-1}, J \cdot C^{-1}$
electric field strength	volt per meter	$V \cdot m^{-1}$	$N \cdot C^{-1}$
electric resistance	ohm	Ω	$V \cdot A^{-1}$
capacitance	farad	F	$A \cdot s \cdot V^{-1}$
magnetic flux	weber	Wb	$V \cdot s$
inductance	henry	H	$V \cdot s \cdot A^{-1}$
magnetic flux density	tesla	T	$Wb \cdot m^{-2}$
magnetic field strength	ampere per meter	$A \cdot m^{-1}$	
magnetomotive force	ampere	A	
luminous flux	lumen	lm	$cd \cdot sr$
luminance	candela per square meter	$cd \cdot m^2$	
illuminance	lux	lx	$lm \cdot m^{-2}$
wave number	1 per meter	m^{-1}	
entropy	joule per kelvin	$J \cdot K^{-1}$	

(continued)

Quantity	Name of unit	Symbol
specific heat capacity	joule per kilogram kelvin	$J \cdot kg^{-1} \cdot K^{-1}$
thermal conductivity	watt per meter kelvin	$W \cdot m^{-1} \cdot K^{-1}$
radiant intensity	watt per steradian	$W \cdot sr^{-1}$
activity (of a radioactive source)	1 per second	s^{-1}
SI Supplementary Units		
plane angle	radian	rad
solid angle	steradian	sr

Definitions of SI Units

meter (m)   The *meter* is the length equal to 1,650,763.73 wavelengths in vacuum of the radiation corresponding to the transition between the levels $2\,p_{10}$ and $5\,d_5$ of the krypton-86 atom.

kilogram (kg)   The *kilogram* is the unit of mass; it is equal to the mass of the international prototype of the kilogram. (The international prototype of the kilogram is a particular cylinder of platinum-iridium alloy which is preserved in a vault at Sèvres, France, by the International Bureau of Weights and Measures.)

second (s)   The *second* is the duration of 9,192,631,770 periods of the radiation corresponding to the transition between the two hyperfine levels of the ground state of the cesium-133 atom.

ampere (A)   The *ampere* is that constant current which, if maintained in two straight parallel conductors of infinite length, of negligible circular cross section, and placed 1 meter apart in vacuum, would produce between these conductors a force equal to 2×10^{-7} newton per meter of length.

kelvin (K)   The *kelvin,* unit of thermodynamic temperature, is the fraction 1/273.16 of the thermodynamic temperature of the triple point of water.

candela (cd)   The *candela* is the luminous intensity, in the perpendicular direction, of a surface of 1/600,000 square meter of a blackbody at the temperature of freezing platinum under a pressure of 101,325 newtons per square meter.

mole (mol)   The *mole* is the amount of substance of a system which contains as many elementary entities as there are carbon atoms in 0.012 kg of carbon 12. The elementary entities must be specified and may be atoms, molecules, ions, electrons, other particles, or specified groups of such particles.

newton (N)   The *newton* is that force which gives to a mass of 1 kilogram an acceleration of 1 meter per second per second.

joule (J)   The *joule* is the work done when the point of application of 1 newton is displaced a distance of 1 meter in the direction of the force.

watt (W)   The *watt* is the power which gives rise to the production of energy at the rate of 1 joule per second.

volt (V)   The *volt* is the difference of electric potential between two points of a conducting wire carrying a constant current of 1 ampere, when the power dissipated between these points is equal to 1 watt.

ohm (Ω)   The *ohm* is the electric resistance between two points of a conductor when a constant difference of potential of 1 volt, applied between these two points, produces in this conductor a current of 1 ampere, this conductor not being the source of any electromotive force.

coulomb (C)   The *coulomb* is the quantity of electricity transported in 1 second by a current of 1 ampere.

farad (F)   The *farad* is the capacitance of a capacitor between the plates of which there appears a difference of potential of 1 volt when it is charged by a quantity of electricity equal to 1 coulomb.

henry (H)   The *henry* is the inductance of a closed circuit in which an electromotive force of 1 volt is produced when the electric current in the circuit varies uniformly at a rate of 1 ampere per second.

weber (Wb)   The *weber* is the magnetic flux which, linking a circuit of one turn, produces in it an electromotive force of 1 volt as it is reduced to zero at a uniform rate in 1 second.

lumen (lm)   The *lumen* is the luminous flux emitted in a solid angle of 1 steradian by a uniform point source having an intensity of 1 candela.

radian (rad)   The *radian* is the plane angle between two radii of a circle which cut off on the circumference an arc equal in length to the radius.

steradian (sr)   The *steradian* is the solid angle which, having its vertex in the center of a sphere, cuts off an area of the surface of the sphere equal to that of a square with sides of length equal to the radius of the sphere.

SI Prefixes   The names of multiples and submultiples of SI units may be formed by application of the prefixes listed in Table 1–1, page 3.

APPENDIX B USEFUL MATHEMATICAL RELATIONS

Algebra

$$a^{-x} = \frac{1}{a^x} \qquad a^{(x+y)} = a^x a^y \qquad a^{(x-y)} = \frac{a^x}{a^y}$$

Logarithims:

$$\log a + \log b = \log (ab) \qquad \log a - \log b = \log (a/b)$$
$$\log (a^n) = n \log a$$

Quadratic formula:

If $ax^2 + bx + c = 0$, $\qquad x = \dfrac{-b \pm \sqrt{b^2 - 4ac}}{2a}$

Binomial theorem

$$(a + b)^n = a^n + na^{n-1}b + \frac{n(n-1)a^{n-2}b^2}{2!} + \frac{n(n-1)(n-2)a^{n-3}b^3}{3!} + \cdots$$

Trigonometry

In the right triangle ABC,
$x^2 + y^2 = r^2$.

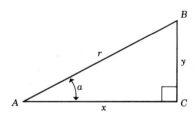

Definitions of the trigonometric functions:

$\sin a = y/r \qquad \cos a = x/r \qquad \tan a = y/x$

Identities:

$\sin^2 a + \cos^2 a = 1 \qquad\qquad \tan a = \dfrac{\sin a}{\cos a}$

$\sin 2a = 2 \sin a \cos a$

$\cos 2a = \cos^2 a - \sin^2 a = 2 \cos^2 a - 1$

$\sin \frac{1}{2}a = \sqrt{\dfrac{1 - \cos a}{2}} \qquad\qquad \cos \frac{1}{2}a = \sqrt{\dfrac{1 + \cos a}{2}}$

$\sin (a \pm b) = \sin a \cos b \pm \cos a \sin b$

$\cos (a \pm b) = \cos a \cos b \mp \sin a \sin b$

$\sin a + \sin b = 2 \sin \frac{1}{2}(a + b) \cos \frac{1}{2}(a - b)$

$\cos a + \cos b = 2 \cos \frac{1}{2}(a + b) \cos \frac{1}{2}(a - b)$

$\sin (-a) = -\sin a \qquad\qquad \cos (-a) = \cos a$

$\sin (a \pm \pi/2) = \pm\cos a \qquad\qquad \cos (a \pm \pi/2) = \mp\sin a$

Geometry

Circumference of circle of radius r: $C = 2\pi r$

Area of circle of radius r: $A = \pi r^2$

Volume of sphere of radius r: $V = 4\pi r^3/3$

Surface area of sphere of radius r: $A = 4\pi r^2$

Calculus

Derivatives:

$$\frac{d}{dx}x^n = nx^{n-1}$$

$$\frac{d}{dx}\sin ax = a\cos ax$$

$$\frac{d}{dx}\cos ax = -a\sin ax$$

$$\frac{d}{dx}e^{ax} = ae^{ax}$$

$$\frac{d}{dx}\ln ax = \frac{1}{x}$$

Integrals:

$$\int x^n\, dx = \frac{x^{n+1}}{n+1}$$

$$\int \frac{dx}{x} = \ln x$$

$$\int \sin ax\, dx = -\frac{1}{a}\cos ax$$

$$\int \cos ax\, dx = \frac{1}{a}\sin ax$$

$$\int e^{ax}\, dx = \frac{1}{a}e^{ax}$$

$$\int \frac{dx}{\sqrt{a^2 - x^2}} = \arcsin\frac{x}{a}$$

$$\int \frac{dx}{\sqrt{x^2 + a^2}} = \ln\left(x + \sqrt{x^2 + a^2}\right)$$

$$\int \frac{dx}{x^2 + a^2} = \frac{1}{a}\arctan\frac{x}{a}$$

$$\int \frac{dx}{(x^2 + a^2)^{3/2}} = \frac{1}{a^2}\frac{x}{\sqrt{x^2 + a^2}}$$

$$\int \frac{x\, dx}{(x^2 + a^2)^{3/2}} = -\frac{1}{\sqrt{x^2 + a^2}}$$

Power series (convergent for range of x shown):

$$\sin x = x - \frac{x^3}{3!} + \frac{x^5}{5!} - \frac{x^7}{7!} + \cdots \qquad \text{(all } x\text{)}$$

$$\cos x = 1 - \frac{x^2}{2!} + \frac{x^4}{4!} - \frac{x^6}{6!} + \cdots \qquad \text{(all } x\text{)}$$

$$\tan x = x + \frac{x^3}{3} + \frac{2x^5}{15} + \frac{17x^7}{315} + \cdots \qquad (|x| < \pi/2)$$

$$e^x = 1 + x + \frac{x^2}{2!} + \frac{x^3}{3!} + \cdots \qquad \text{(all } x\text{)}$$

$$\ln(1 + x) = x - \frac{x^2}{2} + \frac{x^3}{3} - \frac{x^4}{4} + \cdots \qquad (|x| < 1)$$

APPENDIX C THE GREEK ALPHABET

Name	Capital	Lowercase	Name	Capital	Lowercase
Alpha	A	α	Nu	N	ν
Beta	B	β	Xi	Ξ	ξ
Gamma	Γ	γ	Omicron	O	o
Delta	Δ	δ	Pi	Π	π
Epsilon	E	ϵ	Rho	P	ρ
Zeta	Z	ζ	Sigma	Σ	σ
Eta	H	η	Tau	T	τ
Theta	Θ	θ	Upsilon	Υ	υ
Iota	I	ι	Phi	Φ	ϕ
Kappa	K	κ	Chi	X	χ
Lambda	Λ	λ	Psi	Ψ	ψ
Mu	M	μ	Omega	Ω	ω

APPENDIX D NATURAL TRIGONOMETRIC FUNCTIONS

Angle					Angle				
De-gree	Ra-dian	Sine	Co-sine	Tan-gent	De-gree	Ra-dian	Sine	Co-sine	Tan-gent
0°	.000	0.000	1.000	0.000					
1°	.017	.018	1.000	.018	46°	0.803	0.719	0.695	1.036
2°	.035	.035	0.999	.035	47°	.820	.731	.682	1.072
3°	.052	.052	.999	.052	48°	.838	.743	.669	1.111
4°	.070	.070	.998	.070	49°	.855	.755	.656	1.150
5°	.087	.087	.996	.088	50°	.873	.766	.643	1.192
6°	.105	.105	.995	.105	51°	.890	.777	.629	1.235
7°	.122	.122	.993	.123	52°	.908	.788	.616	1.280
8°	.140	.139	.990	.141	53°	.925	.799	.602	1.327
9°	.157	.156	.988	.158	54°	.942	.809	.588	1.376
10°	.175	.174	.985	.176	55°	.960	.819	.574	1.428
11°	.192	.191	.982	.194	56°	.977	.829	.559	1.483
12°	.209	.208	.978	.213	57°	.995	.839	.545	1.540
13°	.227	.225	.974	.231	58°	1.012	.848	.530	1.600
14°	.244	.242	.970	.249	59°	1.030	.857	.515	1.664
15°	.262	.259	.966	.268	60°	1.047	.866	.500	1.732
16°	.279	.276	.961	.287	61°	1.065	.875	.485	1.804
17°	.297	.292	.956	.306	62°	1.082	.883	.470	1.881
18°	.314	.309	.951	.325	63°	1.100	.891	.454	1.963
19°	.332	.326	.946	.344	64°	1.117	.899	.438	2.050
20°	.349	.342	.940	.364	65°	1.134	.906	.423	2.145
21°	.367	.358	.934	.384	66°	1.152	.914	.407	2.246
22°	.384	.375	.927	.404	67°	1.169	.921	.391	2.356
23°	.401	.391	.921	.425	68°	1.187	.927	.375	2.475
24°	.419	.407	.914	.445	69°	1.204	.934	.358	2.605
25°	.436	.423	.906	.466	70°	1.222	.940	.342	2.747
26°	.454	.438	.899	.488	71°	1.239	.946	.326	2.904
27°	.471	.454	.891	.510	72°	1.257	.951	.309	3.078
28°	.489	.470	.883	.532	73°	1.274	.956	.292	3.271
29°	.506	.485	.875	.554	74°	1.292	.961	.276	3.487
30°	.524	.500	.866	.577	75°	1.309	.966	.259	3.732
31°	.541	.515	.857	.601	76°	1.326	.970	.242	4.011
32°	.559	.530	.848	.625	77°	1.344	.974	.225	4.331
33°	.576	.545	.839	.649	78°	1.361	.978	.208	4.705
34°	.593	.559	.829	.675	79°	1.379	.982	.191	5.145
35°	.611	.574	.819	.700	80°	1.396	.985	.174	5.671
36°	.628	.588	.809	.727	81°	1.414	.988	.156	6.314
37°	.646	.602	.799	.754	82°	1.431	.990	.139	7.115
38°	.663	.616	.788	.781	83°	1.449	.993	.122	8.144
39°	.681	.629	.777	.810	84°	1.466	.995	.105	9.514
40°	.698	.643	.766	.839	85°	1.484	.996	.087	11.43
41°	.716	.658	.755	.869	86°	1.501	.998	.070	14.30
42°	.733	.669	.743	.900	87°	1.518	.999	.052	19.08
43°	.751	.682	.731	.933	88°	1.536	.999	.035	28.64
44°	.768	.695	.719	.966	89°	1.553	1.000	.018	57.29
45°	.785	.707	.707	1.000	90°	1.571	1.000	.000	

APPENDIX E PERIODIC TABLE OF THE ELEMENTS

Period	IA	IIA	IIIB	IVB	VB	VIB	VIIB	VIIIB	VIIIB	VIIIB	IB	IIB	IIIA	IVA	VA	VIA	VIIA	Noble gases
1	1 H 1.008																	2 He 4.003
2	3 Li 6.939	4 Be 9.012											5 B 10.811	6 C 12.011	7 N 14.007	8 O 15.999	9 F 18.998	10 Ne 20.183
3	11 Na 22.990	12 Mg 24.312											13 Al 26.982	14 Si 28.086	15 P 30.974	16 S 32.064	17 Cl 35.453	18 Ar 39.948
4	19 K 39.102	20 Ca 40.08	21 Sc 44.956	22 Ti 47.90	23 V 50.942	24 Cr 51.996	25 Mn 54.938	26 Fe 55.847	27 Co 58.933	28 Ni 58.71	29 Cu 63.54	30 Zn 65.37	31 Ga 69.72	32 Ge 72.59	33 As 74.922	34 Se 78.96	35 Br 79.909	36 Kr 83.80
5	37 Rb 85.47	38 Sr 87.62	39 Y 88.905	40 Zr 91.22	41 Nb 92.906	42 Mo 95.94	43 Tc (99)	44 Ru 101.07	45 Rh 102.91	46 Pd 106.4	47 Ag 107.87	48 Cd 112.40	49 In 114.82	50 Sn 118.69	51 Sb 121.75	52 Te 127.60	53 I 126.90	54 Xe 131.30
6	55 Cs 132.91	56 Ba 137.34	57 La 138.91	72 Hf 178.49	73 Ta 180.95	74 W 183.85	75 Re 186.2	76 Os 190.2	77 Ir 192.2	78 Pt 195.09	79 Au 196.97	80 Hg 200.59	81 Tl 204.37	82 Pb 207.19	83 Bi 208.98	84 Po (210)	85 At (210)	86 Rn (222)
7	87 Fr (223)	88 Ra (226)	89 Ac (227)	104 Rf(?) (259)	105 Ha(?) (260)													

58 Ce 140.12	59 Pr 140.91	60 Nd 144.24	61 Pm (145)	62 Sm 150.35	63 Eu 151.96	64 Gd 157.25	65 Tb 158.92	66 Dy 162.50	67 Ho 164.93	68 Er 167.26	69 Tm 168.93	70 Yb 173.04	71 Lu 174.97
90 Th 232.04	91 Pa (231)	92 U 238.03	93 Np (237)	94 Pu (242)	95 Am (243)	96 Cm (247)	97 Bk (249)	98 Cf (251)	99 Es (254)	100 Fm (253)	101 Md (256)	102 No (253)	103 Lr (257)

APPENDIX F UNIT CONVERSION FACTORS

Length

1 m = 100 cm = 1000 mm = 10^6 μm = 10^9 nm
1 km = 1000 m = 0.6214 mi
1 m = 3.281 ft = 39.37 in.
1 cm = 0.3937 in.
1 ft = 30.48 cm
1 in. = 2.540 cm
1 mi = 5280 ft = 1.609 km
1 Å = 10^{-10} m = 10^{-8} cm = 10^{-1} nm

Area

1 cm^2 = 0.155 in^2
1 m^2 = 10^4 cm^2 = 10.76 ft^2
1 in^2 = 6.452 cm^2
1 ft^2 = 144 in^2 = 0.0929 m^2

Volume

1 liter = 1000 cm^3 = 10^{-3} m^3 = 0.0351 ft^3 = 61.02 $in.^3$
1 ft^3 = 0.02832 m^3 = 28.32 liters = 7.477 gallons

Time

1 min = 60 s
1 hr = 3600 s
1 da = 86,400 s
1 yr = 3.156 × 10^7 s

Velocity

1 cm·s^{-1} = 0.03281 ft·s^{-1}
1 ft·s^{-1} = 30.48 cm·s^{-1}
1 mi·min^{-1} = 60 mi·hr^{-1} = 88 ft·s^{-1}
1 km·hr^{-1} = 0.2778 m·s^{-1}
1 mi·hr^{-1} = 0.4470 m·s^{-1}

Acceleration

1 m·s^{-2} = 100 cm·s^{-2} = 3.281 ft·s^{-2}
1 cm·s^{-2} = 0.01 m·s^{-2} = 0.03281 ft·s^{-2}

1 ft·s^{-2} = 0.3048 m·s^{-2} = 30.48 cm·s^{-2}
1 mi·hr^{-1}·s^{-1} = 1.467 ft·s^{-2}

Mass

1 kg = 10^3 g = 0.0685 slug
1 g = 6.85 × 10^{-5} slug
1 slug = 14.59 kg
1 u = 1.661 × 10^{-27} kg

Force

1 N = 10^5 dyn = 0.2247 lb
1 lb = 4.45 N = 4.45 × 10^5 dyn

Pressure

1 Pa = 1 N·m^{-2} = 1.451 × 10^{-4} lb·in^{-2} = 0.209 lb·ft^{-2}
1 lb·in^{-2} = 6891 Pa
1 lb·ft^{-2} = 47.85 Pa
1 atm = 1.013 × 10^5 Pa = 14.7 lb·in^{-2} = 2117 lb·ft^{-2}

Energy

1 J = 10^7 ergs = 0.239 cal
1 cal = 4.186 J (based on 15° calorie)
1 ft·lb = 1.356 J
1 Btu = 1055 J = 252 cal
1 eV = 1.602 × 10^{-19} J
1 kWh = 3.600 × 10^6 J

Mass-Energy Equivalence

1 kg ↔ 8.988 × 10^{16} J
1 u ↔ 931.5 MeV
1 eV ↔ 1.073 × 10^{-9} u

Power

1 W = 1 J·s^{-1}
1 hp = 746 W = 550 ft·lb·s^{-1}
1 Btu·hr^{-1} = 0.293 W

APPENDIX G NUMERICAL CONSTANTS

Fundamental physical constants

Name	Symbol	Value
Speed of light	c	2.9979×10^8 m·s^{-1}
Charge of electron	e	1.602×10^{-19} C
Gravitational constant	G	6.673×10^{-11} N·m^2·kg^{-2}
Planck's constant	h	6.626×10^{-34} J·s
Boltzmann's constant	k	1.381×10^{-23} J·K^{-1}
Avogadro's number	N_0	6.022×10^{23} molecules·mol^{-1}
Gas constant	R	8.314 J·mol^{-1}·K^{-1}
Mass of electron	m_e	9.110×10^{-31} kg
Mass of neutron	m_n	1.675×10^{-27} kg
Mass of proton	m_p	1.673×10^{-27} kg
Permittivity of free space	ϵ_0	8.854×10^{-12} C^2·N^{-1}·m^{-2}
	$\frac{1}{4}\pi\epsilon_0$	8.987×10^9 N·m^2·C^{-2}
Permeability of free space	μ_0	$4\pi \times 10^{-7}$ Wb·A^{-1}·m^{-1}

Other useful constants

Mechanical equivalent of heat		4.185 J·cal^{-1}
Standard atmospheric pressure	1 atm	1.013×10^5 Pa
Absolute zero	0 K	$-273.15°$C
Electronvolt	1 eV	1.602×10^{-19} J
Atomic mass unit	1 u	1.661×10^{-27} kg
Electron rest energy	mc^2	0.511 MeV
Energy equivalent of 1 u	Mc^2	931.5 MeV
Volume of ideal gas (0°C and 1 atm)	V	22.4 liter·mol^{-1}
Acceleration due to gravity (sea level, at equator)	g	9.78049 m·s^{-2}

Astronomical data

Body	Mass, kg	Radius, m	Orbit radius, m	Orbit period
Sun	1.99×10^{30}	6.95×10^8	—	—
Moon	7.36×10^{22}	1.74×10^6	0.38×10^9	27.3 d
Mercury	3.28×10^{23}	2.57×10^6	5.8×10^{10}	88.0 d
Venus	4.82×10^{24}	6.31×10^6	1.08×10^{11}	224.7 d
Earth	5.98×10^{24}	6.38×10^6	1.49×10^{11}	365.3 d
Mars	6.34×10^{23}	3.43×10^6	2.28×10^{11}	687.0 d
Jupiter	1.88×10^{27}	7.18×10^7	7.78×10^{11}	11.86 yr
Saturn	5.63×10^{26}	6.03×10^7	1.43×10^{12}	29.46 yr
Uranus	8.61×10^{25}	2.67×10^7	2.87×10^{12}	84.02 y
Neptune	9.99×10^{25}	2.48×10^7	4.49×10^{12}	164.8 y
Pluto	5×10^{23}	4×10^5	5.90×10^{12}	247.7 y

ANSWERS TO ODD-NUMBERED PROBLEMS

Part II   Electricity and Magnetism, Light, and Atomic Physics

Chapter 24

24-1   625

24-3   0.119 m

24-5   a) 8.34×10^{19} N   b) 8.34×10^5 N

24-7   a) 2.75×10^{26} b) 6.59×10^{15} c) 2.40×10^{-11}

24-9 2.16×10^{-7} N

24-11   a) $q^2/2\pi\epsilon_0 a^2$, vertically upward
b) $q^2 a/2\pi\epsilon_0 (a^2 + x^2)^{3/2}$

24-13   a) 4.04×10^6 b) 4.93×10^{-17}

24-15   a) 5.09×10^5 N   b) 1.44×10^6 N

24-19   a) 5.40×10^{-5} N, -3.45×10^{-5} N;
b) 6.41×10^{-5} N, $32.6°$ below $+x$-axis

Chapter 25

25-1   a) 4.0 N·C^{-1} upward   b) 6.4×10^{-19} N downward

25-3   a) 1010 N·C^{-1} b) 2670 km·s^{-1}

25-5   a) 1.42 cm   b) 9.86 cm

25-7 -24.9×10^{-9} C

25-9   a) 1.80×10^4 N·C^{-1}, $-x$-direction
b) 7.99×10^3 N·C^{-1}, $+x$-direction
c) 3.34×10^3 N·C^{-1}, $70°$ above $-x$-axis
d) 6.36×10^3 N·C^{-1}, $-x$-direction

25-11 $40.8°$

25-13 5.58×10^{-11} N·C^{-1}

25-15 $\sqrt{2}\,\lambda/4\pi\epsilon_0 r$

25-17 8.85×10^{-10} C

25-19   a) 2400 N·m^2·C^{-1} b) Zero
c) 101 N·C^{-1}, $-x$-direction

25-21   b) $q/4\pi\epsilon_0 r^2$

25-25   b) $q/4\pi\epsilon_0 r^2$ or $\rho R^3/3\epsilon_0 r^2$

25-27   a) $Q/4\pi\epsilon_0 r^2$, 0, $Q/4\pi\epsilon_0 r^2$ b) $-Q/4\pi a^2$

Chapter 26

26-1   30 V; positively charged plate

26-3   a) 180 V   b) 180 V

26-5   a) -4.5×10^{-5} J   b) 3.0×10^5 N·C^{-1}
c) -1.5×10^4 V

26-7   b) 0.667 mm

26-9   a) 1798 V   b) Zero   c) 4.49×10^{-5} J

26-11   a) 3.00 m   b) 0.200×10^{-6} C

26-13   b) $q/2\pi\epsilon_0 a$ e) $\pm a\sqrt{3}$

26-15   a) It remains at rest.   b) It oscillates about the origin, along the y-axis.   c) It accelerates away from the origin along the x-axis.

26-17   b) Zero
c) $q/4\pi\epsilon_0[(x^2 + 2ax + 2a^2)^{-1/2}$
$- (x^2 - 2ax + 2a^2)^{-1/2}]$

26-19   c) Zero

26-21 1.03×10^7 m·s^{-1}

26-23   a) $Q/4\pi\epsilon_0(x^2 + a^2)^{1/2}$ b) $Qx/4\pi\epsilon_0(x^2 + a^2)^{3/2}$

26-25   c) $(\lambda/2\pi\epsilon_0) \ln (r_b/r_a)$

26-27   a) Inside $(\rho/4\epsilon_0)(R^2 - r^2)$;
outside $-(\lambda/2\pi\epsilon_0) \ln (r/R)$

26-29   a) -539 V, $+59.9$ V   b) -2.40×10^{-6} J

26-31   a) 4.33×10^{-4} m·s^{-1}
b) 2.74×10^{-4} m·s^{-1} upward

26-33   938 MeV

26-35 2.27×10^{-14} m

26-37   a) 5114 V   b) $0.140c$ c) 9.38 MeV, $0.140c$

26-39   a) 0.704 cm   b) $19.4°$ c) 4.92 cm

Chapter 27

27-1   a) 50 pF   b) 2.82×10^{-3} m^2 c) 400 V
d) 1.0×10^{-6} J

27-3 a) 1000 V b) 2000 V c) 0.5×10^{-3} J
27-5 a) 0.38×10^{-6} C b) 7600 V c) 1.43×10^{-3} J
 d) 1.37×10^{-3} J
27-7 a) 1 μF b) 900 μC c) 100 V
27-9 a) 800 μC, 800 V; 800 μC, 400 V
 b) 533 μC, 533 V; 1067 μC, 533 V
27-11 1.30×10^{-3} J
27-13 0.86 cal
27-15 a) $-(\epsilon_0 A/x^2)\, dx$ b) $(Q^2/2\epsilon_0 A)\, dx$
 c) $Q^2/2\epsilon_0 A = \frac{1}{2}QE$
27-17 a) 0.226 m^2 b) 1250 V
27-19 a) 35.4 pF b) 1.77 nC c) 5000 V·m^{-1}
 d) 4.42×10^{-8} J
 e) 17.7 pF, 1.77 nC, 5000 V·m^{-1}, 8.84×10^{-8} J
27-21 a) $Q^2 x/2\epsilon_0 A$ b) $Q^2(x + dx)/2\epsilon_0 A$
27-23 1770 pF
27-25 a) 26.6 μC·m^{-2} b) 17.7 μC·m^{-2}
27-27 a) 3.43 b) 0.708×10^{-5} C·m^{-2}
27-29 $(Q/4\pi\epsilon_0)(1/r_a - 1/r_b)$

Chapter 28

28-1 a) 20 mA b) 2.74×10^{-6} m·s^{-1}
28-3 1.30×10^5 mm^{-3}
28-5 a) 2.5×10^6 A·m^{-2} b) 0.0430 V·m^{-1}
 c) 5 hr 41 min (2.05×10^4 s)
28-7 a) 2.4×10^{-8} Ω·m b) 20 A c) 3.90×10^{-4} m·s^{-1}
28-9 a) No c) About 1.0 Ω; smaller
28-11 a) 0.106 C b) Zero c) 12.7 A, zero
28-13 0.735×10^{-6} Ω
28-15 a) 99.52 Ω b) 0.0148 Ω
28-17 a) 22.1 Ω b) 0.363 A
28-19 28.5°C
28-21 284°C
28-23 60.9°C
28-25 2.47 mm
28-27 a) 0.50 Ω b) 10 V
28-29 Open: a) 12 V b) 0 c) 12 V
 Closed: a) 11.1 V b) 11.1 V c) 0
28-31 a) 2.22 A b) 1.4 A c) 6.4 V
28-33 90 Ω
28-35 a) VI, I^2R, V^2/R b) 90 Ω
28-37 a) 24 W b) 4 W c) 20 W

Chapter 29

29-3 a) 8 Ω b) 12 V
29-5 9.6 W in 60-W lamp, 14.4 W in 40-W lamp
29-7 a) 141 V b) 4.5 W c) Two in series, connected
 in parallel with two others in series
29-9 a) Two in series, connected in parallel with two
 others in series. b) 1/2 W, if total power is 2 W.
29-11 655 W
29-13 a) -12 V b) 3A c) -12 V
 d) 12/7 A from b to a e) 4.5 Ω f) 4.2 Ω
29-15 a) 18 V b) Point a c) 6 V d) 36 μC
29-17 18 V, 7 V, 13 V

29-19 6 A, 4 A, 2 A, 4 A, 6 A (top to bottom)
29-21 a) 0.8 A, 0.2 A, 0.6 A b) -3.2 V
29-23 a) 1200 Ω b) 30 V
29-25 10.9 V, 109.1 V
29-27 a) 99.0 V b) 0.0526
29-29 4.97 Ω
29-31 2985 Ω, 12,000 Ω; 135,000 Ω; 3000 Ω; 15,000 Ω,
 150,000 Ω
29-35 a) 1450 Ω b) 4500, 1500, 500 Ω c) No
29-37 55 Ω
29-39 2749 Ω
29-41 a) 9.52 V b) No
29-43 7.21 μF
29-45 a) 12 μs b) 2.53 V
29-47 a) 5.31 μC b) 2.0 mA c) 2.0 mA
29-51 a) 5.0×10^6 A·m^{-2} b) 3.34×10^{-10} A·m^{-2}
29-53 a) $i/\pi R^2$ b) $q/\pi\epsilon_0 R^2$ c) $i/\pi R^2$

Chapter 30

30-1 a) qvB along $-z$-axis b) qvB along $+y$-axis
 c) Zero d) $qvB/\sqrt{2}$ parallel to $-y$-axis e) qvB
 in yz-plane at 45° to $-y$- and $-z$-axes
 f) $\sqrt{2}\,qvB/\sqrt{3}$ in yz-plane at 45° to $-y$- and
 $-z$-axes
30-3 3.27 T, perpendicular to direction of v
30-5 a) 1.14×10^{-3} T, into page b) 1.57×10^{-8} s
30-7 Negative
30-9 8 mm
30-11 2.15×10^{-7} Wb
30-13 a) 4.0×10^6 V·m^{-1} b) 20,000 V
30-15 2.13×10^{-2} m
30-17 21
30-19 b) Yes

Chapter 31

31-1 1.20 N, $-z$-direction; 1.20 N, $-y$-direction;
 1.70 N, 45° up, parallel to yz-plane;
 120 N, $-y$-direction; zero
31-3 6.88×10^{28} m^{-3}
31-5 2.4 N
31-7 24 A
31-9 a) 0.16 N, 0.06 N, 0.0083 N·m
 b) 0.16 N, 0.104 N, 0.0048 N·m c) Same torque
31-11 3.6×10^{-6} N·m
31-13 a) 0.181 N·m b) When normal to coil is at 30°
 to field.
31-15 a) 0.5 A b) 4 A c) 108 V d) 60 W e) 48 W
 f) 540 W g) 71%

Chapter 32

32-1 1.6×10^{-5} T; yes
32-3 a) 20.0 μT b) 7.07 μT c) 20.0 μT d) Zero
 e) 5.44 μT
32-5 a) 125 A b) 2.5×10^{-4} T, 1.25×10^{-4} T

32-7 b) $\mu_0 Ia/\pi(a^2 + x^2)$ d) Zero
32-9 a) $\mu_0 I/\pi a$ b) $\mu_0 I/3\pi a$ c) Zero d) $2\mu_0 I/3\pi a$
32-11 a) 1.92×10^{-4} N·m^{-1}, down
　　　b) 1.92×10^{-4} N·m^{-1}, up
32-13 1.02 cm
32-15 7.20×10^{-4} N to left
32-19 $1.91R$
32-21 16
32-23 6.91×10^{-3} T
32-25 a) $\mu_0 I/2\pi r$ b) Zero
32-27 6.0×10^{-4} T
32-29 a) 7958 m^{-1} b) 1250 m
32-31 a) $(1/\epsilon_0 A)(dQ/dt)$

Chapter 33

33-1 a) 5.0 m·s^{-1} b) 2.0 A c) 0.96 N, to left
33-3 a) 3 V b) 3 V c) Upper end
33-5 a) 1.0 V, from B to A b) 1.25 N c) 5.0 W
33-7 0.006 V
33-9 b) FR/B^2l^2
33-11 a) 5.56 rad·s^{-1} b) Zero
33-15 a) 0.2 V b) Zero c) 0.2 V
33-17 a) $\pi r_1^2(dB/dt)$ b) $(r_1/2)(dB/dt)$
　　　c) $(R^2/2r_2)(dB/dt)$ e) $(\pi R^2/4)(dB/dt)$
　　　f) $(\pi R^2)(dB/dt)$ g) $(\pi R^2)(dB/dt)$
33-19 a) Circles, clockwise b) 0.005 V·m^{-1}, 3.14 mV
　　　c) 1.57 mA d) Zero f) 3.14 mV
33-21 b) Zero c) 4 mV d) 2 mA e) Zero
33-23 a) Down b) Up c) Zero
33-25 a) $\mu_0 I/2r$, into plane b) $\mu_0 Il\,dr/2r$
　　　c) $(\mu_0 Il/2)\ln(b/a)$ d) $(\mu_0 l/2)(dI/dt)\ln(b/a)$
　　　e) 1.38×10^{-6} V
33-27 a) a to b b) b to a c) b to a
33-29 a) 1.01×10^{-5} C
33-31 a) 0.737 rev·s^{-1} b) 4320 N·m
33-33 7.96 rad·s^{-1}

Chapter 34

34-1 0.790 mH
34-3 1.0 mH; 2.25 mH or 0.25 mH
34-5 a) 5.0×10^{-4} V; yes b) 5.0×10^{-4} V
34-7 a) 0.100 J b) 2.00 W
34-9 $(1508 \text{ V})\cos(120\pi\text{s}^{-1}t)$; leads by 90°
34-15 387
34-17 a) 0.05 A b) 1.0 A·s^{-1} c) 0.5 A·s^{-1} d) 0.23 s
　　　e) 0.0197, 0.0317, 0.0389, 0.0433 A
34-19 0.326 A to right
34-23 a) 2.39 mH b) 4.41 pF
34-27 a) $4L$ b) L c) $\omega/2$

Chapter 35

35-1 χ_m varies inversely with T.
35-3 a) 2000 A·m^{-1} b) 7.94×10^5 A·m^{-1} c) 397
　　　d) 3.18×10^5 A e) 398

35-5 c) Diamagnetic d) 10 A, opposite
　　　e) It is repelled.
35-9 8 A
35-11 a) 0.045 A b) 0.012 A c) 0.15 A, 0.30 A

Chapter 36

36-1 a) 127 Hz b) 7.96 Hz
36-3 a) 377 Ω b) 2.65 mH c) 2650 Ω d) 2650 μF
36-5 a) 5 mA b) 0.05 A c) 0.5 A
36-7 a) 0.05 A b) 5.0 mA c) 0.50 mA
36-11 a) 583 Ω b) 0.0858 A c) 25.7 V, 42.9 V
　　　d) 59°, lead
36-13 a) 626 Ω b) First larger, then smaller
　　　c) 61.4°, current leads
36-15 16.9, 25.4, 56.4, 31.0, 35.4 V (rms)
36-17 b) 745 rad·s^{-1} or 118 Hz c) 1
　　　d) 35.4, 79.2, 79.2, 0, 35.4 V (rms) e) 745 s^{-1}
　　　f) 0.354 A
36-19 a) 1592 Hz b) 12.4 V (rms)
36-21 0.0184 H
36-23 a) 25 W b) 25 W c) Zero d) Zero e) 0.127
36-25 a) 151 μF b) 917 W
36-27 a) 31.6 b) 3.16 V
36-29 a) 1592 Hz, 10,000 s^{-1} b) 1.0 A c) 1000 V
　　　d) 0.05 J
36-31 a) 150 b) 180 W c) 1.5 A

Chapter 37

37-1 a) 3.0×10^8 m·s^{-1} b) 294 m
　　　c) 4.8×10^{-3} V·m^{-1}
37-3 a) 3.0×10^6 m·s^{-1} b) 3 cm
37-5 a) 1027 V·m^{-1}, 3.42×10^{-6} T b) 3.96×10^{26} W
37-7 About 6×10^4 V·m^{-1}, 2×10^{-4} T
37-9 a) $\rho I/\pi a^2$, parallel to wire
　　　b) $Ir/2\pi a^2$, perpendicular to wire c) $I^2 r\rho/2\pi^2 a^4$
　　　d) $2\pi rlS$
37-11 24.5 V·m^{-1}, 0.0650 A·m^{-1}
37-13 a) $(\mu_0 nr/2)(dI/dt)$, tangent
　　　b) $(\mu_0^2 n^2 r/2)(I\,dI/dt)$, radially inward
37-15 300 m
37-17 a) 10^{10} Hz b) 1.0×10^{-7} T c) 1.19 W·m^{-2}
　　　d) 1.99×10^{-9} N

Chapter 38

38-1 a) 1.5×10^{-9} m, 1.5×10^{-3} μm, 1.50 nm, 15 Å
　　　b) 5.36×10^{-7} m, 0.536 μm, 536 nm, 5360 Å
38-3 16.6 min
38-9 1.732
38-11 a) 32.1° b) Independent of n
38-13 a) 2×10^8 m·s^{-1} b) 333 nm
38-15 1.87
38-19 No
38-21 b) 90°

38-23 1.88
38-27 1.02 cm
38-31 1.75°
38-35 4.54×10^{-5}
38-37 7.07 m
38-39 a) 44 b) 44 kW c) 354 hp

Chapter 39

39-1 Half the observer's height
39-5 a) 9.88 cm b) 5 cm
39-7 a) 3 b) 30 cm behind mirror
39-9 1.71 cm, 0.143
39-11 b) 10 cm behind mirror, 4 cm, erect, virtual
39-13 a) $|s| < 5$ cm b) Erect
39-15 a) 45° b) 1.41 cm
39-17 4.48 cm
39-19 1.35
39-21 30 cm, -1
39-23 0.667 cm

Chapter 40

40-1 $4R$ from center of sphere
40-3 a) The first image b) 30 cm c) Real
d) At infinity
40-5 50 cm
40-7 1.5 cm
40-11

R_1	R_2	f
10 cm	20 cm	40 cm
10 cm	-20 cm	13.3 cm
-10 cm	20 cm	-13.3 cm
-10 cm	-20 cm	-40 cm

40-13 a) 6 cm or 12 cm from object b) -2, -0.5
40-15 60 cm to right of third lens
40-17 -7.21 cm
40-19 a) 20.0 cm b) 18.7 cm
40-21 b) 12 cm
40-23 0.9 cm
40-25 a) 35 mm b) 200 mm
40-27 2.0 cm
40-29 2 cm left of first lens, 2 cm right of second lens
40-31 a) $+3.2$ diopters b) -2 diopters
40-33 a) 50 cm b) 2 m
40-35 a) 7.14 cm b) 3.5 mm
40-37 1/400 s
40-39 a) 120 mm b) 121.4 mm from lens
40-41 a) $842\times$ b) $50\times$
40-43 $380\times$
40-45 $20\times$
40-49 $25\times$, 35 cm from mirror

Chapter 41

41-1 1.0 mm
41-3 1.0 mm

41-5 a) Perpendicular to line joining antenna
b) $0.5I_0$, $0.197I_0$, $0.044I_0$, 0
41-7 a) 5.5 b) zero
41-9 b) 1.1 mm
41-11 0.909 mm
41-13 36.7 fringes·cm^{-1}
41-15 143 nm
41-17 a) Yes b) 89 nm
41-19 2.36 mm
41-21 17.5°, 36.9°, 64.1°
41-23 a) 570 nm b) 43.2°
41-27 0.048 mm
41-29 800 lines·mm^{-1}
41-31 Rainbow-fringed images

Chapter 42

42-1 a) $I_0/4$ b) Parallel to second filter
42-3 a) $(I_0/2) \sin^2 \theta \cos^2 \theta$
42-5 First filter at 45° to polarization of beam, second at 90° to beam; $I_0/4$
42-7 35.5°
42-9 a) 37° b) Horizontal
42-11 a) 58.8° b) 51.1°
42-13 0.005 mm
42-15 a) 1.732 b) 3
42-17 a) Yes b) Dark c) Dark
42-19 c) 0.582 μm
42-21 Linear, rotated 90°
42-23 Left

Chapter 43

43-1 a) 4.3×10^{-8} s b) 10.4 m
43-3 The bolt at A appears to come first.
43-7 $\Delta s' = [(\Delta s)^2 - c^2 (\Delta t)^2]^{1/2}$
43-9 5.77×10^{-9} s
43-11 $v/c = E/mc^2$
43-13 2.60×10^8 m·s^{-1}
43-15 a) 0.5×10^{-4} b) 15.5% c) Factor of 2.29
43-17 1.18×10^{-18} m·s^{-1}
43-19 2.986×10^8 m·s^{-1}
43-21 1.64×10^{-13} J
43-23 a) 4.55×10^{-15} J, 4.92×10^{-15} J
b) 18.2×10^{-15} J, 27.9×10^{-15} J
43-25 a) 1.8×10^{14} J
b) 1.8×10^{20} W
c) 1.84×10^{10} kg
43-27 a) 1.8×10^5 eV b) 6.91×10^5 eV
c) 2.02×10^8 m·s^{-1} d) 2.52×10^8 m·s^{-1}
43-31 a) 0.047 MeV b) 0.294 MeV c) 7.2, 4.0

Chapter 44

44-1 $hf_0 = \phi$
44-3 1.72×10^6 m·s^{-1}

44-5 a) 4.58×10^{14} Hz b) 654 nm c) 1.90 eV
d) 6.62×10^{-34} J·s
44-7 0.34 V greater
44-9 a) 1.04 eV b) 1190 nm
c) 2.52×10^{14} Hz
d) 4.14×10^{-7} eV f) No
44-11 1.41×10^{19} s^{-1}
44-13 a) 6.16×10^{14} Hz b) 486 nm
44-15 1.96 eV or 3.14×10^{-19} J
44-17 a) 2.48×10^4 V b) 0.00062 nm or 0.62 pm
44-19 a) 355 cal·s^{-1} b) 33.8 C° ·s^{-1}
c) High melting temperature and good thermal conductivity
44-21 79,056 K
44-23 4.85×10^{-12} m
44-25 a) 2.481×10^{-11} m b) 2.496×10^{-11} m
c) 49.7 keV

Chapter 45

45-1 a) 13.6 eV b) -26.2 eV c) -13.6 eV d) 13.6 eV
45-3 a) Larger by factor of 4 b) Series ending at $n = 4$, $n = 5$, and one line of series ending at $n = 3$.
45-5 a) -2.80 keV b) 2.57×10^{-13} m or 257 fm
c) 0.59 nm
45-9 97.2 nm, 3.09×10^{15} Hz
45-11 656.3 nm, 364.6 nm
45-13 a) 12,400 eV b) 150 eV
45-15 0.145 nm
45-17 8.67 pm

45-21 a) 2 eV, 621 nm
b) 3.29×10^{-10} eV, 1.02×10^{-7} nm
45-23 2.34×10^{-23} s
45-25

l	0	1	1	1	2	2	2	2	2
m	0	-1	0	1	-2	-1	0	1	2

Chapter 46

46-1 $(1s)^2(2s)^2(2p)^2$, $(1s)^2(2s)^2(2p)^6(3s)^2(3p)^2$
46-3 $(1s)^2(2s)^2(2p)^6(3s)^2(3p)^6(3d)^{10}(4s)^2(4p)^2$
46-5 1.11 nm
46-7 Cesium
46-11 36,480 K
46-13 a) 4.58×10^{-48} kg·m^2
b) 0, 2.43×10^{-21} J, 7.28×10^{-21} J
c) 27.3 μm, 1.10×10^{13} Hz

Chapter 47

47-1 22.7 MeV, 3.64×10^{-12} J
47-3 35.6 MeV, 5.72×10^{-12} J
47-5 17.6 MeV, 2.81×10^{-12} J
47-7 0.137 u, 127.6 MeV, 7.97 MeV
47-9 $\Delta m = 0.000440$ u, 0.410 MeV
47-11 6.05 MeV
47-13 2.76×10^6 s^{-1}, 7.46×10^{-5} curie
47-15 0.782 MeV
47-17 a) 1.32 T b) 4.22 MeV, 2.01×10^7 m·s^{-1}
47-19 23.8 MeV, 3.82×10^{-12} J, 2.3×10^{12} J·mol^{-1}; larger by 10^7
47-21 0.511 MeV (8.19×10^{-14} J), 1.24×10^{20} Hz, 2.43×10^{-12} m

INDEX